PSAM 12
Probabilistic Safety Assessment and Management
22-27 June 2014 • Sheraton Waikiki, Honolulu, Hawaii, USA

Proceedings of the Probabilistic Safety Assessment and Management (PSAM) 12 Conference

Volume 10 - Thursday AM

PSAM 12

Probabilistic Safety Assessment and Management

22 - 27 June, 2014

Sheraton Waikiki, Honolulu, Hawaii USA

CONFERENCE PROCEEDINGS

Volume 10

Thursday AM

Foreword

It is was our honor to welcome you to Honolulu, Hawaii, for the twelfth rendition of the Probabilistic Safety Assessment and Management (PSAM) Conference. The planning for PSAM Honolulu began back in 2007 (before PSAM 9 in Hong Kong), when we looked at several locations around the United States, included Arizona, California, Boston, and even considered locations in Oceania. Based upon the feedback both during and after the conference, PSAM 12 proved to be a great success.

We would like to thank all of the volunteers, those that served before, during, and after the Conference. Members of the Technical Program Committee, the Organizing Committee, the session chairs, and the presenters have our gratitude for making PSAM 12 the most memorable PSAM yet.

This publication represents the technical proceedings for the Conference. Due to the large number of published papers (a total of 391), we have subdivided the technical content (papers) into five volumes, one for each day of the conference.

On behalf of the International Association for Probabilistic Safety Assessment and Management Board of Directors, we hope that this publication will provide a valuable technical resource in addition to a reminder of the memorable stay in the Hawaiian Islands.

Dr. Curtis Smith
Technical Program Chairs

Dr. Todd Paulos
General Chair

Sponsors

Sponsors

PSAM 12 - Probabilistic Safety Assessment and Management
JUNE 22-27, 2014

Technical Program Committee

Technical Program Chair: Curtis Smith, INL USA
Assistant Technical Program Chairs: Steve Epstein, Lloyd's Register Japan
Vinh Dang, PSI Switzerland
Ted Steinberg, QUT Australia

We would like to thank the members of the PSAM 12 Technical Program Committee. These individuals helped to make PSAM 12 a success by reviewing abstracts, technical papers, organizing sessions, and providing technical leadership for the conference.

Technical Committee Members:

- Roland Akselsson
- S. Massoud (Mike) Azizi
- Tito Bonano
- Ronald Boring
- Roger Boyer
- Mario Brito
- Kaushik Chatterjee
- Vinh Dang
- Claver Diallo
- Nsimah Ekanem
- Steve Epstein
- Fernando Ferrante
- Federico Gabriele
- Ray Gallucci
- S. Tina Ghosh
- David Grabaskas
- Katrina Groth
- Seth Guikema
- Steve Hess
- Christopher J. Jablonowski
- Moosung Jae
- Jeffrey Joe

- Vyacheslav S. Kharchenko
- James Knudsen
- Zoltan Kovacs
- Ping Li
- Harry Liao
- Francois van Loggerenberg
- Jerome Lonchampt
- Soliman A. Mahmoud
- Diego Mandelli
- Donoval Mathias
- Zahra Mohaghegh
- Thor Myklebust
- Cen Nan
- Mohammad Pourgolmohammad
- Marina Roewekamp
- Clayton Smith
- Shawn St. Germain
- Ted Steinberg
- Kurt Vedros
- Smain Yalaoui
- Robert Youngblood
- Enrico Zio

Organizing Committee

General Chair: Dr. Todd Paulos
General Vice Chair: Prof. Stephen Hora, USC
Technical Program Chair: Curtis Smith, INL USA
Webmaster, Registration, Support for Papers/Abstracts Submission and Review: Hanna Shapira, TICS

Table of Content

Page	Paper	
10	322	**Effects of Source Term on Off-site Consequence in LOCA Sequence in a Typical PWR** Seok-Jung Han, Tae-Woon Kim, and Kwang-Il Ahn *Korea Atomic Energy Research Institute, Daejeon, South Korea*
21	373	**A Review of U.S. Sodium Fast Reactor PRA Experience** David Grabaskas *Nuclear Engineering Division, Argonne National Laboratory, Argonne, IL, U.S.*
33	527	**Applicability of PSA Level 2 in the Design of Nuclear Power Plants** Estelle C. Sauvage (a), Gerben Dirksen (b), and Thierry Coye de Brunellis (c) *a) AREVA-NP SAS, Paris, France, b) AREVA-NP Gmbh, Erlangen, Germany, c) AREVA-NP SAS, Lyon, France*
41	547	**Use of Corrective Action Programs at Nuclear Plants for Knowledge Management** Pamela F. Nelson (a), Teresa Ruiz-Sánchez (b), and Cecilia Martín del Campo (a) *a) Universidad Nacional Autonoma de Mexico, Mexico City, Mexico, b) Universidad Autonoma de Tamaulipas, Reynosa, Mexico*
53	583	**AES-2006 PSA Level 1. Preliminary Results at PSAR STAGE** A. Kalinkin, A. Solodovnikov, S. Semashko *JSC "VNIPIET", Saint-Petersburg, Russian Federation*
64	262	**Quantitative Risk Assessment for DarkSide 50, a Nuclear Physics Experimental Apparatus Installed at Gran Sasso Nat'l Lab: Results and Technical Solutions Applied** Federico Gabriele (a), Andrea Ianni, Augusto Goretti (b), Michele Montuschi (a), and Paolo Cavalcante (a) on behalf of DarkSide Collaboration *a) Gran Sasso National Laboratory, L'Aquila, Italy, b) Princeton University, Princeton, USA*
76	270	**Safety Analysis and Quantitative Risk Assessment of a Deep Underground Large Scale Cryogenic Installation** Effie Marcoulaki and Ioannis Papazoglou *National Centre for Scientific Research "Demokritos", Athens, Greece*
88	295	**Centrifugal Pump Mechanical Seal and Bearing Reliability Optimization** Peymaan Makarachi, and Mohammad Pourgol-Mohammad *Sahand University of Technology, Tabriz, Iran*
98	463	**A Science-Based Theory of Reliability Founded on Thermodynamic Entropy** Anahita Imanian, Mohammad Modarres *Center for Risk and Reliability, University of Maryland, College park, USA*
108	534	**Quick Quantitative Calculation of DFT for NPP's Repairable Systems Based on Minimal Cut Sequence Set** Daochuan Ge (a,b), Qiang Chou, Ruoxing Zhang (b), Yanhua Yang (a) *a) School of Nuclear Science and Engineering, Shanghai Jiao Tong University, Shanghai, China; b) Software Development Center, State Nuclear Power Technology Corporation, Beijing, China*
117	415	**Air Traffic Controllers' Workload on the Period of ATC Paradigm Shift** Kakuichi Shiomi *Electronic Navigation Research Institute, Tokyo, Japan*
128	416	**Quantification of Bayesian Belief Net Relationships for HRA from Operational Event Analyses** Luca Podofillini, Lusine Mkrtchyan, Vinh N. Dang *Paul Scherrer Institute, Villigen PSI, Switzerland*
140	544	**Task Decomposition in Human Reliability Analysis** Ronald L. Boring and Jeffrey C. Joe *Idaho National Laboratory, Idaho Falls, Idaho, USA*
152	421	**A Comparison of Two Cognition-driven Human Reliability Analysis Processes - CREAM and IDHEAS** Kejin Chen, Zhizhong Li (a), Yongping Qiu and Jiandong He (b) *a) Department of Industrial Engineering, Tsinghua University, Beijing, P. R. China, b) Shanghai Nuclear Engineering Research & Design Institute, Shanghai, P. R. China*
159	203	**Human Reliability in Spacecraft Development: Assessing and Mitigating Human Error in Electronics Assembly** Obibobi K. Ndu (a), Monifa Vaughn-Cooke (b) *a) Space Mission Assurance Group, Johns Hopkins University Applied Physics Laboratory, Laurel, MD, USA, b) Dr. , Department of Mechanical Engineering, Reliability Engineering Program, University of Maryland, College Park, MD USA*

Table of Content

Page	Paper	
172	112	**Use of Bayesian Network to Support Risk-Based Analysis of LNG Carrier Loading Operation** Arthur Henrique de Andrade Melani, Dennis Wilfredo Roldán Silva, Gilberto Francisco Martha Souza *University of São Paulo, São Paulo, Brazil*
184	14	**Probabilistic Analysis of Geological Properties to Support Equipment Selection for a Deepwater Subsea Oil Project** Christopher J. Jablonowski, Edward E. Shumilak, Kenneth F. Tyler (a), Arash Haghshenas (b) *a) Shell Exploration and Production Company, Houston, TX, U.S.A. b) Boots & Coots Services LLC, Houston, TX, U.S.A.*
197	85	**Gas Detection for Offshore Applications** Peter Okoh *Norwegian University of Science and Technology, Trondheim, Norway*
207	165	**BOP Risk Model Development and Applications** Xuhong He, Johan Sörman (a), Inge A. Alme (b), and Scotty Roper (c) *a) Lloyd's Register Consulting, Stockholm, Sweden, b) Lloyd's Register Consulting, Kjeller, Norway, c) Lloyd's Register Drilling Integrity Services Inc., Houston, USA*
215	411	**Determination of the Design Load for Structural Safety Assessment against Gas Explosion in Offshore Topside** Migyeong Kim, Gyusung Kim, Jongjin Jung and Wooseung Sim *Advanced Technology Institute, Hyundai Heavy Industries, Ulsan, Republic of Korea*
225	309	**Propagating Uncertainty in Phenomenological Analysis into Probabilistic Safety Analysis** A. El-Shanawany (a,b) *a) Imperial College London, London, United Kingdom, b) Corporate Risk Associates, London, United Kingdom*
234	349	**A Procedure Estimating and Smoothing Earthquake Rate in a Region with the Bayesian Approach** J.P. Wang *The Hong Kong University of Science and Technology, Kowloon, Hong Kong*
242	512	**Open Conceptual Questions in the Application of Uncertainty Analysis in PRA Logic Model Quantification** Sergio Guarro *ASCA Inc., Redondo Beach, USA*
253	565	**System Initiating Event Frequency Estimation using Uncertain Data** Kurt G. Vedros *NuScale Power, LLC, Corvallis, Oregon, United States*
260	584	**SUnCISTT - A Generic Code Interface for Uncertainty and Sensitivity Analysis** Matthias Behler (a), Matthias Bock (b), Florian Rowold, and Maik Stuke (a) *a) Gesellschaft für Anlagen und Reaktorsicherheit GRS mbH, Garching n. Munich, Germany, b) STEAG Energy Services GmbH, Essen, Germany*
274	312	**BWR-club PSA Benchmarking – Bottom LOCA during Outage, Reactor Level Measurement and Dominating Initiating Events** Anders Karlsson (a), Maria Frisk (b), and Göran Hultqvist (c) *a) Forsmarks Kraftgrupp AB, Östhammar, Sweden, b) Risk Pilot AB, Stockholm, Sweden, c) Havsbrus Consulting, Öregrund, Sweden*
286	371	**Effects of an Advanced Reactor's Design, Use of Automation, and Mission on Human Operators** Jeffrey C. Joe and Johanna H. Oxstrand *Idaho National Laboratory, Idaho Falls, USA*
295	391	**For the Completeness of the PRA Implementation Standard** Yoshiyuki Narumiya (a), Akira Yamaguchi (b), Takayuki Ota, Haruhiro Nomura (a) *a) The Kansai Electric Power Co., Inc, Osaka, Japan, b) Osaka University, Suita, Osaka, Japan*
305	497	**Nuclear Safety Design Principles & the Concept of Independence: Insights from Nuclear Weapon Safety for Other High-Consequence Applications** Jeffrey D. Brewer *Sandia National Laboratories, Albuquerque, NM, USA*
320	498	**Advancing Human Reliability Analysis Methods for External Events with a Focus on Seismic** Jeffrey A. Julius, Jan Grobbelaar, and Kaydee Kohlhepp *Scientech, a Curtiss-Wright Flow Control Company, Tukwila, WA, U.S.A.*

Table of Content

Page   Paper

331   *572*   **Expected Maintenance Costs Model for Time-Delayed Technical Systems in Various Reliability Structures**
Anna Jodejko-Pietruczuk, Sylwia Werbińska-Wojciechowska
Wroclaw University of Technology, Wroclaw, Poland

339   *394*   **Modeling the Reliability and the Performance of a Wind Farm Using Cyclic Non-Homogenous Markov Chains**
Theodoros V. Tzioutzias, Agapios N. Platis, Vasilis P. Koutras
University of the Aegean Department of Financial and Management Engineering, Chios, Greece

349   *273*   **Performance and Reliability of Bridge Girders Upgraded with Posttensioned Near-surface-mounted Composite Strips**
Yail J. Kim (a), Jae-Yoon Kang, and Jong-Sup Park (b)
a) University of Colorado Denver, Denver, CO, USA, b) Korea Institute of Construction Technology, Ilsan, Korea

359   *393*   **A Quantitative Method for Assessing the Resilience of Infrastructure Systems**
Cen Nan (a,b), Giovanni Sansavini (b,c), Wolfgang Kröger (c) and Hans Rudolf Heinimann (a,c)
a) Land Using Group, ETH Zürich, Switzerland, b) Reliability and Risk Engineering, ETH Zürich, Switzerland, c) ETH Risk Center, ETH Zürich, Switzerland

371   *113*   **Use of Reliability Concepts to Support Pas 55 Standard Application to Improve Hydro Power Generator Availability**
Gilberto F. M. de Souza (a), Erick M.P. Hidalgo (a), Claudio C. Spanó (b), and Juliano N. Torres (c)
a) University of São Paulo, São Paulo, Brazil, b) ReliaSoft Brasil, São Paulo, Brazil, c) AES Tietê, Bauru, Brazil

Effects of Source Term on Off-site Consequence in LOCA Sequence in a Typical PWR

Seok-Jung HAN[a], Tae-Woon KIM, and Kwang-Il AHN
[a] Korea Atomic Energy Research Institute, P.O. Box 105, Yuseong, Daejeon, South Korea

---

Abstract: Since the accident of Fukushima, the assessment of source term effects on the environment is a key concern of the nuclear safety. As an effort to take into account the current knowledge of source term in off-site consequence analysis, the effects of the source term according to the containment response simulated by MELCOR code have been examined. In the view of the consequence, the containment response directly affects key features making a shape of plume behaviors to estimate the atmospheric dispersion, which are the release time, duration, and relevant source term features. The source term features for a large break LOCA sequence of a typical PWR plant according to the containment response (failure pressure and break size) have been investigated. In the results of the containment failure pressure, it has been observed that the release time varied 17.4 hour to 52.2 hour according to the containment failure pressure of 4.4 bar to 14.6 bar, respectively. This result potentially affects the radiological emergency strategies such as the public evacuation. Moreover, a considerable amount of the released source term is varied. This is resulted in about twice differences of the radiation exposure dose within the simulation cases. In the break size, it has been observed that the release source term is varied relatively small, but the release features to model the plume behavior are varied according to the break size. In particular, the radiation exposure dose are reduced to 50% according to the plume model approaches (one plume model vs. two plumes model) taking into account the source term release features in this simulation. The obtained insights of source term features will be utilized in an off-site consequence analysis.

Keywords: PRA, Radiological Source Term, Consequence Analysis, MELCOR, MACCS2

---

1. INTRODUCTION

In lessons learned from the Fukushima accident, an improvement of knowledge and understanding of the off-site consequence analysis (CA) became a key concern of the nuclear safety [1]. The CA is to assess an environmental effect of the radiation exposure due to the radioactive materials release to the environment during severe accidents of a nuclear facility. The CA is an integrated analysis including the assessments of radiological source term, atmospheric dispersion, dosimetry according to exposure pathways, health effects of radiation exposure. Among those parts, the radiological source term (shortly, source term)* as a comprehensive technical terminology covering the characteristics of radioactive materials escaped to the environment [2] is a principal part of the CA of nuclear facilities [2].

Because there are a considerable limitation to provide the overall source term features needed in CA and a large degree of uncertainty in their features [3, 4], the simplified source term have been applied in the typical CAs. However, the severe accident analysis codes such as MELCOR [5] and MAAP [6] provide more detailed information for quantifying the source term features. The current state-of-art approaches to the source term estimation in CA are to use these codes. Recently, the US NRC SOARCA report [7] showed an approach to utilize the detailed source term features provided by MELCOR code, of which features are to use a realistic off-site consequence analysis.

---

* This terminology is including the radioactive materials as constituent, radiological characteristics, physicochemical characteristics, relevant phenomenology in their transport, release pathways, amount of their release, etc.

In the present study, as an effort to take into account the current knowledge of source term in CA, the source term features provided by MELCOR code have been utilized. In this work, a large break Loss-Of-Coolant-Accident (LOCA) sequence of a typical large dry containment PWR has been investigated. In a large LOCA sequences, the containment response is a key factor making a shape of the source term behaviors. In the view of the consequence, the containment response directly affects key features making a shape of plume behaviors to estimate the atmospheric dispersion, which are the release time, duration, and relevant source term features. The source term features according to the containment response (failure pressure and break size) simulated by MELCOR code have been examined by MACCS2 code for a CA.

2. SOURCE TERM PROJECTION APPROACHES TO CONSEQUENCE ANALYSIS

There are many features characterizing the source term, but the key features are to determine initial and boundary conditions of an atmospheric dispersion model such as (1) release amounts of source term, (2) release time, and (3) duration during a release phase. For an advanced analysis of atmospheric dispersion, the dispersion features of the source term such as aerosol size or sensible heat of plume are required.

Although a description of dispersion features depends on the atmospheric dispersion models, the typical parts of an atmospheric dispersion model consist of (1) the initial dimension of plumes, (2) plume rise characteristics, (3) deposition characteristics of radioactive materials during the dispersion. Typical information required in CA is shown in Table 1. Among these features, this study focuses on the containment response with the selected accident sequence to make the plume characteristics, release amount, release time, and release duration.

Table 1: Typical information required in off-site consequence analysis

Area	Feature	Element
Accident sequence	Scenario	state of key safety functions
	Phenomenology	progress of severe accident phenomena
	Release pathways	**containment response**
Radioactive materials inventory	Chemical features	classifications
	Radionuclide	total amount
Radioactive materials transport phenomena	Segments	transport (core/RCS/Containment)
	Plume	**characteristics**
	Release	**release amount**
Dispersion features	Aerosol	size distribution
	Release Energy	sensible heat
	Time	**release time**
	Duration	**release duration**

In the view of CA, the source term results provided by the severe accident codes are not directly adopted in CA because of the different modeling techniques. A process utilizing the source term results of the severe accident codes to CA is a kind of the projection technique. To derive the source term features needed in CA, it should assess the atmospheric dispersion model before characterizing the source term features. In this study, the required source term features have been derived based on MACCS2 code developed by Sandia National Laboratories (SNL) in USA [8]. Because the atmosphere is a primary pathway of the radiological dispersion, atmospheric dispersion is a key model to CA. In MACCS2 code, a Gaussian plume model is adopted as a key model to describe the atmospheric dispersion:

$$\chi(x,y,z) = \frac{Q}{\bar{u}} \cdot \frac{e^{-y^2/2\sigma_y^2(x)}}{\sqrt{2\pi}\sigma_y(x)} \cdot \frac{1}{\sqrt{2\pi}\sigma_z(x)} (e^{-\frac{(z+h)^2}{2\sigma_z^2(x)}} + e^{-\frac{(z-h)^2}{2\sigma_z^2(x)}})$$

$$= \frac{Q}{\bar{u}} \cdot \frac{1}{\pi\sqrt{2}\sigma_z(x)\sqrt{2}\sigma_y(x)} \cdot e^{-y^2/2\sigma_y^2(x)} \cdot (e^{-\frac{(z+h)^2}{2\sigma_z^2(x)}} + e^{-\frac{(z-h)^2}{2\sigma_z^2(x)}})$$

(1)

Here χ is the time-integrated concentration of released radiation materials, Q is the total amount of released radiation materials, $\bar{u}$ is the wind speed, σ_y and σ_z are lateral and vertical dispersion coefficients, respectively, and h is the release height. Although the Gaussian plume is a static model, time-dependent features are treated in MACCS2 code using an hourly-based unit-time interval approach for released amounts within the limitation of four plumes. Key factors to represent a plume features using the source term results of the MELOCR code are manipulated considering the MACCS2 plume model features.

In MACCS2, plumes can be modeled upto four plumes, which are specified by a start time and duration. In the typical single plume model, short and long duration approaches are applied in CA according to case by case since a plume shape is determined by duration, of which the release concentrations are varied from high to low because of the conservation of the total amount of released source term (Fig. 1-(a)). One plume model is useful in steady-state estimation such as air pollution effects of normal operating plants, but it is a limitation to investigate an estimation of accident conditions. For the multiple-plume model, the release features of a specific source term could be simulated more realistic (Fig. 1-(b)). Taking into account simulation results, the shapes of the release features could be projected in plume modeling.

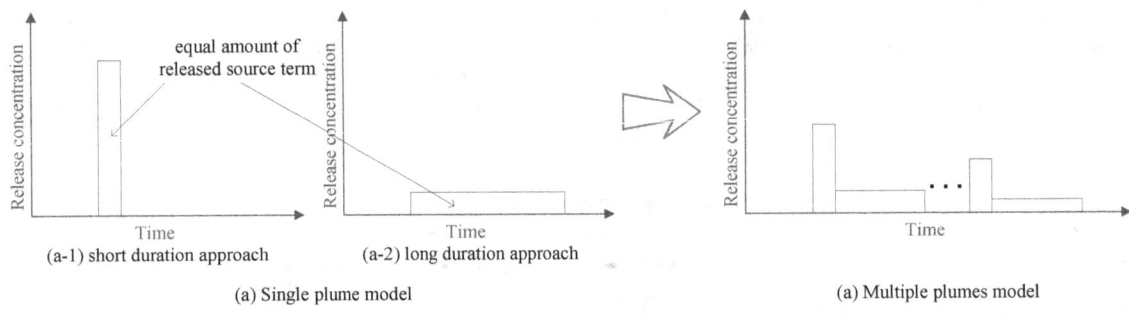

(a) Single plume model   (a) Multiple plumes model

Fig. 1. Plume modeling approaches

3. SOURCE TERM AND CONSEQUENCE ANLYSIS

3.1. Plant Model in MELCOR

An application case, i.e., a loss-of-coolant-accident (LOCA) as a typical sequence reached to severe accident with an over-pressurization containment failure, was selected to investigate the source term behaviors on CA. The containment failure mode due to over-pressurization, although this is the most possible source term release pathway in LOCA sequence, has a large degree of uncertainty to apply the relevant parameters. Most of all, the containment failure pressure and break size are key parameters to determine containment response and the source term behaviors.

In this study, the effects of CA according to the variation of the containment failure pressure and break size have been investigated by MELCOR code (Version 1.8.6 YT). The reference plant for this work was adopted OPR-1000 type plants which are a Korean typical plant [9]. These plants are designed to two-loop (2 steam generator) type PWR with a 2815MW thermal power and housing a large dry containment. The reference plant model in MELOCR is shown in Fig. 2. The containment model adopted four control volumes such as (1) reactor cavity, (2) inner shell, (3) annulus, and (4) upper compartment dome.

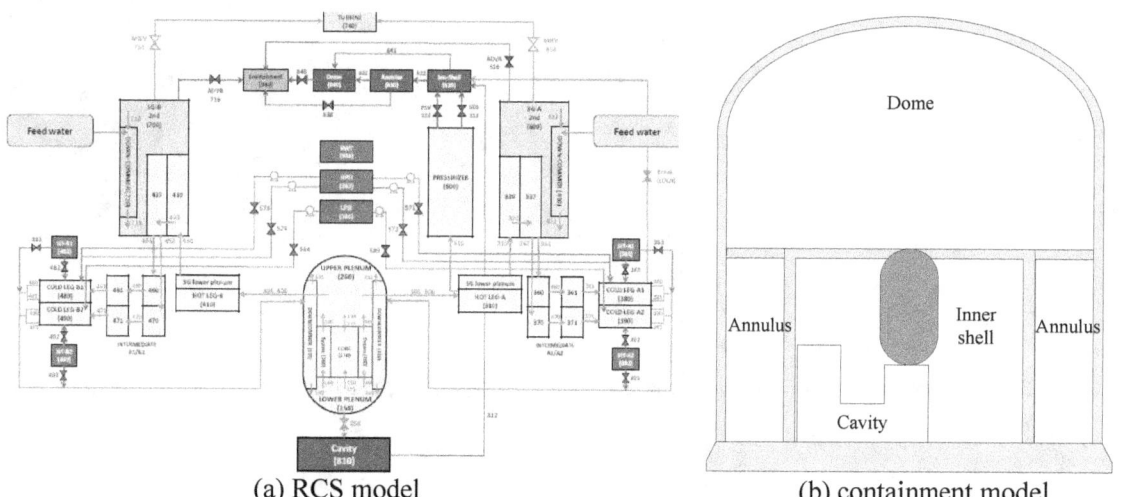

(a) RCS model (b) containment model

Fig. 2. A nodalization diagram of the reference plant

During the severe accident progression initiated from a LOCA, the containment pressure is continuously increasing due to severe accident phenomena, which results in a containment failure. There is a large amount of uncertainty of the containment response. This study focused on key parameters in the containment response, i.e., the containment failure pressure and break size, of which effects on a CA were investigated.

In this study, a six-inch (0.15 meter) break size (break area of 1.82E-2 square-meter) in a cold leg, which is a typical large-break LOCA sequence in the PSA [10], was taken into account. Among the sequences to reach the core damage, a sequence of the recirculation phase failure of safety injection from the containment sump after a dry-out of the water source (RWST) was adopted as a simulation case. This sequence is a highly ranked sequence among the LOCA-induced severe accident sequences [10]. In this sequence, a dominant containment response is that the containment failure occurs by an over-pressurization over the containment design pressure. For this sequence, the cavity state is assumed as a dry state initially. The containment spray did not operate the early phase because the containment pressure did not reach to the operating condition (2.39E5 Pa) and it are assumed not working in the late phase because of the assumption of the recirculation failure. The accident progression of the given case is shown in Table 2.

Table 2. Events of the given accident sequence

Events	Time (hr)
LOCA Started (coldleg break occurred)	0
Main feedwater stopped	0.00
Reactor trip	0.00
MSIV closed	0.00
RCP trip	0.01
Core uncover (-2.28 m)	0.02
SIT-injection started	0.08
LPSI- injection started	0.12
SIT exhausted	0.22
RWST dryout and safety injection fails to operation at recirculation from the sump	1.70
Cladding melt started	2.79
Core dry (-6.09m)	2.87
UO2 relocated to lower head	3.75
RPV lower head penetration	4.92
Cavity dryout	5.25
Containment leak failure start point (4.4 bar (64psi))	18.38

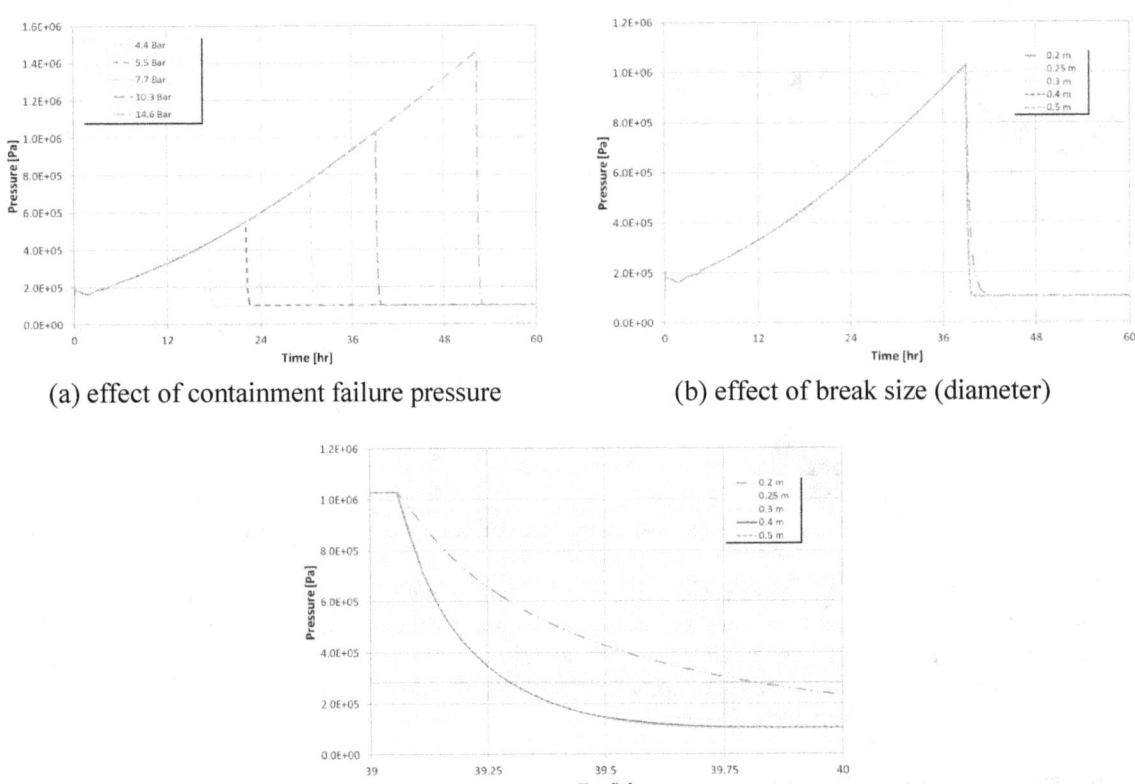

(a) effect of containment failure pressure (b) effect of break size (diameter)

(C) containment pressure responses near failure time for the break sizes
Fig. 3. Containment response according to failure pressure and break size

3.2. Source Term Analysis

The PSA report denoted that the range of containment failure pressure is varied from 4.4 bar (leak failure start point) to 14.6 bar (catastrophic rupture) [10]. For the containment failure pressure, five cases (4.4, 5.5, 7.7, 10.3, and 14.6 bar) were simulated (Fig.3-(a)). In this simulation, the break size of containment is assumed as 0.5 m inner diameter. Containment failure in each case occurs at about 17.4, 22.1, 30.4, 39.1, and 50 hour, respectively. It is noted that this result potentially affects the radiological emergency strategies such as the public evacuation.

The containment break size is another unknown factor making the source term behaviors. In this study, the break size of the containment failure was taken into account from 0.2 to 0.5 meter of hydraulic diameter (Fig.3-(b), (c)). For the containment break sizes, only a containment failure pressure of 10.3 bar was applied.

In these simulations, it was identified that the containment failure pressure affects the containment failure time and it was expected that the containment break size mainly affects the immediate source term behaviors. The source term behaviors (the release fraction and its rate) of noble gases, Cesium and Iodine according to the variation of the containment failure pressure and the containment break size are shown in the Fig.4 and Fig.5. Fig. 6 shows that the variation of the containment failure time (Fig. 6-(a)) and the release fraction of Cesium and Iodine (Fig. 6-(b)) according to the containment failure pressure. It is noted that Fig. 6-(b) delineates that a considerable amount of the release fractions according to the containment failure pressure are reduced to affect the radiological effect on environment. On the other hand, Fig. 7 reveals that the variation of the containment break size affects the source term behaviors, in particular the immediate behaviors near the containment failure time, are drastically changed.

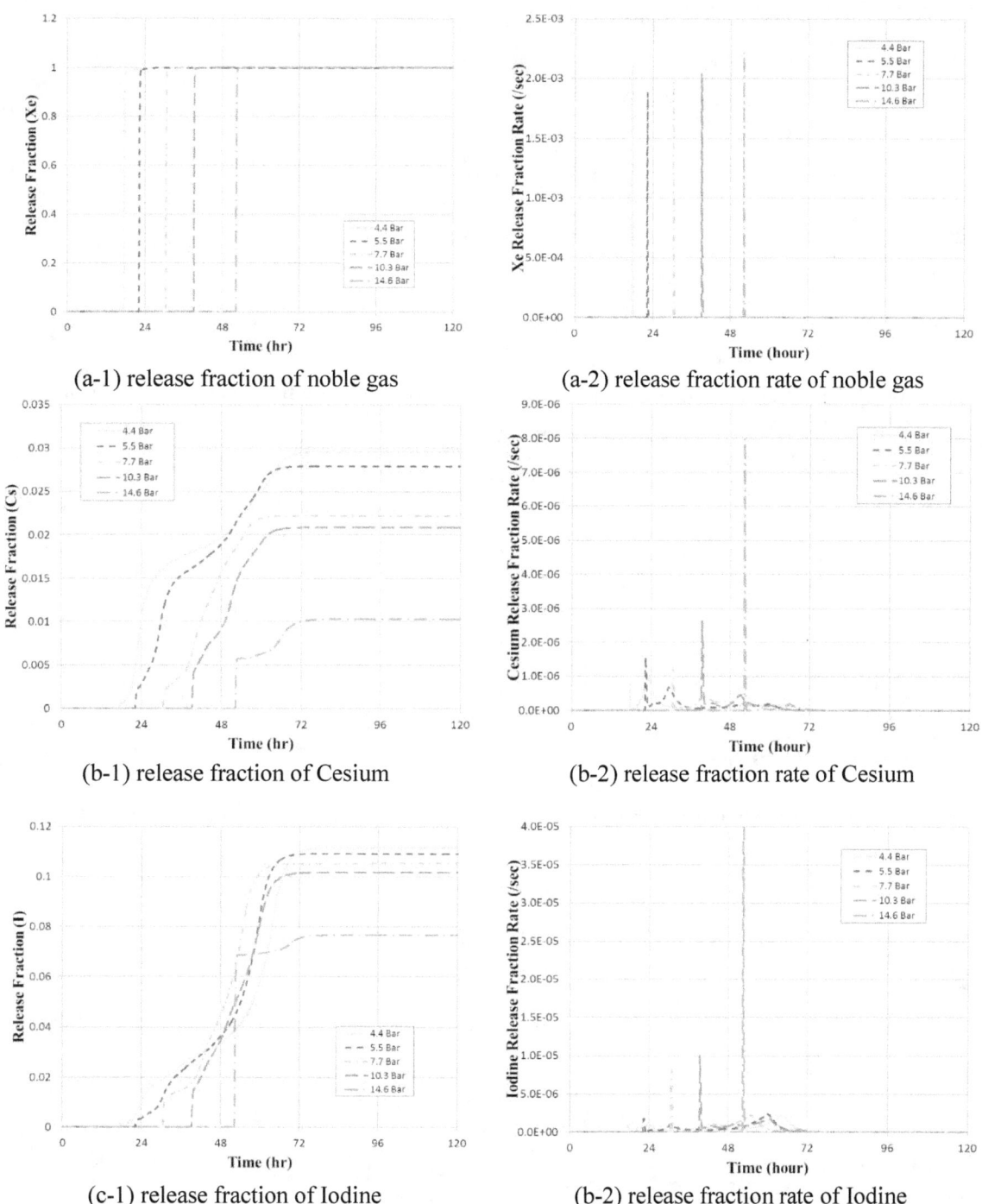

(a-1) release fraction of noble gas
(a-2) release fraction rate of noble gas
(b-1) release fraction of Cesium
(b-2) release fraction rate of Cesium
(c-1) release fraction of Iodine
(b-2) release fraction rate of Iodine

Fig. 4. Source term behaviors according to containment failure pressure

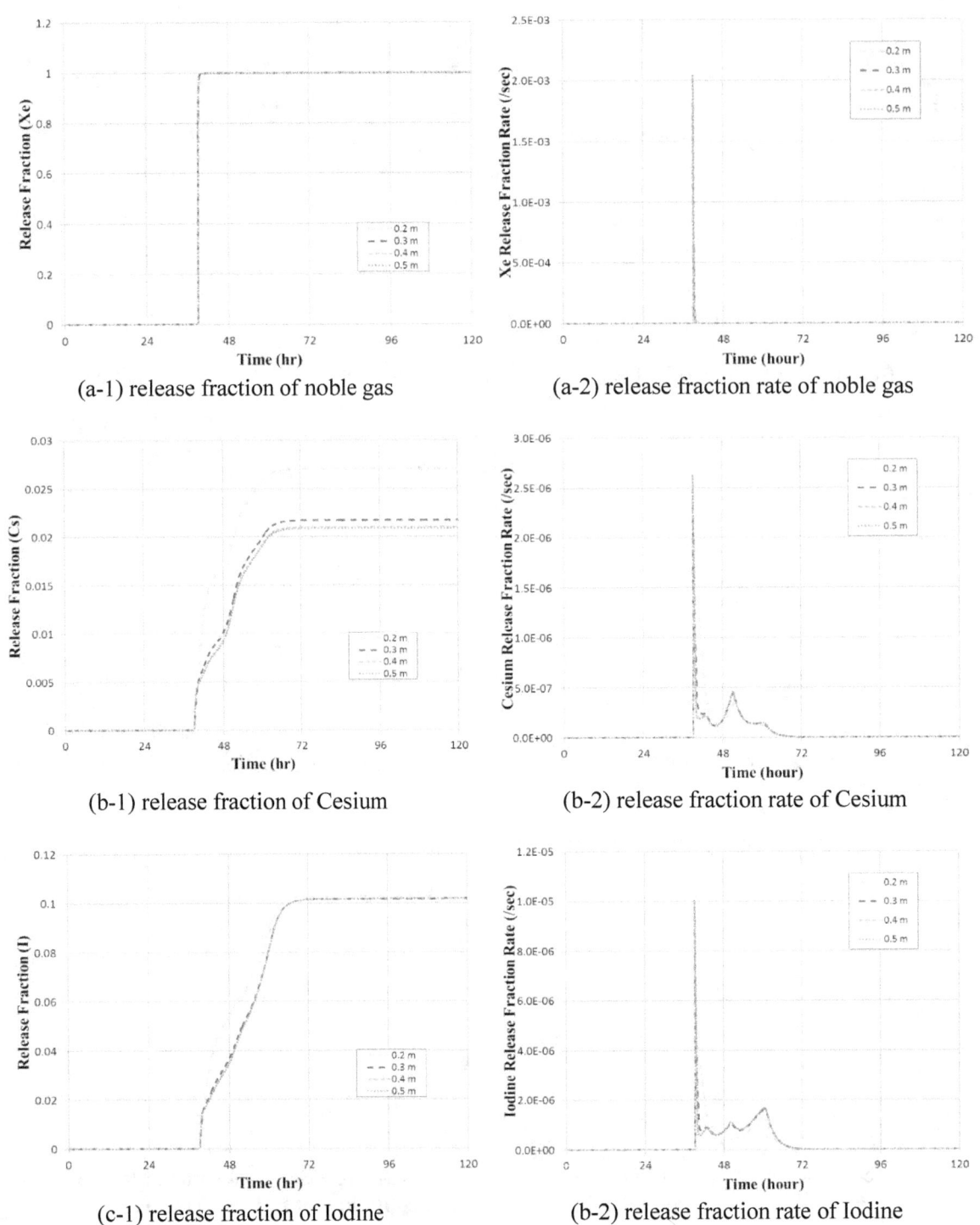

(a-1) release fraction of noble gas
(a-2) release fraction rate of noble gas
(b-1) release fraction of Cesium
(b-2) release fraction rate of Cesium
(c-1) release fraction of Iodine
(b-2) release fraction rate of Iodine

Fig. 5. A source term behavior according to containment failure size

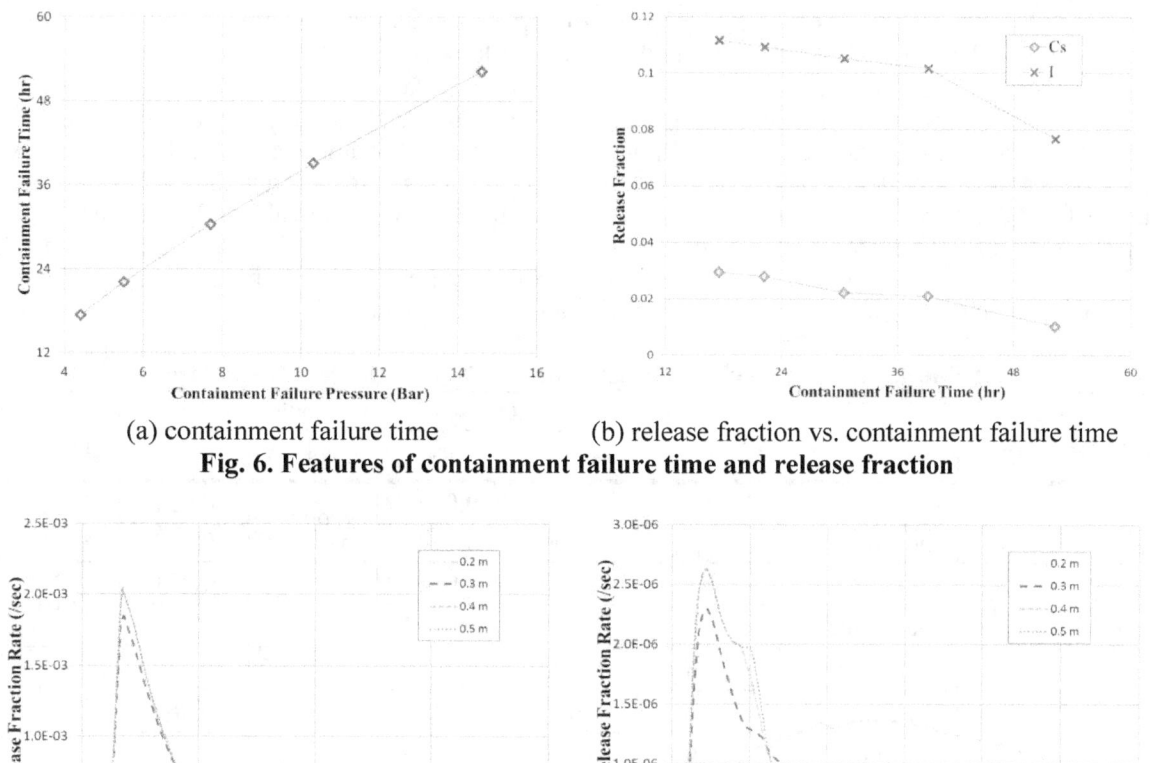

(a) containment failure time (b) release fraction vs. containment failure time
Fig. 6. Features of containment failure time and release fraction

(a) release fraction rate of noble gases (b) release fraction rate of Cesium
Fig. 7. Variation of release fraction rate near failure time

3.3. Source Term Projection

The results of these simulations provide the basis of plume modeling for an atmospheric dispersion. Because this study focused on the effects of the source term according to the containment responses, the different plume models were adopted according to the types of containment response, i.e., containment failure pressure and break size. For the containment failure pressure, one plume model was applied in order to investigate their effects. One-hour duration was applied although a considerable amount of the residual was observed in the simulation results. As the results, Table 3 shows the characterization of this single-plume model. For each chemical group, almost all of the noble gases, maximum 3 % of Cesium and maximum 11 % of Iodine released to the environment.

Table 3. The plumes characterization for the containment failure pressure

Containment failure pressure (Bar)	Failure Time (hr)	Release Fraction of Initial Core Inventory		
		Xe, Kr	Cs	I
4.4	17.4	0.999	0.0296	0.112
5.5	22.1	0.999	0.0279	0.109
7.7	30.4	0.999	0.0222	0.105
10.3	39.1	0.999	0.0210	0.102
14.6	52.2	1.000	0.0103	0.077

For the containment break size, two-plume model was applied as follows:
- Plume 1 : dominant release phase for initial massive release
- Plume 2 : continuous residual phase for assessing additive effects

The first plume modeled taking into account the rapid peak of the release fraction rate near the start point of release. Taking into account the variation of duration as shown in Fig. 7, the durations of the first plume was considered. The second plume modeled taking into account the residual release amount and duration. The characterization of two-plume model is shown in Table 4. It is noted that the total release fraction of isotopes of two-plume case (Table 4) is the same as single-plume case (Table 8, 10.3 Bar). For the noble gases, 99.9% is released to the environment. For the Cesium and Iodine groups, 2.1% and 10.2 % was released to the environment, respectively.

Table 4. The plumes characterizations for the containment break size

Containment Break size (m)	Plumes	Duration of 1st Plume (hr)	Release Fraction of Initial Core Inventory				
			Xe, Kr	Cs	Cs (sum)	I	I (sum)
0.2	1st Plume	7.58	0.999	0.0165	0.0271	0.051	0.103
	2nd Plume		0.000	0.0106		0.052	
0.3	1st Plume	2.83	0.999	0.0065	0.0217	0.021	0.101
	2nd Plume		0.000	0.0152		0.081	
0.4	1st Plume	2.36	0.999	0.0056	0.0210	0.019	0.102
	2nd Plume		0.000	0.0154		0.083	
0.5	1st Plume	1.36	0.999	0.0045	0.0208	0.016	0.102
	2nd Plume		0.000	0.0163		0.086	

3.3. Effects of the source term on the off-site consequence

The effects of the source term according to the characterization of source term aforementioned are simulated by MACCS2 code (WinMACCS Version 3.7). In this study, only three isotope groups (noble gases, Cesium, and Iodine) were considered, although nine isotope groups are treated for the radiological exposure in MACCS2 code[x]. For assessing the specified consequence, weather condition should be fixed. In this case, the following weather condition applied:
- Wind speed: 3.2 m/s
- Atmospheric stability Class: D (neutral)
- Release height: 0 m (ground level release).

To calculate the radiation exposure dosimetry, the peak whole-body dose in the ground centerline under the plume provided by the default output of MACCS2 code were calculated and the default values of dose conversion factors (DCFs) in MACCS2 code were used. In this study, the relative peak whole-body dose comparing with the minimum calculated value was presented. The Fig. 8 shows the relative peak whole-body dose according to distance from a release point for the containment failure pressure. For the simulation cases (the containment failure pressure, 4.4 bar to 14.6 bar), maximum value of the relative peak whole-body dose is about 100% larger than minimum value of them at the same distance, but it is decreased to about 50% at 10 km distance. Revealing the plumes characterization in Table 3, the whole-body dose for the lower containment failure pressure cases are sequentially highly ranked comparing with the higher containment failure pressure cases. This observation shows that higher containment failure pressure reduces the radiation exposure of the environment even except the effects of release start time.

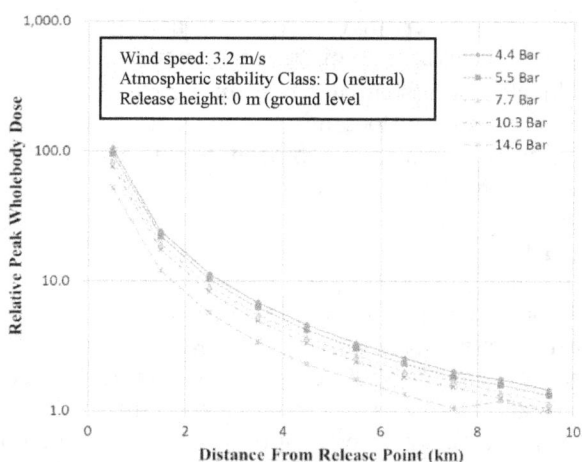

Fig. 8. The relative peak wholebody dose for the containment failure pressure (except the failure time)

Fig. 9 shows the relative peak whole-body dose for the containment break size. In this case, the primary effect is a reduction of the whole-body dose due to two-plume model approach comparing with a similar case of 10.3 bar of the containment failure pressure in Fig. 8. This is due to the split of the amount of source term into two-plume. In particular, it is observed that the effects of first plume are a considerable difference (Fig.9-(b)) although the effect of two-plumes shows a little difference between the cases (Fig.10-(a)). For the break size of 0.2 meter, the primary plume effect which is much higher than other cases is due to the larger portion of the source term in a primary plume as shown in Table 4. As the view of the radiological health effects, this result shows a meaningful effect of the source term because a primary plume is a key contributor to assign an acute effect.

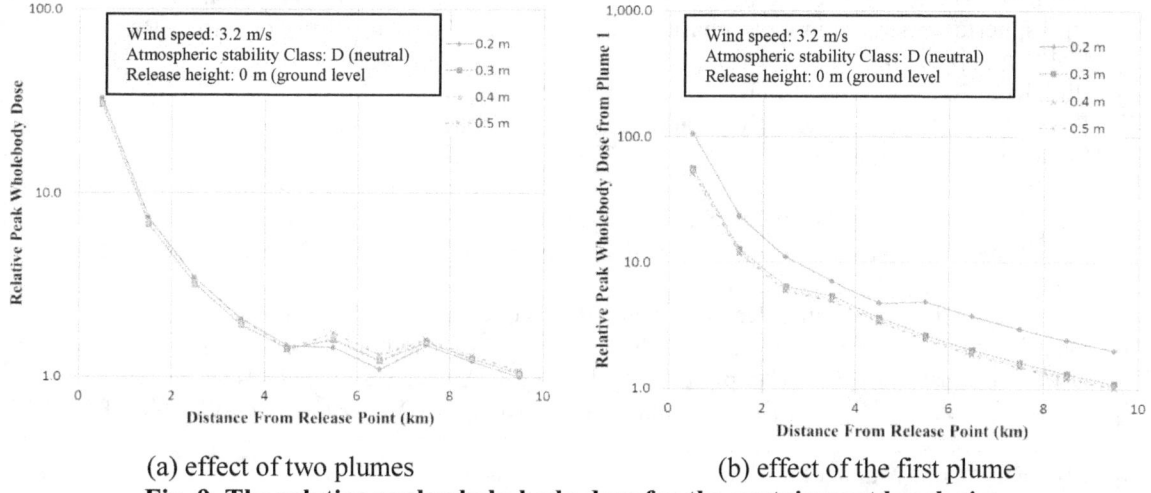

(a) effect of two plumes (b) effect of the first plume

Fig. 9. The relative peak whole-body dose for the containment break size (except the failure time)

From these examinations, it is presented that the characteristics of the containment response to affect the off-site consequence are as follows:

- Increased containment failure pressure delays the source term release time to govern the execution of the emergency plan, so roughly speaking that the better resistance of the containment against the severe accident progression may provide a margin of the execution of the emergency plan.
- In particular, a reduced source term according to the increased containment failure pressure may reduce the consequential health effects.
- A plume model approach to follow the containment response (i.e., release rate instead of cumulative measure of source term) may represent a realistic consequential effect. In the view of

the off-site consequence, the conservative approaches may provide biased insights to reach a different decision making in the execution of the emergency plan.
- In this study, the accident progression and relevant severe accident phenomena has a large uncertainty and the simulation case does not provide overall aspects of these knowledge. To obtain useful insights, a more realistic approach to the accident progression and detailed assessment to show a containment response are required.

4. CONCLUDING REMARK

As an effort to take into account the current knowledge of source term in CA, the effects of the source term according to the containment response simulated by MELCOR code have been examined. The obtained results reveal that the containment response in a large LOCA may affect the off-site consequence. A realistic estimation in the off-site consequence analysis has been a long-lasting issue, due to large uncertainty in the source term estimation. In recent times, however, there were more understandings on severe accident phenomenology and progress in simulation tools such as MELCOR, making it possible to assess more realistically the off-site consequence. The present study examined a containment response focusing on the off-site consequence. Within this simulation case, the useful insights were obtained, but for making a sure insight, further study is recommended.

Acknowledgements

This work was supported by the Nuclear Research & Development Program of the National Research Foundation grant funded by the Ministry of Science, ICT and Future Planning, Korea.

References

[1] Genki Katata et al, "Atmospheric discharge and dispersion of radionuclides during the Fukushima Dai-ichi Nuclear Power Plant accident. Part I: Source term estimation and local-scale atmospheric dispersion in early phase of the accident," Journal of Environmental Radioactivity, Vol. 109, pp.103-113 (2012).
[2] US NRC, "Reactor Safety Study," WASH-1400 (1975).
[3] T. Homma et al, "Uncertainty and Sensitivity Studies with the Probabilistic Accdient Consequence Assessment Code OSCAAR," Nuclear Engineering and Technology, Vol.37, No. 3 (2005).
[4] J.C. Helton et al, "Robustness of an Uncertainty and Sensitivity Analysis of Early Exposure Results with the MACCS Reactor Accident Consequence Model," Reliability Engineering and System Safety, Vol. 48, pp.129-148 (1995).
[5] US NRC, "MELCOR Computer Code Manuals," NUREG/CR-6119 (2005).
[6] Electric Power Research Institute (EPRI), ""MAAP4, Modular Accident Analysis Program User"'s Manual,"" EPRI Report prepared by Fauske & Associates, Inc. (1994).
[7] US NRC, "State-of-the-Art Reactor Consequence Analyses (SOARCA) Report," NUREG-1935, (2012).
[8] US NRC, "Code Manual for MACCS2," NUREG/CR-6613, (1997).
[9] http://www.opr1000.co.kr/
[10] KOPEC, "AOT/STI Relaxation Study for Korean Standard Nuclear Power Plants," the Ministry of Industry and Resources, Korean Language (2004).

A Review of U.S. Sodium Fast Reactor PRA Experience

David Grabaskas[a],
[a]Nuclear Engineering Division, Argonne National Laboratory, Argonne, IL 60439, U.S.

Abstract: The U.S. has a long history of sodium fast reactor (SFR) development. From the almost 30 years of successful EBR-II operation to the competing designs of the Advanced Liquid Metal Reactor (ALMR) project of the early 1990s, much work has been conducted related to SFR safety analysis. Part of this work has involved the creation of PRAs for both operational reactors and those in the conceptual design phase. A review of four of the past U.S. SFR PRAs was conducted, and their strengths and weaknesses were assessed. As part of this review, the past SFR PRAs were compared to the newly issued ASME/ANS Advanced Non-LWR PRA standard, which for the first time offers guidance on the criteria needed for a "complete" advanced reactor PRA. The results of this comparison offer direction for future analyses concerning what methods can be used from the past SFR PRAs, and what new techniques will need to be developed.

Keywords: PRA, Sodium Fast Reactor, ASME/ANS PRA Standard.

1. INTRODUCTION

From the development of the second Experimental Breeder Reactor (EBR-II) in the early 1960s to the Advanced Liquid Metal Reactor project of the 1990s, the U.S. has heavily invested in sodium fast reactor research (SFR) and development. One element of this work has been the development of probabilistic risk assessments (PRAs) for both conceptual designs and operational reactors. In an effort to preserve the knowledge that was gained through these endeavors, this work seeks to review past U.S. SFR PRAs and to assess their capabilities and deficiencies. This includes comparing the past SFR PRAs to the newly issued ASME/ANS PRA Standard for Advanced Non-LWR Nuclear Power Plants [1]. The results of this comparison offer direction for future analyses concerning what methods can be used from the past SFR PRAs, and what new techniques will need to be developed.

The following four PRAs were reviewed for this work:
1) Clinch River Breeder Reactor PRA – CRBRP-4 (1984) [2]
2) Sodium Advanced Fast Reactor (SAFR) PRA (part of PSID[*]) – AI-DOE-13527 (1985) [3]
3) Experimental Breeder Reactor II Level 1 PRA – EBR-II PRA (1991) [4]
4) Power Reactor Innovative Small Module (PRISM) PRA (part of PSID[*]) – GEFR-00793 (1987) [5]

It should be noted that there are many PRA versions and revisions for several of these SFR designs. The versions listed above were chosen since they are openly distributed and widely available, and in the case of SAFR and PRISM, form the basis of a NRC review. The only exception is the SAFR PRA, which is part of the export-controlled Preliminary Safety Information Document (PSID). The SAFR PRA information referenced here is derived from the NRC review in NUREG-1369 [6], which is open access.

2. ASME/ANS PRA Standard for Advanced Non-LWR Nuclear Power Plants

The American Society of Mechanical Engineers (ASME) and American Nuclear Society (ANS) Joint Committee on Nuclear Risk Management recently approved for trial use the new PRA Standard for Advanced Non-LWR Nuclear Power Plants [1]. The source material for the standard was the current ASME/ANS LWR Level-1 PRA standard [7], as well as the draft standards under development for low power and shutdown [8], Level-2 PRA [9], and Level-3 PRA [10]. These documents were modified in order to encompass a technology-neutral field of non-LWR designs. The ASME/ANS

---
[*] PSID – Preliminary Safety and Information Document

non-LWR writing group also worked closely with the Advanced Light Water Reactor (ALWR) writing group [11] to ensure consistency between the two standards. In general, the ASME/ANS PRA standards state what must be done in order to provide an adequate PRA, but not how to do it.

The scope of the advanced Non-LWR PRA standard includes the following [1]:

a) Sources of radioactive material within and outside the reactor core
b) Different plant operating states and shutdown modes
c) Internal and external initiating events (excluding sabotage/terrorism)
d) Different sequence end states, plant damage states, or release categories
e) Evaluation of risk metrics based on sequence end states
f) Quantification of event sequence frequencies, mechanistic source terms, offsite radiological consequences, risk metrics, and associated uncertainties

Unlike the LWR PRA standards, the Non-LWR PRA standard does not use the level-1, level-2, and level-3 terminology. Instead, a single standard encompasses the complete PRA analysis, from initiating event to offsite consequence calculation. This was done since the three-level PRA designation may not be appropriate for some advanced reactor designs, such as those that have liquid fuel. The Non-LWR PRA standard includes the 18 PRA elements seen in Table 1. Like other PRA standards, the Non-LWR PRA standard includes general high level requirements (HLRs) for each PRA element. In order to meet the HLRs, a series of supporting requirement (SRs) must be satisfied. Many of the SRs have different needs depending on the capability category. The capability category is a three-level designation that is determined based on the maturity of the design and the intended application of the PRA.

Table 1: Non-LWR PRA Standard: PRA Elements [1]

A	Plant Operating State Analysis (POS)
B	Initiating Event Analysis (IE)
C	Event Sequence Analysis (ES)
D	Success Criteria (SC)
E	Systems Analysis (SY)
F	Human Reliability Analysis (HR)
G	Data Analysis (DA)
H	Internal Flood PRA (FL)
I	Internal Fire PRA (FI)
J	Seismic PRA (S)
K	Other Hazards Screening Analysis (EXT)
L	High Winds PRA (W)
M	External Flooding PRA (XF)
N	Other Hazards PRA (X)
O	Event Sequence Quantification (ESQ)
P	Mechanistic Source Term Analysis (MS)
Q	Radiological Consequence Analysis (RC)
R	Risk Integration (RI)

There are several HLRs and SRs that are particularly important when reviewing past SFR PRAs. The first topic is the treatment of uncertainty. While the Non-LWR PRA standard does not contain a separate uncertainty PRA element, it embeds the requirements for addressing uncertainty within other elements. For example, the event sequences (ES) PRA element mandates that the sources and assumptions associated with the ES uncertainty must be identified, and the evaluation of those uncertainties must be documented [1]. How this process is carried out is left to the analyst performing the PRA. As with previous PRA standards, capability category I mandates interval estimates for uncertainty, while capability categories II and III require more detailed uncertainty characterization depending on the significance of the uncertainty. Once again, the standard dictates what must be done, but not how to do it.

Next, the mechanistic source term analysis (MS) element describes what must be done to properly model and characterize the source term. The use of a mechanistic source term is a drastic change for advanced Non-LWR designs compared to the current LWR fleet, which uses a postulated source term. As the next section will show, the NRC has criticized past attempts to characterize the source term for SFRs. The Non-LWR standard mechanistic source term HLRs include topics like defining the release categories, establishing the mechanistic analysis method for each category, and the quantitative evaluation of the characteristics of each category. As mentioned above, this includes uncertainty and sensitivity analyses of both parameter and model uncertainty.

For common cause failure analysis, capability category I of the standard mandates the use of the beta-factor approach [12], or equivalent. Capability categories II and III permit the use of the alpha factor model [13], and more sophisticated approaches, like the multiple greek letter model [13]. For human reliability analysis, the standard mandates the use of a systematic method like THERP [14] or ASEP [14] regardless of the capability category.

Lastly, most SFR designs have included passive safety systems, or those systems that require little to no human action and electrical power. How to properly assess the reliability of such systems has been debated for over two decades. The Non-LWR PRA standard does include several references to this issue, including HLR-SC-B5, which mandates the use of mechanistic models supported by empirical data and the characterization of uncertainties associated with safety functions performed by passive means, including those using natural physical processes [1].

3. PAST U.S. SFR PROBABILISTIC RISK ASSESSMENTS

For each of the four past U.S. SFR PRAs, the basic characteristics of the analysis are provided, including the application, scope, and level of detail, which are important inputs when determining the requirements of the new Non-LWR PRA standard. This is followed by any comments from the NRC, if the PRA was reviewed as part of the licensing process.

3.1. Clinch River Breeder Reactor

An overview of the Clinch River Breeder Reactor (CRBR) PRA can be found in Table 2. While the CRBR project was cancelled before the completion of the PRA, there was considerable time and effort dedicated to constructing the analysis. The CRBR was planned as a single 975 MWt, 350 MWe loop-type SFR that was to be built next to Oak Ridge along the Clinch River in Tennessee. CRBR was designed as an oxide-fueled breeder reactor. The project began in 1970, and PRA development started in 1981, two years before the project's termination in 1983.

The development of the CRBR PRA was divided into two phases. Phase I centered on the creation of a tentative list of initiating events and the construction of system fault trees and core damage, core-response, and containment-response event trees. Phase II was to focus on the radionuclide release, health and risk analysis, uncertainty analysis, and common cause failure analysis. However, only the independent common cause failure analysis element of phase II was conducted before the program's termination.

Table 2: CRBR PRA Overview

Metric	CRBR PRA Characteristic
Application	PRA prepared as a supplement for licensing and to aid in design process. Original Preliminary Safety Analysis Report (PSAR) did not include a PRA. Project cancelled before completion of PRA.
Siting	Site specific: Tennessee Valley Authority site on the Clinch River, adjacent to Oak Ridge, Tennessee.
Scope	Power Levels – Full power only Initiating events – Internal initiators (including fire, liquid-metal interactions, internal flood, and missiles) and external events (seismic, tornadoes, wind, and aircraft impact)
End States	Core damage states, core response end states, and containment response end states
Level of Detail	Conceptual design Detailed fault trees for all front-line and essential support systems (except RSS[1]), common cause failure not explicitly modeled in fault trees Accident sequences not modeled mechanistically Human reliability analysis (Combination of THERP [14], OATS [15], and SLIM [16])
Consequence	No radiological consequence analysis or source term calculation
Uncertainty	No detailed uncertainty analysis

[1] RSS – Reactor Shutdown System

The CRBR PRA contained three separate event trees, as shown in Figure 1. The first in the series of event trees, the core-damage tree, consisted of seven top events, including reactor shutdown and short and long-term cooling. However, it is important to note that the accident sequences described in the event trees were not modeled mechanistically, meaning they are postulated sequences that were not derived from a system analysis code or model. Detailed fault trees were created for the systems depicted by top events in the core-damage event tree that had the potential for a significant impact on the frequency of core damage. This included front-line and supporting systems of safety and non-safety-related systems. This analysis also incorporated a detailed human reliability assessment using a combination of THERP [14], OATS [15], and SLIM [16]. As noted earlier, common cause failure was not addressed until phase II of the analysis.

Figure 1: CRBR PRA Structure

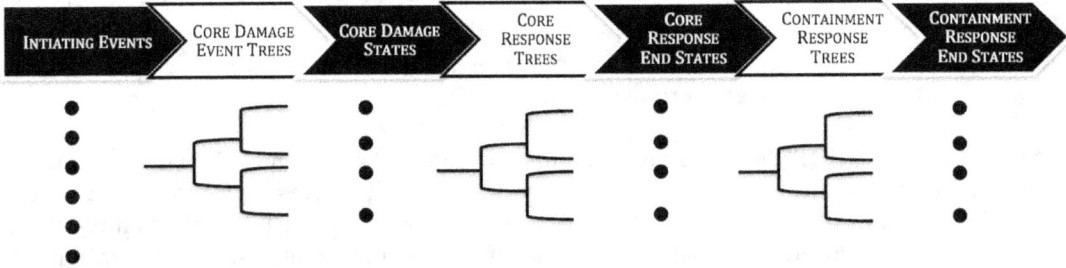

The core damage states were inputs to the core response event trees, which describe core energetics, reactor vessel integrity, and fuel debris coolability. If energetics did occur, or if there was a loss of sodium within the reactor vessel, the plant would be in one of six core response end states where significant core damage occurred with a breach of the primary system boundary. These events were then passed to the containment response event trees, which determined the magnitude of the radiological consequences, and represented such events as containment isolation and liner integrity. In total, there were 180 sequences modeled in the containment-response event tree, which were grouped into 13 bins based on similar release characteristics. As mentioned at the start of this section, the source term for the containment-response tree end states and offsite radiological consequences were not calculated before the termination of the CRBR project. This is also true of the uncertainty analysis.

3.2. Sodium Advanced Fast Reactor (SAFR)

An overview of the SAFR PRA is shown in Table 3. It was provided as part of the Preliminary Safety Information Document (PSID) [3]. SAFR was a planned 900 MWt, 350 MWe pool-type SFR designed by the Rocketdyne Division of Rockwell International. The complete plant site would consist of four independent reactor modules. SAFR was designed to use metallic fuel (although the PRA considers both metallic and oxide fuel) with HT-9 cladding, and would have a breeding ratio greater than one. The SAFR conceptual design PRA was relatively simple compared to other PRAs in this section, since it was only partially completed before the project's termination in 1988. However, it did include an offsite consequence analysis and an attempt at uncertainty quantification. Also, the NRC reviewed the PSID, including the PRA, before the project was cancelled, and documented their findings in NUREG-1369 [6].

Table 3: SAFR PRA Overview

Metric	SAFR PRA Characteristic
Application	PRA provided as appendix to Preliminary Safety Information Document (PSID) for licensing.
Siting	Generic site on Northeastern seaboard of the United States. Population distribution based on licensability for 75% of existing United States LWR sites.
Scope	Power Levels – Full power only Initiating events – Internal initiators (only TOP[1], LOF[2], (U)LOHS[3]), external initiators (limited seismic only)
End States	Plant damage states, release categories, risk measure (latent and acute fatalities)
Level of Detail	Conceptual design No detailed fault trees (some failure state diagrams) Accident sequences not modeled mechanistically No human reliability assessment
Consequence	Source term includes timing of release and core inventory fractions, but based on study of oxide fuel, then scaled to metal fuel. Offsite dispersion calculated using CRAC-2 [17].
Uncertainty	Lognormal distribution used for all input parameter uncertainties based on error factor of 10. Uncertainty propagated by method of moments and discrete probabilistic arithmetic method. Passive reactivity feedback based on engineering judgment and FFTF and TREAT tests.

[1] TOP – Transient overpower
[2] LOF – Loss of flow
[3] (U)LOHS – (Unprotected) Loss of heat sink

As shown in Figure 2, the SAFR PRA consisted of two event trees and an offsite dispersion calculation. However, this description is a bit misleading. Unlike the CRBR PRA, the initiating events include the initial plant response, such as the plant protection system. There were no fault trees developed for any SAFR systems (although some were modeled with failure state diagrams). The SAFR PRA only considered four initiating events, TOP, LOF, LOHS, and ULOHS, but did investigate both metallic and oxide fuel. A very limited examination of seismic initiators was also conducted.

Figure 2: SAFR PRA Structure [6]

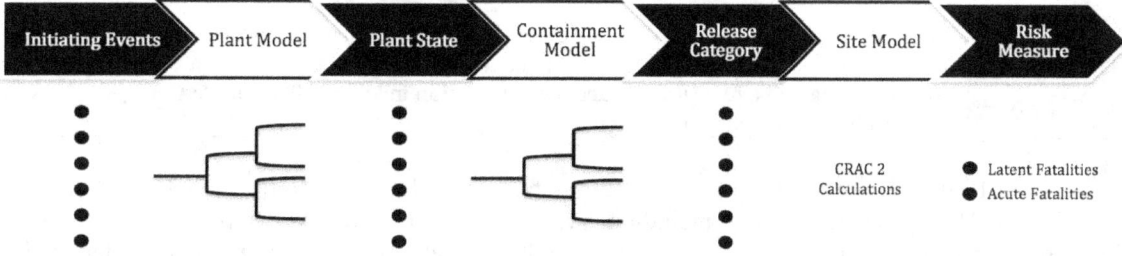

The plant model event tree (also called the core-vessel response) represented events such as energetics, vessel integrity, and in-vessel debris coolability. The containment model event tree included containment isolation, ex-vessel coolability, and transport. However, few containment event tree branches were actually quantified in the analysis before the project's cancellation. The offsite dispersion analysis reviewed 20 release categories, but the source term was not mechanistically derived, and oxide fuel data was scaled to reperesent metal fuel releases. A preliminary uncertainty analysis was also conducted, where a lognormal distribution with an error factor of 10 was assumed for all uncertainties. Then the method of moments and discrete probability arithmetic techniques were used to propagate uncertainties.

In its review of the SAFR PSID, the NRC criticized the simplistic treatment of systems faults, human error, common-cause failures, and passive feedback uncertainty, although it was understood that the fidelity of some of these analyses would have increased without the project's termination. The use of scaled oxide fuel data for the creation of the metal fuel source term was also deemed insufficient by the NRC [6].

3.3. Experimental Breeder Reactor II (EBR-II)

An overview of the EBR-II PRA can be found in Table 4. It is the only reactor of those reviewed in this document that was built and operated. EBR-II is a 62.5 MWt, 20 MWe pool-type SFR that was built in 1965 in Bingham County, Idaho. EBR-II used metal fuel with various clad types, and as the name suggests, was capable of breeding. In 1994, the reactor was permanently shut down after almost 30 years of operation. In 1988, a National Academy of Science study recommended that a PRA be performed for all U.S. Department of Energy (DOE) Class-A reactors, which included EBR-II. A final version of the PRA was released in 1991. The PRA only evaluates to the point of core damage (considered a level-1 PRA in LWR-space).

Table 4: EBR-II PRA Overview

Metric	EBR-II PRA Characteristic
Application	PRA created over 25 years after initial plant operation as part of a DOE study to assess reactor safety.
Siting	Specific site: actual location of reactor in Bingham County, Idaho.
Scope	Power Levels – Full power only (refueling treated independently) Initiating events – Internal initiators (including liquid-metal interactions), external initiators (fires, Na fires, tornadoes, high winds, volcanism, floods, lightning, aircraft impact, industrial accidents, missiles, internal flood), seismic not included in final PRA
End States	Plant transient category, fuel damage category
Level of Detail	As-built plant design Detailed fault trees (beta factor for common cause failure) Accident sequences modeled mechanistically using SAS4A/SASSYS-1 [18] Human reliability analysis included (THERP [14])
Consequence	No radiological consequence analysis or source term calculation (PRA ends at core damage)
Uncertainty	Included data (parameter), model, and completeness uncertainties. Lognormal distribution used for all parameter uncertainties other than reactivity feedbacks, which used screening and a response surface. Quantification involved 5000 Monte Carlo samples. Sensitivity and importance analyses also conducted.

As shown in Figure 3, the EBR-II PRA structure is fairly simple, but overall the EBR-II PRA is the most detailed of those reviewed here, including a vast range of internal and external initiating events. The plant event trees cover such events as scram and decay heat removal. For each system depicted in the plant event trees, a detailed fault tree was constructed, which included both the frontline system and many supporting systems, such as instrument air. Common-cause failure was modeled explicitly in the fault trees through the use of the beta factor technique. Much of the reliability data used to quantify the fault trees was based on the 20+ years of EBR-II operation prior to the PRA development. If fuel damage was encountered, then the event was passed to the fuel damage sub-tree. This tree calculated the time at temperature for the fuel pins and the likelihood of sodium boiling. This information was found by mechanistic modeling using SAS4A/SASSYS-1.

Figure 3: EBR-II PRA Structure

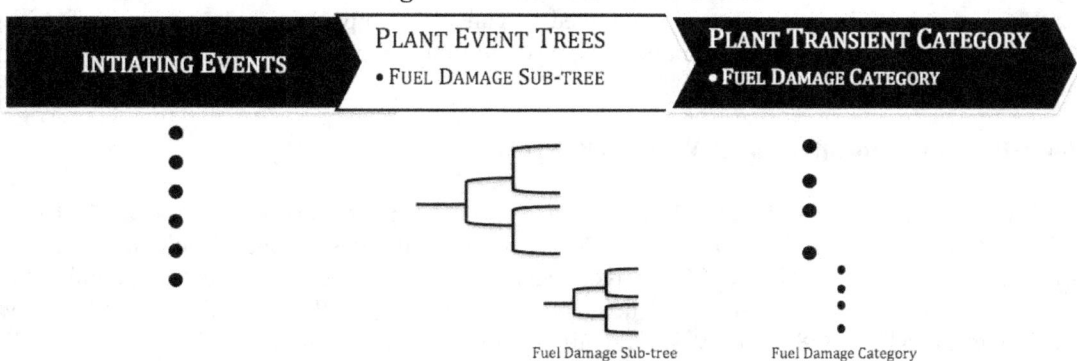

The EBR-II PRA contained a relatively detailed human reliability analysis that centered on human action leading to failure to scram and failure to recover decay heat removal. Since EBR-II was operational at the time of the PRA development, interviews with plant staff were used during the human reliability assessment. The training, procedures, maintenance, control system, and operations information were analyzed using THERP [14]. Human reliability event trees were constructed for actions that could lead to possible system failure.

Several techniques were used to assess the uncertainty within the PRA. First, the uncertainties associated with inherent reactivity feedbacks were addressed through the multistep process seen in Table 5. Due to limited computational resources at the time, parameter screening and response surface methods were used to analyze the reactivity feedback uncertainty. For each unprotected accident sequence, a quasi-static reactivity balance, which included several uncertain parameters, was used to screen initiators. For each uncertain parameter in the reactivity balance, a normal distribution was assumed, and the unprotected accident sequences were investigated. If the uncertainty had negligible impact on the peak cladding temperature of the sequence, it was dropped. If not, a detailed analysis of the sequence was necessary. Next, a SAS4A/SASSYS-1 model was used to screen important uncertain parameters in the reactivity balance for each ATWS sequence. Then, more detailed probability density functions (PDFs) were assigned to the important uncertain parameters. The PDFs of uncertain parameters were used as inputs for a select number of SAS4A/SASSYS-1 calculations, which were used to create a response surface. Finally, roughly one million Monte Carlo sampling calculations were conducted to propagate the important uncertainties using the response surface.

Table 5: EBR-II Reactivity Feedback Uncertainty Analysis Procedure

Step	Description
1	Screen Accident Initiators
2	Calibrate EBR-II Model in SAS4A/SASSYS-1 Transient Code
3	Screen Parameters
4	Quantify Parameter Uncertainties
5	Develop Experimental Constraints
6	Compute Response Surfaces
7	Propagate Parameter Uncertainties Subject to Experimental Constraints
8	Assess Accuracy of Failure Probabilities

For data (parameter), model, and completeness uncertainties, different approaches were used. For data uncertainty, such as the reliability data used in the fault trees, lognormal distributions were assigned to the parameters and 5000 Monte Carlo simulations were used to propagate the uncertainties through the fault and event trees. This process allowed distributions to be found for all output metrics. It should be noted that the uncertainties associated with natural circulation within the reactor vessel were ignored, and natural circulation was assumed to occur. As for completeness uncertainty, the EBR-II PRA states that the PRA is "complete," regarding the scope of the project, as a result of an extensive search of plant records and system models [4]. Model uncertainty was handled through conservative, simplifying assumptions, and separate human error and common mode failure analyses. As part of these analyses, a sensitivity analysis was conducted to gauge the impact of these and other factors. Lastly, an importance calculation was conducted in order to rank the basic events by their contribution to damage frequency.

3.4. Power Reactor Innovative Small Module (PRISM)

An overview of the PRISM can be found in Table 6. PRISM (Mod-A) was designed as a 470 MWt, 180 MWe pool-type SFR by General Electric Co. The conceptual design could include up to nine metal-fueled reactors per site. The PRISM project began in 1984, and the PSID was first submitted in 1986 (with several revisions following). The NRC reviewed the PSID, including the PRA, and responded with NUREG-1368 in 1989 [19], with continued revisions until 1994. However, DOE funding for the program was terminated in the early 1990s.

Table 6: PRISM PRA Overview

Metric	PRISM PRA Characteristic
Application	PRA provided as appendix to Preliminary Safety Information Document (PSID) for licensing.
Siting	Siting envelope using GESSAR II[1], which encompassed the majority of potential reactor sites.
Scope	Power Levels – Full power only Initiating events – Internal initiators (reactivity insertions, heat removal faults, liquid-metal interactions), external events (limited seismic only)
End States	Accident types, core damage categories, release categories
Level of Detail	Conceptual design Simplified fault trees for only three systems (RPS, RSS, and EM pump coastdown), common cause failure modeled using beta factor method Limited or no mechanistic analysis of accident sequences Very limited human reliability analysis
Consequence	Source term includes timing of release and core inventory fractions, but based on study of oxide fuel, then scaled to metal fuel. Offsite dispersion calculated using MACCS [20].
Uncertainty	No uncertainty analysis, best-estimate values only. Several sensitivity studies performed.

[1] The GESSAR (General Electric Standard Safety Analysis Report) II PRA evaluates a BWR plant design, however the siting characteristics are considered applicable to many U.S. locations.

Like the CRBR PRA, the PRISM PRA structure, shown in Figure 4, included three event trees. The system event tree, which was first of the three, included events such as reactor SCRAM, pump coastdown (due to the use of electromagnetic pumps), and decay heat removal. Only three of the top events in system event trees had an associated fault tree: the reactor shutdown system, the reactor protection system, and the electromagnetic pump coastdown. The fault trees are relatively simple and do not include supporting systems or human error, but do use the beta factor approach for common-cause failure. Much of the data used to quantify the trees were derived through judgment, and the NRC believed that many of the system failure probabilities had been understated [19].

Figure 4: PRISM PRA Structure

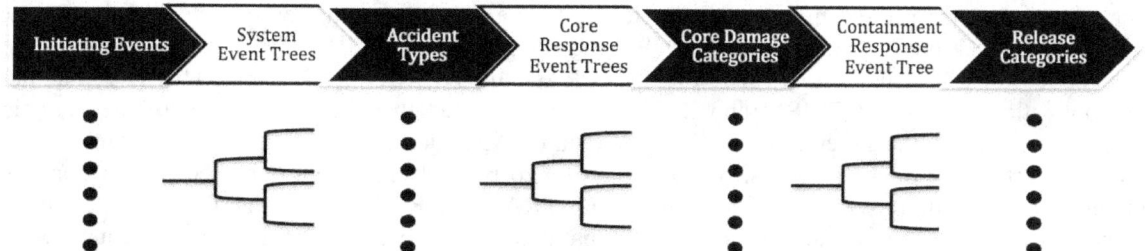

The core response event trees included events such as sodium boiling, clad failure, and energetic release. These sequences were not modeled mechanistically, and as with SAFR, oxide fuel experiments were used as the basis for the analysis, then scaled for metal fuel. The containment response event trees included events such as vessel failure, debris coolability, and late energetics. The output of this analysis was used in the offsite dispersion code MACCS [20], which calculated prompt and latent fatalities.

No comprehensive uncertainty analysis was conducted for the PRISM PRA, although sensitivity studies were performed to asses the impact of initiating event frequencies and the scaling factors used to modify oxide fuel data for metal fuel. More comprehensive uncertainty analyses were planned before the project was cancelled. The NRC found several major sources of uncertainty that would need further analysis in future PRISM PRA revisions. These included the lack of design detail for plant systems (including supporting systems, common mode failures, and human factors), limited test data and experience (related to the electromagnetic pump coastdown, seismic isolators, natural convection decay heat removal systems, and passive feedbacks), human reliability, mechanistic fuel modeling, and metal fuel source term data [19].

4. DISCUSSION

Of the four past SFR PRAs reviewed in this paper, three projects were cancelled before the design reached completion, and one PRA was performed after 25 years of reactor service. Even with those drawbacks, there are lessons that can be learned from these analyses. Table 7 has an overview of three deficiencies seen in past SFR PRAs.

Table 7: Past SFR PRA Deficiencies

Deficiency	Description
Mechanistic scenario modeling	Other than the EBR-II PRA, very little mechanistic modeling (*i.e.*, computer model simulation) was conducted for scenarios depicted in PRA, including the risk dominant sequences. The NRC criticized this deficiency. There was also a lack of mechanistic modeling of passive system performance and reliability during transients, which is now required as part of the ASME/ANS PRA standard.
Mechanistic source term analysis	Of those PRAs that conducted a source term analysis, scaled oxide fuel source terms were used in place of mechanistic metal fuel source terms. The NRC criticized this deficiency, and believed this method may underestimate or mischaracterize the metal fuel source terms.
Uncertainty analysis	Other than the EBR-II PRA, little effort was directed to uncertainty analyses. While this deficiency can be partially accredited to project cancellations, uncertainties were not addressed from the initial stages of the analysis, which is required by the ASME/ANS PRA standard. This was a point of criticism from the NRC. It is also true for the metal fuel performance and passive system reliability analyses.

While all of the SFR designs reviewed here used mechanistic scenario modeling for parts of the safety analysis, there was limited mechanistic modeling of many of the scenarios depicted in the PRAs, including the risk-dominant scenarios, for all of the designs except EBR-II. The NRC highlighted this weakness for both the SAFR and PRISM PRAs. In NUREG-1368 and NUREG-1369, the preapplication reviews of PRISM and SAFR, the NRC stated that due to the lack of mechanistic sequence modeling, the generic assumptions made in the PRA do not accurately represent some of the more important accident sequences, or do not appear to properly represent expected plant responses under specific conditions. In its review of the SAFR PRA, the NRC [6] stated that during fuel damage "mechanistic analyses have not been performed that could otherwise support the generic sequences in this portion of the PRA." A similar criticism was made by the NRC in regard to the accuracy of the metal source term developed for both analyses, with multiple comments directed at the use of scaled oxide fuel data.

The NRC was also critical of both the PRISM and SAFR PRAs for their minimal treatment of uncertainties related to metal fuel behavior, passive system operation, and source term calculations. Even though these projects were cancelled before the PRA and detailed uncertainty analyses could be completed, it does reflect on how uncertainties were to be treated in the analyses. Only point estimate values were used in the PRISM PRA, without identifying the uncertainty interval for input parameters. This would not satisfy even the lowest capability category of the Non-LWR PRA standard, which states that uncertainty intervals must be estimated and the results should include the point estimate and uncertainty bound [1]. The SAFR PRA attempted a simple uncertainty quantification process using a lognormal distribution with an error factor of 10 for all uncertainties, but the level of detail in the PRA was very limited. The NRC questioned the completeness of both the SAFR and PRISM PRAs, and repeatedly stated that the modeling of certain phenomena was not mature enough to forgo a detailed uncertainty analysis. What level of maturity will be adequate for regulatory modelling is an important issue that must be addressed by any future SFR, but it is outside the scope of this work. Even the EBR-II PRA, which is the most detailed of those reviewed here, did not include an uncertainty assessment of the performance of passive decay heat removal, such as the development of natural circulation within the reactor vessel.

Outside of those major deficiencies, there are also shortcomings related to common cause failure and human reliability analysis. For common cause failure analysis, only the EBR-II and PRISM PRAs could meet the capability category I requirement of the Non-LWR standard by using the beta-factor approach. However, the PRISM PRA had simplified fault trees for only three plant systems. The SAFR and CRBR PRAs did not explicitly model common cause failure within the PRA. Both the EBR-II and CRBR PRAs used THERP for their human reliability analysis, which is an acceptable approach in the Non-LWR standard. SAFR and PRISM did not conduct detailed human reliability analyses in the PRA versions reviewed here.

5. CONCLUSION

While the SFR PRAs reviewed here have certain limitations, as described in the previous section, they provide an excellent starting point for future SFR analyses. This is especially true of the EBR-II PRA, which meets many of the Non-LWR standard requirements outlined in Section 2. All of the past SFR PRAs provide information about the types of initiating events that analysts believed were possible for SFRs, and also information regarding how they foresaw the plant response for such transients. The CRBR and EBR-II PRAs give insight into the development of fault trees for sodium specific components and systems, and preliminary data on human reliability analyses. The EBR-II PRA in particular provides very detailed fault trees that included both common cause failure and human actions explicitly.

There are other lessons to be learned from the deficiencies found in these PRAs too. First, the treatment of uncertainty should begin from the initial stages of the analysis. Not only is this now required by the Non-LWR standard, but also is important for the identification of influential uncertainties. Second, demonstrating the reliability of passive safety systems for risk dominant accident sequences is an important task for SFRs, and one which should be completed using mechanistic analyses and taking into account uncertainties. Lastly, the creation of a mechanistic source term is largely an open question for advanced reactors. While none of the reviewed PRAs attempted a mechanistic source term, the strategy used by EBR-II to determine core damage may provide a good starting point. In their analysis, mechanistic simulation was used as a basis to determine time-at-temperature values for regions of the core, and could lead into more detailed fuel damage calculations, if sufficient models exist.

Acknowledgements

Argonne National Laboratory's work was supported by the U.S. Department of Energy, Assistant Secretary for Nuclear Energy, Office of Nuclear Energy, under contract DE-AC02-06CH11357.

References

[1] ASME/ANS Joint Committee on Nuclear Risk Management, *"Probabilistic Risk Assessment Standard for Advanced Non-Light Water Reactor Nuclear Power Plants,"* RA-S-1.4-2013, (2013).
[2] Technology For Energy Corporation, *"Clinch River Breeder Reactor Plant Probabilistic Risk Assessment,"* CRBRP-4, (1984).
[3] U.S. Department of Energy, *"SAFR – Sodium Advanced Fast Reactor, Preliminary Safety Information Document,"* AI-DOE-13527, Export Controlled, (1985).
[4] Argonne National Laboratory, *"Experimental Breeder Reactor II (EBR-II) Level 1 Probabilistic Risk Assessment,"* EBR-II PRA, (1991).
[5] General Electric, *"Power Reactor Innovative Small Module (PRISM) Preliminary Safety Information Document,"* GEFR-00793 UC-87Ta, (1987).
[6] U.S. Nuclear Regulatory Commission, *"Preapplication Safety Evaluation Report for the Sodium Advanced Fast Reactor (SAFR),"* NUREG-1369, (1991).
[7] ASME/ANS Joint Committee on Nuclear Risk Management, *"Standard for Level 1/Large Early Release Frequency Probabilistic Risk Assessment for Nuclear Power Plant Applications,"* RA-Sa-2009, (2009).
[8] ANS Risk Informed Standards Committee, *"Low Power Shutdown PRA Methodology,"* ANS-58.22.
[9] ASME/ANS Joint Committee on Nuclear Risk Management, *"Severe Accident Progression and Radiological Release (Level 2) PRA Methodology to Support Nuclear Installation Applications,"* ANS/ASME-58.24.
[10] ASME/ANS Joint Committee on Nuclear Risk Management, *"Standard for Radiological Accident Offsite Consequence Analysis (Level 3 PRA) to Support Nuclear Installation Applications,"* ANS/ASME 58.25.
[11] ASME/ANS Joint Committee on Nuclear Risk Management, *"Advanced Light Water Reactor (ALWR) PRA Standard,"* ASME/ANS Ra-S 1.5.
[12] U.S. Nuclear Regulatory Commission, *"Procedures for Treating Common Cause Failures in Safety and Reliability Studies,"* NUREG/CR-4780, (1988).
[13] U.S. Nuclear Regulatory Commission, *"Procedures for Modeling Common Cause Failures in Probabilistic Risk Assessment,"* NUREG/CR-5485, (1998).
[14] U.S. Nuclear Regulatory Commission, *"Handbook on Human Reliability Analysis with Emphasis on Nuclear Power Plant Application,"* NUREG/CR-1278, (1983).
[15] J. Wreathall, "*Operator Action Trees, An Approach to Quantifying Operator Error Probability During Accident Sequences,*" NUS Corporation, NUS Report No. 4159.
[16] U.S. Nuclear Regulatory Commission, "*The Use of Performance Shaping Factors and Quantified Expert Judgment in the Evaluation of Human Reliability: An Initial Appraisal,*" NUREG/CR-2986, (1983).
[17] L. T. Ritchie, et al., "*CRAC2 Model Description,*" NUREG/CR-2552 SAND82-0342, (1984).
[18] Editor: T. H. Fanning, "*The SAS4A/SASSYS-1 Safety Analysis Code System: User's Guide,*" Nuclear Engineering Division, Argonne National Laboratory, ANL/NE-12/4, (2012).

[19] U.S. Nuclear Regulatory Commission, "*Preapplication Safety Evaluation Report for the Power Reactor Innovative Small Module (PRISM) Liquid-Metal Reactor,*" NUREG-1368, (1989-1994).
[20] H-N Jow, J. L. Sprung, J. A. Rollstin, L. T. Ritchie, and D. I. Chanin, "*MELCOR Accident Consequence Code System (MACCS), Model Description,*" NUREG/CR-4691 SAND86-1562, (1990).

Applicability of PSA Level 2 in the Design of Nuclear Power Plants

Estelle C. SAUVAGE[a], Gerben DIRKSEN[b], and Thierry COYE de BRUNELLIS[c]

[a] AREVA-NP SAS, Paris, France
[b] AREVA-NP Gmbh, Erlangen, Germany
[c] AREVA-NP SAS, Lyon, France

Abstract:

In the nuclear industry, until recently, the licensing and design of the new Nuclear Power Plants (NPP) were based upon a deterministic approach. The Probabilistic Safety Assessments (PSA) only supported the safety demonstration, mostly by the evaluation of the risks for the population and the environment. The feedbacks in the design of the NPP, when existing, were limited.

Nowadays the use of the PSA becomes more systematic and is extended to the design phase of the new generation of NPP. In this frame the first approach was to develop the concepts of risk based and risk informed decision making to avoid unnecessary burden taking place in the NPP design due to the strong deterministic prescription on low probability events.

Following the development of a new generation of plants, such as the AP1000 or the EPR, which considers the severe accidents in their design, the PSA Level 2 tends to contribute more and more to build the new NPPs. The accident of Fukushima Daishi NPPs even leads to an extended consideration of the severe accident in the design of the nuclear plants and the emergency organization structures.

The interaction between the PSA Level 2 development and the design phase of the NPP became obvious, and part of the safety standards as recommended by the safety authorities and organizations.

This paper assesses how the PSA Level 2 becomes a high visibility topic of the design phase of the NPP. The current safety requirement expectations regarding the use of the PSA Level 2 in the design phase results from this evolution.

Indeed several technical areas can use the insight of a PSA Level 2 to improve the NPP design. It includes the design of hardware and systems (e.g., pipes, valves and tank but also instrumentation and control and civil engineering). It also includes the analysis of human factors, which subject covers the procedures and guidelines, the Human Machine Interface (HMI), the emergency organization, the training and the layout (access to buildings, survivability of the control room...). Two examples of the use of PSA Level 2 for EPR design improvement are provided and reviewed: first the modification of the severe accident spraying system, and second the HMI evaluation for the severe accident.

The use of PSA Level 2 in the design phase depends of the model and the level of detail of the developed probabilistic analysis. Discussions on the areas of improvement regarding the use of the PSA Level 2 in the development of a new NPP are proposed.

Keywords: PSA Level 2, NPP Design, Systems, Severe Accident Management.

1. INTRODUCTION

The severe accident PSA, so called PSA Level 2, is nowadays an integrated part of the safety demonstration. The initial design phase of the new generation of plants, the verification of safety targets, the conception or modification of the mitigation system and the assessment of the human reliability use both the deterministic and the probabilistic approaches.

For the last generations of NPP the verification of the risk based safety criteria have used the results of the PSA Level 1 and 2 to demonstrate the compliance of the NPP with the national requirements provided by the national safety authorities. The International Atomic Energy Agency (IAEA) safety guides (Ref. [1]) state that the overall results of the PSA Level 2 should be compared with the probabilistic safety criteria with the aim to determine whether the risk criteria or targets have been met or whether additional features for prevention or mitigation of accidents need to be provided. Failure to comply with this requirement, as not acceptable on a safety point of view, obviously leads to systems or procedures/guidelines changes.

Indeed, for future nuclear power plants, rather than defining probabilistic criteria, INSAG (Ref. [2]) has proposed the practical elimination of accident sequences that could lead to large early radioactive release, whereas severe accidents that may induce to late containment failure would be considered in the design process with realistic assumptions and best estimate analysis so that their consequences would necessitate only protective measures limited in area and in time. With this approach the design of severe accident mitigation systems becomes a major goal of the deterministic and probabilistic safety assessments.

As part of the IAEA safety standard (Ref. [1]) the recommendation for the use of the PSA for a risk informed approach is strong. The aim of applying a risk informed approach is to ensure that a balanced approach is taken when making decisions on safety issues by considering probabilistic risk insights with any other relevant factors in an integrated manner. It is stated that in any of the applications of the PSA Level 2 described below, the insights from the PSA should be used as part of the process of risk informed decision making that takes account of all the relevant factors when making decisions on issues related to the prevention and mitigation of severe accidents at the plant:
any mandatory requirements that relate to the PSA application being addressed (which would typically include any legal requirements or regulations that need to be complied with);
the insights from deterministic safety analysis;
any other applicable insights or information (which could include a cost–benefit analysis, remaining lifetime of the plant, inspection findings, operating experience, doses to workers that would arise in making necessary changes to the plant hardware, environmental protection concerns, etc.).

The pros and cons require a discussion on the PSA uncertainties that need to be identified, understood and studied to gain confidence in the risk informed approach.

In addition as stated in Ref. [1] the PSA Level 2 report should clearly document important findings including:
- plant specific design or operational vulnerabilities identified;
- key operator actions for mitigating severe accidents;
- potential benefits of various engineered safety systems;
- areas for possible improvement in operations or hardware for the plant and the containment in particular.

Successful application of the PSA Level 2 includes the probabilistic evaluation of plant design to identify potential vulnerabilities in the mitigation of severe accidents, and of the development of severe accident management guidelines that can be applied following core damage. EPR based examples are provided here below.

For generation IV NPP the approach is going further. As detailed in Ref. [3], the Advanced Sodium Technological Reactor for Industrial Demonstration (ASTRID), a demonstration plant to be commissioned in the 2020 decade, is going further in the use of PSA at the conceptual design stage to support the design hypothesis. At this stage, the PSA developed by Commissariat à l'Energie Atomique et aux Energies Renouvelables (CEA) and its partners, AREVA NP and Electricité de France (EDF), aims at providing probabilistic insights to assess design choices and to highlight the weaknesses of the design under safety considerations. Currently only a PSA Level 1 is developed, but a PSA Level 2 is under study in particular to assess the design of the severe accident cooling system.

2. PROBABILISTIC EVALUATION OF A DESIGN CHANGE

2.1. Context of the PSA Level 2 Assessment for Hardware Design Changes

The design changes for the new NPP can benefice from the PSA Level 2 investigations. It can be used to assess the impact of a new systems used in severe accident conditions or the modification of an existing severe accident system. Any change of configuration of an existing system may lead to technical issues that a preliminary impact evaluation on the PSA Level 2 results can identified.

For the standard EPR the connection of the Containment Heat Removal System (CHRS) active flooding line downstream the flooding valve design change was suggested to be evaluated in the frame of the PSA Level 2 study, in order to assess:
- the benefits of this connection modification in term of PSA Level 2 results due the possibility to actuate the active flooding of the spreading area even in case of a failure of the passive flooding,
- the efficiency of measures such as procedures and administrative controls on the Main Control room (MCR) panel to avoid any spurious flooding by an operator error, including for the severe accident scenarios.

In this example the active flooding line connection upstream the passive flooding valves was proposed to be connected downstream of these valves. The impacts on the PSA Level 2 results were assessed.

Currently the suggested risk metrics used to express the frequency results of the PSA Level 2 are expressed in term of Large Release Frequency (LRF) and Large Early Release Frequency (LERF). For the standard EPR a release is large if bigger than 100 TBq of Cesium. A release is early if before or directly concomitant to the vessel rupture. The detailed risk results of the PSA Level 2 are then expressed in term of fraction of initial Cesium, Iodine and Strontium core inventory. But the scope of the study did not include a detailed quantification of the impact of the design modification on the risk results. It was rather limited to the evaluation of the frequency results of the PSA Level 2.

2.2. Probabilistic Assessment of the CHRS Design Change

This evaluation started with the review of all scenarios impacted by the modification. The scenarios of concern cover the normal, incidental, accidental, severe accident and maintenance domains. We consider that any scenarios could evolve to a severe accident situation. In such case phenomenological, system availability and source term impacts on the PSA Level 2 were assessed (qualitative assessment).

Following the design modification an In-containment Refueling Water Storage Tank (IRWST) draining into the core catcher was possible, due to a spurious actuation of active flooding line or an operator error to open the active cooling valves instead of the back-flushing valves. It was also considered that an operator can open the active cooling valves instead of back flushing valves in some sequences, or that the operators could miss the opening of the active cooling valves when needed in some other sequences.

A list of impacted systems by the proposed CHRS design change was set up. This list was indeed derived from this list of impacted scenarios, and includes:

- the CHRS,
- the core catcher:
  The risk is an early flooding of the core catcher. From the core catcher point of view, being flooded is against the functional requirements, and will require draining and cleaning and a plant in cold shutdown.
- the IRWST:
  The status of the active flooding valves and the IRWST water level are available in the control room. If during the accident sequence the active cooling valves are in open position, the operation crew will be aware of the situation and instructs to act in consequence.

Note that the containment isolation function as modeled in the standard EPR RS model is not jeopardized by any spurious opening of the active cooling valves. These valves are part of the containment isolation valves, but the CHRS is included in a bunkered room and is in a closed loop. If the operator would spuriously open the active cooling valve the opening will not lead to releases from the plant.

2.3. Scenario of Concern Detailed Analysis

The IRWST draining into the core catcher due to a spurious actuation of active flooding line or an operator error to opens the active cooling valves instead of the back-flushing valves leads to the presence of water in the spreading area during power operation which required a mandatory shutdown. It also lowers the IRWST inventory available for accident mitigation. The impact on PSA Level 2 considered was an unavailability of the IRWST water that could evolved to core damage in a few sequences.

For the scenarios with an operator error to open active cooling valves instead of back flushing valves in incident or accidental situations, with containment spray and switch to back-flushing required, no impact on PSA Level 2 was found due to the low probability of occurrence of these sequences at power. In shutdown states the increase of the Core Damage Frequency (CDF) could be up to a factor 10.

Finally in severe accident situations a spurious actuation of the active cooling valves may be of concern, leading to an early presence of water in the core catcher. In case of spurious actuation of the active cooling valves before the vessel failure and the corium arrival in the core catcher, the presence of water in core catcher leads to an unavailability to perform its quenching function. The interaction between the water and the corium can lead to several modes of containment failure considered in the PSA Level 2.

To assess the combined frequency of all sequences leading to melt stabilization success, which sequences may be jeopardized by a spurious actuation of active cooling valves, a quantification of the standard EPR PSA Level 2 model was performed. All the severe accident sequences concerned by the analysis include a vessel failure and an early flooded core catcher by spurious actuation of the active flooding valves prior corium arrival.

Note that the quantification of the model is performed by using the software tool RiskSpectrum (RS) PSA Professional developed by Lloyd's Register. Point estimated quantifications were used in the frame of this study for the assessment of containment failure split into different Release Categories (RC).

For these sequences three severe accident phenomena are of concern: first a Fuel Coolant Interaction (FCI) leading to a steam explosion that can jeopardize the containment, second is an incomplete melt transfer leading to Molten Core Concrete Interaction (MCCI) in the transfer channel, third a containment over-pressurization due to the contact of the hot material with the water. The containment failure probability for these sequences was assessed. It was found that the risk of FCI, MCCI and containment pressurization was low enough to be considered as practically eliminated.

Another severe accident scenario of concern includes the failure of both passive flooding valves and the operator missing to open the active cooling valves for cooling the corium in the spreading area with water. In such case the failure of the core catcher system is possible if not garanted, and conservatively considered as equivalent to a failure of the containment. For all states (at power and shutdown) these sequences had very low probability of occurrence.

2.4. Technical Review Committee Conclusion

The impact of the proposed CHRS design change on the probabilistic results is limited for at power, and significantly increases the risk in shutdown by increasing the CDF. The modification was chosen not to be implemented in the standard EPR based on these results.

3. PROBABILISTIC EVALUATION OF HUMAN ACTIONS

3.1. Severe Accident Management Issues in Probabilistic Modeling

With the development of the Severe Accident Management Guidelines (SAMG) the question of the modeling of the human actions in the PSA Level 2 and of the potential feedbacks of the PSA Level 2 results in the SAMG become obvious.

In particular as the severe accident systems are integrated in the design of the new plants, modeling their failure modes become part of the safety demonstration. The human failure of the actions to start or verify the correct actuation of these systems contributes also to the PSA Level 2 results. In addition being in severe accident conditions implies that several equipments or components are failed or unavailable, and maybe repaired. The recovery of failed equipments, when part of the SAMG, should be part of the PSA Level 2.

As stated in Ref. [1] when design improvements are being considered with regard to severe accident management measures, a range of options are often available. The PSA Level 2 may be used to provide an input into the comparison of these options. For example, the PSA Level 2 could provide a basis for determining whether severe accident management measures and guidelines fully address the fourth level of defense in depth as defined in Ref. [4].

Modeling the use of SAMG in a PSA Level 2 faces several issues:
- The SAMG actions are based on guidance as opposed to step-by-step procedures. No verbatim compliance to the SAMG is required. So, depending on the SAMG structure of material, the analysis process of the situation may not be obvious.
- The present Human Reliability Analysis (HRA) techniques used in the PSA Level 1 may not be applicable. Some methods like ASEP mainly remain on time-based HEP calculation for post-fault error, focusing more on decision/diagnosis (very sensitive to time) and less on actions.
- the SAMG are under the responsibility of different actors of the emergency crisis organization. This emergency organization structure adds some complexity and sources of error. The final human or error could result from an overall organization failure,
- The SAMG allows multiple choices to the emergency crisis organization when an evaluation of the situation is performed, including the possibility to decide that no action should be performed following the evaluation.
- The SAMG actions are difficult to define on an accident time line and some are directly dependent on the modeling of system and component recoveries or reparation.

In addition modeling of human errors in a PSA Level 2 may introduce possible new containment challenges that may have unique consequence considerations. For example in the Station Blackout Sequences (SBO) the start of the containment heat removal systems de-inert the containment and allow earlier containment failures due to hydrogen burns. If human recovery actions in severe accident lead to earlier releases in the vicinity of the plant the off-site emergency planning actions may be ineffective in the population protection. This aspect of the HRA was studied at the function event level in the RS model of the PSA Level 2.

3.2. HRA Methodology for EPR

One of the major obstacles to be overcome in the development of a method to consider SAMG in the PSA Level 2 is the modeling of the human reliability associated with SAMG. For the standard and Chinese EPR a new methodology for the HRA model in the PSA Level 2 was developed. The objective was the selection of an appropriated process and quantification method of the human error in severe accident conditions. The state-of-the-art of the EPR SAMG, so called Operating Strategies for Severe Accident (OSSA), were used as an input.

Due to the specificity of the organization and of the nature of the tasks performed under severe accident conditions (multiple actors, complex chain of command, complex diagnosis, use of different procedures…), the Standardized Plant Analysis Risk-Human Reliability Analysis (SPAR-H) methodology was chosen. It provided simplified assessment of the failure mode, action vs. diagnosis failures. Furthermore SPAR-H leads to a simplified but conservative Human Error Probability (HEP) assessment via the Performance Shaping Factor (PSF) rating.

Besides the HRA process developed follow the good practices prescribed in the NUREG-1792 Ref. [5].

The methodology models the dedicated organization set up to face the severity of the accident. The needed organization relies on interacting local and national teams with different levels of knowledge and responsibilities. The chain of command involves multiple actors using different procedures or dedicated guidelines to cope with an evolving situation.

The methodology assessed the role of the emergency organization in the decision process, and if a Technical Support Center (TSC) evaluation was required. If these human actions correspond to human actions realized on the mitigation path or performed when a challenge exists the TSC evaluation was judged obvious.

3.3. Actions and Tasks Analysis for the OSSA

Actions, in MCR and performed locally, necessary for the operation of systems/functions credited in the PSA Level 2 were screened and classified in four categories depending when they were performed during the accident (at the OSSA entrance, as immediate actions, as intermediate actions or in the long term).

Then the actions were divided in tasks. A task modeling so called Hierarchical Task Analysis (HTA) and a Task Analysis (TA) allowed identifying the needed data and cues necessary to perform the actions and the required controls to actuate systems/functions. The involved required HMI could be identified. As a feedback to the HMI design the missing or inadequate interfaces can be identified during the process. Note that a TimeLine Analysis (TLA) supported the evaluation of the time available for the operators to perform the actions for the reference scenarios.

The RS software was used to develop action by action Fault Tree (FT) including all tasks, and to calculate the intermediate and final HEP; the lower level of HEP being calculated with SPAR-H. Some values were given to the defined human recovery tasks in the fault trees. Human recovery is used in severe accident as a safety mean or level of defense to enhance the human operator tasks reliability. The recovery could consist in a recovery of an operator by another one or by a member of the emergency organization. When an action HEP was high, consideration can be given to the recovery of one or several tasks linked to the action. Finally the FT models the organization reliability and not only individual human error.

According to the SPAR-H methodology dependencies between human failures were also modeled. It was chosen to value the recovery tasks trough the fault trees and not trough the work process PSF. This modeling puts the emphasis on the work organization retained in order to enhance the whole organization reliability/resilience.

Note that the consideration of recovery is more difficult with the current static PSA Level 2 model. In the development of the generation IV NPP the use of PSA Level 2 dynamic models intends to develop time/state dependant models. Currently the Research and development (R&D) programs on the Petri nets or on Monte Carlos tries coupled to a RS models allows developing PSA Level 2 models that cover the equipment recovery and the component maintenance.

3.4. Improvement of the PSA Model and OSSA following the Human Reliability Analysis

The study emphasized the impact of the dependency between actions. As an example the action to connect the severe accident batteries had not a high HEP, but by the combination of the dependencies between actions it lead to impossibility to isolate the containment, and to major impact on the LRF/LERF.

Recovery was also a requirement for some actions. The OSSA entrance initially modeled with no recovery from any member of the safety organization had a strong contribution to the early release frequencies. Monitoring of the OSSA entry condition by several operators, safety engineer or any other member of the safety organization decreased the delayed entry into the severe accident guideline and the lack of initial response to the severe accident situation.

4. CONCLUSION

It is a strong IAEA recommendation for the PSA Level 2 to be used to provide inputs into design evaluation throughout the lifetime of a NPP or during the design process for a new plant.

Currently the use of the PSA Level 2 in the design of the EPR showed some interesting feedbacks of the probabilistic safety into the design, both for the hardware aspects and the severe accident management. It emphasized the major benefits of the emergency organization in the severe accident management.

The studies can cover with a good confidence in the frequency and the source term results the impact of the modified design. However a further step is the knowledge of the uncertainty and their propagation in the PSA Level 2 can improve the PSA Level 2 and the conclusions.

Following the Fukhusima Daiichi Nuclear Power Plant (NPP) accident in Japan resulted from the combination of two correlated extreme external events (earthquake and tsunami). The consequences went beyond what was initially considered in the design of the NPP. The concept of extended PSA is reviewed. It may give in the future a new dimension for the use of PSA Level 1 and 2 in the concept and design phase of NPP.

In the future PSA Level 2 taking into consideration the reparation of the failed or malfunctioning system will change the approaches and results.

REFERENCES

[1] "Development and Application of Level 2 Probabilistic Safety Assessment for Nuclear Power Plants", IAEA Safety Standards, Specific Safety Guide n° SSG-4, 2010
[2] "Basic Safety Principles for Nuclear Power Plants", 75-INSAG-3 Rev. 1, International Nuclear Safety Advisory Group INSAG-12, IAEA, 1999
[3] Use of simplified PSA Studies to support the ASTRID design process", ICAPP 2014, USA, April 6-9th, 2014
[4] "Safety of Nuclear Power Plants Design", IAEA Safety Standards, Series No. NS-R-1, IAEA, 2000
[5] "Good Practices for Implementing Human Reliability Analysis", NUREG-1792, 2005
[6] "Severe Accident management Plan (SAMP) for Nuclear Power Plants", IAEA Safety Guide NS-G-2.15, Draft 2014
[7] "The SPAR-H Human Reliability Analysis Method", NUREG/CR-6883, 2005

ACRONYMS

ASTRID	Advanced Sodium Technological Reactor for Industrial Demonstration
CDF	Core Damage Frequency
CEA	Commissariat à l'Energie Atomique et aux Energies Renouvelables
CHRS	Containment Heat Removal System
EDF	Electricité de France
FCI	Fuel Coolant interaction
FT	Fault Tree
HEP	Human Error Probability
HMI	Human Machine Interface
HRA	Human Reliability Analysis
HTA	Hierarchical Task Analysis
IAEA	International Atomic Energy Agency
IRWST	In-containment Refueling Water Storage Tank
LERF	Large Early Release Frequency
LRF	Large Release Frequency
MCCI	Molten corium concrete interaction
MCR	Main Control Room
NPP	Nuclear Power Plant
OSSA	Operating Strategies for Severe Accidents
PSA	Probabilistic Safety Assessment
PSF	Performance Shaping Factor
R&D	Research and Development
RC	Release Category
RS	RiskSpectrum
SAMG	Severe Accident Management Guideline
SAMP	Severe Accident Management Program
SBO	Station BlackOut
SPAR-H	Standardized Plant Analysis Risk-Human Reliability Analysis
TA	Task Analysis
TLA	TimeLine Analysis
TSC	Technical Support Center

Use of Corrective Action Programs at Nuclear Plants for Knowledge Management

Pamela F. Nelson[a*], Teresa Ruiz-Sánchez[b], and Cecilia Martín del Campo[a]
[a] Universidad Nacional Autonoma de Mexico, Mexico City, Mexico
[b] Universidad Autonoma de Tamaulipas, Reynosa, Mexico

Abstract: Due to the uncertainty of many of the factors that influence the performance of the humans in nuclear power plant maintenance activities, we propose using Bayesian networks to model this kind of system. In this study, several models are built from the information contained in the Condition Reports from the Corrective Action Program at a nuclear power plant. This first study, using actual nuclear power plant data, includes a method for data processing and highlights some potential uses of Bayesian networks for improving organizational effectiveness in the nuclear power industry. The tool described in this paper is designed to provide a systematic approach to assist in managing an organization's knowledge base and support improvements in organizational performance. This paper describes the utilization of cause codes recorded in the Corrective Action Program for determining their effect on consequential events.

Keywords: Bayesian Networks, Corrective Action Program, Knowledge Management, Organizational Performance.

1. INTRODUCTION

Almost every organization manages data in some way. Data is a major corporate resource; however it is frequently poorly documented. Descriptions of data and other resources are metadata, which are part of the corporate memory for the organization, and preserving corporate memory is one of the basic features of knowledge management. At present, many countries are experiencing a large percentage of the personnel at nuclear power plants reaching retirement age. As a result, recording the memories of these workers, including the meaning of data, is increasingly important. Preserving metadata is crucial for understanding data years after the data were created. Human error and organizational performance is of special interest in any industry, and the nuclear industry has developed methods for performing Human Reliability Analysis (HRA) to calculate the contribution of human error to accidents. There have been attempts to collect data to inform quantification in HRA, starting with the work done for the THERP methodology by Swain [1]. These efforts have continued to the present time, with efforts like the US Nuclear Regulatory Commission's Human Event Repository and Analysis (HERA) system [2] and the UK's CORE Database [3]. However, many experts in HRA have related the opinion that there should have been more effort on collecting human error data for the purpose of quantifying the probabilities of human error [4]; however, there does exist a wealth of information in the Condition Reports (CRs), products of the Corrective Action Programs (CAPs), at most nuclear plants. The corrective action process includes formal mechanisms to report, capture, assess, and correct organizational failures or shortcomings. Typically the focus is placed on identifying root causes and implementing corrective actions to ensure organizational learning and improvement [5].

In fact, in the nuclear industry, we asseverate that the CAPs are a source of this metadata. The information contained in the CRs at every nuclear power plant is invaluable, and while the reports for each event may be lengthy, there should be an efficient way to record, store and retrieve data and feedback continually. The statistics of the data can tell us much about the trends in failures, whether system or human failures; however, if the information is not codified to work for the intended database, the results may be inaccurate. For this reason, this paper describes the review and work done to extract benefit from the root cause analysis done on any abnormal occurrence at a nuclear plant, and

---

[*] *pnelson_007@yahoo.com*

presents a model to include this wealth of information in a structure that furthers the knowledge management about human errors in nuclear power plants.

The cause coding together with study of written descriptions in the failure and repair work orders helps to identify candidates of human errors related to maintenance activities from the failure and maintenance history. From a sample containing thousands of human related labeled condition reports it was possible to define a model of the factors that influence the occurrence of these events. The model provides the structure of the information necessary in the database to be able to utilize it for knowledge management. Once validated, the model can serve to predict events as well as risk-inform the procedures to reduce the reoccurrence of the events as well as avoid consequential events.

2. DEVELOPING A MODEL FROM STATISTICS

While some of the HRA methods have made attempts to collect data to inform quantification of HRA, these efforts have not been completely successful in being able to provide useful and complete data for human error quantification. Even if we were to be able to derive the quantitative HRA data, which focus entirely on the human error probability (HEP), this does not necessarily provide the information about the performance measures that can be used to track weakness in human factors. For example, knowing the error likelihood does not actually tell the human factors researcher the expected performance or the level of performance degradation that may precede an actual error. One of the principal motivations for this research is the lack of data in the human as well as organizational areas.

In order to obtain the information necessary to quantify the human errors and obtain the performance measures necessary to identify risk significant process steps for frequently performed activities (e.g., surveillance procedures) and interpret the degraded barriers at a nuclear plant, it is necessary to study and understand the events that actually occur at the plants. In the United States, the Nuclear Regulatory Commission requires a Problem Identification and Reporting program. Compliance with this requirement is performed through station specific Corrective Action Programs. These Corrective Action Programs generally use a reporting mechanism that is available to the general station employee population to identify and record problems, issues, or actions that need to be performed to accomplish the station's business and operational missions. Typically, a standard Condition Report form is used to document and enter this information into station databases. Thus, the CRs are significant and objective source of events and metadata. The information contained in the Condition reports is invaluable and while the reports can be lengthy and be highly variable from CR to CR due to the many "authors" at a plant for a CR, it is important that the CR information be processed and evaluated to generate important data and analyses relative to plant and human performance. This offers a significant opportunity to associate consequential station events to those processes and activities that were being performed by station personnel at the time of the event as well as their associated causes. This provides important opportunities to develop human performance models from objective plant specific data that has the potential to reveal organizational weaknesses and those station activities that are risk significant relative to consequential human failures (not just core damaging events). There needs to be a process model that provides an efficient and consistent way to record the data, store it, and have it provide the basis for follow-on technical analyses related to human performance. The statistics of the data can tell us much about the trends in failures, whether system or human failures; however, if the information is not codified to work for the intended database, the results may be inaccurate and uncertain. For this reason, the cause coding was reviewed at a pilot plant in order to use the data to develop a model.

2.1 Factor Analysis

A Factor Analysis (FA) was conducted on the database from the pilot plant to test the methodology described in the work reported in Groth [6]. In particular, the Factor Analysis in the XLSTAT computer program [7] was applied in order to extract the factors and determine the grouping of the variables for the causal model network that was developed. A sample size of 95 events was used, which are the significant events reported in the seven year period from 2004 - 2011. While strict rules regarding sample size for exploratory factor analysis have mostly disappeared, studies have revealed that adequate sample size is partly determined by the nature of the data. It is considered adequate to have 5 or 10 times the number of samples as variables, thus the 95:11 ratio of samples to variables is considered sufficient for the analysis [8].

Factor analysis is one of a family of techniques for taking high-dimensional data, and using the dependencies between the variables to represent it in a more tractable, lower-dimensional form, without losing too much information. FA is one of the simplest and most robust ways of doing such dimensionality reduction.

The variables considered in this study were taken from the causes indicated for each event reported in the Condition Reports and are listed in Table 1. We can observe the name and the description of the cause code variables, with some examples to demonstrate the types of causes considered in this study. Table 2 presents a sample of the first 10 lines of the data used for the FA; there are 95 entries in the complete matrix. In order to conduct the FA, the data was converted to binary, which means that the 0 represents "*Adequate*" for that variable, and a 1 means "*Less than Adequate*" for that variable. And, in this way a binary matrix is formed for the data for the significant events reported in the Condition Reports.

Table 1: Variables Considered

Variable	Description	Examples
DE	Equipment Design/Manufacture/Performance Monitoring	Predictive Maintenance Program Inadequacy, Preventive Maintenance Program Inadequacy
HF	Human Factors/Work Environment	Human Factors Not Properly Addressed in Work Area/Equipment
LS	Job Leader/Supervisory Methods	Pre-job Preparation or Briefing Inadequate, Prioritization of Work Activities Inadequate
MA	Management Assessment/Corrective Action	Organization Not Sufficiently Self-Critical, Cause Analysis for Known Problem Inadequate
MC	Change Management	Need for Change Not Recognized, Change Not Implemented in a Timely Manner
MP	Management Practices	Communication Within an Organization Inadequate/Untimely, Communication Between Organizations Inadequate/Untimely, Management Practices Promote/Allow Unacceptable Behaviors
MR	Management Resources	Prioritization/Scheduling of Activities Inadequate (Management level)
PA	Procedure Adherence	Procedure/Instruction/Step Implemented Incorrectly (Intent Not Met)
TR	Training	Necessary Initial/Refresher Training Not Provided
WI	Work Instructions	Document Contents Incorrect or Missing
WP	Work Practices	Slip or Lapse

Table 2: Binary Matrix for 11 Variable FA

Condition Report	DE	HF	LS	MA	MC	MP	MR	PA	WI	WP	TR
1	0	0	1	0	0	1	0	0	0	0	0
2	1	0	0	0	1	0	0	0	0	0	0
3	1	0	0	1	0	0	0	0	1	1	0
4	0	0	0	0	0	0	0	0	1	0	0
5	0	0	0	0	1	0	0	0	1	0	0
6	1	0	0	0	0	0	0	0	0	0	0
7	0	0	1	1	0	1	0	0	0	1	0
8	1	0	0	0	0	0	0	0	0	0	0
9	1	0	1	1	0	1	0	0	1	0	1
10	0	0	1	1	0	1	1	0	0	1	0

The default in most statistical software packages is to retain all factors with eigenvalues greater than 1.0, corresponding to the first five factors in our analysis; however, there is broad consensus in the literature that this is among the least accurate methods for selecting the number of factors to retain [8]. Many researchers describe that the best choice for researchers is the scree test. This method is described in every textbook discussion of factor analysis. The scree test involves examining the graph of the eigenvalues and looking for the natural bend or break point in the data where the curve flattens out. The number of data points above the "break" (i.e., not including the point at which the break occurs) is usually the number of factors to retain. Figure 1 presents the Scree plot where we can see that the first four factors explain almost 60% of the variance.

Figure 1: Scree Plot

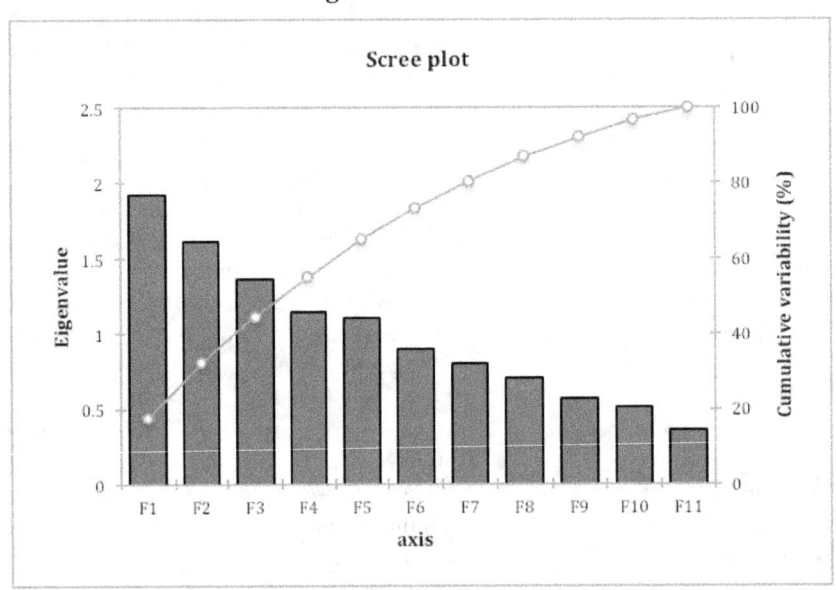

Table 3 illustrates the loading of the variables on the four extracted factors. As we can see, DE loads on F1, while the management variables load on F2 (MA, MC, MP, MR), etc. We can interpret this by observing that the factors divide the human performance difficulties into four categories: maintenance programs, management issues, work practices and supervision, and training, procedures and instructions.

Table 3: Factor Loadings

Variables	F1	F2	F3	F4
DE	0.871			
HF		0.436		
LS			0.366	
MA		0.436		
MC		0.456		
MP		0.504		
MR		0.276		
PA				0.583
WI				0.415
WP			0.538	
TR				0.736

Table 4 shows the correlations among the variables, which in turn are used to define the links or arcs between the variables in the causal model that was developed and is presented in Section 4.

Table 4: Correlations among Variables

Variables	DE	HF	LS	MA	MC	MP	MR	PA	WI	WP	TR
DE	1	0.216	-0.147	-0.001	-0.184	-0.248	-0.197	-0.137	-0.155	**-0.297**	0.008
HF	0.216	1	0.156	0.106	0.164	0.201	-0.062	-0.043	-0.131	-0.127	-0.072
LS	-0.147	0.156	1	0.075	0.044	**0.369**	0.243	0.086	0.034	0.254	-0.029
MA	-0.001	0.106	0.075	1	0.129	0.199	0.152	-0.143	0.030	-0.107	0.152
MC	-0.184	0.164	0.044	0.129	1	0.091	0.123	-0.115	-0.073	-0.228	-0.025
MP	-0.248	0.201	**0.369**	0.199	0.091	1	0.152	0.106	0.040	0.060	0.153
MR	-0.197	-0.062	**0.243**	0.152	0.123	0.152	1	-0.047	0.061	-0.036	0.076
PA	-0.137	-0.043	0.086	-0.143	-0.115	0.106	-0.047	1	0.043	-0.097	-0.055
WI	-0.155	-0.131	0.034	0.030	-0.073	0.040	0.061	0.043	1	0.126	**0.282**
WP	**-0.297**	-0.127	**0.254**	-0.107	-0.228	0.060	-0.036	-0.097	0.126	1	-0.072
TR	0.008	-0.072	-0.029	0.152	-0.025	0.153	0.076	-0.055	**0.282**	-0.072	1

Another result from the Factor Analysis is a biplot, as shown in Figure 2. As used in FA, the axes of a biplot are a pair of extracted factors. These axes are drawn in black and are labeled F1, F2 in this case. There is another plot for F2, F3, etc. which are not shown here.

A biplot uses vectors to represent the coefficients of the variables on the factors. Both the direction and length of the vectors can be interpreted. Vectors point away from the origin in some direction. A vector points in the direction that is most like the variable represented by the vector. This is the direction which has the highest squared multiple correlation with the factors. The length of the vector is proportional to the squared multiple correlation between the fitted values for the variable and the variable itself. For example, in Table 3 DE is loaded on Factor 1 with 0.871, thus the vector representing the DE variable has a value of 0.871 on the F1 axis in Figure 2.

The fitted values for a variable are the result of projecting the points in the space orthogonally onto the variable's vector (to do this, you must imagine extending the vector in both directions). The observations whose points project furthest in the direction in which the vector points are the observations that have the most of whatever the variable measures. Those points that project at the other end have the least. Those projecting in the middle have an average amount. For example, the perpendicular line from MP to the F2-axis intersects in 0.541, while MR in 0.276.

Thus, vectors that point in the same direction correspond to variables that have similar response profiles, and can be interpreted as having similar meaning in the context set by the data.

Figure 2: Biplot Vectors

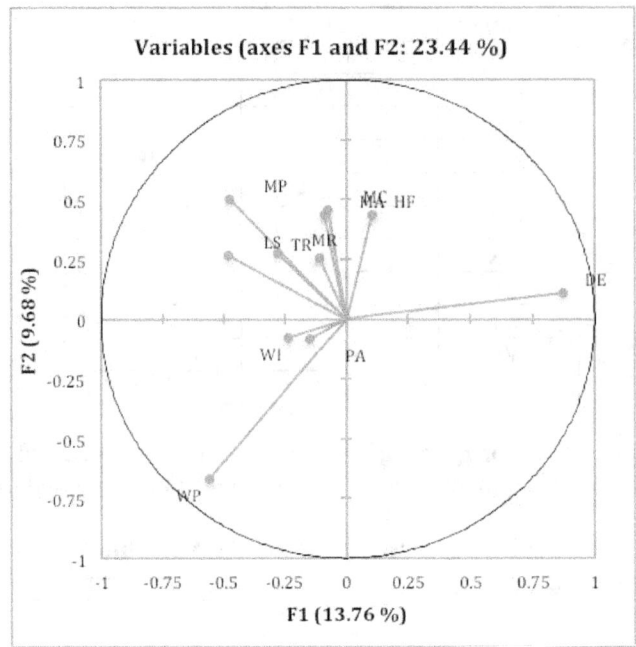

The biplot uses points to represent the scores of observations on the factors, and in Figure 3, each numbered point represents one of the condition reports, and the vectors represent the causes (variables). The relative location of the points can be interpreted in the following manner: points that are close together correspond to observations that have similar scores on the factors displayed in the plot. To the extent that these factors fit the data well, the points also correspond to observations that have similar values on the variables. In this example, events that are close together are ones that have similar profiles of causes.

Figure 3: Scatter Plot of Significant CAP Events

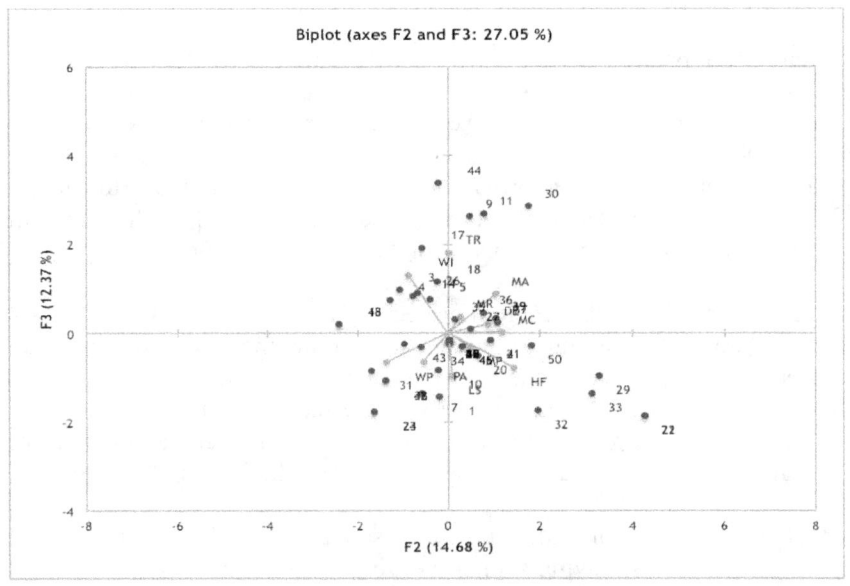

Probabilistic Safety Assessment and Management PSAM 12, June 2014, Honolulu, Hawaii

3. GRAPHICAL MODELS

3.1 Causal Models

A causal model can be used as an estimating approach based on the assumption that future value of a variable is a mathematical function of the values of other variable(s). It is used where sufficient historical data is available, and the relationship (correlation) between the dependent variable to be forecasted and associated independent variable(s) is well known [9]. Sufficient data can be considered as having ten times the data samples as the number of variables that describe the samples, as mentioned in Section 2.1.

Among the different data mining algorithms, probabilistic graphical models, in particular Bayesian Networks (BN) is a sound and powerful methodology grounded on probability and statistics, which allows building tractable joint probabilistic models that represent the relevant dependencies among a set of variables. The resulting models allow for efficient probabilistic inference. In this work, a BN represents the probabilistic relationships between the causes and the incidents reported at a nuclear power plant during routine maintenance and surveillance activities, providing a new methodology for probabilistic downscaling: i.e. allowing to compute the probability of a certain type of event, including reactor trip: Pr (reactor trip | a certain cause combination), as well as the decreased probability given a decrease in the probability of a given cause. For instance, the improvement in procedures (in this case, we assume perfect procedures) can decrease the probability of reactor trip.

Formally, BNs are directed acyclic graphs (DAGs) whose nodes represent variables, and whose arcs encode conditional independencies between the variables. The graph provides an intuitive description of the dependency model and defines a simple factorization of the joint probability distribution leading to a tractable model that is compatible with the encoded dependencies. Here we present two models derived from the same set of data, with two purposes: (1) to predict types of events, given different less than adequate performance in different areas, (2) to predict the probability of undesired consequences in routine operation at a nuclear power plant. Efficient algorithms exist to learn[†] both the graphical and the probabilistic models from data, thus allowing for the automatic application of this methodology in complex problems. Generalizations of Bayesian networks that can represent and solve decision problems under uncertainty are called influence diagrams, an example of which is presented in Section 4.

3.1.1 Probabilistic Networks

A Bayesian network is a graphical model that encodes probabilistic relationships among variables of interest. Bayesian networks can be used to learn causal relationships and be used to gain understanding about an area of interest and to predict the consequences of intervention.

Building a probabilistic network for a domain of application involves three tasks:

- Identify variables that are of importance and their possible values

- Identify the relationships between variables and express in a graphical structure

-Obtain the probabilities that are required for the quantitative part

Basically we can understand how a Bayesian network is used for quantification by observing the following simple example in Figure 4. If we have four variables that are used to describe cases, we can derive the graphical model from the data and quantify the probability of the network from the following equation 1.

[†] The verb *learn* is used to mean build the graphical model and calculate the conditional probability table directly from the data, in the computer science lexicon.

Figure 4: Example Bayesian Network

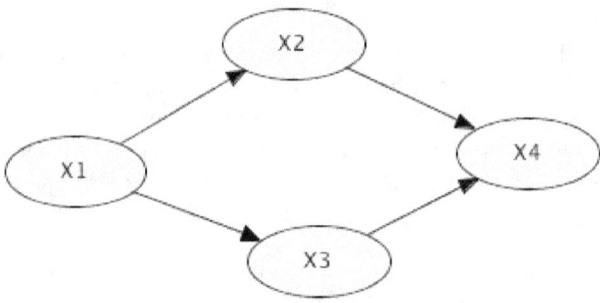

$$Pr\ (X1,X2,X3,X4) = Pr\ (X1) * Pr\ (X2|X1) * Pr\ (X3|X1) * Pr\ (X4|X2,X3) \tag{1}$$

There are two approaches to developing a BN: using the factor analysis results, as described in the previous section or learning the graphical structure from data. The graphical model resulting from the first approach is presented in Figure 5, while the learned structure is presented in Figure 6. Once the graphical structure has been established, assessing required probabilities is straightforward and involves studying subsets of data that satisfy various conditions [10].

Concentrating on the task of obtaining the probabilities, the most common sources are statistical data, literature, and human experts. Although there is abundant information, these sources seldom provide all the numbers required for a real application; however, we propose that the use of the data available in the Corrective Action Programs, once put into ontological terms, could provide the statistics necessary for producing extremely accurate probabilistic models for human error, including organizational influence.

Figure 5 illustrates the model created from the data analyzed in the Section 2.1. From the factor analysis, four factors were maintained. Thus, we can observe that the model identifies the variables loaded on the four factors, previously identified in the factor loadings and the arcs represent the correlations (greater than 0.240) between the variables. The arc between MC and TR was added, since the author has seen cases of correlation, despite the low correlation found in the FA.

Figure 5: Bayesian Network Designed from Pilot Plant CAP Data

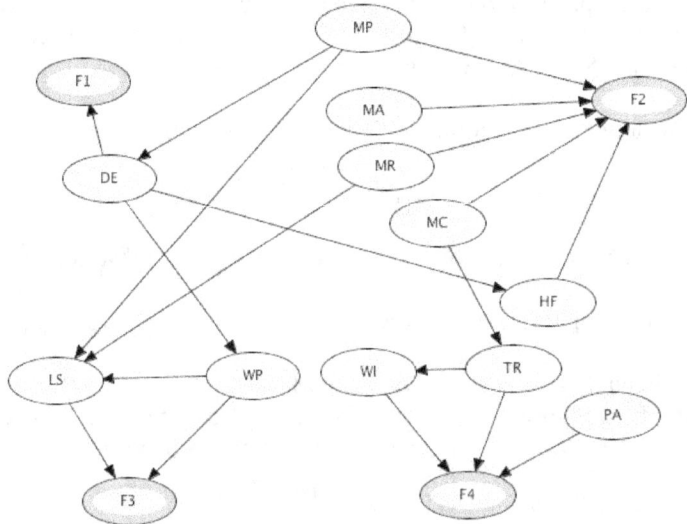

From this BN it is possible to determine how to reduce the probability of the four types of events (F1, F2, F3, F4 as described in Section 2.1) by reducing the probability of a less than adequate performance of the variable (considered to be one of the causes of the event). These shadowed nodes can also be considered as error forcing contexts, as described by Groth [6]; however, for the purpose of this study we use the former definition because it enables us to calculate the probability of an F3 type event given less than adequate performance in WP and LS, for example.

3.2. Predicting Reactor Trip

The next step is to add the undesirable consequences as nodes to the model. For this paper, we add only one node, in this case, reactor trip (RT), to simplify the explanation of the method. We will use pink boxes to represent "treatments" or "aids" in order to reduce the probability of the undesired event. These treatments are considered barriers or defenses used in nuclear power plants to aid human performance, such as 3-way communication, process and procedure approvals, pre-job brief, etc. However, before we can get develop the influence diagram, we must develop the graphical representation of the relationship between the variables and the consequences, defined as reactor trip in this example.

We can also correlate the causes or variables to the occurrence of consequential events, such as reactor trip. The structure of this next model was learned directly from the data. While the sample size is sufficient for the statistical analysis, it may scarce to directly determine all the arcs (correlations between the variables) directly using the HUGIN [11] program. The node HF that is not linked in this model evidences this effect. A column was added to the data table for another variable called RT (reactor trip) and the 1 indicates when the event caused or terminated in reactor trip. Thus, this BN is used to determine the effect of reducing adverse impacts of the underlying causes of the events on the probability of reactor trip. One result shows that the elimination of procedural adherence errors would decrease the probability of reactor trip by one third. This BN was learned using the Greedy algorithm [12] and is presented in Figure 6.

Figure 6: Learned Structure of the BN

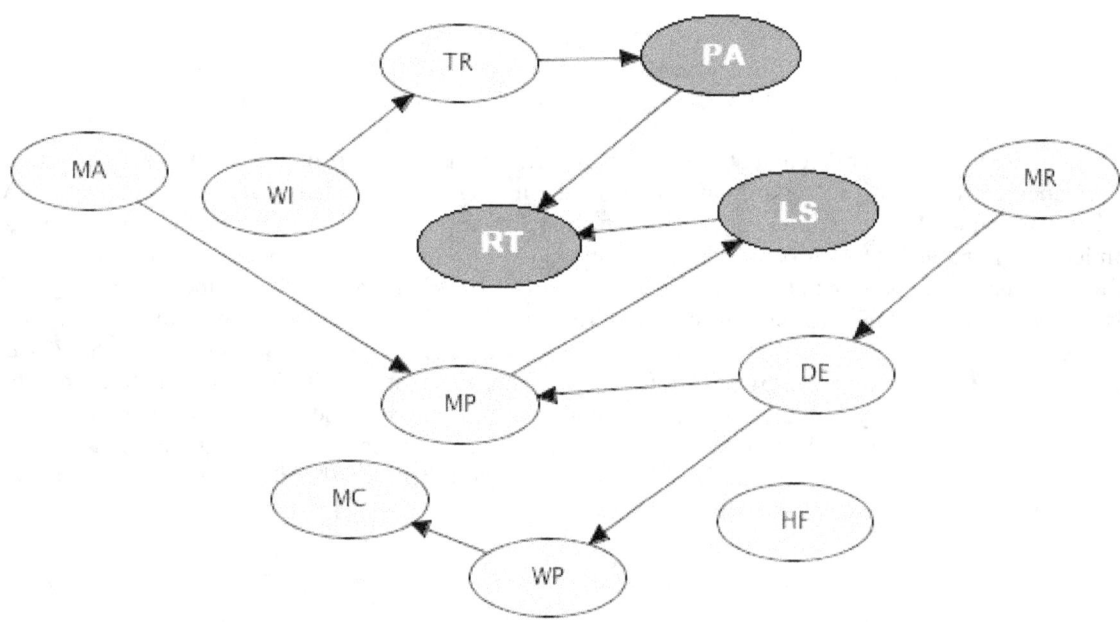

4. RESULTS

Now we have the information necessary to be able to construct the influence diagram; however for the purposes of this paper, we shall concentrate on only one part of the diagram. The probability of reactor trip (RT) depends directly on procedure adherence (PA) and on Supervision (LS), as shown in Figure 6 (nodes in red). So, we will amplify this section of the network. We add a decision node, that is, should plant management implement a human performance barrier to reduce the probability of reactor trip from these cause sources. In this case we are referring to human factors/organizational barriers, not actual physical barriers. The pink box in Figure 7 represents this decision node. We can also add a cost node and a utility node (savings acquired from avoiding a reactor trip), represented by the green diamonds in this same figure.

Figure 7: The Influence Diagram Propagated with RT=100%

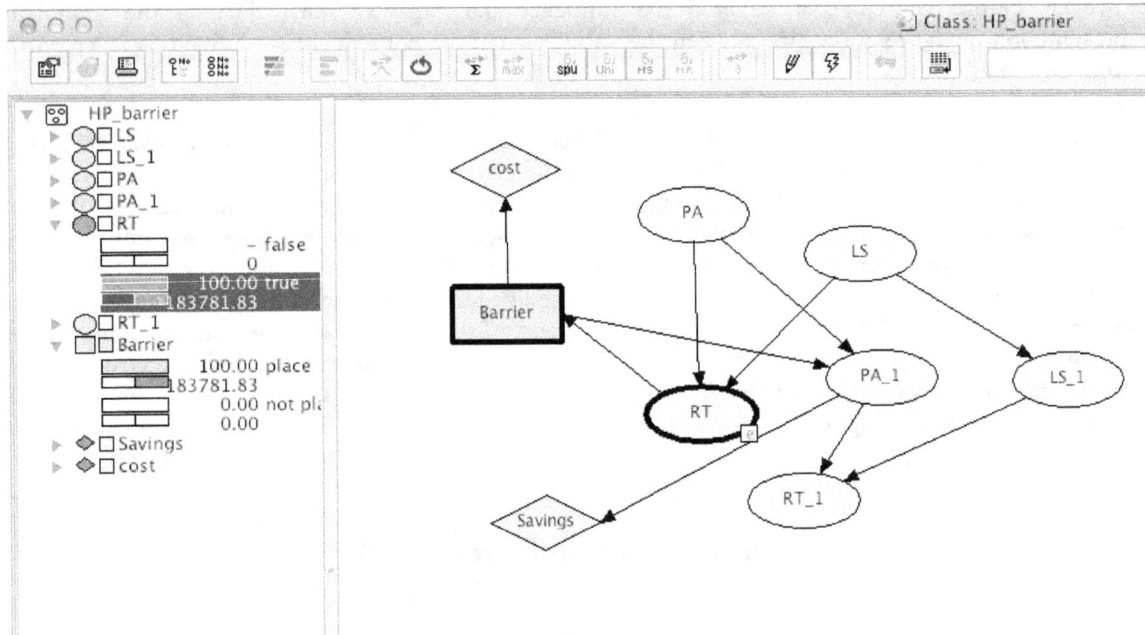

For this paper it is postulated that operating experience has shown an unacceptable level in the occurrence of reactor trips due to human errors. The model indicates that procedure adherence (PA) and supervision (LS) are key contributors to reactor trip, thus for this example, we propose implementing a barrier to PA (e.g., additional procedure approvals). The process shown in this paper allows the proposed barrier to be evaluated in terms of cost and savings to determine its viability and, also provides a means to determine its effectiveness in reducing the occurrence of future reactor trips. In this hypothetical situation, the barrier was determined to have an expected utility of 183,781.83 (i.e., benefit) and an associated organizational implementation cost of 40,000. Thus, due to high value relative to cost, the analysis indicates it is advantageous to implement the barrier. This demonstrates a key benefit of this approach in that focused barriers can be specifically structured to target and improve specific human and organizational performance activities relative to consequential historical plant events such as reactor trips.

5. CONCLUSIONS

Since a small set of data was used for the development of the models relative to the amount of data contained in Corrective Action Programs, the quantitative results are preliminary. However, the models developed in this paper are functional and the results are promising for several reasons: (1) the methodology enables the incorporation of operational experience into the model by using information from the Corrective Action Programs at nuclear power plants; (2) the models make it possible to identify and incorporate organizational factors into the probabilities of human error in a meaningful way; and (3) the influence diagrams, developed from the Bayesian networks, enable the user to evaluate the utility of adding human performance barriers or other organizational effectiveness initiatives and calculate their effect on undesirable consequences in a nuclear power plant caused by human error during routine operation and maintenance activities.

It is important to emphasize that the purpose of these models is to illustrate the type of insights that can be gathered through the model development process and data collection effort, as well as to provide a road map for future model development and data collection process improvements. Despite the many limitations of the data, the models are useful and the uncertainty in the results will be reduced by additional data collection and associated screening. Additional work will be performed to develop a more comprehensive model and data screening process to support the development of a database and associated data processing specification (e.g., to define the fields necessary, etc.) for Corrective Action Programs that would further support and facilitate an analysis such as described in this paper. This work will contribute to the development of a method for trending and tracking human and organizational performance events, as well as associated causes to support efforts in improving knowledge management and organizational effectiveness.

Acknowledgements

We would like to acknowledge the participating plant personnel for spending time to perform the searches we requested, without which, this work would not be possible.

References

[1] Swain, A.D., & Guttman, H.E. *Handbook of human reliability analysis with emphasis on nuclear power plant applications*. Final report. NUREG/CR-1278. Washington, DC: US Nuclear Regulatory Commission. (1983).

[2] Hallbert, B., Whaley, A., Boring, R., McCabe, P., & Lois, E. *Human Event Repository and Analysis (HERA):The HERA Coding Manual and Quality Assurance*, NUREG/CR-6903, Vol. 2. Washington, DC: US Nuclear Regulatory Commission. (2007).

[3] Kirwan, B., Basra, G., and Taylor-Adams, S.E. *CORE-DATA: A computerized human error database for human reliability support*. Proceedings of the 1997 IEEE Sixth Conference on Human Factors in Power Plants, 9-7 – 9-12. (1997).

[4] Boring, R.L., *Fifty Years of THERP and Human Reliability Analysis*, PSAM11 (2012).

[5] IAEA, *Comparative Analysis of Methods and Tools for Nuclear Knowledge Preservation*, IAEA Nuclear Energy Series No. NG-T-6.7, Vienna (2011).

[6] Groth, K., *A Data-informed model of Performance Shaping Factors for use in Human Reliability Analysis*, Ph.D. dissertation, University of Maryland, College Park, MD, (2009).

[7] XLSTAT (computer program) Available at www.xlstat.com (Accessed October 2013).

[8] Costello, A.B. and Osborne J.W., *Best Practices in Exploratory Factor Analysis: Four Recommendations for Getting the Most From Your Analysis*, Practical Assessment, Research & Evaluation, Vol. 10 No. 7, July (2005).

[9] http://www.businessdictionary.com/definition/causal-model.html#ixzz2vgnD4wnN, (Accessed March 2014).

[10] *Brief introduction to graphical models and Bayesian Networks* http://www.cs.ubc.ca/~murphyk/Bayes/bnintro.html#appl (Accessed November 2013).

[11] S.K. Andersen, K.G. Olesen, F.V. Jensen, and F. Jensen. *HUGIN* — a Shell for Building Bayesian Belief Universes for Expert Systems. In Proc. of the 11th IJCAI, pages 1080–1085, 1989. Available at http:www.*HUGIN.com* (Accessed November 2013).

[12] Chickering, D., *Optimal structure identification with greedy search*. Journal of Machine Learning Research, 3(3):507{554, (2002).

AES-2006 PSA LEVEL 1.
PRELIMINARY RESULTS AT PSAR STAGE

A. Kalinkin[a*], A. Solodovnikov[a], S. Semashko[a]
[a] JSC "VNIPIET", Saint-Petersburg, Russian Federation

Abstract: This report represents PSA level 1 results for AES-2006 project at LAES - 2 first unit configuration on PSAR stage. Report contains short description and base composition of LAES - 2 project, composition and basic requirements of normative documents regulating process of PSA level 1 implementing, list of operating conditions considered in PSA, list of initiating events selected for analysis, brief description of most significant accident sequences leading to fuel damage, characteristics of input data used, results for fuel damage frequency assessment.

Keywords: AES-2006, PSA level 1 results, Leningrad NPP – 2, Baltic NPP.

1. AES-2006 PROJECT SHORT DESCRIPTION

AES-2006 with VVER -1200 reactor is the base project for construction of Leningrad NPP- 2 (LAES -2), Baltic NPP and the Belarusian NPP. Currently, AES - 2006 project is being finalized to meet STUK requirements for NPP «Hanhikivi» construction in Finland. Development of LAES - 2 project was carried out using experience of design, construction and industrial operation for Tianwan NPP. Currently in commercial operation since 2007, are two blocks of Tianwan NPP.

AES-2006 project at Baltic NPP configuration passed Volume 2 EUR [1] requirements conformity assessment by WorleyParsons Nuclear Services JSC. Results of conformity assessment presented at Table 1.

Table 1: AES-2006 project Volume 2 EUR requirements conformity assessment results

Assessment category	Amount	Statistic
Conformity	3424	87,6%
Conformity with requirement objectives	220	5,6%
Nonconformity	42	1,1%
Not applicable	176	4,5%
Not considered	44	1,2%
Total	3906	100%

LAES -2 project utilize following approaches:
- Maximum use of solutions and studies for already developed projects with VVER type reactors;
- Risk minimizing and system performance improving through usage of proven technical solutions and reference equipment;
- Improving systems and equipment performance by optimizing design margins;
- Providing required level of safety, including beyond design basis accidents, based on the choice of a rational configuration of safety systems with the combination of active and passive elements to implement the principles of diversity and functional redundancy, as well as to reduce the human impact on the basis of the principle of reasonable sufficiency;
- Operating and capital costs reducing;
- NPP`s cost and construction time reducing through usage of existing groundwork for documentation.

---

[*] a_kalinkin@nio.spbaep.ru

LAES-2 project safety concept based on:
- Defense-in-depth principle;
- Deterministic safety principles;
- Target probabilistic criteria;
- Radiation safety criteria.

Following target probabilistic criteria are established for the project [2, 3]:
- Total frequency of severe fuel damage for all accident sequence must not exceed 10^{-6} 1/reactor per year;
- In order to avoid evacuation of population outside of emergency measures planning zone, determined in accordance with the regulatory requirements for NPP placement, design should strive to ensure, that estimated value for limited accidental release probability, does not exceed 10^{-7} 1/reactor per year.

Following safety barriers are established at project:
- 1st barrier: fuel matrix, preventing release of fission products to fuel cladding;
- 2nd barrier: fuel cladding, preventing release of fission products to main circulation path;
- 3rd barrier: main circulation path, preventing release of fission products to containment;
- 4th barrier: containment, preventing release of fission products to environment.

Communication between plant operational states, goals and levels of defense-in-depth, as well as states of security barriers, associated with failures of defense levels, presented in Table 2.

Table 2: Communication between plant operational states, goals and levels of defense-in-depth, and states of security barriers associated with failure of defense levels

Defense-in-depth levels	Level 1	Level 2	Level 3	Level 4	Level 5
Defense-in-depth goals	Terms of NPP location and prevention of normal operation violations	Preventing design accidents by normal operation systems	Preventing beyond design basis accidents by safety systems	Beyond design basis accidents management	Preparation and implementation of emergency measures plans at site and beyond
Plant operational states	Normal operation	Normal operation violations	Design basis accident	Beyond design basis accidents	Severe post-accident situation
Strategy	Accident prevention	Accident prevention	Accident mitigating	Accident mitigating	Accident mitigating
Management	Normal operation control system	Normal operation control system, including limiting function	CSS, including RTS & ESFAS	Normal operation control system, CSS, BDBA management system	Emergency management
Procedures	Instructions and guidance for normal operation	Technological regulations for safe operation Normal operation violations management procedures	DBA management procedures	Symptom-based emergency procedures	Emergency measures plans
Reaction	Normal operation systems	Normal operation systems	Engineering protective measures	Special design measures	External emergency preparation

Defense-in-depth levels	Level 1	Level 2	Level 3	Level 4	Level 5
Security barriers state, associated with failure of defense level	Fuel element damage within limits of radiation safety, operability of safety barriers 3, 4	Fuel element damage within limits of radiation safety, impact on safety barriers limitation	Limited 2, 3 safety barriers damage. Impact on containment limitation	1, 2, 3 safety barriers damage. Functionality of containment	1, 2, 3, 4 safety barriers damage

2. COMPARISON WITH REFFERENCE PROJECT

Comparison of main characteristics and parameters for Tianwan NPP with reactor plant V-428 and LAES-2 NPP with reactor plant V - 491 (see Table 3) shows that the - reactor plant V - 491 has some advantages, in particular:
- Tanks of borated water storage system were moved out of the safety building to containment and their functions were combined with those of pit-tanks. This allowed to simplify the water flow scheme to reactor from the ECCS systems, solve problem of pit - tanks valves failure-to-open during LPIS and HPIS pumps transition to recycling, and realize water return scheme from leak or because of a sprinkler system work, back to the pit- tanks.
- Feed and boron regulation system is able to perform the functions of emergency boron injection during ATWS accidents and function of 1 circuit feeding at shutdown modes and "small" leaks of 1 circuit;
- Cooling for responsible consumers process water supply system, using spray ponds;
- Refusal of EDG water cooling in favor of air cooling;
- Presence of passive safety systems;
- Service life for equipment has been increased from 40 to 60 years.

Table 3: Main differences between NPP projects with V-428 and V-491 reactor plants

Design data	Reactor plant V-428	Reactor plant V-491
Service life of main equipment, years	40	60
Pressure in the reactor (nominal) at core outlet, MPa	15,7	16,2±0,3
Coolant temperature at core outlet, °C	321	328,9±5
Coolant temperature at core inlet, °C	291	298,2
The temperature difference (heating) in the reactor, °C	30	30,7
Coolant flow rate through reactor, m³/hour	86000	86000
Reactor internal diameter (cylindrical portion), mm	4150	4150
Thickness of the reactor wall, mm	192,5	197,5
Thermal power (nominal), MW	3000	3200
Time spent (campaign) fuel in the core, year	3 – 4	4
Average fuel burnup (stationary fuel cycle), MW × day/kg U	43	До 70
Operating time at full capacity during the year (effective), hour	7000	8400
2nd circuit design excessive pressure, MPa	7,84	8,1
Steam generators type	PGV -1000M horizontal	PGV-1000MKP horizontal
Steam capacity, ton/hour	1470	1602
Generated steam pressure at SG steam collector outlet (at rated load) MPa	6,27	7,00
Generated steam temperature (at rated load), °C	278,5	287,0±1,0
Feedwater temperature (at rated load), °C	218	225±5
Coolant flow rate through loop, m³/hr	21500	21200
Pressure in pressurizer, MPa	15,7	16,2
Turbine type	K-1000-60/3000	K-1200-6,8/50

Design data	Reactor plant V-428	Reactor plant V-491
Turbine power, MW	1060	1170
Containment inner shell height (from inside), m	~63,0	67,85
Containment inner shell free volume, m^3	69169,0	75000
Containment outer shell height (from inside), m	~66.0	77
Containment outer shell vertical concrete wall thickness, mm	600,0	800,0
Residual heat removal scheme	RHRS + LPIS+ Sprinkler system	RHRS+ LPIS
Borated water storage tanks accommodation	Safety building	Under containment
Passive heat removal system through steam generators for beyond design basis accidents management	-	+
Passive heat removal system from containment for beyond design basis accidents management	-	+
Quantity and power of EDG, pcs. × kW	4×5500	4×6300
Quantity and power of normal operation reliable power supply DG, pcs. × kW	2×5000	1×6300

3. PSA IMPLEMENTATION

In accordance with specification requirements for LAES-2 PSA level 1 [2, 3], reactor core, fuel assemblies in the spent fuel pool and assemblies undergoing handling operations, are regarded as a source of radioactivity.

Following regulatory requirements were used during analysis:
- Regulations on main recommendations for development of PSA level 1 for internal initiating events for all NPP operational modes [4];
- Key recommendations for NPP PSA implementation [5];
- Procedures for Conducting PSA of NPP (Level 1) [6].

In the course of PSA following groups of plant operational states were considered:
- Power operation;
- "Hot" state during power shutdown, for planned and unplanned shutdowns;
- Cooldown through 2nd circuit within 1st circuit temperature range of 255 - 135 ° C with disabled ECCS accumulator tanks, for planned and unplanned shutdowns;
- Cooldown through 1st circuit within 1st circuit temperature range of 135 ° C to 60 ° C, for planned and unplanned shutdowns;
- «Cold» state during cooldown, for planned and unplanned shutdowns. Unscheduled repair of equipment;
- Preparation for reactor disassembly, reactor disassembly, for planned and unplanned shutdowns. Scheduled and unscheduled repair of equipment;
- Refueling. Scheduled maintenance of equipment;
- Revision of RP equipment. Scheduled maintenance of equipment;
- Reactor assembly at planned and unplanned shutdowns. Scheduled maintenance of equipment;
- «Cold» state during unit startup, for planned and unplanned shutdowns. Scheduled maintenance of equipment;
- Warming up before RHRS pumps shutdown, for planned and unplanned shutdowns;
- Hydraulic tests for 1 and 2 circuits;
- Warming up within 1 circuit temperature range 135-220 ° C with disabled ECCS accumulator tanks, for planned and unplanned shutdowns;
- "Hot "state during unit startup, for planned and unplanned shutdowns.

For selected groups of plant operational states, a list of events that could disturb normal operation of NPP was compiled. For this study the recommendations of the IAEA, experience of PSA implementation for similar units, and consistent deductive analysis for undesirable processes causes, were used. For further analysis were selected events, directly or indirectly affect NPP normal operation, characterized by estimated frequency of occurrence not less than 10^{-7} 1/year and contribution for total fuel damage frequency less than 1%.

Selected initiating events are grouped on the basis of similarity of accident process percolation paths and final states, caused by these IE, as well as on the basis of similarity success criteria. The aim of such grouping is to limit the number of accident sequences models.

List of IE groups for unit on-power-states shown at Table 4, for shutdown states at Table 5.

Table 4: IE groups for unit-on-power states

Name & content of IE groups
1. IE group, leading to loss of 1 circuit coolant
Compensated 1 circuit leak inside containment
Small 1 circuit leak inside containment
Middle 1 circuit leak inside containment without safety systems dependent failure
Middle 1 circuit leak inside containment with safety systems dependent failure
Large 1 circuit leak inside containment without safety systems dependent failure
Large 1 circuit leak inside containment with safety systems dependent failure
Small 1-to-2 circuit leak
Middle 1-to-2 circuit leak
Large 1-to-2 circuit leak inside containment
Reactor vessel rupture
SG tube rupture, caused by steam line rupture, outside of containment in non-isolated from SG part
Small 1 circuit leak outside of containment
Compensated 1 circuit leak outside of containment
2. IE group, leading to loss of 2 circuit coolant
Steam line/feed water pipes rupture in non-isolated from SG part
Steam line rupture in isolated from SG part
Feed water pipes rupture in isolated from SG part
3. IE group, leading to transient processes
3.1 Transient processes for 1 circuit
Scram
Loss of feed water flow rate caused by control system failure or partial loss of feed water flow rate at one SG
Automatic reactor shutdown with SG isolation by operator
Administrative shutdown (with different configuration of safety system, concerning safe shutdown)
Small feed water/ condensate system leak
3.2 Transient processes for 2 circuit
Loss of normal heat removal
Spontaneous closure of MSIV at one loop
Spontaneous closure of MSIV at all loops
4. IE group leading to failure of support systems
LOOP
Partial loss of own needs power supply
Loss of responsible consumers cooling

Table 5: IE groups for shutdown states

Name & content of IE groups
1. Termination of heat removal from the core because of primary coolant leaks
Compensated 1 circuit leak inside containment
Small 1 circuit leak inside containment
Large & middle 1 circuit leak inside containment without safety systems dependent failure
Middle 1 circuit leak inside containment with safety systems dependent failure
Large 1 circuit leak inside containment without safety systems dependent failure
Large & middle 1 circuit leak inside containment with safety systems dependent failure
Small 1-to-2 circuit leak
Middle 1-to-2 circuit leak
Large 1-to-2 circuit leak inside containment
Reactor vessel rupture
SG tube rupture, caused by steam line rupture, outside of containment in non-isolated from SG part
Small 1 circuit leak outside of containment
Compensated 1 circuit leak outside of containment
2. Termination of heat removal from the core because of 2 circuit leaks
Steam line/feed water pipes rupture in non-isolated from SG part
Steam line rupture in isolated from SG part
Feed water pipes rupture in isolated from SG part
3. Termination of residual heat removal due to support systems failures
LOOP
Termination of residual heat removal due to equipment failures
Termination of residual heat removal through 2 circuit
Termination of residual heat removal through 1 circuit
Spontaneous closure of MSIV at one loop
4. 1 circuit brittle strength conditions violation due to "cold" overpressure
1 circuit coolant withdrawal lines spontaneous closure
5. Termination of heat removal from SFP
Failure of one SFP heat removal system chanel (including failure of support systems)
6. Fuel damage during handling operations
SFA damage due to heavy objects falling at SFP or reactor pit
SFA damage by refueling machine due to refueling machine faults, personnel errors, violations of normal operating conditions, loss of refueling machine power supply

On the basis of thermal-hydraulic calculations, as well as on results of engineering evaluations performed for PSA, for each initiating events in each operational state group, success criteria were established. Success criteria was supposed as need to find the unit in a controlled state in the final state of accident with absence of threats to leave this state not associated with random equipment failures.

Analysis object modeling and calculation of model characteristics performed, using Risk Spectrum 1.2.1 software. System models takes into account existing equipment interdependence and interconnections between systems that could affect its functions performance. The models also take into account the possibility of common cause failures due to implicit dependencies.

During personnel reliability analysis human errors, which could take place both before and after initiating event (latter subdivided into errors in response to initiating events and errors in the commission of recovery actions), are simulated. Personnel reliability analysis used THERP and SAIC-TRC approaches.

Within data analysis were evaluated frequencies of initiating events, as well as equipment reliability parameters, including the parameters of the common causes for failure.

IE group frequency assessment for operational states groups, in order to quantify contribution of each operational state at total fuel damage frequency, performed using:
- OKB Gydropress data, obtained on the basis of statistic for nuclear power plants with VVER-1000 type reactors operation and results of the probabilistic analysis for destruction of RP equipment and pipelines, performed as part of AES-2006 technical design[7, 8];
- Results of earlier performed PSA [9];
- Operating experience for nuclear power plants with VVER type reactors in Russia and Ukraine [10 - 12];
- Quantitative analysis for nuclear fuel handling operation safety at Unit 1 LAES -2 [13];
- Calculation of reliability for polar crane 360 (205)/32 +10-41, 5 - UHL4 [14].

Equipment reliability parameters estimations were made using:
- Operational data for Novovoronezh, Kalinin and Balakovo NPP equipment within period 1986 – 2010 [15, 16];
- Results of earlier performed PSA [9];
- FRAMATOM reliability data for control safety systems [17];
- Reliability data for Alfa Laval plate heat exchangers [18];
- IAEA-TECDOC- 478 data [19];
- IAEA-TECDOC- 508 data [20];

For common cause failures modeling, α and β - factor models are used. US NRC data used for common cause failures quantitative estimates [21].

4. PSA RESULTS

Quantification results for major accident sequences leading to fuel damage for on-power states are shown in Table 6, for shutdown states in Table 7. Major initiating events contribution at total fuel damage frequency for on-power states are shown in Figure 1, for shutdown states in Figure 2.

Table 6: Quantification results for major accident sequences leading to fuel damage for on-power states

AS code	FDF (1/rpy)	Relative contribution (%)
VS-JND_FM-JNG_GM	$3,80 \cdot 10^{-8}$	29,5
1_2S-P1_2-SPOT_S	$2,62 \cdot 10^{-8}$	20,3
NISP-H-N4	$1,32 \cdot 10^{-8}$	10,2
S-LOCA-JND_FM-JNG_GM	$1,16 \cdot 10^{-8}$	9,0
RR	10^{-8}	7,74

Explanations to Table 6 AS codes:
- VS - compensated 1 circuit leak inside containment;
- JND_FM –1 circuit feed by 1 of 4 HPIS channels;
- JNG_GM - 1 circuit feed by 1 of 4 LPIS channels;
- 1_2S – small 1-to-2 circuit leak inside containment;
- P1-2 - BRU-A works in cooldown mode on 1 of 3 non-emergency SG and water supply in these SG from corresponding channel of emergency feedwater system;
- SPOT_S - work of three channels of SG PHRS at non-emergency SG;
- NISP - steam line/feed water pipes rupture in non-isolated from SG part;
- H – continuous heat removal through 2nd circuit
- N4 – emergency SG feedwater isolation;
- S-LOCA - small 1 circuit leak inside containment.

Table 7: Quantification results for major accident sequences leading to fuel damage for shutdown states

AS code	FDF (1/ rpy)	Relative contribution (%)
LSFPHR-FAK-SFP_FEED-DNU	$1,54 \cdot 10^{-7}$	34,45
LOOP_HR1_5-R_HUM-JNG_HUM	$7,60 \cdot 10^{-8}$	17,00
LNHR-HUM_EHRS-D_1/4	$4,80 \cdot 10^{-8}$	10,74
LHR1-R_HUM-JNG_HUM-KBA_HUM	$2,80 \cdot 10^{-8}$	6,26
LOOP_HR1-3,4,8,9,10-RE2-R_HUM-JNG_HUM-KBA_HUM	$2,17 \cdot 10^{-8}$	4,85
LOOP_HR1-5-RE2-R_HUM-JNG_HUM	$1,74 \cdot 10^{-8}$	3,89

Explanations to Table 7AS codes:
- LSFPHR - failure of one SFP heat removal system chanel;
- FAK – SFPHRS reserve line work;
- SFP_FEED – SFP feed by emergency heat removal tanks &SFP feed pump;
- DNU - SFP feed by diesel-pump unit;
- LOOP_HR1_5 - LOOP under residual heat removal through 1st circuit condition, reactor lid removed;
- LOOP_HR1-3, 4, 8, 9, 10 - LOOP under residual heat removal through 1st circuit condition, reactor lid maintained;
- RE2-R – LOOP duration within 2 -8 hours;
- R_HUM – RHRS startup by operator;
- JNG_HUM – LPIS pumps startup by operator;
- LNHR - termination of residual heat removal through 2nd circuit;
- HUM_EHRS - BRU-A cooldown mode startup by operator;
- D_1/4 - BRU-A works in cooldown mode on 1 of 4 SG and water supply in these SG from corresponding channel of emergency feedwater system;
- LHR1 – termination of residual heat removal through 1st circuit;
- KBA_HUM – 1st circuit feed from feed and boron regulation system, startup by operator.

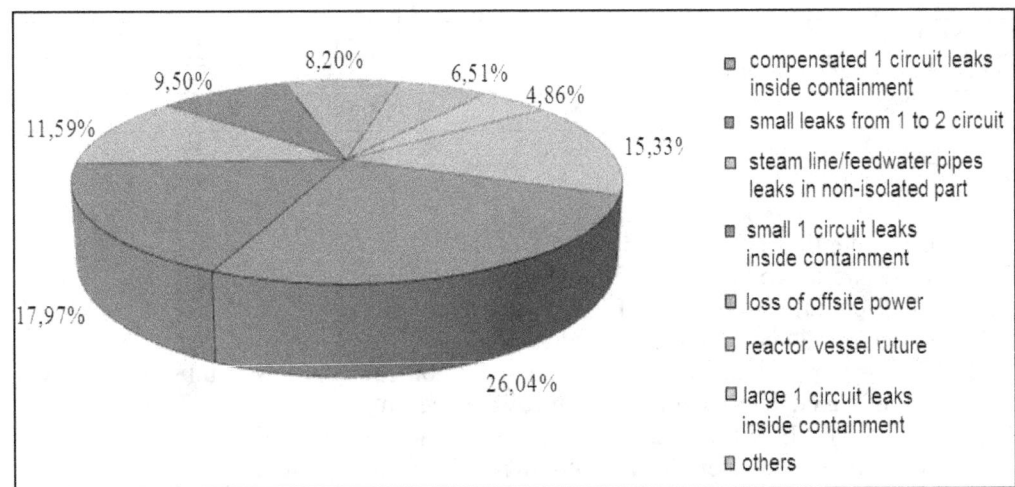

Figure 1 – Major initiating events contribution at total fuel damage frequency for on-power states

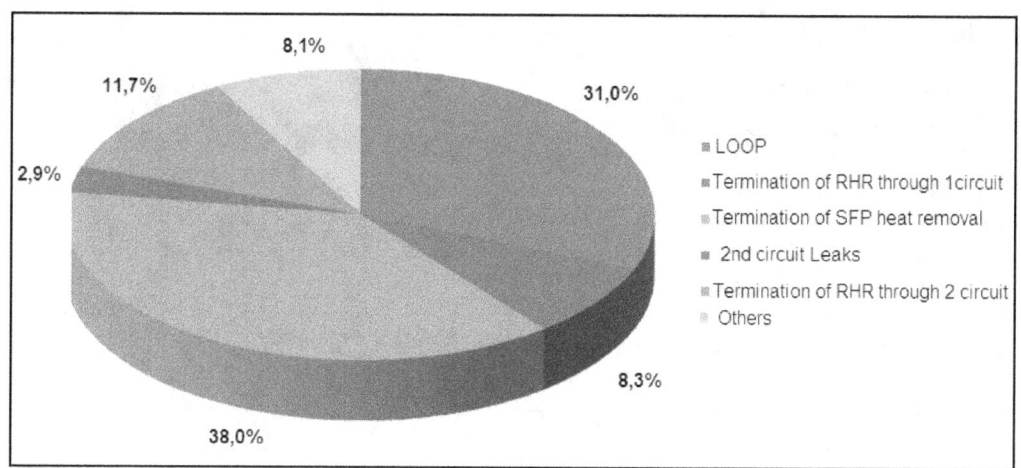

Figure 2 – Major initiating events contribution at total fuel damage frequency for shutdown states

Following results for fuel damage frequency obtained during PSA [22, 23]:
- Average value for total fuel damage frequency at plant-on-power states is $1,29 \cdot 10^{-7}$ 1/ reactor per year;
- Average value for total fuel damage frequency at plant shutdown states is $4,47 \cdot 10^{-7}$ 1/ reactor per year.

As a result of the uncertainty analysis performed for FDF at on-power states, following 90% confidence interval border obtained:
- Lower (5 %) - $2,23 \cdot 10^{-8}$ 1/ reactor per year;
- Median (50 %) - $7,98 \cdot 10^{-8}$ 1/ reactor per year;
- Upper (95 %) - $4,15 \cdot 10^{-7}$ 1/ reactor per year.

As a result of the uncertainty analysis performed for FDF at shutdown states, following 90% confidence interval border obtained:
- Lower (5 %) - $1,69 \cdot 10^{-7}$ 1/ reactor per year;
- Median (50 %) - $3,47 \cdot 10^{-7}$ 1/ reactor per year;
- Upper (95 %) - $9,52 \cdot 10^{-7}$ 1/ reactor per year.

Below are some recommendations made on the basis of PSA:
- Operational procedures improving: remove feed and boron regulation system outage to beginning of refueling stage;
- Technical solutions: consider possibility of automatic startup for LPIS backup channel, while executing normal operation function; consider possibility of automatic startup for SFPHRS backup channel; envisage backup reactor feed line under reactor dismantling condition to reduce LOOP and termination of residual heat removal through 1 circuit -related IE contribution at total FDF.

4. CONCLUSION

Results of PSA level 1 at PSAR stage for AES-2006 (first unit of LAES-2) shows that project meets its target probabilistic criteria. Further work should be directed to conservatism degree reduction and model detailed elaboration at FSAR stage.

AES-2006 project Volume 2 EUR requirements conformity degree, high level of safety and its possibility to be reconfigured to meet customer specific requirements, allows it worthily compete with other NPP projects at international market.

List of abbreviations

ATWS - Anticipated Transient Without Scrams.
BDBA - Beyond Design Basis Accident.
BRU-A - Fast acting atmosphere steam release installation.
CSS - Control Safety System.
DBA - Design Basis Accident.
DG - Diesel Generator.
ECCS - Emergency Core Cooling System.
EDG - Emergency Diesel Generator.
ESFAS - Engineering Systems Fast Actuation System.
FDF - Fuel Damage Frequency.
FSAR - Final Safety Analysis Report
HPIS - High Pressure Injection System.
IAEA - International Atomic Energy Agency.
IE - Initiating Event.
LOOP - Loss Of Offsite Power.
LPIS - Low Pressure Injection System.
MSIV - Main Steam Isolation Valve.
NPP - Nuclear Power Plant.
PSA - Probability Safety Analysis.
PSAR - Preliminary Safety Analysis Report.
RHR -Residual Heat Removal.
RHRS - Residual Heat Removal System.
RP - Reactor Plant.
rpy – reactor per year.
RTS -Reactor Trip System.
SFA - Spent Fuel Assembly.
SFP - Spent Fuel Pool.
SFPHRS - Spent Fuel Pool Heat Removal System.
SG -Steam Generator.
SG PHRS - Passive Heat Removal System from Steam Generators.

References

[1] European Utility Requirements for LWR nuclear power plants. Volume 2, revision C, April 2001.
[2] LAES-2 units № 1 & № 2. Technical task for development of project documentation. LN2O.B.051.&.&&&&&&.&&&&&.000.MB.0001K, Rosatom, 2012, Moscow.
[3] AES-2006. LAES-2 units № 1 & № 2. Project. Technical task. PSA level 1 for shutdown and low power modes, FGUP «SPBAEP», 2008, Saint-Petersburg.
[4] RB - 024- 11. Regulations on main recommendations for development of PSA level 1 for internal initiating events for all NPP operational modes, Federal Agency for Environmental, Technological and Nuclear Supervision, Moscow, 2011.
[5] RB - 032- 04. Key recommendations for NPP PSA implementation, Federal Agency for Environmental, Technological and Nuclear Supervision, Moscow, 2004
[6] Procedures for Conducting Probabilistic Safety Assessment of Nuclear Power Plants (Level 1), IAEA, 1992, Vienna.
[7] AES-2006. Baltic NPP. Reactor plant V-491. Initiating event frequency analysis for PSA. Part 1. Power operation modes. Work material, OKB «Gydropress», 2013, Podolsk.
[8] AES-2006. Baltic NPP. Reactor plant V-491. Initiating event frequency analysis for PSA. Part 2. Shutdown modes. Work material, OKB «Gydropress», 2013, Podolsk..
[9] Novovoronezh NPP units № 1 & № 2. PSA. Consolidated database for PSA. Volume 1. NW2O.C.165.&.&&&&&&.01&&&.022.HH.0001, Russian Academy of Sciences, 2007, Moscow.

[10] Gathering and organizing initiating event data for NPPs with VVER-1000 and initiating events frequencies calculation, VNIIAES, 2010, Moscow.

[11] IAEA information system IRS (International Reporting System for Operating Experience) https://websso.iaea.org.

[12] IAEA information system IRS (Power Reactor Information System) http://www.iaea.org/PRIS/CountryStatistics/CountryStatisticsLandingPage.aspx.

[13] LAES-2. Quantitative safety analysis for nuclear fuel handling operations at Unit №1 LAES-2. Report. LGM1.01-01.01, JSC «Diakont», 2011, Saint - Petersburg

[14] Roadway electric circular actions crane 360(205)/32+10-41,5-UHL4. Explanatory note. FZ1 5290 029 PZ. Redaction 1, JSC «Sybtyazhmash», 2011, Krasnoyarsk.

[15] Gathering and organizing equipment reliability data for NPPs with VVER-1000 and equipment reliability parameters calculation. VNIIAES, 2010, Moscow.

[16] Gathering and organizing equipment data for NPPs with VVER-1000, affecting important to safety functions performing, normal heat removal system, for using at LAES-2 safety justification, VNIIAES, 2010, Moscow.

[17] Work report FRAMATOM ANP "Safety systems reliability analysis – part RTS", 2001, Erlangen.

[18] Reliability Report "Alfa Laval Plate Heat exchangers" T100077-20, 2001;

[19] INTERNATIONAL ATOMIC ENERGY AGENCY, Component reliability data for use in probabilistic safety assessment. IAEA-TECDOC-478, Vienna, (1988);

[20] International Atomic Energy Agency, Survey of ranges of component reliability data for use in probabilistic safety assessment. IAEA-TECDOC-508, Vienna, (1989);

[21] NUREG/CR-5497 (update 2008) Common-Cause Failure Parameter Estimation

[22] AES-2006. LAES-2. PSA level 1. PSA level 1 for low power and shutdown modes. LN2O.B.110.&.&&&&&.04&&&.022.HH.0002. Book 4, VNIPIET, 2013, Saint-Petersburg.

[23] AES-2006. LAES-2. PSA level 1. PSA level 1 for power operation modes. LN2O.B.110.&.&&&&&.04&&&.022.HH.0001. Book 4, VNIPIET, 2013, Saint-Petersburg.

Quantitative Risk Assessment for DarkSide 50, a Nuclear Physics Experimental Apparatus Installed at Gran Sasso Nat'l Lab: Results and Technical Solutions Applied

Federico Gabriele[a1], Andrea Ianni[b], Augusto Goretti[b], Michele Montuschi[a], and Paolo Cavalcante[a] on behalf of DarkSide Collaboration

[a] Gran Sasso National Laboratory, L'Aquila, Italy
[b] Princeton University, Princeton, USA

Abstract: DarkSide 50 (DS50) is a two phase argon Time Projection Chamber designed to search for dark matter at the Gran Sasso National Laboratory (LNGS). As in most rare event experiments hosted at the LNGS, the challenge of DS50 is to reduce the background due to natural radioactivity. To meet this challenge, DS50 has the unique feature of underground depleted argon and uses an active veto to account for neutrons, the most important source of background for WIMP searches.

In this paper we report the Quantitative Risk Analysis (QRA) of the whole apparatus that we implemented and developed in order to bring the failure rates in the range of the LNGS acceptable matrix.

Due to the complexity of the experimental apparatus the analysis takes a variety of accident scenarios into account.

Even though QRA is widely used internationally for many purposes, the peculiarity of this application makes the involved issues, interpretation and results extremely interesting for risk assessment in the application of low background experiments in a confined underground space.

Keywords: QRA, Low-background, Gran Sasso Nat'l Lab, Confined Space, Dark Matter.

1. INTRODUCTION

A wide range of astronomical evidence suggests the existence of dark matter, but as yet the nature of this major component of the Universe is completely unknown. A leading candidate explanation is that dark matter is composed of weakly interacting massive particles (WIMPs) formed in the early universe and gravitationally clustered together with the standard baryonic matter.

Such WIMPs could in principle be detected through their collisions with ordinary nuclei in a sensitive target, giving observable low-energy (<100 keV) nuclear recoils. The predicted collision rates are very low and require ultra-low background detectors with large (0.1–10 tonnes) target masses, located in deep underground sites to reduce the background produced by neutrons from cosmic ray muons [1].

Several technologies have been developed for direct detection of dark matter WIMPs. These detectors all share the common goal of achieving the low background and low threshold energy required to detect the nuclear recoils that are possibly produced by the extremely rare collisions of WIMPs with target nuclei.

Among the variety of detector technologies, noble liquid Time Projection Chambers (TPCs), which detect both the scintillation light and the ionization electrons produced by recoiling nuclei, are particularly promising.

DarkSide 50 is a Liquid Argon Time Projection Chamber (LAr-TPC) with a projected sensitivity of 2×10^{-45} cm^2 for a WIMP mass of 100 GeV.

What makes this notable is that DS50 is designed to deliver this sensitivity via a long run with expected background well under 1 event, making a dark matter detection plausible.

---

[1] federico.gabriele@lngs.infn.it

2. GRAN SASSO NATIONAL LABORATORY

The Gran Sasso National Laboratory (LNGS), Figure 1, is one of the most important worldwide underground laboratories. The LNGS facility is part of the Italian National Institute of Nuclear Physics (INFN) and is made up of two main areas:
- the External Operations Centre (External Lab) in the town Assergi,
- and the Underground Laboratory.

Figure 1: The LNGS Location and a 3D Underground Lab Image

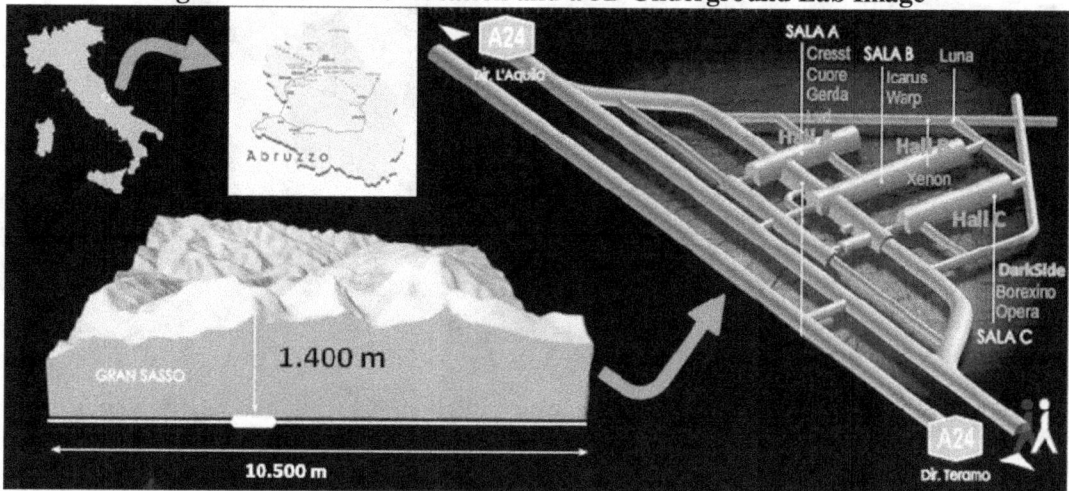

The whole experimental research centre is located in the heart of Gran Sasso and Monti della Laga National Park. The underground laboratory is located under a rock layer about 1,400 m thick, acting as a shield against cosmic radiation and currently housing about 20 experiments of different sizes. The underground cavity in the middle of a huge reservoir along the Gran Sasso highway tunnels (a double-tunnel 10,500 m long). The Underground Lab consists of three experiment halls: Hall A, Hall B and Hall C (about 100x20x20 m each), and a series of interconnecting smaller tunnels.

From the "safety point of view", beyond health and safety regulation in the work place the LNGS is classified, according to the European Directive Seveso III (2003/105/CE), as a major accident hazard plant because of experiments using and storing a large amount of substances classified dangerous to the environment [2].

In this environment and in compliance with Seveso Law, the satisfactory demonstration of acceptable risk levels is a requirement for approval of an experimental apparatus at LNGS. For that reason, we performed a Safety Risk Analysis in order to evaluate the likelihood of occurrence of possible events and to improve the safety standards in a complex system such as the LNGS.

3. THE DARKSIDE 50 APPARATUS

The DarkSide 50 Detector apparatus consists of three nested detectors, as illustrated in the left panel of Figure 2.
From the center outward, the three detectors are:
- the DarkSide 50 Detector, (right panel of Figure 2),
- the Liquid Scintillator Detector Neutron Veto (right panel of Figure 2),
- the Water Čerenkov Detector Muon Veto (left panel of Figure 2).

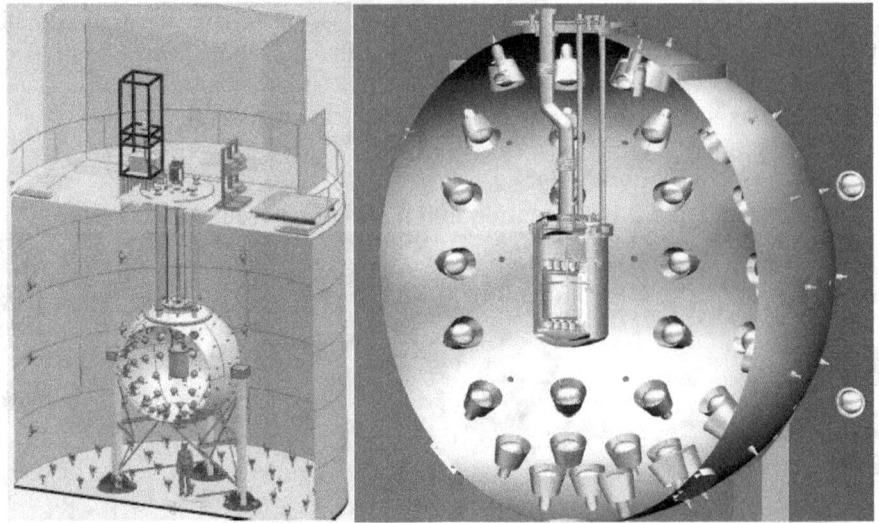

Figure 2: On the right the Nested Detector System of DarkSide 50, on the left a 3D Image.

3.1. The Water Čerenkov Muon Veto

The outermost detector is an 11 m diameter, 10 m high cylindrical tank (variable thickness in height from 12 to 8 mm) filled with high purity and deionized water, 1000 tonnes, (this tank is known as the CTF). An array of 80 photomultiplier tubes (PMTs) is mounted on the side and bottom of the water tank to detect Čerenkov photons produced by muons traversing the water. The inside surface of the tank is covered with a laminated Tyvek-polyethylene-Tyvek reflector to enhance Čerenkov light detection efficiency. At the top of water surface a Gas Nitrogen (GN_2) blanketing with slight overpressure is present in order to prevent contamination by air inlet.

3.2. The Liquid Scintillator Neutron Veto

A 30 tonnes borated Liquid Scintillator Detector is contained in a 4 m diameter (6 mm thickness) stainless steel sphere (Liquid Scintillator Vessel, or LSV) inside the CTF water tank. The scintillator consists of a mixture of 2 hydrocarbons: Pseudocumene (PC) and Trimethylborate (TMB), in equal amounts, with the wavelength shifter Diphenyloxazole (PPO) at a concentration of 3 g/liter.

An array of 110 PMTs is used to detect the scintillated photons. The inside surface of the sphere is covered with a Lumirror reflecting foil. Each PMT is fixed to the external liquid scintillator vessel wall through 30 mm single o-ring welded supports. A mixture level indicator in scintillator expansion vessel is also provided. The Liquid Scintillator external walls have been tested with helium in order to detect leaks.

On the top of the LSV there are two flanges: a 2,000 mm flange used to seal the scintillator top dome and a 900 mm flange used to hold the seven tubes and the DS50 Detector. Finally, on the scintillator external wall there are three blind covered flanges, used for optical fibers inlet.

3.3. The DarkSide 50 Detector

The DarkSide 50 Detector is a two-phase Liquid Argon Time Projection Chamber (TPC), right panel of Figure 3, designed to be sensitive to nuclear recoils possibly associated to dark matter interactions.

Figure 3: On the right the Cryostat, on the left the TPC.

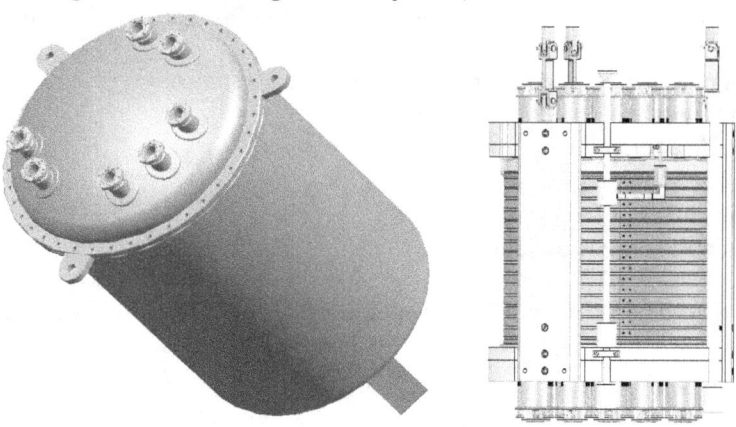

The Cryostat (or DAr) containing the TPC, as shown in the Figure 2 and in the left panel of Figure 3, is supported at the center of the LSV and operates at liquid argon (LAr) temperature immersed in the Liquid Scintillator Detector Neutron Veto. The total amount of LAr contained in the cryostat is about 150 kg.

The active volume of the TPC, 53.6 cm high and 47 cm wide, is contained laterally by a PTFE cylinder and at the top and bottom by fused silica windows. The volume is read out by 38 PMTs, nineteen each on the top and the bottom which view the active LAr.

The cylindrical surface (made of PTFE) is a reflector that is coated with tetra-phenyl-butadiene (TPB) wavelength shifter which is used to shift the UV scintillation photons to the visible spectrum for detection. The windows at the top and bottom of the cylinder are also coated with the wavelength shifter on the inner surfaces, and on both surfaces with transparent ITO conductive layers.

The Cryostat is a stainless steel, double walled, vacuum-insulated and super-insulated cylinder with two-to-one elliptical dished heads composed by two different vessels: the DAr vessel and the DAr vacuum jacket, both connected on the bottom of the top flange of the CTF and lowered inside the LSV through its north-side opening called the N2 Flange.

As shown in Figure 3, DAr has seven ports on top:
- LAr inlet port;
- vacuum port;
- 2 cable inlet/outlet ports (GAr oulet in normal conditions);
- GAr safety outlet port (GAr outlet in emergency conditions) [Blind Flange];
- 2 high voltage (HV) ports.

The LAr inlet, GAr safety outlet, cable inlet/outlet and HV ports are welded to the DAr inner vessel top dome and flanged with a ConFlat flange to the DAr vacuum jacket while the vacuum port, located in the center of cryostat top dome, is welded to the DAr vacuum jacket top dome.

The seven inlets listed above are provided through corrugated stainless steel-made tubes. The LAr inlet is vacuum insulated and the five others are with single wall (the Gar safety outlet port, vacuum port, the HV and cable inlet/outlet ports).
Both vessels are closed on top with two flanges:
- a 584 mm wire-seal type DAr vessel top flange,
- a 710 mm o-ring type DAr vacuum jacket top flange.

3.4. The DarkSide 50 Cryogenic System

The DarkSide-50 Cryogenic System is designed to continuously recirculate GAr through the purification getter and re-condense it for return to the detector, while precisely maintaining the working pressure of the argon in the cryostat under a range of heat loads during commissioning and operation. The design also insures that the system is safe in the case of loss of electric power.

The system (see Figure 4) can be divided in the following subsystems:
- Purification System,
- Charcoal (Radon) Trap,
- Ar Condenser,
- Liquid Nitrogen Supply System

and two closed-circulation cryogenic loops.

Figure 4: The DarkSide 50 Cryogenic System: on the left the System, on the right a Schematic Sketch

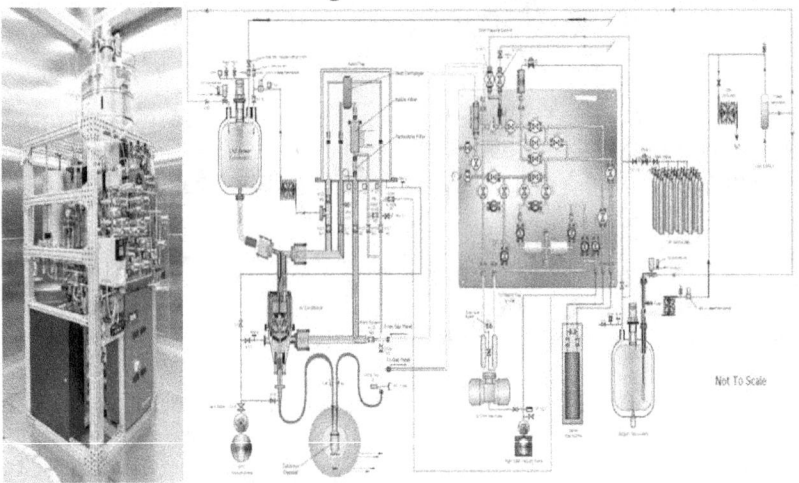

3.4.1. The Purification System

The Purification System (PS) is located in the so-called Hanoi Cleaning Room (CRH) placed on the top of the CFT tank, as illustrated in the Figure 2. The PS is directly connected to the Liquid Scintillator Detector and to the DS50 Detector through the N1 and N2 Flanges. The system is comprised of the Gas Panel, the Heated (SAES) Getter and the Circulation Pump (P-1). Its aim is to purify the GAr coming from the DAr and send it clean to the Radon Trap. All the above components are connected together through 0.5" DN tubes in which the GAr is circulated.

The gas panel itself is comprised of:
- Two filters (F-3, F-4) usually by-passed, except when the radioactive source connection between them is in use (only for calibration).
- Twelve pneumatic valves (VG01-12) which in case of emergency allow argon to bypass the gas panel itself, letting argon flow from the detector to the Ar condenser.

The Heated (SAES) Getter component is comprised of:
- the Getter, that is used to purify argon gas from H_2O and N_2 traces;
- the Heat Exchanger;
- the Filter (F-G).

The volumetric pump (P-1) is used to circulate GAr inside the purification system in order to overcome pressure drops in the purification loop. P-1 operates in resonance mode and the oscillation amplitude is related to input power, so it does not have a fixed compression ratio and it has a designed burst pressure of 27 bar, but the Maximum Allowable Working Pressure (MAWP) can be chosen according to particular needs and it will be dependent on input power and inlet delivery pressure (usually set equal to gas panel MAWP of 3 barg); the system is designed in order to allow only the set MAWP to be generated, even if the valves are blocked downstream.

P-1 is also set for a 100 sl/min maximum flow value but it normally operates at 50 sl/min. P-1 pressure and flow values are monitored by a Pressure Transmitter (PT P1), connected to a Pressure Relief Valve (PRV-P1) and a by-pass, activated by a pneumatic (V-G04) or a manual valve (V-S04).

3.4.2. Charcoal (Radon) Trap

The Charcoal (Radon) Trap is part of the gas circulation loop and its aim is to remove the radioactive contaminant Radon from the argon steam, using a vertical cold charcoal column: cold Nitrogen coming out from the Ar condenser head flows through a heat exchanger to cool down clean warm argon coming from the gas panel to a temperature just above the argon condensation temperature.

Cold argon gas then passes through the charcoal trap, it cools down and enters the Ar condenser, while warm Nitrogen coming out from the heat exchanger flows back to the liquid Nitrogen (LN_2) production dewar where it is condensed. A pump (P-20) is needed to overcome the pressure loss in the Mass Flow Control (MFC 20) and to force the Nitrogen back into the production dewar.

The vessel of the Charcoal Trap is a flat-headed vertical cylinder that includes the main filter, the heat exchanger and an additional filter (F-55) used to remove other gas traces. It is directly connected and located inside a stainless steel vacuum jacket, operating at temperature of 20 °C and a Maximum Allowable Working Pressure of 1.1 bar. Pressure values inside the Radon Trap are monitored by a pressure transmitter (PT 22) located on the Radon filter by-pass connection and a pressure indicator (PI 21) connected with the GAr outlet tube after control valve V-C2. Inside the Radon Trap, four temperature elements are also provided:
- TE N2-3, located on heat exchanger LN_2 inlet line;
- TE AR-1, located on heat exchanger GAr outlet line;
- TE RD-1/2, connected to the radon filter.

3.4.3. Argon Condenser

The argon condenser condenses clean cold Ar coming from the Radon trap and delivers the LAr to the cryostat through the above vacuum insulated transfer line. The cold head of the condenser condenses incoming gaseous argon using liquid N_2 through a copper-made exchanger.

The cold GN_2 generated during the argon condensation leaves the condenser through a vacuum insulated transfer line and feeds the Radon Trap heat exchanger that pre-cools the argon before entering the filter. The argon condenser is provided with stainless steel vacuum jacket, operating at a Temperature of 20 °C and a Maximum Allowable Working Pressure of 1.1 bar. Temperature inside the argon condenser is monitored by four temperature elements:
- TE N2-1, located on copper made exchanger LN_2 inlet line;
- TE N2-2, located on copper made exchanger LN_2 outlet line;
- TE AR-2/3, inside the argon condenser.

3.4.4. Liquid Nitrogen Supply System

The Liquid Nitrogen Supply System or Nitrogen Loop provides cooling for the liquefaction of argon. Starting at the LN_2 reservoir (a 160 l Dewar), a significant amount of LN_2 is stored as a buffer in the event of a power failure for an estimated time of about 24 hours. At the LAr condenser, LN_2 is used to liquefy the argon for the detector. This process of cooling the argon changes the phase of the LN_2 to N_2 cold gas. The cold gas return goes through a heat exchanger that is used to pre-cool the argon going to the radon trap and then to the condenser. The N_2 then travels to a room temperature heat exchanger that warms up the nitrogen gas before going through the MCF 20. Once the N_2 passes through the MFC 20 it returns back to the LN_2 Dewar where it is condensed to LN_2 completing the loop. The re-condensation of N_2 to LN_2 is done by the 300 W GM cryocooler mounted on the top of the LN_2 dewar. The above system allows high system stability and controllability during both normal and abnormal including emergency conditions.

Moreover the nitrogen system can be operated directly by a transfer line connected to the lab's large liquid nitrogen reservoir. There are the following control/monitoring devices connected to the LN_2 production dewar:
- the above MFC 20, on the Nitrogen gas line, which controls the cooling power: the MFC is also connected to the Distributed Control System (DCS), so that in the case of a malfunction, an alarm is sent from the control system;
- A pressure transmitter (PT 20) which measures the GN_2 line pressure value;
- 3 Pressure Relief Valves.

3.4.5. Argon Recovery System

The Argon Recovery System allows the recovery and storage of the Gas Underground Argon (GUAr) coming from the DS50 Detector in case of evacuating of the LAr. It is connected directly to the above gas panel and composed of a 600 W GM cryocooler head-cold installed on the top of the LAr dewar. The system has two modes of operation, one by the local liquefaction of the 600 W GM cryocooler installed on top of the liquid argon dewar, and the other by the lab's liquid nitrogen reservoir. In the second mode, the Nitrogen return gas line provides another source of LN_2 System.

4. THE QUANTITATIVE RISK ANALYSIS

We began the analysis with a (PRA) Preliminary Risk Analysis (April 2012) [3, 5, 6, 7] in which we performed a FMECA for the above DS50 Apparatus. After this first evaluation and method application, we carried out a (QRA) Quantitative Risk Analysis [4, 5, 6, 7] (June 2013).

In the following table we report the Top Events (TE) and the Events (E) that emerged from the FMECA and then that were enhanced in the Fault Tree Analysis (FTA).

Table 1: Top Events and Events

	Top Events
A	Cryostat (DAr) overheating/overpressure
B	DarkSide 50 overheating/overpressure
	Other Events
C	Argon direct release
D	Water release in Hall C
E	Nitrogen direct release
F	Liquid Scintillator direct release in Hall C

We performed the FTA by the FTPlus software and we assessed the calculations assuming a DS50 life-time of 20 years. We consider component and safety system failures according to specifications and operating modes:
- Immediately revealed failures (Rate model);
- Dormant failures (Dormant model);
- Fixed model.

Safety system failures are typically detected when their availability is requested (dormant failures). In order to estimate unavailability when activation is requested ("No opening/working when requested") we apply the "dormant" model.

To assess non-time-dependant unavailability, failure rates, and probabilities (e.g. "human error" in normal operations) we shall use the "fixed" model, according to scientific literature. In this model, output value ω stands for the frequency of a human operating error based on the fact that the operation is done "n" times in a fixed time interval (e.g. 1 year).

4.1. Top Events A and B

In the assessment of Top Event A, we consider all initiator events which could lead to the condition of possible overheating/overpressure in DAr and consequent cryostat breakdown.

The Top Event A "Cryostat overheating/overpressure" main initiator events are:
- DAr vessel failures:
  - Hole or leak from DAr vessel external wall and top dome;
  - Hole or leak from n.10 DAr vessel welded points;
  - Hole or leak from n.7 DAr vessel top flanged ports for inlets/outlets;
  - Leak from DAr vessel top flange;
- Dar vacuum jacket failures:
  - Hole or leak from DAr vacuum jacket external wall and top dome;
  - Hole or leak from n.2 DAr vacuum welded points;
  - Hole or leak from n.7 DAr vacuum jacket top flanged ports for inlets/outlets;
  - Leak from DAr vacuum jacket top flange;
- Hole or leak from stainless-steel inlet/outlet tubes:
  - LAr inlet corrugated stainless steel tube;
  - DAr vacuum jacket vacuum line.
- Hole or leak from flex inlet/outlet lines:
  - n.2 cables flex lines;
  - DAr cryostat inner vessel vacuum line.
- Leak from gate valve GV-2.

The Top Event A main enablers are:
- DAr cryostat vacuum providing system unavailability;
- No Darkside-50 system rupture disks (RD-1, RD-3) opening when requested.

In the assessment of Top Event B, we consider all initiator events which could lead to the condition of possible overheating/overpressure in Darkside-50 system, including all auxiliaries (e.g. Argon condenser, LN2 production dewar, etc.) and also the "Purification with pump" configuration, that correspond to Darkside-50 normal operating conditions.

The Top Event B "Darkside-50 overheating/overpressure" initiator events are:
- Overpressure/vacuum loss in DAr cryostat (see Top Event A);
- Overpressure/vacuum loss in Argon condenser due to:
  - Vacuum gap failure;
  - Hole or leak from LN2 inlet tube;
  - Hole or leak from GAr inlet welded port;
  - Hole or leak from GN2 outlet welded port;
- Overpressure/vacuum loss in gas panel/Radon trap due to:
  - Vacuum gap failure;
  - Pneumatic valves (V-21, V-22) stuck closed;
  - Leak from filters (Radon filter, F-5, F-G);
  - Hole or leak from GAr inlet tube;
  - Leak from heat exchanger external shell;
  - n.8 pneumatic valves (V-G01, V-G05, V-G07, V-G08, V-G11, V-G12, V-C1 and V-C2) stuck closed.

Top Event B main enablers are:
- All enablers considered in Top Event A simulation (see Top Event A);
- Radon trap/Argon condenser vacuum providing system unavailability;
- No pump P-1 by-pass/stop (e.g. PRV-P1/PT-PI unavailability, P-1 failure, etc.);
- No Pressure Relief Valves (PRV-2, PRV-4, PRV-21, PRV-11) opening when requested or failure;
- No LN2 production dewar Rupture Disk (RD-2) opening when requested.

In order to follow a conservative approach, holes/leaks considered in the analysis are micro-leaks due to operational error during welding, stainless-steel imperfections, etc. which have higher failure rates but also effects (e.g. Pressure increase, vacuum loss, etc.) that could be mitigated by vacuum providing system.

We also apply a high performance super insulation layer on the vacuum jacket internal walls in order to slow temperature/pressure increase down and help maintain the vacuum.

Figure 6: Top Event A: Gate 1

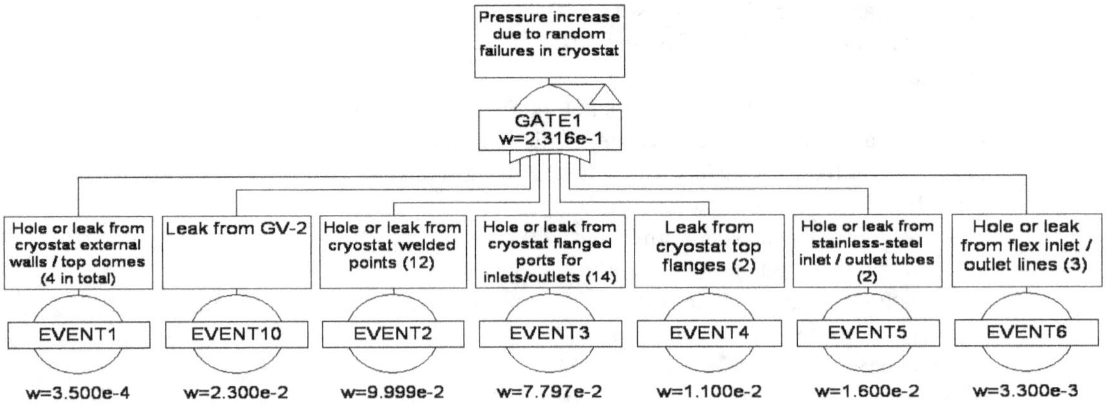

Pressure increase due to random failures in cryostat

GATE1 w=2.316e-1

Hole or leak from cryostat external walls / top domes (4 in total)	Leak from GV-2	Hole or leak from cryostat welded points (12)	Hole or leak from cryostat flanged ports for inlets/outlets (14)	Leak from cryostat top flanges (2)	Hole or leak from stainless-steel inlet / outlet tubes (2)	Hole or leak from flex inlet / outlet lines (3)
EVENT1	EVENT10	EVENT2	EVENT3	EVENT4	EVENT5	EVENT6
w=3.500e-4	w=2.300e-2	w=9.999e-2	w=7.797e-2	w=1.100e-2	w=1.600e-2	w=3.300e-3

4.2. Events C, D, E and F

In the assessment of Event C, we consider all initiator events which could lead to the condition of possible Argon direct release which could be localized in the PS inside the CRH or directly in Hall C, depending on release of the starting point position.

The main initiator events in the PS inside the CRH are:
- Leak from pump P-1;
- Leak from n.2 sample valves (SVs);
- Leak from n.4 manual valves (V-Ss);
- Leak from n.4 vacuum valves (V-Vs and vacuum valve on P-1 line);
- Leak from manual valve V-PI3;
- Leak from n.14 pneumatic valves (V-Gs, PRV-1, PRV-3);
- Leak from filters (F-3, F-4, F-G).

The main initiator events with releases directly in Hall C are:
- Leak from gate valve GV-1;
- Leak from n.5 pneumatic valves (V-Cs);
- Pneumatic valve V-C6 stuck open/partially open;
- Leak from GAr outlet tube in Radon trap;
- Leak from GAr inlet port in Argon condenser;
- PRV-11 or Rupture Disks (RD-1, RD-3) undue openings.

The Event D does not lead to safety related consequences, but we have evaluated it regardless because it could lead to Hall C flood in case of relevant water release from Darkside-50 water tank.

The Event D "Water release in Hall C" initiator events are:
- Leak from water tank lateral manhole;
- Hole or leak from water inlet (water purification plant) or outlet (water drain) tubes;
- Hole or leak from n.8 blind covered flanges;
- Hole or leak from water tank external wall;
- Leak from manual valves (V-W1, V-W2).

In the assessment of Event E, we consider all initiator events which could lead to the condition of possible Nitrogen direct release.

The Event E "Nitrogen direct release" main initiator events are:
- Leak from n.6 pneumatic valves (V-21, V-22, V-23, V-24, V-25, V-26);
- Leak from n.3 manual/by-pass valves (V-BP1, V-BP2, Valve on PI 20 line);
- Manual valve V-25 stuck open/partially open;
- Hole or leak from Radon trap GN2 inlet/outlet tubes;
- Hole or leak from LN2 production dewar GN2 inlet tube;
- Pressure Relief Valves (PRV-21, PRV-2, PRV-4) and Rupture Disk (RD-2) undue opening;
- Hole or leak from Argon condenser GN2 outlet port;
- Hole or leak from LN2 production dewar outlet port;
- Leak from pump P-20.

In the assessment of Event F, we consider all initiator events which could lead to the condition of possible Liquid Scintillator directly in Hall C, depending on release starting point position.

The Event F initiator events are:
- Leak from the manual gate valves on the CTF pumping station, the draining lines and from the filter F-502;
- Leak from 3 solenoid valves on the CTF pumping station and draining lines;
- Leak from needle valve DS-514;
- Hole or leak due to buffer (HT 5302) failures;
- Hole or leak from scintillator inlet/outlet tubes;
- Hole or leak from CTF tank scintillator inlet/outlet ports;
- Leak from pumps in both CTF pumping station and draining lines;
- Leak from filter F-502;

It is important to underline that all component failure rates used for the analysis, and therefore the simulation performed, follow a conservative approach mainly because:
– All values derive from component failure rate records based on laboratory specific tests and previous experiences on each component: for that reason, all failure rates could be further specified, for example, considering more precise failure rates provided by component producers/suppliers.

– All component failure rates used in the analysis include all possible type of hole/leaks, from micro to relevant ones. For that reason, the above Events could lead to dangerous safety related consequences only if a relevant hole/leak occurred: these types of holes/leaks have a lower frequency of occurrence than small ones.

Following the same conservative approach, it is important to underline that no enablers can be considered in the analysis.

It is also important to note that, for the Events C and E, the Hall C ventilation system is provided, and for the Events D and F, the Hall C retaining basin is provided.

5. CONCLUSIONS

In this section we report remarkable technological solutions adopted, a summary of results, and the application of ALARP methodology.

First of all, it must be emphasized that a possible risk scenario leading to a Rapid Phase Transition (RPT) due to contact between LAr and liquid scintillator inside the LSV is not part of the analysis because of the high improbability of its occurrence due to the presence of a super-insulated layer which represents a further safety "containment" in case of a hole or leak in addition to the double-walled cryostat explained above.

That super-insulated layer represents also a significant reduction of heat exchange between two above liquids. The following technological solutions have been installed as results of analysis in order to reduce the likelihood of occurrence:
- Pressure Relief Valve (PRV) and Rupture Disc (RD) release points have been located in safe position and connected to the vent.
- A pressure regulator on gaseous Nitrogen line before LN_2 production dewar has been installed.
- A Slow Control System has been developed and installed in order to monitor and control the Cryogenic System and the Ar and N_2 flows.
- A specific maintenance/monitoring plan for pumps, filters, mass flow controllers, sensors, rupture disks and valves has been defined.
- A pressure comparison measuring system has been actuated between pressure transmitters and indicators.

In Table 2, we summarize the QRA results in term of probability.

Table 2: Summary of QRA Results

	Top Events	Probability (ev/year)*
A	Cryostat (DAr) overheating/overpressure	$<10^{-7}$
B	DarkSide 50 overheating/overpressure	$<10^{-7}$
	Other Events	**Probability (ev/year)***
C	Argon direct release	$5.4e^{-1}$
D	Water release in Hall C	$8e^{-2}$
E	Nitrogen direct release	$3.7e^{-1}$
F	Liquid Scintillator direct release in Hall C	$3.4e^{-1}$

The events listed are evaluated and reported in the LNGS Acceptable Matrix, which defines the following categories: N=Unacceptable, T=Tolerable and A=Acceptable.

We created this matrix by taking into account the following 2 parameters:
- Frequency: in terms of event frequency of occurrence (ev/year or ev/hour).
- Consequence: in terms of effects on human health and safety. Consequences have been estimated on a quality level therefore a more precise consequence evaluations could be assessed in order to have more accurate results.

Table 3: Acceptable Matrix

Consequence / Frequency	Lethal / Irreversible Effects	Major Effects	Serious Effects	Minor Effects	No Relevant Effects
Frequent >1 ev/year $<1.1e^{-4}$ ev/hour					
Probable $e^{-1} - 1$ ev/year $1.1e^{-4} - 5.7e^{-6}$ ev/hour				Event C Event E Event F	
Occasional $3e^{-2} - e^{-1}$ ev/year $5.7e^{-6} - 2.8e^{-6}$ ev/hour					Event D
Remote $3e^{-3} - 3e^{-2}$ ev/year $2.8e^{-7} - 2.8e^{-6}$ ev/hour					
Improbable $3e^{-4} - 3e^{-3}$ ev/year $2.8e^{-7} - 2.8e^{-6}$ ev/hour					
Extremely Improbable $<3e^{-4}$ ev/year $<5.7e^{-8}$ ev/hour	TE A TE B				

It is important to emphasize that we adopted the same conservative approach for all Top Events and for Events (C, E and F); we placed them in the Tolerable Area, the first one, for their frequency of occurrences which are much lower than credibility threshold (10^{-8} ev/year), and the second one, in terms of their potential effects on human safety which are much less dangerous, therefore the associated risks could be considered almost Acceptable (A).

In conclusion, the assessment altogether represents, from the point of view of the approach and modality to face the analysis, a relevant reference for the Risk Analysis in the application of the low background Experimental Apparatus of DarkSide 50 in the LNGS Undergound Lab. Moreover, the results and the technological solutions emerged from the above analysis are extremely interesting and very useful for this application.

Acknowledgements

We would like to acknowledge all the personnel of the DarkSide 50 Collaboration. Thanks to those without whom the Experiment would be impossible, and in special way, to the DS50 Engineer Operative Group and to the Nier Ingegneria for the huge activities and efforts achieved during the 2 years it took in order to carry out the Risk Analysis on which this paper is based.

References

[1]   DarkSide 50 Collaboration. *"DS-50 NSF PNA Nov 2012"*, Proposal Experimental Collaboration (2012).
[2]   LNGS. *"Rapporto di Sicurezza"*, 2011.
[3]   DarkSide 50 Collaboration and Nier Ingengeria. *"FMECA DarkSide 50 rev.2 FINAL"*, Preliminary Risk Analysis (2012).
[4]   DarkSide 50 Collaboration and Nier Ingegneria. *"DarkSide 50 QRA rev.2"*, Quantitative Risk Analysis (2013).
[5]   RIAC. *"NPRD – Nonelectronic Parts Reliability Data"*, (2011).
[6]   D. Swain, H.E. Guttmann, *"Handbook of human reliability analysis with emphasis on nuclear power plant applications"*, NUREG/CR-1278, U.S. Nuclear Regulatory Commission, Washington, DC (1983).
[7]   G. Mulé, *"Guida alla scelta e dimensionamento delle valvole di sicurezza e dei dischi di rottura"*, AIDIC Servizi S.r.l., ISBN 88-900775-6-5. First-edition 2005.

Safety analysis and quantitative risk assessment of a deep underground large scale cryogenic installation

Effie Marcoulaki[*] and Ioannis Papazoglou
National Centre for Scientific Research "Demokritos", Athens, Greece

Abstract: This work considers the safety analysis and quantitative risk assessment of a deep underground cryogenic installation intended for neutrino physics. The neutrino detector equipment will be submerged in 50ktons fiducial mass of purified liquid argon, stored in a specially designed heat insulated tank located inside a deep underground cavern. The conditions inside the tank and the cavern, and the purity of argon will be maintained using appropriate systems for cooling, heating, pressurization and filtration. Smaller adjacent caverns will host the process unit equipment (process unit caverns). The caverns for the tank and the process units are planned to be excavated inside a mine at about 1400 meters underground. The quantitative results presented here provide incentives for improvements on the current process design of the installation that can reduce significantly the expected frequencies of accidental argon release due to tank overpressure.

Keywords: safety assessment, cryogenic argon, loss of containment, underground installation, tank overpressure, neutrino detectors

1. INTRODUCTION

Advances in neutrino physics, low energy neutrino astronomy and direct investigation of Grand Unification require the construction of very large volume underground observatories. Many European national underground laboratories with high technical expertise are currently operated with forefront smaller-scale underground experiments, and there is currently a lot of activity worldwide on the construction of a large-scale facility. The heart of these neutrino observatories is a detector consisting of specialized sensitive detection devices submerged into a detector medium. The detector medium on which the neutrino interact can be purified water, organic scintillators or in the present case liquid argon. The measurements of the particles produced are analyzed to establish the characteristics of the neutrinos that generated the interactions. There is significant evidence that large-scale observatories will enable fundamental discoveries in the field of particle and astroparticle physics [1].

The safety analysis presented here is part of an excessive multiyear design study elaborated by a large consortium of neutrino physicists and European construction companies, within the projects LAGUNA (2008-2011) and LAGUNA-LBNO[†] (2011-2014) funded by the European Commission. Among other issues, the study considered (a) the selection of an appropriate location and the conduction of geomechanical investigations, (b) the design and costing of the excavation of underground caverns and tunnels, (c) the design and costing of a tank and its construction plan to host the liquid argon, (e) the design, construction and costing of the detector equipment, (f) the design and costing of on ground and underground processes to fill the tank with argon and (g) to maintain argon at the desired conditions etc. Safety issues have been granted significant attention from the beginning of this study. Several types of risks have been identified and registered, and specific measures were proposed to prevent and mitigate them. The authors were responsible for assessing the safety of the underground installation at normal operation and quantifying the risks of cryogenic argon releases.

---

[*] Contact person: Dr. Effie Marcoulaki, email: emarcoulaki@ipta.demokritos.gr, tel.: +302106503743

[†] LAGUNA is the abbreviation for Large Apparatus for Grand Unification and Neutrino Astrophysics. LBNO stands for Long Baseline Neutrino Observations

2. GENERAL DESCRIPTION OF THE INSTALLATION

In the present work, the neutrino detector is a tank with gross capacity 52,300 m^3 filled with cryogenic argon at 87K. The argon tank and associate processes considered here will be located in specially excavated caverns 1400m underground (see Figure 1a). The location considered is a copper and zinc mine at Pyhäsalmi in central Finland (see Figure 1b). The location selection was based on various criteria, including the excellent rock characteristics, the absence of nearby nuclear plants and hence of neutrons, the depth of the mine, the low seismicity and the distance from CERN. The latter is ideal for the study of accelerated neutrinos generated at CERN and aimed towards a detector at Pyhäsalmi. Information on the site selection and characteristics is publicly available in the LAGUNA project deliverables [1].

Figure 1a: Pyhäsalmi mine, Central Finland   **Figure 1b: Present mine layout [1]**

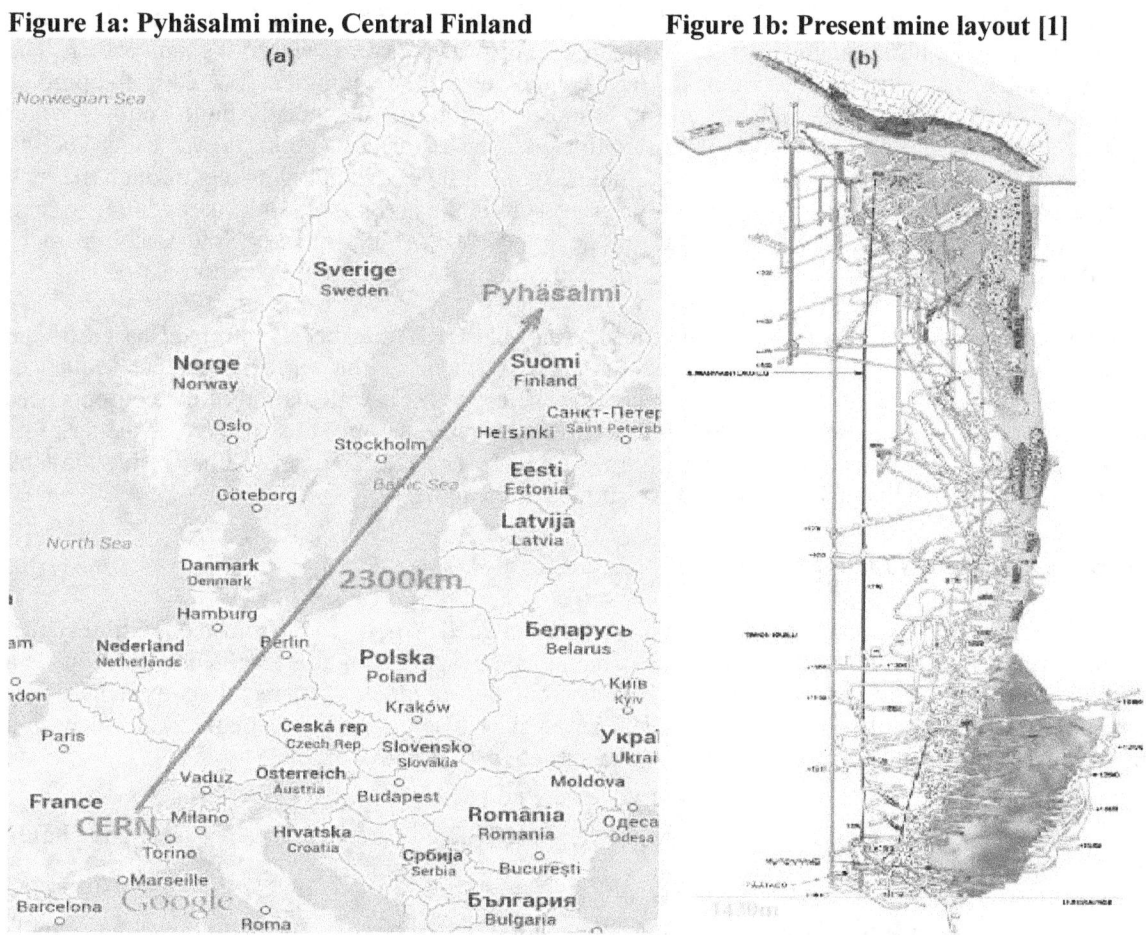

At -1400m (see Figure 1b) three large caverns will be excavated, suitable for two 50kton f.m. liquid argon tanks and one 50kton f.m. liquid scintillator tank (tank caverns). Smaller adjacent caverns will host the process unit equipment (process unit caverns). The tank caverns will be isolated from the process unit caverns and the rest of the mine using lock doors.

The LAGUNA-LBNO project team has provided detailed designs on the location, sizing, cost and equipment types of the on ground facilities and the shafts used for technical infrastructure (like piping, electricity wires). This information is published at [1] in strict confidence. The process flow sheet, the argon tank and process unit sizes, the temperature and pressure conditions etc used in the present analysis are also found in [1]. In particular, the present QRA is based on the geotechnical designs and rock mechanical analysis conducted by Kalliosuunnittelu Oy Rockplan Ltd, the tank design and calculations elaborated by Technodyne International Ltd and Rhyal Engineering Ltd, and the liquid handling process design proposed by Sofregaz SA.

The detector medium is argon, kept at cryogenic conditions and vapour-liquid equilibrium at 87K and 1265mbar in specially designed and heat insulated tank. A detector network and appropriate instrumentation to measure neutrino interactions with argon atoms will be immersed in the argon liquid inside the tank. A ventilation system (pressurizer) and a cavern heating system will work in parallel to maintain the conditions in the tank cavern at 22°C and 1250mbar.

The neutrino experiments require argon liquid of particularly high purity, so argon liquid will be taken from the tank and processed in a set of filtration units. The argon boil-off produced inside the tank due to heat influx in the tank will be re-condensed in heat exchanger units that use nitrogen as coolant. The argon cooling system will also include expansion and compression units arranged in loops that aim to cool down, re-condensate and reuse the nitrogen coolant, and a nitrogen storage vessel.

The ventilation and heating system units, the pumps, motors and turbines for fluid transmission and processing, and the filtration units will be located in the processes cavern, next to the tank cavern. The heat generated by the instrumentation in the tank cavern and all the equipment located in the process cavern will be rejected finally into the surface atmosphere through a specially built Cooling Water System. Presently the design of this system comprises a motor driven cooling tower using cooling water from the nearby Lake Pyhäjärvi, and a pipeline running several hundred meters through the rock to transmit the water underground. In case of overpressure in the argon tank, the water cooling system is able to transmit the released argon gas to the mine surface atmosphere. Note that, the pipes transmitting argon between the surface and -1400m are empty during normal operation.

A safety analysis along with a Quantitative Risk Analysis (QRA) has been performed based on the existing design details. The QRA includes: (a) an assessment of the various accident sequences leading to Ar release; (b) an estimation of their frequencies; and (c) a calculation of the consequences for three types of Ar release using CFD tools. This paper presents only the accident sequences that lead to Ar release owning to overpressure in the LAr tank. The release can be either inside the tank cavern, or in the process units cavern, or at the surface of the mine.

3. HAZARD IDENTIFICATION

The main concern of storing such a large quantity of liquefied Argon stems from the possibility that Liquid Argon (LAr) may be released from its containment, evaporate and (a) reach concentrations in a particular confined space that can expel oxygen from the air and/or (b) result to extremely low temperatures in the confined space. Two main areas exist where argon is contained and there is a possibility of a Loss of Containment (LOC) and a subsequent release of argon:
   a. Tank cavern: Main cavern with the cryogenic argon tank
   b. Process unit cavern: Process area where argon is cooled / condensed (if it is in gas phase), filtrated and returned to the argon tank.

A LOC and hence an argon release may occur as a result of a number of immediate causes. A number of safety measures either engineered features or procedural are employed to prevent the occurrence of each of the immediate causes of LOC. This section presents the immediate causes for LOC due to overpressure and the associated safety measures.

All different type of containment failure may be divided in two major categories:
   a. Structural Failures
   b. Containment Bypassing

A structural failure of the containment occurs if the stress employed on the containment by the various operating conditions is larger than the stress of the containment. This inequality may happen when the strength of the containment is as designed but the stress exceeds the design limits. Alternatively, an inequality may also occur if the strength of the containment becomes lower than the normally expected stresses.

Containment Bypassing occurs whenever an engineered opening in the containment (like a valve) opens inadvertently when it is supposed to be closed. Such failures are mainly caused by human actions either during normal operation or during test and maintenance activities. Assessment of causes of containment bypass requires a detailed final design, and the detailed relevant operating, test and maintenance procedures. At this stage of the system design these details are not available, hence the corresponding potential causes of argon release are not included. Immediate causes of structural failure of the liquid argon containment are determined with the help of the Master Logic Diagram [2].

4. OVERPRESSURE AND ASSOCIATE SAFETY MEASURES

Increase of the internal pressure can cause an increase of the stress on the containment exceeding its design limits and resulting in LOC. In the present installation the critical design feature that might cause a LOC owing to overpressure, is the roof of the tank. Its design requires that the overpressure inside the tank with respect to the pressure in the cavern does not exceed the 25mbarg. It is assumed that, causes that result in a gauge pressure greater than 25mbarg will result in LOC and in particular in a small break at the gaseous phase of the argon. Furthermore, if the pressure increase is relieved through the pressure relief valves of the tank, Ar is transferred outside the tank in the Ar-processes cavity, from where it is transferred to the mine surface through an appropriate system.

This section presents the safety functions and the safety measures incorporated in the design to prevent pressure difference beyond the design limit, and the mitigating measures to relieve overpressure and preserve the containment integrity. Systems that directly serve a safety function are called *Primary Safety Systems* (PSS). For successful operations, the Primary Safety Systems sometimes depend on other systems that provide support services to them, called *Support Safety Systems* (SSS). Their operation is also safety significant, but only through the impact they have on the operation of the PSS.

Argon is kept at vapor-liquid equilibrium (VLE) at cryogenic conditions using a cooling system, to remove the heat influx from the cavern air to the argon contained in the tank. Any heat imbalance can cause an increase in the tank internal pressure, leading to LOC and argon release. The heat imbalance may be caused by either an increase in the heat influx or a decrease in the heat removal capacity. Heat flux into the cryogen from the environment will vaporize the liquid and potentially cause pressure build-up in cryogenic containment vessels and transfer lines. Cryogenic fluids have small latent heats and expand 700 to 800 times to room temperature, so even small heat inputs can cause significant pressure increase.

4.1. Safety Function F1: Maintain Heat influx into LAr within design limits

To maintain the heat influx down to acceptable levels, the tank is equipped with an appropriate insulation. Provided that the cavern air conditions are at the required levels, the insulation is designed to maintain the heat influx rate from the cavern air at 43kW.

Safety system PS1: Tank Insulation

Additional support systems to maintain the integrity of the insulation are systems to detect loss of insulation, and the design considerations allowing insulation replacement [1]. Given the preliminary stage of the design, in section 5 it is simply assumed that given a loss of insulation it will take approximately one month to detect the loss and repair it to the design specifications.

4.2. Safety Function F2: Maintain pressure within Tank below the maximum acceptable level

Heat influx into the tank must be removed, to keep the pressure inside the tank below the acceptable level. There are two sources of heat influx: (a) 43kW from the cavern air (through the insulation); and (b) 84kW from the detector instrumentation.

System PS2: The re-condensation system

This system removes the Ar boil-off generated by the heat influx, maintaining the tank pressure at 1265mbara. The boil-off circulates with the help of a motor driven pump through a Heat Exchanger where it is re-condensed. The process uses liquid N_2 converted to N_2 gas. Heat is removed through the N_2 cycle and eventually is rejected into the surface atmosphere through the Service Water Cooling system (section 4.5/SS1). There are two N_2 loops to convert N_2 gas back to liquid N_2. One loop is sufficient to remove the 127kW produced during normal operation of the LAr tank The success criteria for the re-condensation system depend on the size of the extra heat load generated by the loss of insulation and on whether the instrumentation has been turned off (as presented later in Table 5).

System PS3: Operator's action to switch off Instrumentation

Operators must diagnose the increased heat flux and switch off the instrumentation and other heat producing loads. Two possible states are considered: (a) Instrumentation loads are switched off (success); and (b) Instrumentation loads are not switched off (failure).

4.3. Safety Function F3: Maintain Cavern Conditions

Cavern air pressure should be maintained at 1250mbar or higher, given that the Tank pressure is increasing and the tank gauge pressure should remain lower than 25mbarg. Furthermore, the cavern air temperature should be maintained at 22°C.

System PS4: Ventilator system

This system takes air from the tank cavern exterior and injects it into the tank cavern to maintain the required ambient pressure of 1250mbar. There are two loops, and each one is capable to maintain the cavern pressure at the required level even if the Cavern Heating system (PS5) has failed.

System PS5: Cavern Heating System

This system is heating the cavern to maintain the air temperature at the required 22°C. If the cavern-air temperature decreases, this system can heat it up to the required 22°C. Therefore, this system alone can maintain the temperature of the cavern air and hence the air pressure at its required levels. If the heating system is not available, the cavern temperature will decrease, thus reducing the air pressure, unless PS4 is available. Based on design information available at the time of the analysis it is assumed that, if PS5 is available then it can maintain the cavern air at the required temperature. Systems PS4 and PS5 are thus completely redundant.

4.4. Safety Function F4: Relieve Extra Pressure

If the gauge pressure in the tank exceeds 25mbarg, argon gas must be relieved from the tank and released into the surface atmosphere via the Service Water Cooling system (section 4.5/SS1).

System PS6: Set of Tank Relief Valves

The argon tank is equipped with a set of safety relief valves. The pressure relief system typically comprises at least 4 valves in operation. The same applies for vacuum relief valves [1]. The released argon is then lead to the surface by the Service Water Cooling system (section 4.5/SS1).

4.5. Support safety systems

System SS1: Service Water Cooling System

Heat generated in the cavern and the associated process systems is removed by SS1 and transferred to the surface atmosphere through a cooling tower-type process. Hot water carrying the heat from all systems in the installation (including heat generated by motors and turbines) is fed into an evaporating system where the heat is deposed into water ultimately coming from the surface and the nearby lake. This water is evaporated and sent to the surface through forced circulation powered by a ventilator.

The safety significance of this system is extremely high. Lack of cooling signals inability of heat transfer from the tank, and failure of all motors, pumps and turbines in the process equipment. Given the current level of information, it has been assumed that failure of this system (as a whole) will result in failure of all PSSs and SSSs that need cooling for their continuing function. In case of SWCS failure, success of the pressure relief (PS6) results in Ar release in the area of the SWCS. If PS6 fails then tank overpressure occurs and Ar is released in the tank cavern.

System SS2: Electric Power

Electric power is another essential service needed by almost all PSSs and SSSs systems for successful operation. Loss of AC and/or DC power for prolonged period of time would result in failure of all PSSs, as well as SS1. Consequently, the safety significance of the electricity supply system is extremely high. At this stage of the design there is no detailed information on the design of the electricity supply system. This analysis assumes two sources of electric power: an offsite source (from the grid) and an Emergency Diesel Generator (DG). Loss of both these sources for more than certain time periods (to be determined for each initiating event) will cause a total blackout, which in turn will cause failure of all equipment depending on electricity.

5. DELINEATION OF ACCIDENT SEQUENCES

Accident sequences are sequences of events that lead to LOC integrity and hence to argon release. An accident sequence consists of an initiating event that challenges the safety functions/systems of the installation, and of additional events representing either hardware failures and/or human actions that collectively result in a LOC. For the safety systems described above, accident sequences are obtained using Functional Block Diagrams and the associated Event Trees [3]. All systems are operating during normal operation. Failure of any major component of the safety systems constitutes an initiating event.

5.1. Loss of Offsite Power

With the exception of the Ar-N_2 Heat Exchanger, some turbine driven pumps in the Re-condensation system PS2 and the relief system PS6, all safety related systems depend on electric power to operate. Consequently, following Loss of Offsite Electric Power (LOOP), if the DG does not start, the installation gets into blackout conditions. Unless power is restored on time, over-pressurization occurs. Then, if the relief valves open we have release of Ar gas in the processes cavern, otherwise we have release of Ar gas in the tank cavern.

During blackout, the SWCS, the systems maintaining cavern pressure and the Ar re-condescension system do not operate. Instrumentation is turned off since there is no power, and the heat influx towards the Ar tank is equal to the 43kW from the cavern air. Since boil off is not re-condensed, the pressure inside the Tank increases at a rate of 0.25mbar/hr [1]. Furthermore, since there is no cabin air pressurization or heating, the predicted cavern pressure drop is 4.3mbar/hr. As a result, the gauge pressure in the Tank will reach the tank design limit of 25mbarg within 2.2 hours. Consequently there is a time period of about 2.2 hours within which either the offsite power or the DG ought to be repaired to avoid argon release in the cavern. The events considered in this Event Tree are given in Table 1. The calculations of component availabilities and failure/repair probabilities within a given time period are based on classical reliability techniques.

A reduced Event tree depicting the accident sequences following a LOOP is given in Figure 2.
- Accident sequence #1 is a successful sequence since following the initiation of LOOP, offsite power is restored within 2.2hrs. The state of the DG and its repair is indifferent in this sequence.

- Accident sequence #2 is also successful since although offsite power is not restored within 2.2 hours, the DG starts and it runs for more than 2.2 hours.
- Accident sequences #3 & #4 result in an overpressure because offsite power is not restored within 2.2 hours, the DG starts but it fails to run for 2.2 hours. In sequence #3 the relief valves open releasing Ar in the Ar-processes cavern. Sequence #4 results in Ar release in the tank cavern since the relief valve system fails.
- In sequences #5, #6 & #7 the offsite power is not restored within 2.2 hours and the DG does not start. Sequence #5 is successful since the DG is repaired and started within 2.2 hours.
- In sequences #6 & #7 the DG is not is repaired and started within the 2.2 hour window. Then, if the relief valve system operates there is Ar release in the Ar-processes cavern (#6). If the relief valve system fails there is Ar release in the tank cavern due to overpressure (#7).

Table 1: Events considered in Loss of Offsite Power Event Tree

EVENT	PROBABILITY
Loss of offsite Power	10^{-5}/hr (Frequency) [4]
Mission Duration	2.2 hours
Failure to recover Offsite Power within 2.2 hours (MTTR=2hours)	0.37
Emergency Diesel Generator starts on Demand	0.98
Emergency Diesel Generator is repaired within 2.2 hours	1.54×10^{-1}
Given that Emergency Diesel Generator starts, mean availability over 2.2 hours	9.11×10^{-1}
Pressure Relief System failure (on demand)	4×10^{-3}

Figure 2: Loss of Offsite Power Event Tree

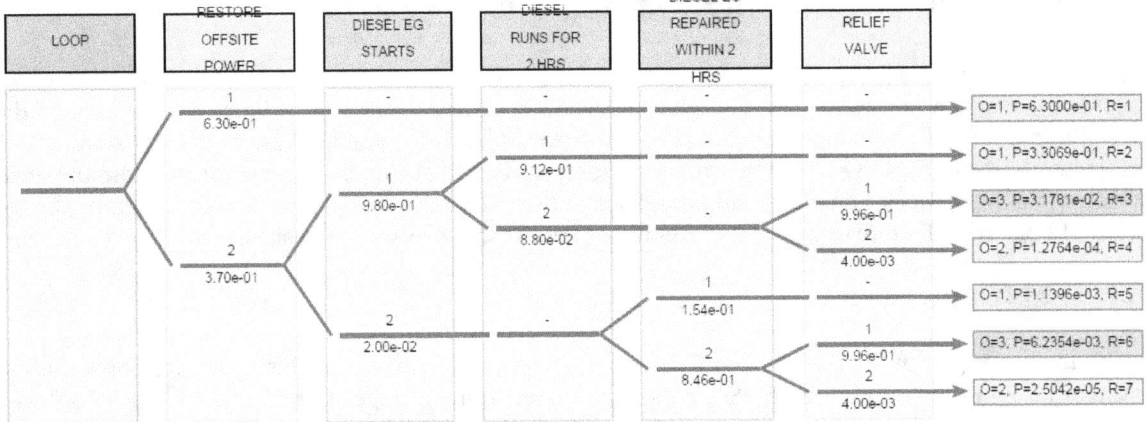

5.2. Loss of Service Water Cooling System

The initiating event here is a failure of the Service Water Cooling System (SWCS) for reasons other than lack of electric power. Without cooling, all safety systems except the human action switching off the instrumentation (PS3) will fail. If instrumentation is switched off, then the situation is exactly as in the station blackout (section 5.1), so there is a period of 2.2 hours before argon release due to overpressure. If the instrumentation is ON then the heat input rate at the Ar in the tank is 126.6kW resulting at an internal pressure increase rate of 0.74mbar/hr [1]. The cavern pressure still drops at a rate of 4.3mbar/hr, but in this case the differential pressure limit will be exceeded in just 2 hours. Consequently, quantification of this event should include the probability of not recovering the system within 2 hours.

There are two general ways for losing the SWCS: either by failure of one of the motor drive pump or by failure of the motor driven fan. The events considered in this Event Tree are given in Table 2. An additional failure mode consists in loss of cooling water transported from the nearby lake. This failure mode has not been considered since the design of this part system did not exist at the time of the analysis.

A reduced Event Tree for the loss of SWCS is depicted in Figure 3. Two initiating events have been considered, corresponding to: (a) the loss of a pump and (b) a loss of a Heat exchanger. The logic is simple resulting in success if the failed component is repaired within 2.2 hours (if instrumentation is turned off) and in 2 hours (if instrumentation is not turned off). If the failed component is not repaired within the available grace period the sequences result in failure due to overpressure.

Sequences #2 & #5 are accident sequences, as they involve failure of the SWCS pump with failure to recover it within the available time. PS6 is successful and Ar is released in the SWCS area, Sequences #3 & #6 involve Ar release in the tank cavern since PS6 has failed. Similar accident sequences #8 & #11 and #9 & #12 are calculated for shorter grace period since the Instrumentation is left ON.

Table 2: Events considered in Loss of Service Water Cooling System Event Tree

EVENT	PROBABILITY
Loss of the operating fan	3.1×10^{-4}/hr (Frequency)
Loss of the Heat exchanger (ventilator)	3.6×10^{-5}/hr (Frequency)
Failure to recover pump within 2 hours (MTTR=12hours)	0.85
Failure to recover pump within 2.2 hours (MTTR=12hours)	0.83
Failure to recover Heat Exchanger within 2 hours (MTTR=36hours)	0.95
Failure to recover Heat Exchanger within 2.2 hours (MTTR=36hours)	0.94
Failure to switch Instrumentation OFF	0.1

Figure 3: Event Tree for Loss of Service Water Cooling System (SWCS)

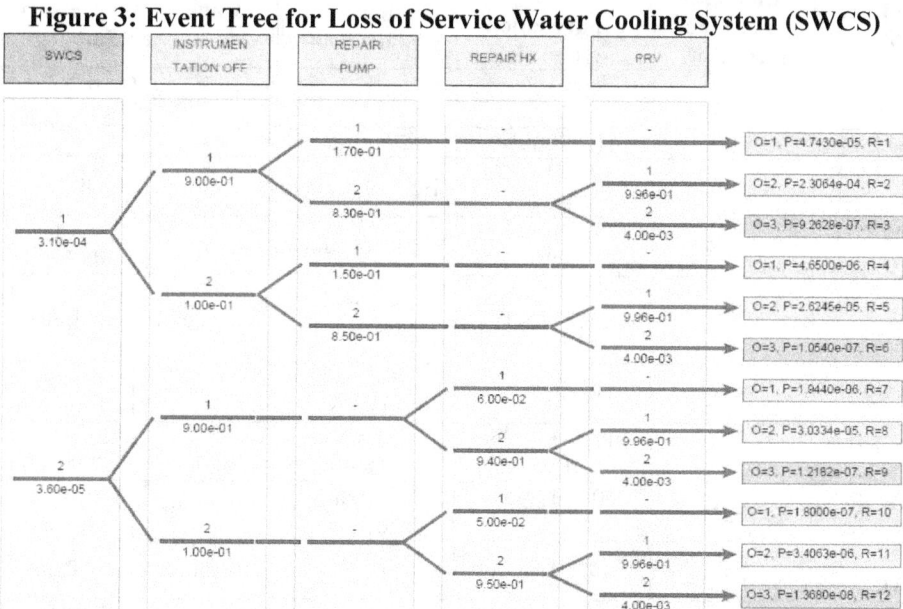

5.3. Loss of Cavern Heating System

Loss of the cavern heating system, for causes other than a station blackout or loss of service water cooling system, does not create an immediate problem provided that the cavern ventilator (pressurizer) is available. This is because it has been assumed that the cavern ventilator is capable to maintain the cavern pressure at the required levels. The analysis is similar to subsections 5.1 and 5.2 and the results are given in Table 7.

5.4. Loss of one Cavern Ventilator (Pressurizer) Train

Cavern air is maintained at a constant pressure of 1250mbar through the Ventilator system which is continually operating during the operational phase of the LAGUNA LAr detector. The system has two trains each capable of pressurizing the cavern. Failure of the operating train constitutes the initiating event. The standby train has to start and continue operation. However, even if the second train does not start, the Cavern heating system is capable to maintain the cavern air at the required pressure. The analysis here is similar to subsections 5.1 and 5.2 and the results are given in Table 7.

5.6. Loss of Ar boil off re-condensation pump

Loss of the argon boil-off re-condensation pump results in loss of the boil-off re-condensation system. Depending on whether instrumentation has been turned off or not the available grace period is 10 hours and 40 hours, respectively. The analysis is similar to subsections 5.1 and 5.2 and the results are given in Table 7.

5.7. Loss of Operating Re-condensation Train Event Tree

Loss of the Operating Re-condensation Train results in a challenge to the boil off re-condensation system. For this system to fail, the standby train or the Ar pump must fail before restoring the online failed N_2 train. Depending on whether instrumentation has been turned off or not the available grace period is 10 hours or 40 hours, respectively. The analysis is similar to subsections 5.1 and 5.2 and the results are given in Table 7.

Table 5: Success criteria for Re-condensation system given Loss of insulation

Extra Heat load (H) due to Loss of Insulation	Instrumentation and other loads turned off	Total heat load to Re-condensation system (H)	Re-condensation requirement	Time window before Tank – Cavern overpressure exceeds 25mbarg		
				2 N_2 trains	1 N_2 train	0 N_2 trains
LI1: H ≤ 87kW	YES	H ≤ 130kW	1/2 N_2 trains	∞	∞	> 13 hrs
LI2: 87kW < H ≤ 130kW	YES	130 < H ≤ 173 kW	Both N_2 trains	∞	>40 hrs	10 < t < 13 hrs
LI3: 130kW < H ≤ 217kW	YES	173 < H ≤ 260 kW	Both N_2 trains	∞	13 < t < 40 hrs	6.6 < t < 10 hrs
LI4: 217kW < H	YES	260kW < H	System inadequate	Depends on size of LI	t < 13hrs	t < 6.6 hrs
LI1: H ≤ 87kW	NO	H ≤ 217kW	Both N_2 LOOPs	∞	t > 20 hrs	t >8 hrs
LI2: 87kW < H ≤ 130kW	NO	217 < H ≤ 260 kW	Both N_2 LOOPs	∞	13 < t < 20 hrs	6.6 < t < 8 hrs
LI3: 130kW < H ≤ 217	NO	260 < H ≤ 347 kW	System inadequate	t > 20 hrs	8 < t < 13.6 hrs	5 < t < 6.6 hrs
LI4: 220kW < H	NO	347kW < H	System inadequate	t < 20 hrs	t < 8 hrs	t < 5hrs

5.8. Loss of Tank Insulation

Loss of insulation (LI) results in higher heat influx to the tank, increasing the Ar boil off rate and hence the required re-condensation rate. Under normal operating conditions (with insulation intact) the heat load that must be removed through the re-condensation of the boil-off is about 127 kW. If there is an abnormal situation it is possible to switch off the instrumentation and other heat producing functions, thus reducing the necessary heat removal rate to 43kW.

In order to include all possible combinations of insulation loss and whether instrumentation is switched off, the following four different types of insulation loss are considered:
- LI1: the additional resulting heat influx to the LAr Tank is H ≤ 87kW
- LI2: the additional resulting heat influx to the LAr Tank is 87kW < H ≤ 130kW
- LI3: the additional resulting heat influx to the LAr Tank is 130kW < H ≤ 217kW
- LI4: the additional resulting heat influx to the LAr Tank is H > 217kW

Since each train of the N_2 cooling system can remove up to 130kW, the requirements for the N_2 cooling system are given in Table 5. Loss of insulation is an event that occurs randomly in time. It is assumed that it will take, on the average, one month or 720 hours to completely repair the insulation. Since all the safety systems are online (continuously operating), the probability of failing given a loss of insulation event (and within the time that this event has not been recovered) is a second order effect, can therefore be neglected. In other words, the failure of the various safety systems has been taken into consideration when their failure has been considered as an initiating event. This is true for the all safety systems with the exception of the LAr boil-off-re-condensation system. Successful operation of this system depends on the amount of heat that it has to remove from the tank and hence it is directly affected by the occurrence of reduction in the insulation. For this reason, in calculating the conditional probability of argon release due to overpressure given a Loss of Insulation initiating event, only the LAr-re-condensation system is considered. All other systems are assumed to operate successfully. The frequency for loss of insulation is taken equal to 1×10^{-2}/yr. The conditional probabilities that, given a loss of insulation, the loss will be according to LI1, LI2, LI3 and LI4, are suggested equal to 0.8, 0.1, 0.07 and 0.03, respectively.

Loss of insulation resulting in a Heat Influx less than 87kW

The calculation for the frequency of Loss of insulation resulting in a Heat Influx less than 87kW take into account the state of operation of the N_2 cooling system. If this system is completely failed then with the instrumentation OFF there is a heat input equal to 43kW + 87kW = 130kW (maximum) hence the pressure inside the tank will increase by 25mbar within 13 hours (for 130kW). If the instrumentation is ON then the maximum heat input is about 217kW and the available time is 8 hours. If one N_2 train is operating, then with the instrumentation OFF it can remove the net heat input of 130kW. If the instrumentation is ON then net heat input is about 217kW the operating train removes 130kW, hence the second train is needed and it has to be available within 20 hours. The events included in the Event Tree are given in Table 6.

Loss of insulation resulting in a Heat Influx between 87kW & 130kW

With the added heat influx due to LI, the available time for recovering failed parts of the LAr re-condensation system changes. If this system is completely failed then with the instrumentation OFF there is a heat input between 130kW and 173kW (maximum) hence the pressure inside the tank will increase by 25mbar within 10 to 13 hours. If the instrumentation is ON then the heat input is between 217kW and 260kW and the available time is between 6.6 hours and 8 hours. If one N_2 train is operating, then with the instrumentation OFF there is an extra heat input between 0 and 43kW (maximum) hence the pressure inside the tank will increase by 25mbar in more than 40 hours. If the instrumentation is ON then the extra heat input is between 87kW and 130kW and the available time for the second train to be recovered is between 13 hours and 20 hours. Similarly to LI1, the frequency for Loss of insulation resulting in a Heat Influx between 90kW & 130kW is set at 1.14×10^{-7}/hr.

Table 6: Events considered in the for LI resulting in a Heat Influx less than 87kW

EVENT	PROBABILITY
Loss of insulation resulting in a Heat Influx less than 90kW	9.12×10^{-7}/hr (Frequency)
Failure of Ar Pump (mean failure probability for 720 hours)	0.1
Ar Pump not recovered within 13 hours (Instrumentation OFF)	0.338
Ar Pump not recovered within 8 hours (Instrumentation ON)	0.513
Failure of the Operating Re-condensation N_2-Train (ORTr) (mean failure probability for 720 hours)	0.16
Standby Re-condensation N_2-Train fails to start	2×10^{-2}
Probability of not recovering failed ORTr within 20 hours (MTTR=12hours) (Instrumentation ON and one train operating)	0.189
Probability of not recovering failed ORTr within 8 hours (MTTR=12hours) (Instrumentation ON and no train operating)	0.513
Probability of not recovering failed ORTr within 13 hours (MTTR=12hours) (Instrumentation OFF and no train operating)	0.338

Loss of insulation resulting in a Heat Influx between 130kW & 217kW

With the added heat influx, the available time for recovering failed parts of the LAr re-condensation system changes. If this system is completely failed then with the instrumentation OFF there is a heat input between 173kW and 260kW (maximum) hence the pressure inside the tank will increase by 25mbar within 6.6 to 10 hours. If the instrumentation is ON then the heat input is between 260kW and 347kW and the available time is between 5 hours and 6.6 hours. If one N_2 train is operating, then with the instrumentation OFF there is an extra heat input between 43 and 130kW (maximum) hence the pressure inside the tank will increase by 25mbar within 13 hours and 40 hours. If the instrumentation is ON then the extra heat input is between 130kW and 217kW and the available time for the second train to be recovered is between 8 hours and 13 hours. Similarly to LI1, the frequency for Loss of insulation resulting in a Heat Influx between 130kW & 220kW is set at 7.98×10^{-8}/hr.

Loss of insulation resulting in a Heat Influx Higher than 217kW

If insulation is lost to such an extent that the extra heat input exceeds 217kW, then the total heat input with the instrumentation OFF exceeds 260kW. Since the existing Ar re-condensation system (even if both N_2 trains are available) is able to remove only 260kW, the extra heat will eventually lead to Tank overpressure. The time until tank over-pressurization depends on the extent of the heat input in excess of 260kW. The conditional probability of this happening is therefore equal to unity and whether there will be Ar release into the surface atmosphere or inside the tank cavern depends on the successful operation of the pressure relief system. The estimated frequency of this event occurring is 3.42×10^{-8}/hr.

5.8. Summary of argon release owning to overpressure and recommendations

The results of the event tree analysis are summarized in Table 7. In the case of overpressure in the argon tank, argon gas can be released to the mine surface atmosphere through the PSVs and the SWCS. The frequency of such an event has been estimated at 5.74×10^{-1}/year or once every 21 months. This mainly comprises the frequencies of loss of the argon pump in the re-condensation system (by 99%), and the loss of the operating nitrogen loop (about 0.4%), and finally the loss of insulation (0.6%). Increasing the reliability of the argon gas pumping system can reduce significantly the expected frequency of this type of release.

Perusal of Table 7 indicates that the frequency of argon release inside the tank cavern owing to overpressure, is almost once every seven months. This frequency is notably high, but only 17% of this value is attributed to initiating events other than the failure of the SWCS. If the SWCS is not considered, the main failure cause is the loss of the argon pump in the re-condensation system (99%). This is due to the complete lack of redundancy in this system. It is recommended that the final design incorporates redundancy in the SWCS, as well as in the Ar side of the re-condensation system. This could reduce the expected frequency of LOC due to overpressure by two to three orders of magnitude. It is noteworthy that, the literature for double walled refrigerated LNG containments suggests a frequency for failure of the roof and vapor release in the order of 3.5×10^{-9}/hr [5]. This is the order of magnitude for the contribution of those systems that contain redundancy, like the cavern atmosphere regulating systems, the N_2 side of the re-condensation system and the loss of insulation events.

7. CONCLUSIONS

This work quantifies the risk for loss of containment in a deep underground installation involving large quantities of a cryogenic substance. The work is part of a very extensive study to select the site, design the excavation, construction and deployment phases, design of the process units etc, undertaken by a pan-European team of highly specialized and experienced engineers. The engineers collaborated closely with distinguished scientists, so that the proposed installation meets the requirements for next generation particle and astroparticle physics experiments.

Table 7: Frequencies for Loss of Containment owning to overpressure

INITIATOR = LOSS OF:	Frequency (hr^{-1})	Conditional Probability of Argon Release in			Frequency (hr^{-1}) of Argon Release in		
		Tank Cavern	Process Area	Surface of Mine	Tank Cavern	Process Area	Surface of Mine
LOOP	1.00x10^{-5}	1.5x10^{-4}	3.8x10^{-2}	--	1.50x10^{-9}	3.80x10^{-7}	--
SWCS Pump	3.10x10^{-4}	3.65x10^{-3}	8.3x10^{-1}	--	1.13x10^{-6}	2.57x10^{-4}	--
SWCS HX	3.60x10^{-5}	3.76x10^{-3}	9.4x10^{-1}	--	1.35x10^{-7}	3.38x10^{-5}	--
Operating Pressurizer Train	1.00x10^{-5}	2.00x10^{-8}		5.00x10^{-6}	2.00x10^{-13}		5.00x10^{-11}
Cavern Heating System	3.10x10^{-4}	6.80x10^{-10}		1.70x10^{-7}	2.11x10^{-13}		5.27x10^{-11}
Ar Pump	3.10x10^{-4}	8.60x10^{-4}		2.10x10^{-1}	2.67x10^{-7}		6.51x10^{-5}
Operating N$_2$ train	5.00x10^{-5}	2.60x10^{-5}		6.40x10^{-3}	1.30x10^{-9}		3.20x10^{-7}
Insulation: LI1	9.12x10^{-7}	2.60x10^{-5}		3.90x10^{-2}	2.37x10^{-11}		3.56x10^{-8}
Insulation: LI2	1.14x10^{-7}	2.30x10^{-4}		5.70x10^{-2}	2.62x10^{-11}		6.50x10^{-9}
Insulation: LI3	7.98x10^{-8}	4.50x10^{-4}		1.10x10^{-1}	3.59x10^{-11}		8.78x10^{-9}
Insulation: LI4	3.42x10^{-8}	4.00x10^{-3}		9.96x10^{-1}	1.37x10^{-10}		3.41x10^{-8}
TOTAL					**1.54x10^{-6}**	**2.92x10^{-4}**	**6.55x10^{-5}**

The analysis presented here generates and investigates accident sequences leading to overpressure in the cryogen (argon) containment and estimates their frequencies of occurrence. The results based on the present state of the design identify weak spots. Incorporating design with higher reliability for these parts of the system would greatly decrease the associated risks.

Note that, the present study is a summary of a detailed analysis, including CFD dispersion simulations, and assessment of adverse consequences on personnel health and the cavern structural integrity following a potential argon release in the experimental facility. The analysis concluded that there is no risk for personnel present in the tank cavern when the release occurs or the tank cavern integrity [1].

Acknowledgements

The authors acknowledge financial support from the European Commission, project LAGUNA-LBNO "Design of a pan-European Infrastructure for Large Apparatus studying Grand Unification, Neutrino Astrophysics and Long Baseline Neutrino Oscillations", European Commission, Grant Agreement No. 284518, FP7-INFRASCTRUCTURES-2011-1.

References

[1] Deliverables of LAGUNA (GA 212343) and LAGUNA-LBNO (GA 284518) European Commission projects. Available at: http://laguna.ethz.ch:8080/Plone
[2] I. A. Papazoglou and O. N. Aneziris. "*Master Logic Diagram. Method for hazard and Initiating event identification in process plants*", Journal of Hazardous Materials, A97, pp. 11-30 (2003)
[3] I. A. Papazoglou. "*Functional block diagrams and automated construction of event trees*", Reliability Engineering & System Safety, 61(3), pp. 185–214 (1998)
[4] Offshore Reliability Data (OREDA). Available at http://www.oreda.com
[5] Health and Safety Executive. "*Failure rate and event data for use within land use planning risk assessments*", document ID: HSE PCAG chp_6K Version 12 – 28/06/12. Available at: http://www.hse.gov.uk

CENTRIFUGAL PUMP MECHANICAL SEAL AND BEARING RELIABILITY OPTIMIZATION

Peymaan Makarachi [a], and Mohammad Pourgol-Mohammad [a*]
[a] Sahand University of Technology, Tabriz, Iran

Abstract: Centrifugal pumps are used in a wide range of field and industrial applications and as significant rotating equipment, incurred high real life costs. The earlier researches illustrate that the main cost is borne by the seals and bearings as critical components of the pump. Most of the pump maintenance work is initiated by the failure of a mechanical seal or bearing as well. Reliability allocation is developed for the early design stage of a system to apportion the system reliability requirement to its individual subsystems. This article examines possible approaches to allocate the reliability values to the components of the mechanical seals and bearings such that the total cost is minimized. The cost of increasing reliability of these components is considered as an exponential function that contains four parameters of component reliability, feasibility factor, maximum achievable reliability and minimum reliability, which is estimated by Monte Carlo simulation. The Genetic Algorithm (GA) optimization is applied to the reliability allocation topic for a typical mechanical seal and bearing components. Optimization process yield optimum values of the components reliabilities, while considering the cost function as an objective in the GA method.

Keywords: Centrifugal pump, Mechanical seal, Bearing, Reliability allocation, Genetic algorithm, Monte Carlo

1. INTRODUCTION

Before a pump can be selected or a prototype designed, the application must be clearly defined. Whether a simple recirculation line or a complex pipeline is needed, the common requirement of all applications is to move liquid from one point to another. As pump requirements must match system characteristics, analysis of the overall system is necessary to establish pump conditions. This is the responsibility of the user and includes review of system configuration, changes in elevation, pressure supply to the pump, and pressure required at the terminal. Relevant information from this analysis is passed on to the pump manufacturer in the form of a pump data sheet and specification.

Centrifugal pumps are extensively used in different industries and in some instances number of utilized pumps could easily count to hundreds of pumps. A pump is usually classified into two general classes of centrifugal and positive displacement. The centrifugal pump has two main parts: a rotating element which includes an impeller and a shaft, and stationary elements made up of a casing, the mechanical seal, and the bearings. With centrifugal pumps, the energy is added continuously by increasing the fluid velocity with a rotating impeller while reducing the flow area.

Centrifugal pumps are the most common type of kinetic pumps and these pumps are used in a wide range of field and industrial with moderate to high flow and low head applications. Mechanical seals are used in centrifugal pumps to provide a leak proof seal between the component parts. There are many different designs for mechanical seals to meet specific applications. Mechanical seal is compromised of the primary and mating rings. When in contact they form the dynamic sealing surfaces that are perpendicular to the shaft. The primary ring is flexibly mounted in the seal head assembly, which usually rotates with the shaft. The mating ring with a static seal, forms another assembly that is usually fixed to the pump gland plate. Each of the sealing planes on the primary and mating rings is lapped flat to eliminate any visible leakage. The basic components of mechanical seal of a centrifugal pump are shown in the Fig. 1 [1].

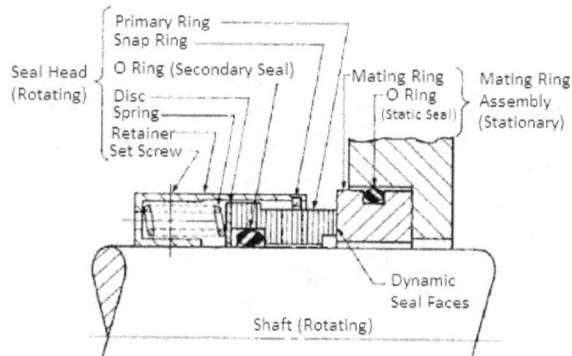

Figure 1: Schematic of Pump Mechanical Seal [1]

Bearings are manufactured to take pure radial loads, pure thrust loads, or a combination of the these two kinds of loads. The nomenclature of a ball bearing is illustrated in Fig. 2, which also shows the four essential parts of a bearing. These are outer ring, inner ring, balls or rolling elements, and separator.

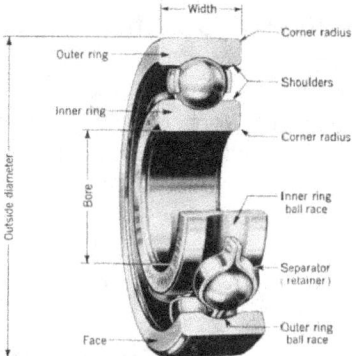

Figure 2: Schematic of Pump Bearing [2]

More than eighty percent of root causes of rotating equipment outages are related to failures of mechanical seals and bearings. Recently, there is great attention on mechanical seal and bearing failures and reliability. For instance, failures of mechanical seals are evaluated in [3-[4], failure modes are analyzed in [5] and improving reliability of seals is discussed in [6-[9]. In addition for increase the reliability of the pumps, researches have been conducted on bearings [10-[13]. A designer needs to achieve the target reliability while minimizing the total cost. Intuitively, some of the lowest reliability components may need special attention to raise the overall reliability level. Such an optimization problem may arise while designing complex system. The cost is formulated as a function of reliability and it has an exponential behavior. It is assumed that the cost function satisfy three conditions. Cost function is a positive definite function, non-decreasing and increases at a higher rate for higher values of reliability. This mathematical formulation depends on certain parameters that they are calculated in this article.

Here, the reliability of the seal and bearing are allocated to the components with optimum value to achieve the minimum cost of increasing their reliability. The minimum required reliability for each component of a seal and a bearing are approximated in order to achieve a system reliability goal with minimum cost and this minimum required component reliability will be achieved via fault avoidance [14]. Monte Carlo method is used for the evaluation of minimum reliability. Feasibility parameter is evaluated for application in the cost function. The maximum achievable reliability of each component is considered 99.99%. The problem of reliability allocation and optimization has been widely treated by many authors. A number of studies have examined these problems for last several decades [15].

2. SCOPE AND OBJECTIVE

This research is aimed to determine optimum reliability value for mechanical seal and bearing components subject to minimization of the specific cost function. Total cost is sum of each component cost. Cost is a function of the components minimum reliability and Monte Carlo method is used for estimating the minimum reliability value. Feasibility parameter is evaluated according to the different indexes like state of the art, complexity, environment and operating time.

Here, GA model is developed based on a binary coding that can easily deal with variables for finding minimum cost of mechanical seal and bearing components of a centrifugal pump. Components reliabilities are used as random variables in the optimization model based on GA. Optimized parameters are resulted as output of the GA.

Mechanical seal and bearing failure modes of a centrifugal pump are defined in section 3. Monte Carlo method which is used in the calculation of minimum reliability of components is described in section 4. The details of the GA programming methodology are given in section 5. Defining the cost function that is used as an objective function in the GA method and determining of the unknown parameters of this function are explained in section 6. The results of the program are presented in section 7 and these results are discussed in section 8. Concluding remarks are provided in section 9.

3. MECHANICAL SEAL AND BEARING FAILURE MODES

Failure mode and effect analysis (FMEA) is a powerful technique for reliability analysis. This method is inductive in nature. The FMEA analysis describes inherent causes of events that lead to a system failure, determines their consequences, and devises methods to minimize their occurrence or recurrence. Here, the information about the critical failure modes and the related failure causes with effects of the pump mechanical seal and bearing are in Table 1 and 2. When mechanical seals are properly applied, there should be no static leakage and, under normal conditions, the amount of dynamic leakage should range from none to just a few drops per minute. If excessive leakage occurs, the cause must be identified and corrected. Causes for seal leakage with possible corrections are listed in Table 1.

Table 1: The FMEA of Mechanical Seal [16-20]

No.	Potential Failure Mode	Potential Cause(s)/Mechanisms of Failure	Potential Effect(s) of Failure
1	Leakage for secondary seals	Installation problems	Seal drips steadily
		Overaged Oring	
		Chemical attack	
		Poor maintenance	
2	Excessive clearance around the seal	Weakness in distortion resistance	Stuffing box leaks abnormally
		Excessive preloads on seal faces	
		Excessive vibration	
		Excessive flush flow	
3	Leakage between rotary and stationary ring	Vibration	Seal life is short
		Poor maintenance	Seal leaks
4	Faces not flat	Foreign particles between seal faces	Seal faces blistered and distorted
		Problems of gland gasket for proper compression	
		Improper material	Seal drips steadily
		Chemical attack	
		Improper cooling of flush lines	Seal life is short
		Incorrect installation	
5	Seal fluid vaporizing	Bypass flush line	Seal spits and sputters
		Problems in gland plate orifices	Seal life is short
6	Inadequate amount of liquid to lubricate seal faces	Bypass flush line	Seal squeals during operation
		Problems in gland plate orifices	Seal life is short

In order for bearing to operate properly, the equipment must be in good condition. The main failures of a bearing are related to mounting, vibration, dirt and improper lubrication. Table 2 lists common troubles that affect the bearing life.

Table 2: The FMEA of Bearing [19-22]

No.	Potential Failure Mode	Potential Cause(s)/Mechanisms of Failure	Potential Effect(s) of Failure
1	Improper mounting	Not observing the basic concepts	Bearings do not give good service life
		Improper workmanship during installation of bearings	Pump operates with noise or vibrations, or both
			Excessive radial or axial load
2	Vibration	Cavitation	Balls and rollers to jam into the
		Bent shafts	
		Unbalanced rotary assemblies	
		Shock thrust loads	The surfaces of the balls and rollers begin breaking away
		Slapping v-belts	
		Improper foundation	
3	Dirt and Abrasion	Careless handling during storage and assembly	Contamination between the balls and races can start a round of false brinelling
			Pump operates with noise or vibrations, or both
			Mechanical seal and stuffing box fails prematurely
4	Inadequate Lubrication	Wrong type of lubricant	Too much friction
			High heat
			Metal-to-metal contact between rolling and stationary elements
			Pump draws higher amps than specified
5	Excessive Lubrication	Too much lubrication	Forming the foam and froth mixed with air
			Overheating
			Pump draws higher amps than specified

4. MONTE CARLO SIMULATION AND ERROR BOUNDS

Monte Carlo Simulation (MCS) is a method that presents the following characteristics: it is applied to many practical problems allowing the direct consideration of any type of probability distribution for the random variables; it is able to compute probability of failure with desired precision; and it is easy to implement.

In reliability analysis the Monte Carlo simulation is used when the analytical solution is not attainable and the failure domain can neither be expressed nor approximated by an analytical form. A reliability problem is formulated using a failure function, $g(X_1, X_2, X_3, ... X_n)$, where $X_1, X_2, X_3, ... X_n$ are random variables. Violation of the limit state is defined by the condition $g(X_1, X_2, X_3, ... X_n) \leq 0$ and the probability of failure, $\hat{p}_f$ is expressed by the following expression:

$$\hat{p}_f = \frac{\sum_{i=1}^{N_T} I(X_1, X_2, X_3, ... X_n)}{N_T} \qquad (1)$$

Where $I(X_1, X_2, X_3, ... X_n)$ is a function defined as:

$$I(X_1,X_2,X_3,...X_n) = \begin{cases} 1 & \text{if } g(X_1,X_2,X_3,...X_n) \leq 0 \\ 0 & \text{if } g(X_1,X_2,X_3,...X_n) > 0 \end{cases} \quad (2)$$

Here, NT independent sets of values are obtained based on the probability distribution for each random variable and the failure function is computed for each sample. Using direct simulation Monte Carlo, an estimate of the reliability of component is obtained by:

$$\hat{R} = \frac{N_S}{N_T} \quad (3)$$

where, N_S is total number of successful trials in the simulation [23].

Monte Carlo estimates have associated error bounds. The Monte Carlo trials are discrete events and independent of each other. Consequently, their outcome follows the binomial distribution [24]; the beta inverse cumulative distribution function is used to determine the lower and upper confidence limits on the reliability predicted by the Monte Carlo simulation method at desired confidence levels. Here, the desired confidence level is considered 95%.

5. GENETIC ALGORITHM

A genetic algorithm generates the initial population of solutions. This population evolves over successive generations based on the survival of fitness. The operations such as reproduction, cross over and mutation are performed on the populations and the fitness of each individual is evaluated. Based on the new fitness of each individual, the population of next generation is produced probabilistically and the individuals with poor fitness will disappear, and the individuals with high fitness will survive. The genetic algorithm can search huge space rapidly. The GA starts with a group of chromosomes known as the population and a matrix of uniform random numbers between zero and one is generated.

In evaluation process, only the best candidate solutions are selected to continue, while the rest are deleted. These elite individuals are passed to the next population.

Mating is the creation of offspring from the parents selected in the pairing process.

Random mutations alter a certain percentage of the bits in the list of chromosomes and change the characteristics of a gene. Mutation is the second way a GA explores a cost surface. A single point mutation changes a 1 to a 0, and vice versa [25].

The number of generations that evolve depends on a set number of iterations is exceeded. The best string seen up to the last generation provides the solution to the problem. Here, after 1000 generations the algorithm is stopped.

6. COST FUNCTION

There is always a cost associated with changing a design due to change of vendors, use of higher-quality materials, retooling costs, administrative fees, etc. The cost as a function of the reliability for each component is quantified before attempting to improve the reliability. The preferred approach would be to formulate the cost function from actual cost data. In many cases however, this data is not available and is hard to obtain. For this reason, a general behavior model of the cost versus the component's reliability was developed for performing reliability optimization. The proposed cost function is [14]:

$$C = \sum_{i=1}^{n} c_i(R_i) = \sum_{i=1}^{n} \exp\left[(1-f_i)\frac{R_i - R_{i,\min}}{R_{i,\max} - R_i}\right] \quad (4)$$

where, the constraint is $R_s \geq R_G$ and each variable range is $R_{i,\min} \leq R_i \leq R_{i,\max}, i = 1,2,...,n$ and this function is for a system consisting of n components. $R_{i,\min}$ is minimum reliability of a component, $R_{i,\max}$ is maximum reliability of a component, f is the feasibility (or cost index) of improving a component's reliability relative to the other components in the system, R_G is goal reliability and R_s is system reliability.

The cost increases as the allocated reliability departs from the minimum or current value of reliability and it increases as the allocated reliability approaches the maximum achievable reliability. The cost is a function of the range of improvement, which is the difference between the component's initial reliability and the corresponding maximum achievable reliability. It is easier to increase the reliability of a component from a lower initial value.

6.1. FEASIBILITY

The feasibility parameter is a constant, which represents the difficulty in increasing component reliability relative to the rest of the components in the system. Depending on the design complexity, technological limitations, etc., certain components can be very hard to improve, relative to other components in the system [13]. Weighting factors for allocating reliability have been proposed by [26], are used to quantify feasibility. This parameter is given by

$$f_i = \frac{I_i}{\sum_{j=1}^{n} I_j} \quad (5)$$

For any component:

$$I = A(C + E + T) \quad (6)$$

Where A is state of the art index, C is complexity index, E is environment index, and T is operating time index. The state of the art index is given by:

$$A = K^{\upsilon} \; ; \; \upsilon = a^{-T_w} \quad (7)$$

T_w is number of years during which work has been done on the component, a=0.9842 for bearing and mechanical systems and K factor is defined as below:

$$K_i = \lambda_i K_{bi} \Big/ \sum_{j=1}^{n} \lambda_j K_{bj} \quad (8)$$

$$K_{bi} = 10 n_{bi} / n_{bc} \quad (9)$$

Where λ_i is failure rate for component i, n_{bi} is number of parts in component i and n_{bc} is number of parts in the most complex component.

The complexity index is given by:

$$C = 1 - e^{-K_b + 0.6 K_p} \quad (10)$$

K_b is described in the previous paragraph and K_p is defined as below:

$$K_{pi} = 10 n_{pi} / n_{pc} \quad (11)$$

where n_{pi} is number of redundant parts in component i and n_{pc} is number of redundant parts in the most complex component.

The environment index is given by:

$$E = 1 - \frac{1}{f'} \quad (12)$$

where f' is unit stress. The stress level at which complete failure is expected, a value of 100 is assigned and at which no failure is expected, a value of 0 is assigned.

The operating time index is given by:

$$T = \frac{T_m}{T_u} \quad (13)$$

where T_m is total mission time of the item and T_u is operating time of the component.

The assessment of this parameter is summarized in Tables 3 and 4 for ten years. Failure rates of the components are estimated in accordance with [20] and are in fails/ million hours.

Table 3: Data for feasibility evaluation for mechanical seal

No.	Components	n_b	K_{bi}	λ_i	f'	A	C	E	I	F
1	Disc	1	1.666	0.0567	35	2.86E-02	0.811	0.971	5.26E-02	3.63E-02
2	Flush lines	2	3.333	0.0486	30	5.38E-02	0.964	0.966	1.09E-01	7.49E-02
3	Gland plate	6	10.0	0.0567	35	2.34E-01	0.999	0.971	5.22E-01	3.60E-01
4	O-ring (secondary seal)	1	1.666	0.1135	70	6.45E-02	0.811	0.986	1.21E-01	8.34E-02
5	O-ring (static seal)	1	1.666	0.0973	60	5.38E-02	0.811	0.983	1.00E-01	6.92E-02
6	Primary & Mating ring	2	3.333	0.0973	60	1.21E-01	0.964	0.983	2.53E-01	1.74E-01
7	Retainer	1	1.666	0.0243	15	1.06E-02	0.811	0.933	1.93E-02	1.33E-02
8	Set screw	3	5.0	0.0162	10	2.39E-02	0.993	0.9	4.82E-02	3.32E-02
9	Snap ring	1	1.666	0.073	45	3.84E-02	0.811	0.978	7.10E-02	4.90E-02
10	Spring	2	3.333	0.065	40	7.54E-02	0.964	0.975	1.54E-01	1.06E-01

Table 4: Data for feasibility evaluation for bearing

No.	Components	n_b	K_{bi}	λ_i	f'	A	C	E	I	F
1	Balls	8	10	0.00375	60	0.62	0.999	0.983	1.854	0.7457
2	Rings	2	2.5	0.00375	73.3	0.12	0.864	0.986	0.349	0.1402
3	Lubricant	1	1.25	0.00375	40	0.05	0.632	0.975	0.141	0.0568
4	Cage	1	1.25	0.00375	71.6	0.05	0.632	0.986	0.142	0.0571

6.2. MAXIMUM ACHIEVABLE RELIABILITY

In reliability allocation, a limiting reliability value is defined. The cost function near this value is high and it is influenced by technological and financial constraints. The maximum achievable reliability acts as a scale parameter for the cost function. By decreasing $R_{i,max}$, the cost function is compressed between $R_{i,min}$ and $R_{i,max}$. In this paper, the maximum achievable reliability is considered 99.99% for each component of a mechanical seal and a bearing.

6.3. MINIMUM RELIABILITY OF COMPONENTS

The cost is a function of minimum reliability of each component. To estimate this parameter of cost function, Monte Carlo method is used. Here, minimum reliability of 10 main components of a centrifugal pump mechanical seal and 4 components of a bearing is predicted using this method.
In a mechanical seal, the spring load is applied on primary ring to keep the seal faces in contact and this load is distributed uniformly by a metal disc. The reliability of the disc, flush line and gland plate are predicted based on state, cooling capability and safety factor, respectively. Design parameters of dynamic O-ring seal are considered pressure, leakage, seal size, hardness, surface finish, temperature and PV coefficient. Design parameters of static O-rings are like the dynamic O-rings. In static seals, surface finish has a different value and PV coefficient is not applicable for this kind of O-rings.
The basic components of a mechanical seal are the primary and mating rings. Together they form the dynamic sealing surfaces, which are perpendicular to the shaft. The primary ring is part of the seal head assembly, while the mating ring and static seal form a second assembly, making a complete installation for a pump [16]. PV coefficient and heat transfer in a mechanical seal are considered as design factors for this assembly. A metal retainer locked to the shaft and provides a positive drive through the shaft and to the primary ring. The reliability of retainer is predicted based on safety factor of applied stress and its strength. The function of set screws is to restrict or control motion. The reliability of the set screw is predicted based on its strength. The snap ring retains the assembly on the

shaft and the reliability of the snap ring is predicted based on groove deformation. The reliability of the spring is predicted based on its strength.

In a bearing, the balls are inserted into the grooves by moving the inner ring to an eccentric position. The balls are separated after loading, and the separator is then inserted. The use of a filling notch in the inner and outer rings enables a greater number of balls to be inserted, thus increasing the load capacity. The thrust capacity is decreased, however, because of the bumping of the balls against the edge of the notch when thrust loads are present. The angular-contact bearing provides a greater thrust capacity. The minimum reliability of these components is predicted based on strength and applied stress. The calculations results are summarized in the Table 5 and 6.

Table 5: Minimum Reliability Evaluation for Mechanical Seal

Components	R_{min}	Lower limit	Upper limit	Iterations
Disc	98.25%	98.22%	98.27%	1,000,000
Flush line	92.6%	92.54%	98.64%	1,000,000
Gland plate	95.95%	95.91%	95.99%	1,000,000
Dynamic seal	90.1%	89.91%	90.28%	100,000
Static seal	91.85%	91.67%	92.18%	100,000
Retainer	97.98%	97.95%	98%	1,000,000
Setscrew	93.67%	93.63%	93.72%	1,000,000
Snap ring	96.7%	96.66%	96.73%	1,000,000
Spring	96.34%	96.3%	96.37%	1,000,000

Table 6: Minimum Reliability Evaluation for Bearing

Components	R_{min}	Lower limit	Upper limit	Iterations
Balls	95%	94.87%	95.96%	1,000,000
Rings	90%	89.93%	90.05%	1,000,000
Lubricant	97.52%	97.49%	97.55%	1,000,000
Cage	95.86%	95.83%	95.9%	1,000,000

6.4. GOAL RELIABILITY

Cost function of mechanical seal satisfies goal reliability as a constraint. In this paper, goal reliability is determined according to [22] for 10 years operation of the pump.

7. RESULTS

Reliability allocation optimization calculation of the mechanical seal and bearing of a centrifugal pump are shown in Figures 3 and 4:

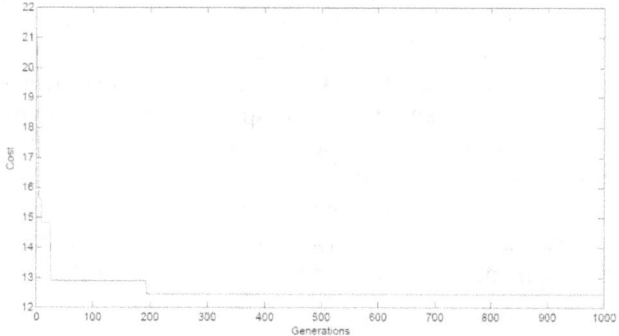

Figure 3: Convergence procedure for mechanical seal

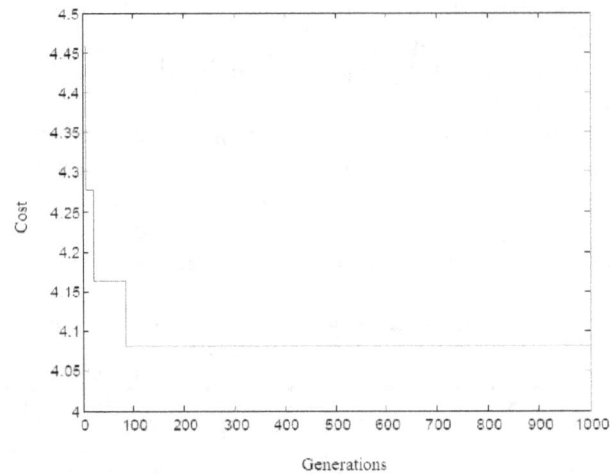

Figure 4: Convergence procedure for mechanical seal

Outputs for the reliability allocation of a mechanical seal are calculated as: $R_{disc} = 98.81\%$, $R_{flush\_lines} = 94.4\%$, $R_{gland\_plate} = 96.25\%$, $R_{dynamic\_seal} = 90.4\%$, $R_{static\_seal} = 93.13\%$, $R_{primary\_mating} = 94.95\%$, $R_{retainer} = 98.21\%$, $R_{setscrew} = 93.71\%$, $R_{snap\_ring} = 97.64\%$, $R_{spring} = 97.46\%$.

Outputs for the reliability allocation of a bearing are calculated as: $R_{balls} = 95.47\%$, $R_{rings} = 90.06\%$, $R_{separator} = 95.98\%$, $R_{lubricant} = 97.57\%$.

8. ANALYSIS OF RESULTS

In accordance with the previous results, the minimum cost of mechanical and bearing are calculated 12.45 and 4.08, respectively. The ranking of the optimized reliabilities shows that for mechanical seal, the minimum increase of reliability is related to the set screw and the maximum increase of reliability is related to the flush line. In addition, the ranking of the optimized reliabilities shows that for bearing, the minimum increase of reliability is related to the lubricant and the maximum increase of reliability is related to the balls. These values can be used at least for the initial designing of the mechanical seal and bearing components. Optimizing design with respect to the reliability is a step to design a reliable centrifugal pump.

9. CONCLUDING REMARKS

In this article the mechanical seal of and bearing of a centrifugal pump were evaluated for reliability optimization problem through reliability allocation at the component level. A general cost function with estimated parameters is used as an objective function for the optimization with GA method. These parameters can be altered and different allocation scenarios are investigated. GA is utilized to solve constrained optimization problems effectively. The analyzed results show that the genetic algorithm can be used as a useful decision-supporting tool to optimize the design of a pump. Fundamental techniques for performing a Monte Carlo simulation have been explained. This tool can be applied to any system that compromises smaller components with a known or at least determinable failure distribution. Monte Carlo method is applied for estimation of minimum reliability of mechanical seal and bearing components. Further research can be concentrated in obtaining such functions based on actual cost data and this procedure is applicable for the other components of the pump. The results of the analysis can then be used to provide economic justification for reliability improvements to existing equipment or to purchase new equipment for the system.

References

[1] Val S. Lobanoff, "*Centrifugal pumps: design & application*", Gulf Publishing Company, Houston, TX, (1992).
[2] J.E. Shigley, C.R. Mischke, "*Mechanical Engineering Design*", McGraw-Hill Book Company, NY, (1989).
[3] S. Shiels, "*Failure of mechanical seals in centrifugal pumps: Part two*", Stan Shiels on Centrifugal Pumps, pp. 257-263, (2004).
[4] S. Shiels, "*Failure of mechanical seals in centrifugal pumps*", World Pumps, Volume 2002, Issue 429, pp. 20-22, (2002).
[5] J. Singh, S. Angra and V.K. Mittal, "*Failure mode analysis of mechanical seals*", Journal of Engineering and Technology, vol. 2, issue 2, (2012).
[6] K.-D. Meck, G. Zhu, "*Improving mechanical seal reliability with advanced computational engineering tools, part 1: FEA*", Sealing Technology, Volume 2008, Issue 1, pp. 8-11, (2008).
[7] K.-D. Meck, G. Zhu, "*Improving mechanical seal reliability with advanced computational engineering tools, part 2: CFD and application examples*", Sealing Technology, Volume 2008, Issue 2, pp. 7-10, (2008).
[8] C. Watkinson, "*Improving the reliability of mechanical seals in ethylene oxide applications*", Sealing Technology, Volume 2007, Issue 12, pp. 8-12, (2007).
[9] R. Gabriel, "*API 610 and API 682: A powerful combination for maximum pump/mechanical seal reliability*", World Pumps, Volume 1996, Issue 360, pp. 56-60, (1996).
[10] S. Shiels, "*Troubleshooting centrifugal pumps: Rolling element bearing failures*", Stan Shiels on Centrifugal Pumps, pp. 241-247, (2004).
[11] T. Sahoo, "*Making centrifugal pumps more reliable*", World Pumps, Volume 2009, Issue 513, pp. 32-36, (2009).
[12] R. Sehgal, O.P. Gandhib, S. Angra, "*Reliability evaluation and selection of rolling element bearings*", Reliability Engineering and System Safety, 68, pp. 39–52, (2000).
[13] H.P. Bloch, F.K. Geitner, "*An introduction to machinery reliability assessment*", New York: Van Nostrand Reinhold, 1990.
[14] A. Mettas, "*Reliability allocation and optimization for complex systems*", Proceedings Annual Reliability and Maintainability Symposium, Los Angeles, CA, 216-221, (2000).
[15] W. Kuo, VR. Prasad, "*An Annotated Overview of System-Reliability Optimization*", IEEE Transactions on Reliability, 49, 176-187, (2000).
[16] J. Karassik, P. Messina, P. Cooper, C. Heald, "*Pump Handbook*", McGraw-Hill, NY, (2001).
[17] J. Sun, X. Hea, L. Weib, X. Feng, "*Failure Analysis and Seal Life Prediction for Contacting Mechanical Seals*", International Conference on Experimental Mechanics (ICEM), (2008).
[18] J. Singh, S. Angra, V. Mittal, "*Failure Mode Analysis of Mechanical Seals*", Journal of Engineering and Technology, Vol. 2, Issue 2, (2012).
[19] P. Girdhar, O. Moniz, "*Practical Centrifugal Pumps Design, Operation and Maintenance*", IDC Technologies, (2005).
[20] OREDA Participants, 4th ed.,"*OREDA handbook*", Trondhim: OREDA Participants, (2002).
[21] L. Bachus, A. Custodio, "*Know and understand centrifugal pumps*", Elsevier Ltd., UK, (2003).
[22] P. Makarachi, P., M. Pourgolmohammad, "*Optimization of Failure Rate of Centrifugal Pumps Using Genetic Algorithm*", ASME2012 International Mechanical Engineering Congress & Exposition, (2012).
[23] T.A. Cruse, "*Monte Carlo Simulation: Reliability-based Mechanical Design*", pp. 123–46, Marcel Dekker, NY, (1997).
[24] D. Kececioglu, "*Robust Engineering Design-by-Reliability with Emphasis on Mechanical Components & Structural Reliability*", Vol. 1, DEStech, (2003).
[25] R. L. Haupt, S. E. Haupt, "*Practical genetic algorithms*", 2nd ed., John Wiley & Sons, Inc., Hoboken, New Jersey, (1998).
[26] W. Adams, L. Waling, R. Dingman, J. Parker, "*The Role of Off-Design Pump Operation on Mechanical Seal Performance*", Proceedings of the 11th International Pump User Symposium, (1994).

A Science-Based Theory of Reliability Founded on Thermodynamic Entropy

Anahita Imanian, Mohammad Modarres
Center for Risk and Reliability
University of Maryland, College park, USA

Abstract: Failure data-driven stochastic and probabilistic techniques that underlie reliability analysis of components and structures remain unchanged for decades. The present study relies on a science-based explanation of damage as the source of material failure, and develops an alternative approach to reliability assessment based on the second law of thermodynamics. The common definition of damage, which is widely used to measure the reliability over time, is somewhat abstract, and varies at different geometric scales and when the observable field variables describing the damage change. For example, fatigue damage in metals has been described in several ways including reduction of elasticity modules, variation of hardness, cumulative number of cycle ratio, reduction of load carrying capacity, crack length and energy dissipation. These descriptions are typically based on observable changes in the physical or spatial properties, and exclude unobservable and highly localized damages. Therefore, the definition and measurement of damage is subjective and dependent on the choice of observable variables. However, all damage mechanisms share a common feature at a far deeper level, namely *energy dissipation*. Dissipation is a fundamental measure for irreversibility that, in a thermodynamic treatment of non-equilibrium processes, is quantified by *entropy generation*. Using a theorem relating entropy generation to energy dissipation via generalized thermodynamic forces and thermodynamic fluxes, this paper presents a model that formally describes the resulting damage. This model also contains cases where there is a synergy between different irreversible fluxes, such as in corrosion-fatigue damage where the mechanical deformation rate leading to fatigue is coupled with the electrochemical reaction rate leading to corrosion. Employing thermodynamic forces and fluxes to model the damage process, not only enables us to express the entropy generation in terms of physically measurable quantities including stress diffusion and electrochemical affinities, but also provides a powerful technique for studying the complex synergic effect of multiple irreversible processes. Having developed the proposed damage model over time, one could determine the time that damage accumulates to a level where the component or structure can no longer endure and fails. Existence of any uncertainties about the parameters and independent variables in this thermodynamic-based damage model leads to a time-to-failure distribution. Accordingly, such a distribution can be derived from the thermodynamic laws rather than estimated from the observed failure histories.

1 Introduction

The definition of damage due to the physical mechanisms varies at different geometric and scales. For example, the definition of fatigue damage can vary from nano-scale through the macro-scale. At the atomic level the grain boundary is a likely location where atoms are more loosely packed. At the micro-scale damage is the accumulation of micro-stresses in the neighborhood of cracks. At the meso-scale level, damage might be defined as growth and coalescence of micro-cracks to meso-cracks. However, measuring damage is subject to the physically measurable variables (i.e., observable marker) when dealing with specific failure mechanisms. That means damage in this context is a characterization of the observable symptoms in the form of measurable field variables such as crack length, amount of wear, and degree of deformation (Singpurwalla, 2010). In fact, Arson (2012) states that damage prediction relies on the field variables chosen to describe the anticipated degradation or aging. Singpurwalla (2010) refers to these subjective choices of observable field variables as "observable markers." For example, in the corrosion-fatigue mechanism material weight loss, change of impedance, density of pits, accumulated number of cycles-to-failure, and crack length may be used as "observable markers" that measure the damage. Therefore, defining a consistent and broad definition of damage is necessary and plausible. To reach this goal, we elaborate on the concept of material damage within the thermodynamic framework.

Thermodynamically, all forms of damage share a common characteristic, which is the dissipation of energy. In thermodynamics, dissipation of energy is the basic measure of irreversibility, which is the main feature of the degradation processes in materials [1]. Chemical reactions, release of heat, diffusion of materials, plastic deformation, and other means of energy production involve dissipative processes. In turn, dissipation of energy can be quantified by the *entropy generation* within the context of irreversible thermodynamics. Therefore, dissipation (or equivalently entropy generation) can be considered as a substitute for characterization of damage.

The common practice in damage analysis and prediction of structural life and integrity is through the application of the traditional reliability and Physics-of-Failure (PoF) methods. The traditional generic handbook-based reliability prediction methods such as those advocated in MIL-HDBK-217F [2], Telcordia SR-332 [3], and FIDES [4] rely on analysis of the field data (with incoherent operating and environmental conditions), with the assumption of the constant failure rates. Numerous studies have shown that misleading and inaccurate results from applications of these handbooks can lead to poor designs, incorrect reliability prediction and operating decisions [5, 6, 19, 20]. The PoF models such as the Coffin-Manson model [5], Norris-Landzberg model [6] and Bayerer's model [7] offer more rigorous and improved reliability estimation approach. However, these empirically-based methods are limited to simple failure mechanisms and are hard to model multiple competing and common cause failure mechanisms.

In contrast, with the empirically-based PoF approach to reliability prediction, which considers only the most predominant failure mechanisms ([8, 9]), the definition of damage in the context of the thermodynamic entropy allows for the incorporation of all underlying dissipative processes. For example in the case of corrosion-fatigue, consider the physically measurable quantities such as stress and electrochemical affinity of the oxidation-reduction electrode reaction (Me$\Leftrightarrow$Me^{z+}+ze) of a metal. The entropy as a state function is independent of the path of the failure (which commonly depends on factors such as geometry, load, frequency of load, etc.) from the initial state to the final failed state of the material, considering a known failure threshold (endurance limit) [10]. Entropy provides a science-based approach to model a wide range of damage processes with favorable results in fracture mechanics, fatigue damage analysis [10, 11] and tribological processes such as friction and wear [12, 13]. Additionally, it provides a powerful technique for studying the synergistic effects arising from interaction of multiple processes [18].

However, lack of detailed knowledge about the independent variables that generate entropy such as the exact loading conditions applied to a structure, materials-to-materials variability, environmental and seasonal factors means there would be uncertainties about the entropic-based trajectory of cumulative damage. While in the absence of such uncertainties an exact time of failure can be calculated, existence of such uncertainties will lead to estimation of a time-to-failure distribution.

The remainder of this paper is organized as follows. Section 2 presents the features of material damage. Section 3 describes our construction of the entropy model. Section 4 describes the relationship between the damage and entropy generation. Section 5 links the entropy, as an index of damage, to the reliability assessment and section 6 offers concluding remarks.

2 A Characterization of Material Damage
Damage can be viewed as surface or volumetric deterioration of materials. For example, fatigue and creep are processes that cause volumetric damage in structures, while corrosion and wear cause surface damage. At the macroscopic level, realization of material damage becomes difficult and significantly dependent on

the type of volumetric or surface damage. Lemaitre and Chaboche [14] state that unless any macroscopic discontinuity or permanent distortion can be observed, it would be very difficult to assess integrity and health of structural materials. The common practice of damage evaluation is through the quantification of the symptoms or observable markers of damage (e.g., crack size, density of crack, depth of the pit, weight loss) and other mechanical markers (e.g., reduction of the elastic modulus, accumulation of plastic strain, or change in viscoplastic properties). Difficulties to develop a consistent definition of damage from physical and mechanical points of view have compelled researchers to look for a microscopically consistent definition in the context of the continuum damage mechanics [14]. For example, in the continuum damage mechanics the damage, D, as an internal variable is defined as the effective surface density of microdefects:

$$D = \frac{S_D}{S} \tag{1}$$

where S_D is the damage surface area and S is the initial cross section area. Due to the difficulty in direct measurement of the density of defects on the surface or volume of materials, Lemaitre [15] used the strain equivalent principle to correlate between other measurable properties of material (e.g. variation of elastic strain, module of elasticity, micro-hardness, density, and plastic strain) and damage. However, these different damage indexes do not provide a consistent measure of damage, including all the observable and unobservable damages.

At the micro-level, however, material damage can be defined in a more coherent way. In fact, at the nano-scale, damage may refer to breaking and reestablishment of the interatomic bonds in crystalline metals and polymers ([16]). Based on this fundamental definition of damage, meso-scale characteristics of damage such as dislocations, slips, micro-cavities, and micro-cracks can be quantified.

As discussed above, several definitions and measures of damage exist, however, the concept of damage is somewhat abstract, and definitions are relative. All damage mechanisms share a common characteristic at a much deeper level, i.e., the "dissipation" of energy. Dissipation can be described well within the context of non-equilibrium thermodynamics using the second law of thermodynamics. In a thermodynamics treatment of non-equilibrium irreversible processes, dissipation is quantified by the "entropy generation". We consider this characterization of damage highly general, consistent and scalable. In the following section the focus is on the formulation of the entropy generation caused by dissipative mechanisms, using the corrosion-fatigue as a demonstration example.

3 Total Entropy Produced in a System

Consistent with the second law of thermodynamics, entropy does not obey a conservation law. Therefore, it is essential to relate the entropy not only to the entropy crossing the boundary between the system and its surroundings, but also to the entropy produced by the processes taking place inside the system. Processes occurring inside the system may be reversible or irreversible. Reversible processes inside a system may lead to the transfer of the entropy from one part of the system to other part of the interior, but do not generate entropy. Irreversible processes inside a system, however, result in generation of the entropy, and hence in computing the entropy they must be taken into account.

Using the second law of thermodynamic, it is possible to express the variation of total entropy flow per unit volume, dS, in the form of

$$dS = d^r S + d^d S \tag{2}$$

where $d^r S$ is the entropy supplied to the system by its surroundings through transfer of matters and heat (e.g., in an open system where wear and corrosion mechanisms occur). The rate of exchanged entropy is obtained as

$$\frac{d^r S}{dt} = -\int^\Omega J_s \cdot n_s dA \tag{3}$$

where J_s is a vector of the total entropy flow per unit area, crossing the boundary between the system and its surroundings, and n_s is a normal vector. Similarly, $d^d S$ is the entropy produced inside of the system, which can be obtained from Eq. 4,

$$\frac{d^d S}{dt} = \int^V \sigma dV \tag{4}$$

where, σ is the entropy generation per unit volume per unit time. The second law of thermodynamics states that $d^d S$ must be zero for reversible transformations and positive ($d^d S > 0$) for irreversible transformations of the system.

The balance equation for entropy shown in Eq. 5 can be derived using the conservation of energy and balance equation for the mass.

$$\frac{dS}{dt} + \nabla J_s = \sigma \tag{5}$$

This gives us an explicit expression for total entropy in terms of reversible and irreversible processes [14, 17, 18]

$$\frac{dS}{dt} = -\nabla \cdot \left(\frac{J_q - \Sigma \mu_k J_k}{T}\right) - \frac{1}{T^2} J_q \cdot \nabla T - \Sigma_{k=1}^n J_k \left(\nabla \frac{\mu_k}{T}\right) - \frac{1}{T}\Sigma_{f=1}^n F_k J_f - \frac{1}{T}\Pi : \nabla V - \frac{1}{T}\Sigma_{j=1}^r v_j A_j \tag{6}$$

where, T is the temperature, μ_k the chemical potential, J_q the heat flux, J_k the diffusion flow, v_i the chemical reaction rate, Π the stress tensor, ∇V the velocity gradient (equal to strain rate $\dot{\epsilon}$), $A_j = -\Sigma_k \mu_k v_{jk}$ the chemical affinity or chemical reaction potential difference, F_k the force due to external field, and J_f the corresponding flux. Each term in Eq. 6 is derived from various mechanisms involved, which define the macroscopic state of the complete system. External forces may be resulted from different factors including electrical field, magnetic field, gravity field, etc., where the corresponding fluxes are electrical current, magnetic current and velocity.

By comparing Eq. 6 with Eq. 5 we can make the identifications

$$J_s = \frac{J_q - \Sigma \mu_k J_k}{T} \tag{7}$$

$$\sigma = -\frac{1}{T^2} J_q \cdot \nabla T - \Sigma_{k=1}^n J_k \left(\nabla \frac{\mu_k}{T}\right) - \frac{1}{T}\Sigma_{f=1}^n F_k J_f - \frac{1}{T}\Pi : \nabla V - \frac{1}{T}\Sigma_{j=1}^r v_j A_j \tag{8}$$

where, Eq. 7 shows the entropy flux resulted from heat and material exchange. Equation 8 represents the total energy dissipation terms from the system that from left to the right include heat conduction energy, diffusion energy, external force energy, mechanical energy, and chemical energy. Equation 8 is fundamental to non-equilibrium thermodynamics, and represents the entropy generation σ as the bilinear form of forces and fluxes as

$$\sigma = \Sigma_{i=1}^m X_i J_i \tag{9}$$

It is through this form that the contribution from the applicable thermodynamic forces and fluxes are expressed. For example, in the case of a chemically reactive system, the chemical affinity A_j drive the chemical reaction with velocities v_j, the mechanical stress cause the deformation rate, the concentration gradient cause the diffusion rate, and the temperature gradient generate the heat flow.

In the linear non-equilibrium thermodynamics (LNT) theory whereby the variation of thermodynamic forces and fluxes are small, the components of the thermodynamic fluxes, J_i, are assumed to be a linear combination of the components of the thermodynamic forces, X_i, so that [19]

$$J_i = \sum_{k=1}^{n} L_{ik} X_k \tag{10}$$

where, L_{ik}'s are the phenomenological coefficients. According to the Onsager reciprocity theorem [19], the equation of motion for each individual particle is time reversible as in classical dynamics or quantum mechanics. The macroscopic result obtained from this assumption is that $L_{ik} = L_{ki}$. It is worth to note that, the LNT application is limited to linear systems with no general means of minimizing entropy generation, which is implied by Prigogine's theorem[1] as a special case of Zeigler principle [20]. Further, the phenomenological coefficients of LNT must be determined experimentally. An advantage is that it can relate the fluxes and their coupling in a very fast process by using a linear relationship [21, 22]. The Seebeck effect, Furiers's law of heat conduction, Fick's law of diffusion, Ohm's law of electrical condition, and the Navier-Stokes equations for fluids may be formally derived from the LNT [17].

For example, damage in the corrosion-fatigue damage mechanism is produced by the synergy between two irreversible fluxes of anodic dissolution current density and the plastic deformation [23]. Gutman [23] first identified the influence of plastic deformation on anodic dissolution rate and vice versa (i.e., the mechanochemical effect), where the entropy generation is the summation of entropy generation due to electrochemical reaction and plastic deformation. Summing the contributions of the mechanical and electrochemical processes, we can write the total entropy generation for combined effect of plastic deformation and anodic and catholic dissolution as:

$$T\sigma = \Pi : \dot{\varepsilon}_P + \tilde{A} i_{corr} \tag{11}$$

where $\tilde{A}$ is the electrochemical and mechanochemical potential losses (over-potential). Employing LNT theory, we can write the following system of phenomenological equations to take the cross effects of two processes into consideration

$$\dot{\varepsilon}_P = L_{11}\Pi + L_{12}\tilde{A} \tag{12}$$

$$i_{corr} = L_{21}\Pi + L_{21}\tilde{A} \tag{13}$$

4   Physical Damage versus Total Entropy Generation

To validate the entropy as an appropriate parameter representing the physical or mechanical damages (e.g. weight loss, crack size, young modulus), Bryant et al. link quantitative and observable markers of degradation processes leading to damage (e.g. wear volume, crack size and corrosion mass loss) to the associated entropy generation [24]. They derived the damage rate $\dot{w} = \Sigma_i \dot{w}_i = \Sigma_i \Sigma_j Y_i^j J_i^j$ as a linear combination of the components of entropy generation σ_i produced by the dissipative processes.

$$\dot{w} = \Sigma_i B_i \sigma_i \tag{14}$$

where, B_i represents the degradation coefficient that relates the generalized degradation force, Y_i, to generalized thermodynamic force, X_i, so that $Y_i = B_i X_i$. In an application of this approach, they also show

---

[1] According to this theory the entropy production given in Eq. 9 takes a minimum on stationary states.

that in wear, the rate of weight loss derived from the concept of entropy generation is in agreement with Archard's sliding wear model [25], and Doelling et al. experimental results [26]. Amiri et al. showed a linear relationship between the dissipative energy and wear volume [12]. Ontiveros et al., denoted a linear correlation between the cumulative plastic strain energy and cycles-to-crack initiation in high-cycle fatigue of Aluminum alloys [11].

5 Reliability Assessment Using Entropy as an Index of Damage

It was stated earlier that damage caused through a degradation process could be viewed as the consequence of dissipation of energy that can be measured and expressed by entropy such that:

$$\text{Damage} \equiv \text{Entropy}$$

In the earlier discussion in this paper it was shown (through Eq. 5) that one could express the total entropy per unit time per unit volume for the individual dissipation processes resulting from the corresponding failure mechanisms. Therefore, the evolution trend of the damage, D, is obtained from

$$D|t \sim \int_0^t [\dot{S}|X_i(u), J_i(u)] du \qquad (15)$$

where $D|t$ is the monotonically increasing cumulative damage starting at time t from a theoretically zero value or practically some initial damage value. When D reaches a predefined (often subjective) level of endurance it may be assumed beyond that point the component or structure fails. It is worth to note that failure in this context is the point when an item becomes effectively nonfunctional (but possibly still operational). That is failure is considered as the point where the item is no longer meeting a functionality requirement (e.g., an acceptable performance level or an endurance limit such as a given level of thermodynamic efficiency).

Because entropy as a parameter of degradation includes all observable damage markers (cracks, wear debris and pit densities) and unobservable damages such as subsurface dislocations, slip and micro-cavities, definition of a single failure threshold might not be possible due to long stretch of damage measurement from nono-scale to macroscopic scale. In this case, the cumulative damage and alternatively entropy endurance level can be estimated through the measurement of certain observable damage markers. The correlation between the observable damage markers and entropy, justified by several studies [10, 11, 12], enables the definition of failure threshold on the basis of observable markers. In the other word, the damages grow, coalesce and eventually the weakest link among all coalesces damages manifests itself as an observable damage which causes failure.

Materials, environmental, operational and other variability in degradation forces impose uncertainties on the cumulative damage, D. Existence of such uncertainties leads to confidence intervals around the mean value and about the time that a failure occurs (as depicted in Figure 1). This figure shows how the interfaces of the accumulated damage and endurance level result in the probability density function (PDF) of the time-to-failures, which is the variable of interest in most reliability analysis. It is self-evident that the probability that the random variable, $D|t$, (i.e., the cumulative damage at time, t) exceeds the constant endurance level, D_f (thus causing the failure), must be equal to the probability that the random variable time-to-failure is less than t. Accordingly, the magnified section in Figure 1 depicts the cumulative probability of damage as the diagonal shaded area, and the cumulative time to failure probability as the solid shaded area.

Accordingly, assuming the constant endurance limit, D_f, one can derive the time-to-failure distribution, $g(t)$, from the thermodynamically-based damage relation expressed by Eq. 15,

$$\int_0^t g(x)dx = \int_{D_f}^{\infty} f(D|t)dD \qquad (16)$$

where $f(D|t)$ is the PDF of the damage at time t. The corresponding time to failure PDF, $g(t)$, would be

$$g(t) = \frac{d}{dt}\int_{D_f}^{\infty} f(D|t)dD \qquad (17)$$

Obviously, the reliability function can be expressed as

$$R(t) = \int_t^{\infty} g(t)dt = 1 - \int_{D_f}^{\infty} f(D|t)dD \qquad (18)$$

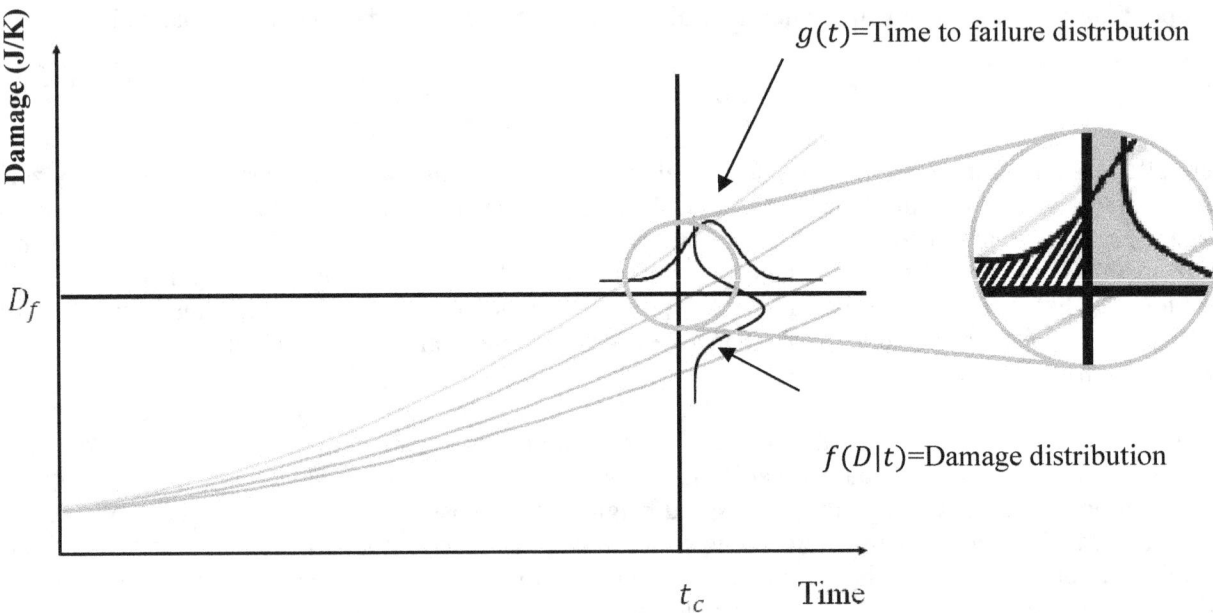

Figure 1: Damage- endurance modeling.

Similar to the uncertainties about the amount of damage, the endurance limit, D_f, may also be uncertain because composition of materials and thus their strengths varies from sample to sample. In this case, a failure occurs when the damage in a component exceeds its endurance level. The probability that no failure occurs is equal to the probability that the random variable, D, is less than the random variable, D_f, describing the component endurance level.

$$R(t) = \Pr(D|t < D_f|t) \qquad (19)$$

Where $R(t)$ is the relaibilty of component at time t. Knowing the PDFs of the random variables D and D_f expressed by $f(D)$ and $h(D_f|t)$, the PDF of the time-to-failure distribution can be obtained by

$$R(t) = 1 - \int_0^{\infty} h(D_f|t)dD_f \int_{D_f}^{\infty} f(D|t)dD, \qquad (20)$$

and

$$g(t) = -\frac{d}{dt}R(t) \tag{21}$$

It is possible that the PDF of the endurance limit is independent of time (i.e., $D_f|t \approx D_f$). If multiple dissipative forces are at work then using the weakest link principle

$$R(t) = \prod_{i=0}^{n}\left\{1 - \int_0^\infty h(D_{f_i}|t)dD_{f_i}\int_{D_{f_i}}^\infty f(D_i|t)dD_i\right\} \tag{22}$$

Again Eq. (20) serves as the basis to derive the PDF of the time-to-failure.

6 Conclusions

This paper presents a thermodynamic framework for the degradation level assessment using entropy generation as a measure of damage. It suggests that a unified measure of damage can be defined based on the entropy generation concept. Applying the entropic method can improve our understanding of the degradation mechanism and the quantification of damage. Entropy, as a state function, is independent of the failure path, and provides a formal means to analyze the synergy between different degradation mechanisms and forces. The general entropy generation function is derived in terms of energy losses due to heat conduction, diffusion losses, mechanical dissipations, chemical losses and external force field effects (e.g. magnetic, electrical and gravity fields). It is shown that entropy generation and damage are related. However, the entropy generation function is subject to various stochastic variations of forces that cause damage. The reliability model is built off of the relationship between damage PDF, endurance limit distribution and time-to-failure PDF.

This paper discusses a fundamental foundation for a science-based explanation of damage as a source of material failure and thus materials reliability. As such we offer an alternative approach for reliability assessment based on the second law of thermodynamics. As the next step, which is validating the proposed framework, we are now studying the entropy growth rate as a degradation parameter for the corrosion-fatigue mechanism in materials. This approach could open the window for further exploration of the applications of thermodynamics for reliability assessment and analysis of materials and prognosis and health management of critical components and structures.

Acknowledgement

This work is part of an ongoing research through grant number N000141410005 from the Office of Naval Research (ONR).

Reference

[1] H. Tang and C. Basaran, "*A Damage Mechanics-Based Fatigue Life Prediction Model for Solder Joints*", Transaction of ASME, Journal of Electronic Package, vol. 125, pp. 120–125, 2003.

[2] Military Handbook for Reliability Prediction of Electronic Equipment, Version A, U.S. Department of Defense, 1965.

[3] Reliability Prediction Procedure for Electronic Equipment, Telcordia Technologies. Special Report SR-332, Telcordia Customer Service, Piscataway, NJ, 2001.

[4] Reliability Methodology for Electronic Systems, FIDES Guide Issue, FIDES Group, 2004.

[5] S. S. Manson, "*Thermal Stress and Low Cycle Fatigue,*" New York: McGraw-Hill, 1966.

[6] K. C. Norris and A. H. Landzberg, "*Reliability of Controlled Collapse Interconnections*", IBM Journal of Research and Development, vol. 13, pp. 266–271, 1969.

[7] R. Bayerer, T. Hermann, T. Licht, J. Lutz, and M. Feller, "*Model for Power Cycling Lifetime of IGBT Modules – Various Factors Influencing Lifetime*", Integrated Power Electronics Systems (CIPS) Conference, Nuremberg, Germany, pp. 11-13, 2008.

[8] M. Held, P. Jacob, G. Nicoletti, P. Scacco, and M. H. Poech, "*Fast Power Cycling Test of IGBT Modules in Traction Application*", International Conference on Power Electronics and Drive Systems, vol. 1, pp. 425-430, 1997.

[9] M. Ciappa, "*Selected Failure Mechanisms of Modern Power Modules*", Microelectronic Reliability, vol. 42, pp. 653-667, 2002.

[10] M. Naderi, M. Amiri, and M.M. Khonsari, "*On the Thermodynamic Entropy of Fatigue Fracture*", Proceedings of the Royal Society A – Mathematical Physical and Engineering Sciences, vol. 466, pp. 423 - 438, 2010.

[11] V. Ontiveros, M. Amiri, and M. Modarres, "*Fatigue Crack Initiation Assessment based on Thermodynamic Entropy Generation*", Journal of Risk and Reliability, (Submitted, 2013).

[12] M. Amiri and M. M. Khonsari, "*On the Thermodynamics of Friction and Wear—A Review*", Entropy, vol. 12, pp. 1021-1049, 2010.

[13] M. Nosonovsky and B. Bhushan, "*Thermodynamics of Surface Degradation, Self-organization, and Self-healing for Biomimetic Surfaces*", Transaction of Royal Society, vol. 367, pp. 1607–1627, 2009.

[14] J. Lemaitre and J. L. Chaboche, "*Mechanics of Solid Materials*", 3rd edition; Cambridge University Press: Cambridge, UK, 2000.

[15] J. Lemaitre, "*A Course on Damage Mechanics*", Springer, France, 1996.

[16] C. W. Woo and D. L. Li, "*A Universal Physically Consistent Definition of Material Damage*", International Journal of Solids Structure, vol. 30, pp. 2097-2108, 1993.

[17] S. R. de Groot and P. Mazur, "*Non-Equilibrium Thermodynamics*", Wiley, New York, 1962.

[18] D. Kondepudi and I. Prigogine, "*Modern Thermodynamics: From Heat Engines to Dissipative Structures*", Wiley, England, 1998.

[19] L. Onsager, "*Reciprocal Relations in Irreversible Processes*", Journal of Physics, Revision 37, vol. 405, 1931.

[20] H. Ziegler and C. Wehrli, "*On a Principle of Maximum Rate of Entropy Production*", Journal of Non-equilibrium Thermodynamics vol. 12, pp. 229–243, 1997.

[21] H. W. Haslach, "*Maximum Dissipation Non-Equilibrium Thermodynamics and its Geometric Structure*", New York, Springer, 2011.

[22] Y. Demirel and S. I. Sandler, "*Non-equilibrium Thermodynamics in Engineering and Science*", Journal of Physical Chemistry, vol. 108, pp. 31–43, 2004.

[23] E. M. Gutman, "*Mechanochemistry of Materials*", Cambridge International Science Publishing, Cambridge, UK, 1998.

[24] M.D. Bryant, M. M. Khonsari, and F. F. Ling, "*On the thermodynamics of degradation*", Proceeding of the Journal of Royal Society, vol. 464, pp. 2001-2014, 2008.

[25] J. F. Archard, "*Wear Theory and Mechanisms*", in Wear Control Handbook ASME, pp. 35-80. New York, 1980.

[26] K. L. Doelling, F. F. Ling, M. D. Bryant, and B. P. Heilman," *An Experimental Study of the Correlation between Wear and Entropy Flow in Machinery Components*", Journal of Applied Physics vol. 88, pp. 2999-3003, 2000.

Quick Quantitative Calculation of DFT for NPP's Repairable Systems Based on Minimal Cut Sequence Set

Daochuan Ge[a,b,*], Qiang Chou[b], Ruoxing Zhang[b], Yanhua Yang[a]

a. School of Nuclear Science and Engineering, Shanghai Jiao Tong University, Shanghai, China;
b. Software Development Center, State Nuclear Power Technology Corporation, Beijing, China

Abstract: The quantitative calculations of Nuclear Power Plant (NPP)'s repairable system are mainly based on Markov model. However, with the increase of the system's size, the system's state space increases exponentially, which makes the problem hard or even not to be solved. This paper proposes a method about quick calculation of Dynamic Fault Tree (DFT) for NPP's repairable system based on Minimal Cut Sequence Set (MCSS), which divides a complex DFT into individual failure chain defined by MCSS. For each failure chain, the Markov model is applied. Then the unavailability of system is obtained synthesizing the result of each failure chain. This approach decreases the system's size increasing from exponentially to linearly and reduces the computation complexity. As to the NPP's dynamic systems with low failure rate and high repair rate, this approach can give a solution with a high-precision and conservative result and has practical value.

Keywords: Failure Chain, Quick Quantitative Calculation, Unavailability, Repairable Systems, NPP.

1. INTRODUCTION

To ensure the operating safety and design balance of NPP, it is essential to make reliability assessment of critical safety-critical systems. The real-life safety- critical systems of NPP often exhibit dynamic failure mechanisms, i.e., sequence- and functional- dependent failure behaviours, which make it hard to model and analyze the systems' reliability. For the powerful modelling ability and intuitiveness of DFT, NPPs often adopt DFT to describe the failure behaviours of the systems. The commonly-used methods for quantifying a DFT are mainly Markov-based [1,2], multi-integration-based [3,4,5] and Monte Carlo simulation-based methods [6,7]. However, each of these approaches has its own shortcomings: For Markov-based method, it is only applicable for exponential components time-to-failure distribution systems. Moreover, this method often confronts the problem of "state space explosion"; as to the multi-integration-based method, this method is only applied to non-repairable systems; as to the Monte Carlo simulation-based method, it may be very time-consuming, especially when the solutions with high degree of accuracy are desired. In addition, a new simulation procedure must be implemented whenever a component's failure parameters value changes. Considering some components existing in NPP are repairable, it is necessary to develop a practical approach for evaluating the reliability of repairable systems of NPP which should be easily to be implemented and computed. In this paper, an approach used to evaluate the reliability of repairable system of NPP based on MCSS is proposed.

The reminder of this paper is organized as follows: Section 2 gives some related models and concept, including the MCSS model, the proposed model, and the uniqueness property of DFT's MCSS. Section 3 implements numerical experiments of the proposed model. Section 4 presents two cases study. Section 5 gives the final conclusions.

2. RELATED MODELS AND CONCEPTS

2.1. The MCSS Model

It is well known the occurrences of a DFT's top event not only depend on the combination of basic events but also depend on their failure orders. To characterize this failure behaviour, the researchers [8] propose a new concept of Minimal Cut Sequence (MCS) which is used to express what minimal basic events combination and in what failure orders that can lead to the occurrence of a DFT's top event. A MCS comprises several capital letters characterizing the failure behaviour of basic events and some temporal connecting symbols ($\prec$) expressing specific failure sequence. For an illustrative purpose, an example is shown in Fig.1.

Fig.1: An Illustrative Example

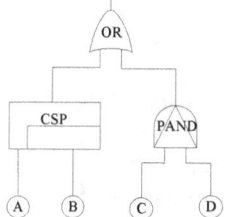

The OR gate, i.e., top event, fires if any input event occurs. As to the CSP gate, it fires only if all input events fail. Considering the cold standby component B never fails before primary B fails, the MCS of the CSP gate is written as: $A \prec B$. As to the PAND gate, it fires only if all input events fail in a left to right failure order, thus the MCS of the PAND gate is expressed as: $C \prec D$. Since the failure behaviour of the example is expressed by the logic OR of the two dynamic gates, the Minimal Cut Sequence aggregate of the system is $\{A \prec B, C \prec D\}$. How to obtain the complete MCS of a general DFT is beyond the scope of this paper, the interested readers can refer to Refs. [9,10,11] for more information.

A DFT generally have more than one MCS and all these MCS compose a set, i.e., Minimal Cut Sequence Set (MCSS). So the MCSS captures the complete failure information of a DFT. Suppose a DFT has n MCSs, and then the MCSS of the DFT can be written as:

$$MCSS = MCS_1 \cup MCS_2 \cup \cdots \cup MCS_n \qquad (1)$$

The occurrence probability of the top event can be expressed by

$$P_r(system\ failure) = P_r(MCSS)$$
$$= P_r(MCS_1 \cup MCS_2 \cup \cdots \cup MCS_n) \qquad (2)$$

To solve the Eq. (2), an Inclusion-Exclusion Principle [12] is applied as follows:

$$P_r(system\ failure) = P_r(MCS_1 \cup MCS_2 \cup \cdots \cup MCS_n)$$
$$= \sum_{i=1}^{n} P_r(MCS_i) - \sum_{1 \leq i < j \leq n} P_r(MCS_i \cap MCS_j) + \cdots$$
$$+ (-1)^{n-1} P_r(MCS_1 \cap MCS_2 \cap \cdots \cap MCS_n) \qquad (3)$$

Apparently, the MCSS model is an algebraic approach. it avoids the notorious problem of "state space explosion". Yet this approach is becoming unavailable when the system is repairable.

2.2. The Uniqueness of A DFT's MCSS

The uniqueness of a DFT's MCSS means that the MCSS is unique once the system's DFT is modelled determinately. That to say the MCSS is independent on whether the system is repairable or not. It is well know the occurrences of a DFT's top event are determined jointly by the combinatorial and sequential constraints. For a DFT, the combinatorial and sequential restrictions are uniquely decided. As a result, whether a system is repairable or not, the MCSS of the system's DFT is unique. Suppose a system has n components and its corresponding DFT has m MCSs, then the Eq. (4) must hold.

$$\begin{cases} MCS_i(b_1(\lambda_1,\mu_1),\cdots,b_n(\lambda_n,\mu_n)) = MCS_j(b_1(\lambda_1),\cdots,b_n(\lambda_n)) & 1 \le (i,j) \le m \\ MCSS_R = MCSS_{NR} = \sum_{j=1}^{m} MCS_j \end{cases} \quad (4)$$

Where $b(\lambda)$ denotes a non-repairable component, $b(\lambda,\mu)$ expresses a repairable component, λ, μ is the failure rate and repair rate of a component, $MCSS_R$ is the MCSS of the repairable system and $MCSS_{NR}$ is the MCSS of the non-repairable system. Therefore, the MCSS of a repairable system can be obtained using the approaches mentioned in [9,10,11].

3. THE PROPOSED MODEL

As to a repairable system's DFT, the top event fails if any MCS occurs, and vice versa. Therefore, the reliability of a repairable system is closely related to the MCSS of its corresponding DFT. Assume a system components has low failure rate and high repair rate, then the time deviating from the normal state, especially the time in failure state is much less than in normal state. For an illustrative purpose, we suppose a system state set {S0, S1, S2, S3, S4}. S0 is the initial state, S1, S3, S4 are the degraded states, S2 is the failure state. The system states transition schematic is shown in Fig. 2.

Fig.2: The Schematic of System States Transition

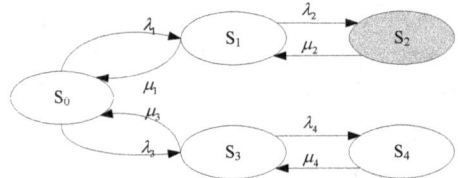

Assume the system runs for T hours and the time staying at every state are T_{S0}, T_{S1}, T_{S2}, T_{S3}, T_{S4}. Then the unavailability Q of the system can be calculated by

$$Q = \frac{T_{S2}}{\sum_{i=0}^{4} T_{Si}} = \frac{T_{S2}}{T} \quad (5)$$

To reduce the model scale, we directly adopt the failure chain ($S0 \rightleftarrows S1 \rightleftarrows S2$), i.e., approximate model, to express the system states. Meanwhile we suppose the time at each state are T'_{S0}, T'_{S1}, T'_{S2}, and then the approximate unavailability Q_{app} of the system can be computed by

$$Q_{app} = \frac{T'_{S2}}{\sum_{i=0}^{i=2} T'_{Si}} = \frac{T'_{S2}}{T} \quad (6)$$

Considering the approximate model gives up two degraded states, i.e., S3 and S4, it increase the duration that the system staying at failure state, i.e., $T'_{S2} > T_{S2}$. Combining the Eq. (5) and (6), we have

$$Q = \frac{T_{S2}}{T} < \frac{T'_{S2}}{T} = Q_{app} \quad (7)$$

The Eq. (7) indicates the solution obtained by the approximate model is comparatively conservative. Let $\lambda = \max\{\lambda_1,\lambda_2,\lambda_3,\lambda_4\}$, $\mu = \min\{\mu_1,\mu_2,\mu_3,\mu_4\}$, and then the following Equation holds.

$$\lim_{\substack{\lambda \to 0 \\ or \\ \mu \to \infty}} (Q_{app} - Q) = 0 \quad (8)$$

Apparently, as λ is small or μ is large, we can adopt an approximate model to evaluate the reliability of a repairable system, i.e., $Q \approx Q_{app}$. As to the repairable system with small failure and high repair rate, the failure chains capture the most failure information which can be understood as the failure chains contribute significantly to the system failure. In this point of view, we propose a generalized approximate model for evaluating the reliability of a repairable system. Suppose a repairable system has n failure chains, i.e., $\overline{L_1}, \overline{L_2}, \cdots, \overline{L_n}$ determined by the MCSS, and then the approximate model is expressed by

$$Q(repairable\ system) = Q\left(\overline{L_1}, \overline{L_2}, \cdots, \overline{L_n}\right)$$
$$\approx Q_{app}\left(\overline{L_1}\right) + Q_{app}\left(\overline{L_2}\right) + \cdots + Q_{app}\left(\overline{L_n}\right) \quad (9)$$

4. NUMERICAL EXPERIMENT

4.1. Experiment Design

To validate the proposed model, a numerical experiment is implemented. First, we define the ratio of a DFT's failure chains as $N_{\overline{L}} / N$. Where, $N_{\overline{L}}$ is the number of the failure chains, N is the total number of the chains. For $1 \leq N_{\overline{L}} \leq N$, and then we can get $1/N \leq \varphi \leq 1$. Without loss of generality, we choose PAND gate ($\varphi = 1/N$) and WSP gate ($\varphi = 1$) as our experimental subjects. The corresponding approximate models are shown in Fig.3 and Fig.4 separately.

Fig.3: Approximate Model for PAND gate (n=2)

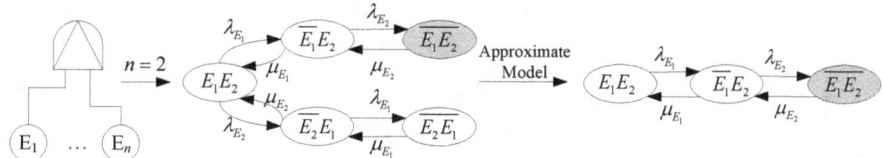

Fig.4: Approximate Model for WSP gate (n=2)

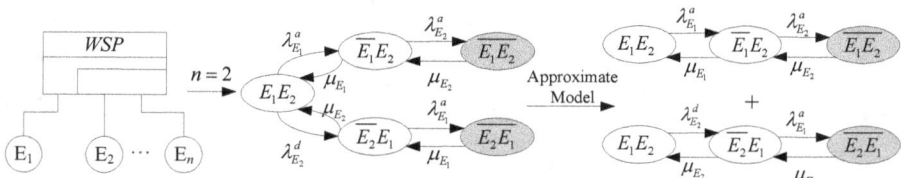

The experiment designs are described in Table 1.

Table 1: The Experiment Designs

The Gate Type	The Schemes	Design Parameters	Design Points	Mission Time(h)
PAND	Scheme 1: n=2	$\lambda_1 = \lambda_2 = \cdots = \lambda_n$ $\mu_1 = \mu_2 = \cdots = \mu_n$	$(\lambda_i^{(1)}, \mu_i^{(1)})$	T=10⁶
	Scheme 2: n=3		$(\lambda_i^{(2)}, \mu_i^{(2)})$	
	Scheme 3: n=4		$(\lambda_i^{(3)}, \mu_i^{(3)})$	
WSP	Scheme 1: n=2	$\lambda_2^d = \lambda_3^d = \cdots = \lambda_n^d = \alpha\lambda_1$ $\mu_1 = \mu_2 = \cdots = \mu_n$	$(\lambda_i^{(1)}, \mu_i^{(1)})$	
	Scheme 2: n=3		$(\lambda_i^{(2)}, \mu_i^{(2)})$	
	Scheme 3: n=4		$(\lambda_i^{(3)}, \mu_i^{(3)})$	

Note: the right superscript "d" of design parameters presents component in standby state; α is a dormant factor, and $\alpha = 0.1$; n is the total number of the input events; $\mu_i^{(1)} = 0.5\mu_i^{(2)} = 0.25\mu_i^{(2)} = 0.5$, $\{\lambda_1^{(1)}, \lambda_2^{(1)}, \cdots, \lambda_n^{(1)}\}$ = {1.0E-2, 5.0E-3, 2.5E-3, 1.0E-3, 5.0E-4, 2.5E-4, 1.0E-4, 5.0E-5, 2.5E-5, 1.0E-5, 5.0E-6, 1.0E-6, 5.0E-7, 1.0E-7}, $\lambda_i^{(1)} = \lambda_i^{(2)} = \lambda_i^{(3)}$.

4.2. Experiment Results and Analysis

The experiment results are shown in Fig.5-Fig.10, and the relative error between approximate model and exact model are shown in Fig.11-Fig.14. The relative error is defined as: $\varepsilon_{er} = (Q_{app} - Q_{exc})/Q_{exc}$, where the Q_{exc} is the exact solution.

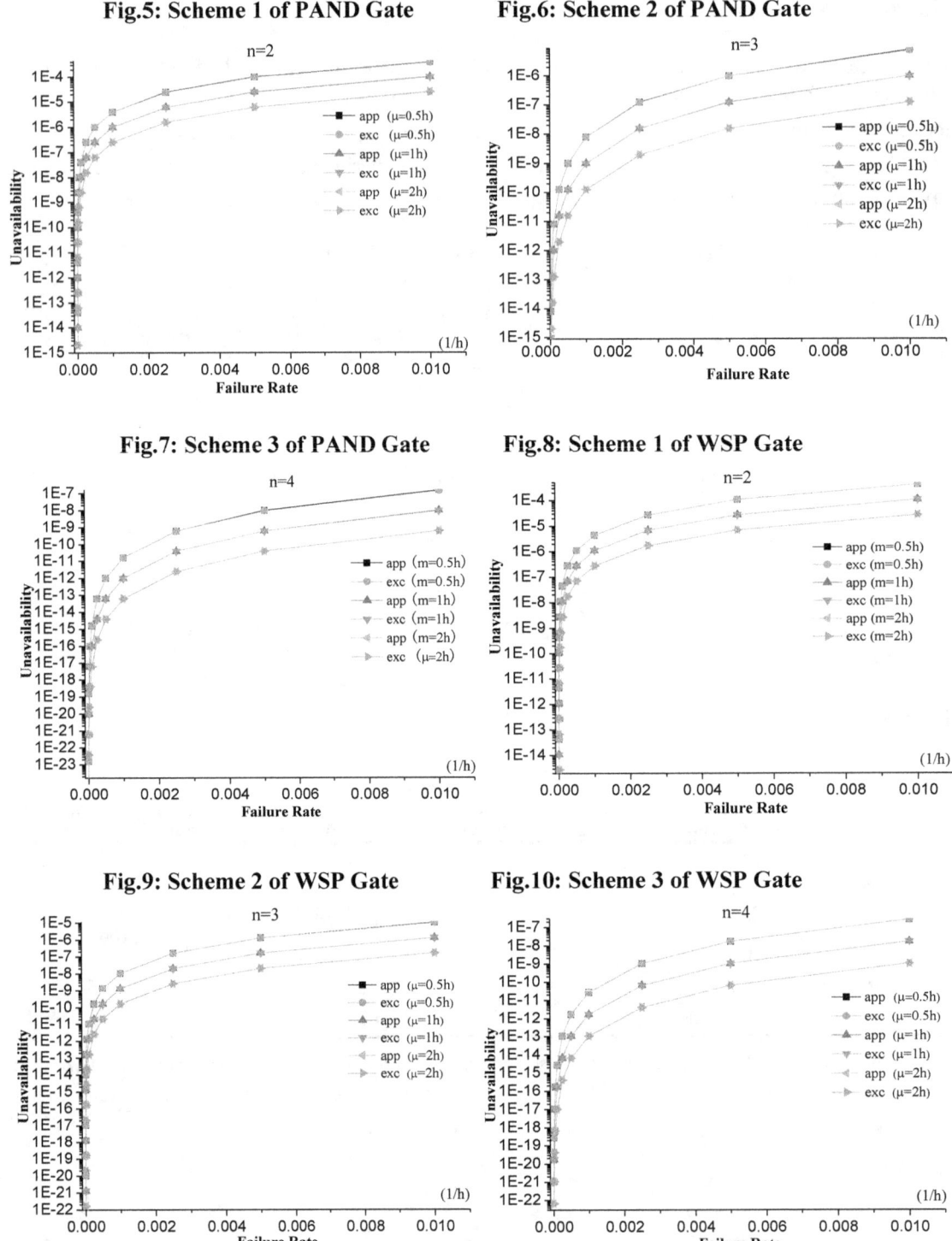

Fig.5: Scheme 1 of PAND Gate

Fig.6: Scheme 2 of PAND Gate

Fig.7: Scheme 3 of PAND Gate

Fig.8: Scheme 1 of WSP Gate

Fig.9: Scheme 2 of WSP Gate

Fig.10: Scheme 3 of WSP Gate

Fig.11: the ε_{er} for Scheme 1 of PAND Gate   Fig.12: the ε_{er} for Scheme 1 of WSP Gate

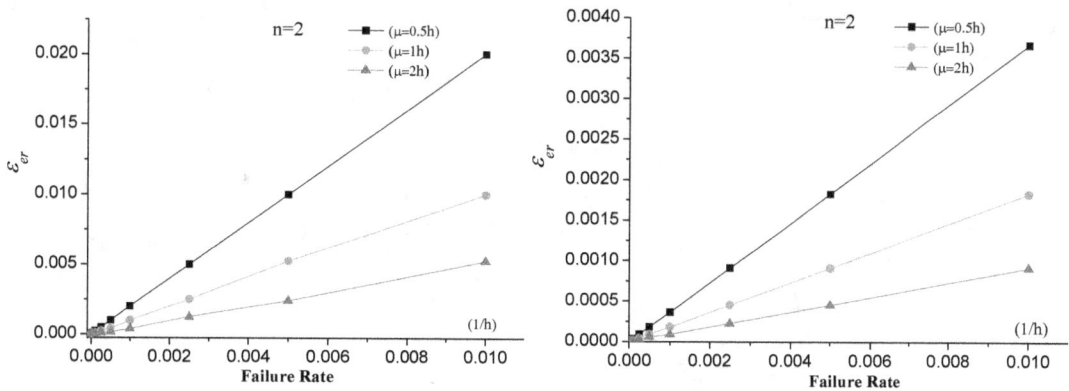

Fig.13: the ε_{er} for Scheme 3 of PAND Gate   Fig.14: the ε_{er} for Scheme 3 of WSP Gate

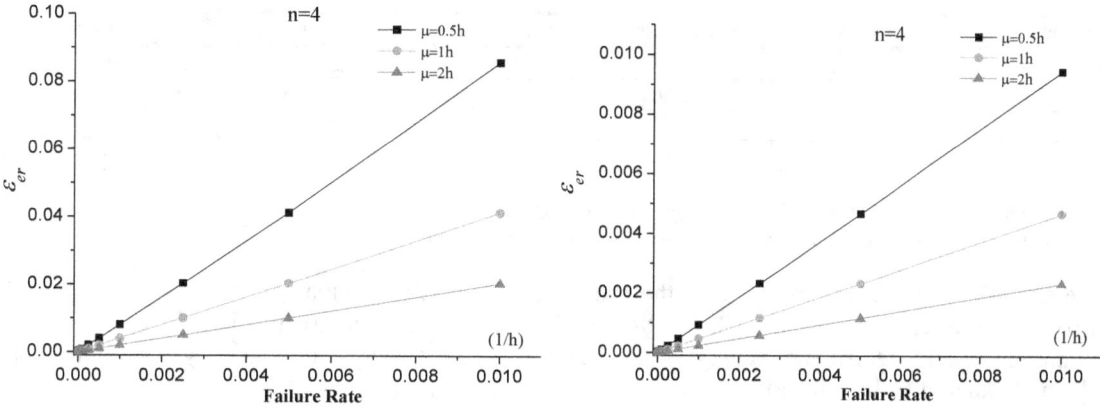

Fig.5-Fig.10 shows at each design point, the result obtained from the approximate model is in good agreement with that from the exact model (Markov-based). In addition, Fig.11-Fig.14 demonstrates the results calculated by the approximate model are conservative compared with the exact solutions. Moreover, with the decrement of the failure rate or the increment of the repair rate, the value ε_{er} is becoming smaller and smaller, and even can be neglected. It is found the value ε_{er} from WSP gate is smaller than that from PAND gate at the same design point, which can be interpreted as that with the augment of the failure chain ration (φ), the accuracy of the approximate model is becoming higher and higher. Considering $\varphi \in [1/N, 1]$ and most components in NPP with high repair rate and low failure rate, it is reasonable that the proposed model is valid and conservative for NPP's dynamic repairable systems.

5. CASE STUDY

5.1 Case Study 1

For model validation purpose, a case study is analyzed which is from a partial safety system of one Chinese NPP. The system's DFT model is shown in Fig.15 and its corresponding approximate model is shown in Fig.16.

Fig.15: The Simplified DFT Model **Fig.16: The Approximate Model**

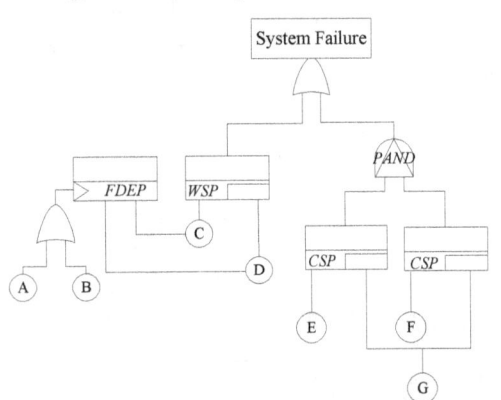
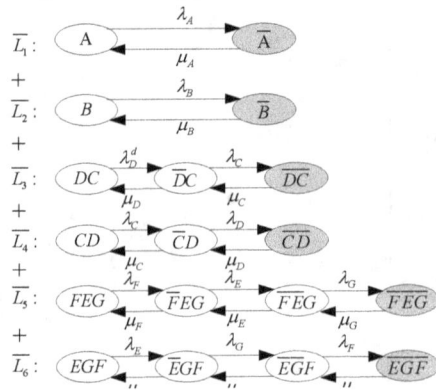

The reliability parameters of the components contained in the case 1 are listed in table 2.

Table 2: Reliability Parameters for Case 1

Component	Failure Rate	Repair Rate	Component	Failure Rate	Repair Rate
A	1.0e-7	0.25	D_d	2.0e-4	1.00
B	5.0e-7	1.20	E	1.4e-3	2.00
C	1.0e-7	1.50	F	2.5e-3	3.00
D_a	5.0e-7	1.00	G	2.0e-3	0.50

Note: the symbol D_a denotes the component D in working state; D_d denotes the component D in standby state; the time-to failure and time-to-repair of all components are following exponential distributions.

We suppose that the mission time of the system is 5000h. For comparison purpose, we apply the approximate model and exact model, i.e. Markov-based model, respectively to analyze the system's unavailability. The steady and average unavailability of the system calculated by the approximate model and exact model are listed in Table 3.

Table 3: Results of Case 1

	Approximate solution	Exact Solution	Relative Error (ε_{er})
Average Unavailability	1.93036E-6	1.93031E-6	2.5903e-5
Steady Unavailability	1.95397E-6	1.95367E-6	1.5356e-4

Analysis: As to the Markov-based model, i.e., converting the whole DFT into Markov Chain, the number of the system states would grow up to 2^7. It is a hard and error-prone job. By contrast, the max number of the states defined by the longest failure chain involved in our proposed model is only 4. Therefore, the proposed method is more efficient than the Markov-based approach. In addition, the results calculated by the proposed method are highly matched with those obtained by the Markov-based method.

5.2. Case Study 2

For further model validation purpose, a more complex case is analyzed, which is from a partial I&C safety system of one Chinese NPP. The system's simplified DFT model is shown in Fig.17.

Fig.17: The Simplified DFT Model   **Fig.18: The Approximate Model**

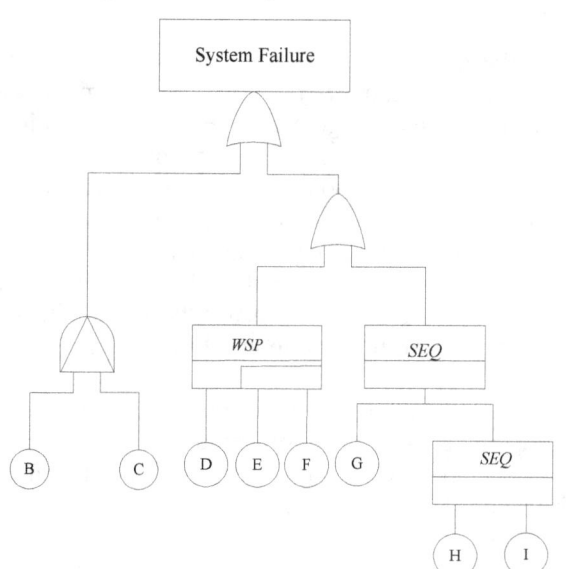
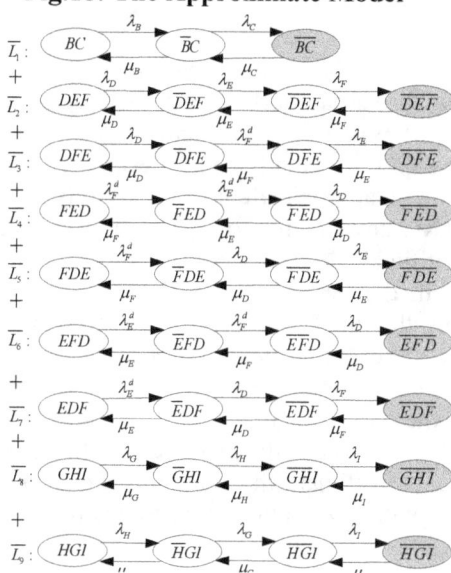

The reliability arguments of the system's components are shown in Table 4.

Table 4: The Reliability Parameters for Case 2

Component	Failure Rate	Repair Rate	Component	Failure Rate	Repair Rate
B	8.5e-4	0	F_a	1.0e-2	3
C	1.0e-4	12	F_d	2.0e-3	3
D	1.0e-2	3	G	1.0e-2	2.5
E_a	1.0e-2	3	H	6.0e-3	3.5
E_d	2.0e-3	3	I	5.0e-3	2

Similarly, we adopt the proposed method and Markov-based method to analyze this system's unavailability separately. Assume the failure time and repair time of system's components follow exponential distribution and the mission time of the system is 5000h, and then the solutions of the system's unavailability obtained by the two methods are listed in Table 5.

Table 5: Results of Case 2

	Approximate solution	Exact Solution	Relative Error (ε_{er})
Average Unavailability	8.42951E-6	8.42191E-6	9.024E-4
Steady Unavailability	6.49651E-6	6.48887E-6	1.200E-3

Obviously, the results obtained by our proposed method are very close to those derived by the Markov-based method.

6. Conclusions

As to the repairable systems modelled by DFTs, the quantitative analyses of these systems are mainly based on Markov approach. Although this approach can offer an exact solution, it may confront the notorious problem of "state space explosion". For a large-scale DFT, the conventional Markov-based method would become hard to be implemented. To solve this problem, this paper proposes an approximate method to analyze DFT with repairable components. This method divides the whole Markov Chain into separate failure chain and neglects the successful chains. Each failure chain is quantified by Markov-based method, and then the results of the separate failure chains are integrated to obtain the system's unavailability. Therefore, in contrast to the conventional state space-based method, this method gets over the problem of "state space explosion". The results of experimental design and cases analysis demonstrate, as to a system with high repair rate and low failure rate, this method can offer a solution with a high accuracy.

For the NPP's repairable systems, most components involved have low failure rate and high repair rate. It is reasonable the proposed method has a high engineering application value in NPP, which can be used to estimate the reliability of safety-critical system quickly. However, as to the repairable systems with high failure rate and low repair rate, it still needs further research.

References

[1] M. Alam, and U.M. Al-Saggaf, "Quantitative reliability evaluation of repairable phased-mission systems using Markov approach." *IEEE Trans. Rel.,* vol.R-35, n.5, pp. 498-503. Dec. 1986.
[2] L. Xing, K.N. Fleming, and W.T. Loh, "Comparison of Markov model and fault tree approach in determining initiating event frequency for systems with two train configurations." *Reliab. Eng. Syst. Saf.,* Vol. 53, n. 1, pp. 17-29. Jul. 1996.
[3] W. Long, Y. Sao, and M. Horigome, "Quantification of sequential failure logic for fault tree analysis." *Reliab. Eng. Syst. Saf.,* vol. 67, no. 3, pp. 269-274.Mar. 2009.
[4] S.V. Amari, G. Dill, and E. Howald, "A new approach to solve dynamic fault tree." In *Proc. Annu. Reliab. Maintainability Symp.,* PP. 1-7. 2003.
[5] D. Liu, , C. Zhang, , W. Xing, R. Li, and H. Li, "Quantification of Cut Sequence set for Fault Tree Analysis." in Proc. *HPCC,* vol. 4782, *Lecture Notes in Computer Science,* 2007, pp. 755-765.
[6] K.D. Rao, V. Gopika, V.V.S.S. Rao, H.S. Kushwaha, A.K. Verma, and A. Srividya, "Dynamic fault tree analysis using Monte Carlo simulation in probabilistic safety assessment." *Reliab. Eng. Syst. Saf.,* vol. 94, no. 4, pp. 872-883. Apr. 2009.
[7] P. Zhang, and K.W. Chan, "Reliability Evaluation of Phasor Measurement Unit Using Monte Carlo Dynamic Fault Tree Method." IEEE Trans. Smart Grid., vol. 3, n. 3, pp. 1235-1243. Sep. 2012.
[8] Z. Tang, and J.B. Dugan, "Minimal Cut Set /Sequence Generation for Dynamic Fault Trees." In *Proc. Annu. Reliab. Maintainability Symp.,* pp.207-213. Jan. 2004.
[9] D. Liu, W. Xing, C. Zhang, R. Li, and H. Li, "Cut Sequence Set Generation for Fault Tree Analysis." in Proc. *ICESS,* vol. 4523, *Lecture Notes in Computer Science,* 2007, pp. 592-603.
[10] G. Merle, J.-M.Roussel, J.-J. Lesage, Algebraic determination of the structure of Dynamic Fault Trees. *Reliab. Eng. Syst. Saf.,* vol. 96, no. 2, pp. 267-277. Feb.2011.
[11] J. Liu, w. Tang, and Y. Xing, "A Simple Algebra for Fault Tree Analysis of Static and Dynamic Syatems" *IEEE Trans. Rel.,*vol. 62, n. 4, pp. 846-861. Dec. 2013.
[12] J.B. Dugan, and S.A. Doyle, "New results in fault-tree analysis." In *Proc. Annu. Reliab. Maintainability Symp., Tutorial Notes,* pp.1-23. 1997.

Air Traffic Controllers' Workload on the Period of ATC Paradigm Shift

Kakuichi Shiomi
Electronic Navigation Research Institute, Tokyo, Japan

Abstract: The real time simulation was performed in order to investigate the influence of the introduction of CPDLC into Japanese domestic ATC operations. The simulation was carried out with the participation of retired ATC controllers. Based on the results, it was considered that the introduction of CPDLC would be effective to reduce total communication time for ATC instructions, but it was also confirmed that the ATC workload is not simply dependent on the length of communication time. It is impossible to avoid the increase of workload in the transient situation, and it is not desirable from the viewpoint of the operational workload of ATC controllers that the introductory situation of 30% of aircraft with CPDLC capability continues for a long term.

Keywords: ATC, CPDLC, workload

1. INTRODUCTION

It is expected that the introduction of CPDLC (Controller-Pilot Data Link Communication) into the Air Traffic Control (ATC) operations will improve ATC safety and reliability by eliminating incorrect hearing in the present analogue voice communication, etc., since the quality of current analogue VHF radio communication is not sufficient.

In Japan, CPDLC has been introduced into the ATC operations of the Pacific Oceanic area, since the ATC instruction interval is relatively large in the oceanic ATC area, and ATC controllers have sufficient time to carry out CPDLC operations. In the U.S. and Europe, CPDLC of the domestic air route ATC operations have already been introduced now, but the introduction is not yet in Japan.

It is clear that the transition to data communication from analogue voice communication greatly shortens spatial occupancy time of radio waves in ATC air-ground communication. Therefore the introduction of CPDLC significantly improves the use efficiency of radio wave resources. It is also considered that a wider introduction of CPDLC into Japanese domestic ATC operations is necessary to improve the efficiency of utilization of radio wave resources, and to keep sufficient radio wave resources required in the near future.

The Japanese Civil Aviation Bureau (JCAB) is now seeking to carry out their introduction plan of CPDLC as an important part of overall future planning, named CARATS (Collaborative Actions for Renovation of Air Traffic Systems). The JCAB is expecting the improvement of the convenience of ATC operations, the efficiency of air transportation, and the total air safety with the introduction of CPDLC.

2. THE FIRST STEP FOR CPDLC INTRODUCTION

The wider introduction of CPDLC into Japanese domestic air route ATC operations will be inevitable in the near future.

And now, there are two ways to introduce CPDLC in Japan; one way is to introduce a CPDLC system similar to that introduced in the U.S. and Europe domestic air route ATC operations, and the other way is to develop the Japanese CPDLC system corresponding to the requests from Japanese ATC controllers.

But no one knows which way is better for the first step of the introduction of the Japanese domestic air route ATC operations, since Japanese is currently in a unique situation. Japanese airspace is not as overcrowded as Europe, and VHF radio wave resources are also not in short supply at present.

When the Japanese CPDLC is introduced into the Japanese domestic air route ATC operations, it has to have minimum compatibility with the CPDLC of the U.S. and Europe, since the Japanese CPDLC must be valid for international flights from the U.S. and Europe in Japanese airspace.

However, the most important thing in the first step of the introduction of CPDLC into the Japanese domestic air route ATC operations is to minimize ATC controllers' contradictory feelings towards the introduction of CPDLC by minimizing the impact of its introduction on the current ATC operation environment.

In the current Japanese situation that there is no severely congested airspace except in the Kanto-South sector (Southern side of the Tokyo Metropolitan area) in which many aircraft are cruising to Tokyo/Haneda International Airport, and there are few pressing problems in the current ATC operation environments. Due to this situation, the introduction of CPDLC will not be expected to be greatly welcomed. It is then very important that the introduction of the CPDLC will never annoy or trouble the ATC controllers.

The study on the CPDLC is aiming to make clear the way to introduce CPDLC into the Japanese domestic ATC operation as the first step of the next CPDLC paradigm, and also aiming to establish a firm foothold for the second step by giving answers to the following questions.
In the first step of the introduction of CPDLC,
a) What kinds of CPDLC messages will have to be handled?
b) How much will it reduce the time for ATC communication?
c) How will ATC controllers' workloads change?
We were required to answer as quantitatively as possible the above questions.

If the basic data such as contents of domestic air-ground ATC communication, their frequency, etc. exist, it is possible to estimate the expected length of the communication time that will be reduced by the introduction of CPDLC. It will be possible to calculate the relationship between the expected reduced ATC communication time and the contents of introduced CPDLC messages. Since the performance of VHF radio communication media used for CPDLC is given as statistical data, the relationship between the reduced time and the CPDLC messages is able to be calculated more precisely by making use of fast-time simulators, etc.

However, it is impossible to obtain any information about behavior change of an ATC controller, who has "humanity" associated with changes in operation configuration, occurrence of new-type human errors, the sense of the ATC workloads, etc. from the approximate calculation or the fast-time simulation, because it assumes a simplified CPDLC model that eliminates the human factors.

Since the subjective evaluation of the first step of the introduction of CPDLC into the ATC operations has to be made by ATC controllers as the users of the CPDLC system after the introduction, the provider of the system must prepare to make the CPDLC system acceptable by them, and to get higher evaluation scores. Since the good usability of handling CPDLC messages strongly depends on the design of the user interface, it is essentially necessary for the system provider to repeat the evaluation and improvement of the user interface for ATC controllers, such as GUI, custom input devices, etc.

The aim of this study on the introduction of CPDLC into the Japanese domestic ATC operations was to obtain useful data for preparation of the introduction as the results of the real-time ATC simulations that were carried out with the cooperation of ATC controllers. As it was possible to perform proper CPDLC/ATC simulations without being affected by unnecessary dissatisfaction such as due to poor operational capabilities when handling CPDLC messages, we had prepared the CPDLC/ATC simulator which looks sophisticated enough, and which had enough capability for handling the messages.

Our study aimed to obtain quantitative answers to the questions mentioned above, and also aimed to obtain some additional information for thinking about the second step of the introduction of CPDLC and the following steps.

3. CPDLC/ATC SIMULATION

The long-term aim of the introduction of CPDLC into world global ATC is to correspond to the needs in the future paradigm of the fourth dimensional trajectory management of aircraft. In the future paradigm of 4D-trajectory management, it will be necessary for the ATC controller to give very complex instructions, which is difficult to communicate by the current analogue voice, and then he or she has to use CPDLC to give instructions. A set of CPDLC messages is already designed to be able to handle the complex, high-level, and advanced ATC instructions.

With the exception of exceptional circumstances, the ATC controller will no longer command flight altitude, flight direction and flight speed to individual aircraft in the future paradigm of 4D-trajectory management. However, in the introductory period of CPDLC, it can be thought that there will be many situations where the ATC controller has to make CPDLC messages to command flight altitude, flight direction and flight speed to individual aircraft.

Figure 1: Custom Input Device for CPDLC Messages

The data input device shown in Figure 1 was one of the custom input devices designed and developed for quick change of aircraft's flight altitude, flight direction and flight speed in CPDLC messages. However, these special input devices were not used in the CPDLC/ATC simulation, since the ATC controllers who participated in the simulation used a regular wheel-mouse for handling CPDLC messages. Fortunately or unfortunately for us, the combination of the wheel-mouse and the pop-up menu was enough for the ATC controllers' message handling in the simulation.

3.1. CPDLC/ATC Simulator

Figure 2 shows the CPDLC/ATC simulator prepared for simulation of the introduction of CPDLC into Japanese domestic ATC operations. The simulator has two large LCDs and one small LCD. One of the large LCDs displays the information for the participant who plays the role of an ATC controller in the simulation. The other large LCD displays the information for the participant who plays the role of an aircraft pilot. The small LCD displays the GUI to control the simulation for the simulation supervisor.

Figure 2: Appearance of the CPDLC/ATC Simulator

The displayed information on the large LCD is designed according to the current digitalized ATC radar image display. The LCD displays the positions of aircraft on an airspace sector map, which consists of air routes, waypoints, etc. Figure 3 shows the displayed image provided by the simulator for ATC operations and a real radar image provided for the ATC controller by the current actual ATC information system. Though the position of an aircraft is indicated with a triangle symbol in the current actual ATC system, the position of an aircraft which has CPDLC capability is indicated with a circle symbol, as shown in Figure 4, in the simulator to discriminate between conventional aircraft without CPDLC capability and new aircraft with the capability.

Figure 3: Displayed Images for the ATC Controller

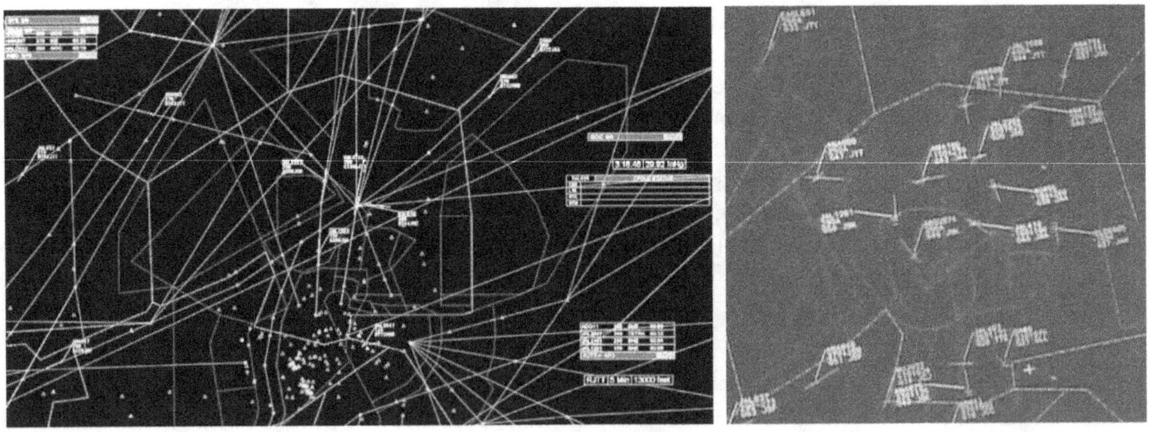

The left image is a simulated display image, and the right image is a picture of an actual radar display.

Both large LCDs display almost the same information and look almost the same, since the participant who plays the role of an aircraft pilot has to control the same number of aircraft that were instructed by the ATC controller.

The ATC controller can create a typical CPDLC message by selecting the pop-up menu according to the procedure shown in Figure 5. After creating the CPDLC message as shown in Figure 5 (4), if the "OK" button is clicked, the CPDLC message is sent to the target aircraft, and a symbolized up-link icon is displayed until the acknowledgment signal from the aircraft is received.

Figure 4: Aircraft's Symbols displayed in the ATC Information Display

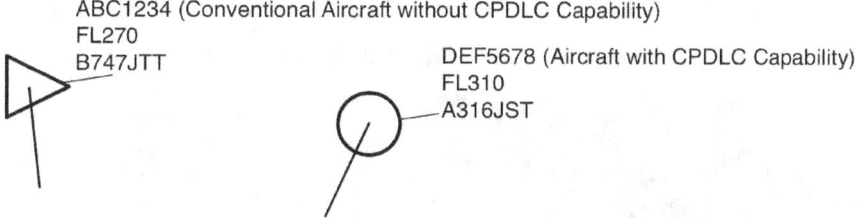

Figure 5: Procedure to create CPDLC Message

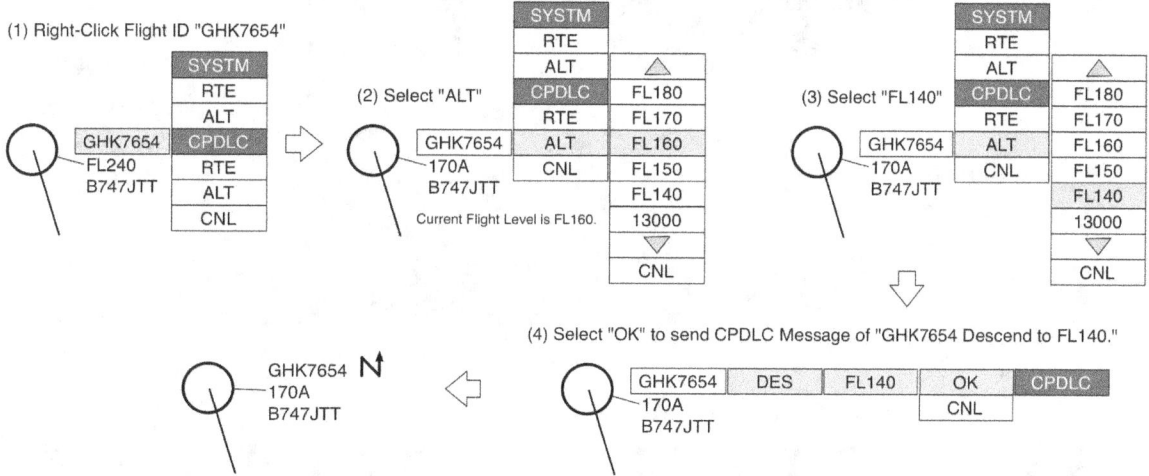

Figure 6: Pop-Up Menu for Creating CPDLC Messages

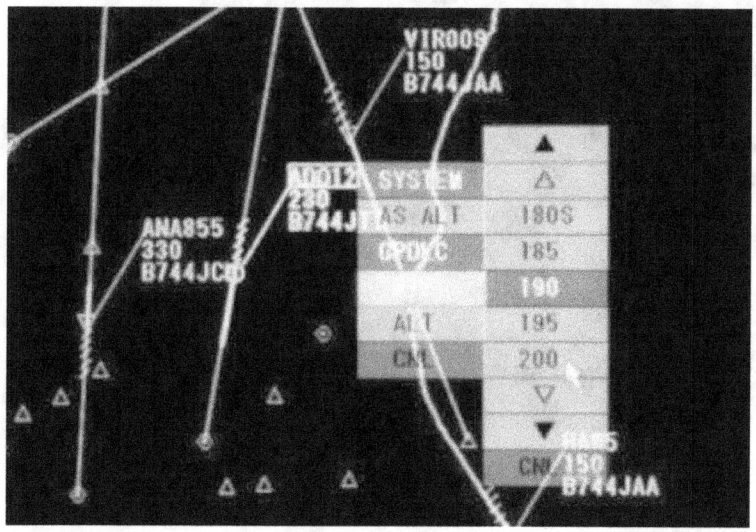

Figure 6 shows the situation when the ATC controller creates a CPDLC message to change the flight direction. In the case shown in Figure 6, item "190" is highlighted in the pop-up menu, since the current flight direction of the selected aircraft is 190 degrees.

3.2. CPDLC Simulation Scenarios and Conditions

Figure 7 shows a picture of the CPDLC/ATC simulation being carried out. Since the participant playing the pilot role should frequently coordinate with the ATC controller and the ATC coordinator, they carried out the simulation sitting next to each other. Since this seating resembles the actual ATC environment as shown in Figure 8, it is considered that this would not likely become a factor in increasing the workload of the participant playing the role of ATC controller.

Figure 7: Configuration of the CPDLC/ATC Simulation

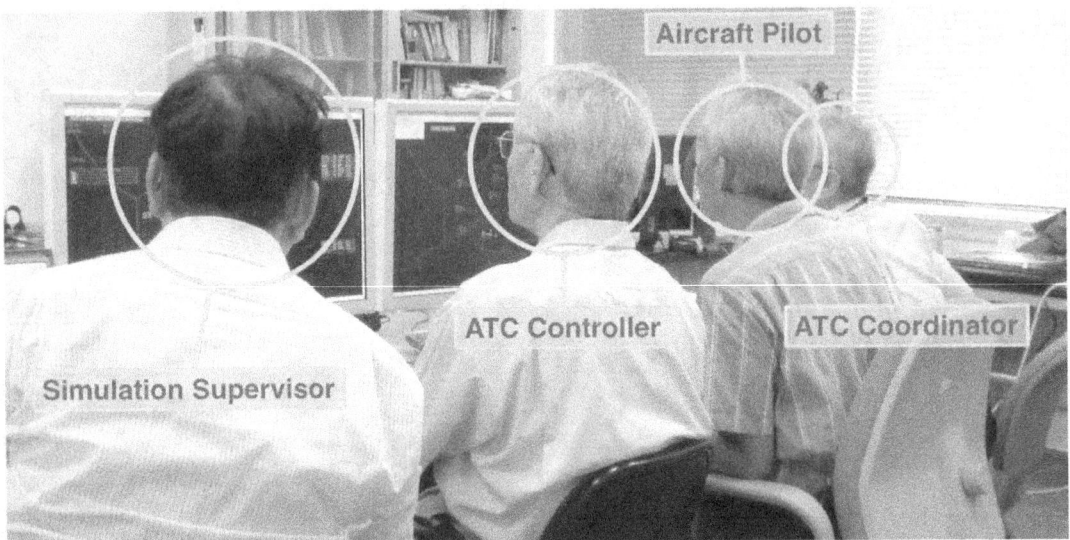

Figure 8: Configuration of Actual Japanese ATC Environment in 2008

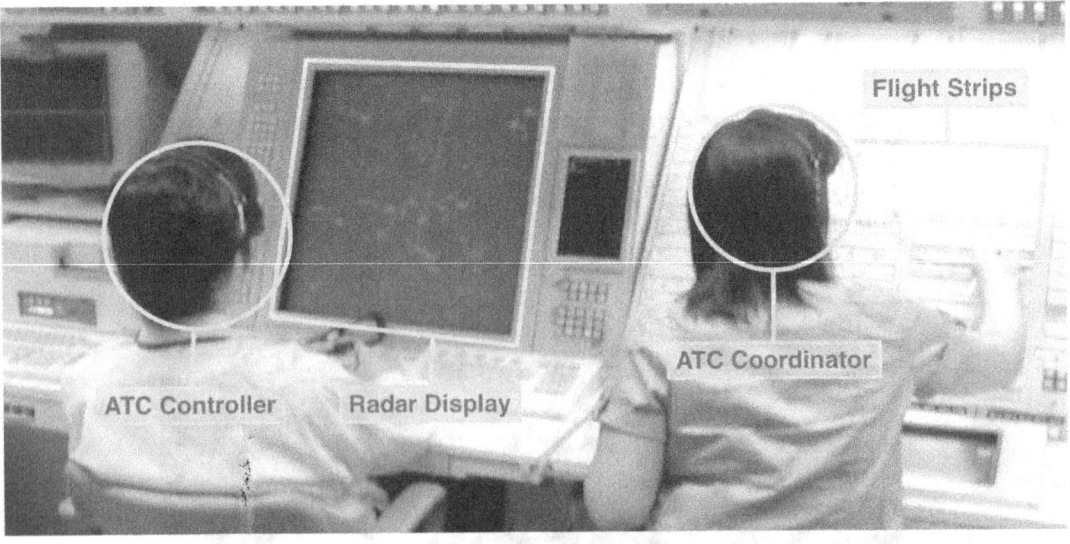

Figure 9: Layout of Radar Display and Flight Strips

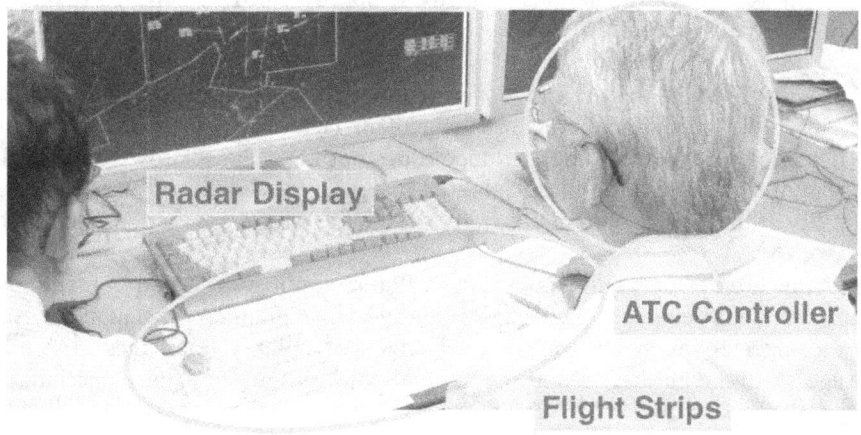

Flight strips representing flight information of aircraft were preliminarily printed according to the current Japanese ATC format, and arranged in front of the Radar display as shown in Figure 9.

The Kanto-North (Northern side of Tokyo) sector was set as the airspace for CPDLC/ATC simulation. Though the Kanto-North sector is not as congested as Kanto-South, it has a rich variation of flights. The Kanto-North sector adjoins the Japanese major terminal airspace of two major airports, Haneda/Tokyo and Narita/Tokyo. And the ATC controllers deal with passing flights from the Pacific Ocean, take-off and landing flights of small aircraft at Fukushima airport, and military aircraft at Hyakuri/Ibaraki, Yokota/U.S. Air-Force, etc.

The Kanto-North sector has various air routes at an altitude of between 9,000 feet and 37,000 feet. This structure of airspace makes it possible to carry out complex ATC operations including changing flight altitude. And it is suitable for the simulation with the expectation of some or more additional results, which cannot be obtained without good luck.

One of the important things that must be revealed by this simulation is to confirm the validity of the contents of the set of CPDLC messages that will be introduced into the Japanese domestic ATC operation at the introductory phase. The contents of the set of CPDLC messages must correspond to the characteristics of the CPDLC communication media and ATC operational environment.

At first, in this simulation, it was assumed that it would be best to use VDL (VHF Digital Link) Mode-2 as air-ground communication media, since the JCAB had already decided to use VDL Mode-2 as the data link media for CPDLC in Japanese domestic airspace. VDL Mode-2 will be useful enough to send complex messages, such as Pre-Departure Clearance, the CPDLC message for setting 4D-trajectory of aircraft, etc. in the future paradigm of 4D-trajectory management, since ATC instructions consist of complex contents which do not require a quick response from an aircraft pilot. However, VDL Mode-2 will never provide the real-time feature like analogue voice communication. In the data link communication on VDL Mode-2, every message will be delayed by about three seconds, as averaged delay time, and some messages will be delayed by more than 10 seconds. It is impossible to carry out real-time communication like daily conversation by using VDL Mode-2. Also, in the data link communication on VDL Mode-2, a message sent later may arrive earlier than another message sent before it.

The experiments for the evaluation of the VDL Mode-2 function and performance were carried out, and consequently the function for simulating the VDL Mode-2 performance was implemented in the CPDLC/ATC simulator used in this simulation.

The functions of the CPDLC application were implemented according to the ATN (Aeronautical Telecommunication Network) regulations that were approved by the ICAO (International Civil Aviation Organization) [1].

Since the CPDLC application has the function to confirm message delivery, the application repeats sending a message until an automatic acknowledgement is received from the objective destination aircraft, even if the lower communication protocol does not have the acknowledgment function. The logical delivery confirmation function implemented in the CPDLC application is considered essential to ensure reliability of interactive communication on the VDL Mode-2 with poor data-link capability.

The data link characteristics of VDL Mode-2, the CPDLC application function mentioned above, the corresponding functional level on the side of the aircraft, the current Japanese ATC operation environment, etc. interfere with the use of some kinds of CPDLC messages. A utilizable set of CPDLC messages will be limited according to both the aircraft cruising functions and the total environments of ATC operations.

Though a limitation in the set of CPDLC messages existed in the CPDLC evaluation tests carried out in Europe and the U.S., the limitation itself had not always been considered a severe problem. However, it is also true that a limitation in the message set would seem unnatural to the ATC controllers, if CPDLC messages were simply regarded as an alternative to voice instructions.

In this simulation experiment, the following message services were set to be able to be used for ATC operations: frequency change instruction messages as the ACM (ATC Communications Management) service message, and both flight altitude change instruction messages and flight direction change instruction message as the ACL (ATC Clearances) service messages.

But flight speed change instruction messages were not available, because the ATC controller had no way to know the current flight airspeed. In order to accurately instruct flight speed change it is necessary for the ATC controller to know current flight airspeed affected by wind before giving the instruction. There is no other way to know the current flight airspeed of aircraft than by the ATC controller questioning the pilots, and it is suggested not to do so from the point of view of operational workload increase due to interactive message exchange, under the situation of larger transmission delays occurring on the VDL Mode-2 data link. Therefore the flight speed change instruction does not allow the ATC controllers to use it as long as the current airspeed data is not distributed automatically from aircraft through DAPs (Downlink Aircraft Parameters), etc.

It is also considered that utilization of the ACL service messages with frequent corrections given by the "disregard" message should be limited.

Though more than 300 kinds of CPDLC messages have been defined in the ICAO SARPs (Standards and Recommended Practices) [1], it is assumed to use the message set size of only about 1% in this simulation. However, this setting is not unrealistic, since the frequency change instruction in area control ATC operation has been executed for all the handling aircraft, and the flight altitude change instruction has been frequently executed for the taking off and landing aircraft, and it is considered that these message services are expected to show good effects as the result of the introduction of CPDLC. The flight direction change instruction as the ACL service message is also used frequently for spacing control between aircraft.

Three kinds of simulation scenarios, A, B and C, were prepared for this CPDLC/ATC simulation experiment. Each scenario can be processed in about 30 to 40 minutes. Aircraft density of 45 aircraft per hour is assumed in these scenarios, since allowable process capacity of the Kanto-North sector is 50 aircraft per hour. The characteristics of each simulation scenario are as follows:
Scenario A – relatively uniform aircraft density
Scenario C – temporarily high aircraft density
Scenario B – intermediate between Scenario A and C

Each scenario can be processed without difficulty, if the participant of ATC controller role had ATC service experience at the Kanto-North sector.

In all three scenarios, the percentages of aircraft which have CPDLC capability were set at 0%, 30% and 80%. 0% means the current ATC situation before the introduction of CPDLC into ATC operations, 30% means the introductory phase of CPDLC, and 80% means that all aircraft except military aircraft have CPDLC capability.

As a result of setting the conditions of this simulation, it was decided that each participant playing the ATC controller role should execute nine simulation experiments. Each participant, as a matter of course before the actual simulation, had performed the exercise practice to experience the CPDLC operation carefully.

4. SIMULATION RESULTS AND CONCLUSION

Ten retired ATC controllers belonging to ATCAJ (Air Traffic Control Association, Japan) participated in the role of ATC controller in the CPDLC/ATC simulation. And valid results were obtained from eight participants.

Figure 10: Total Communication Time vs. Rate of Aircraft with CPDLC Capability

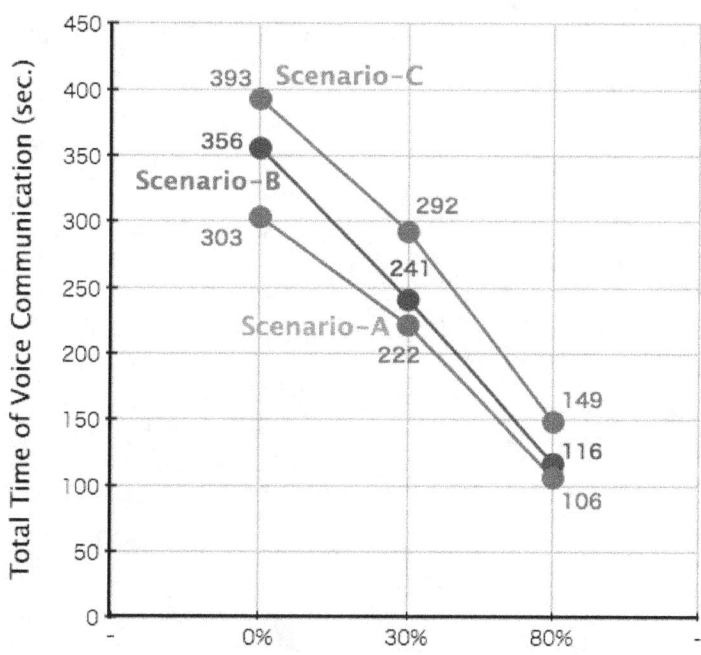

Rate of Aircraft with CPDLC Capability

Figure 10 shows the decreasing effect of total voice communication time per scenario by the average value of 8 valid participants.

When the simulation results of each participant playing the ATC controller role is observed respectively, there was a case that the difference of total communication time was more than 100 seconds depending upon the scenario. The average of total time of voice communication decreased as naturally expected according to the increase of the rate of aircraft with CPDLC capability. In particular, the frequency change instruction as the ACM service message was effective to decrease communication time. Conventionally, in analog radio voice communication, it took more than several seconds to give the instruction of the communication frequency change. After the introduction of

CPDLC into domestic ATC operations, the frequency change instruction is instantly completed with a key pack and/or pointing device such as a mouse, etc.

When the percentage of aircraft with CPDLC capability was 80%, there were many positive comments and opinions about the introduction of CPDLC after the simulation experiments, such as follows:
1. Communication time became shorter and the margin occurred during thinking time.
2. It was easier for ATC operations to concentrate.

When it was 30%, the instructions were given almost by conventional analogue radio voice, and there were many negative comments and opinions, such as follows:
1. There were many ATC instructions given by voice to aircraft with CPDLC capability.
2. There was some confusion about the discrimination between aircraft with CPDLC capability and those without the capability.
3. It was not easy to distinguish between aircraft with CPDLC capability and aircraft without the capability, only according to the slight difference in displayed aircraft symbols.

It can be thought that some ATC controllers think "It's useful if air traffic volume is small, but it is necessary to limit the use of ACM service to only when traffic volume is large." No one can tell how the rate of aircraft with CPDLC capability will change in the future, after the introduction of CPDLC into air route ATC operations. It does not seem to be desirable from a viewpoint of an ATC operation workload that the situation of 30% of aircraft with CPDLC capability continues for a long term.

Figure 11: CEM vs. Rate of Aircraft with CPDLC Capability

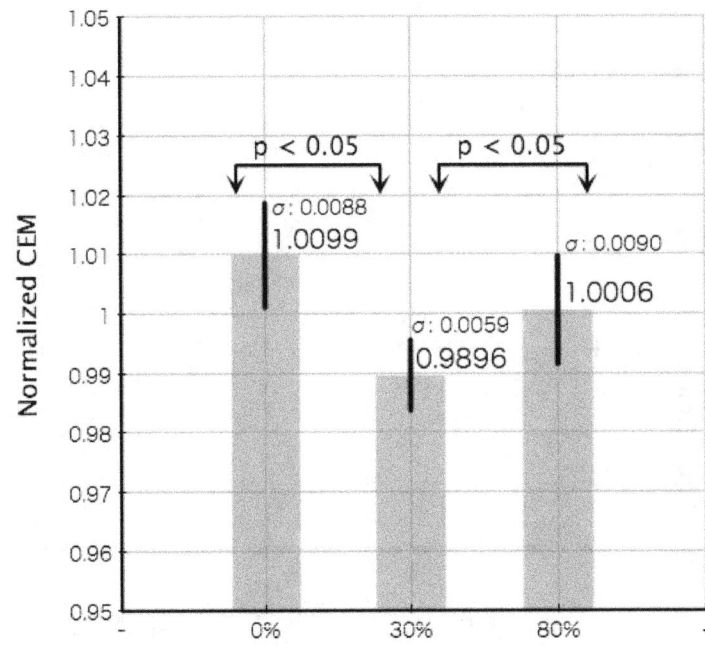

Rate of Aircraft with CPDLC Capability

Figure 11 shows the relationship between the percentages of aircraft with CPDLC capability and the average of CEM (Cerebral Exponent Macro) values that were calculate from a human voice and correlate with mental load of the speaker [3, 4]. Every CEM value was calculated from every two seconds ATC communication voice by using the software of implementation of SiCECA (Shiomi's Cerebral Exponent Calculation Algorithm) [4, 5].

In 1998, S. Hirose and K. Shiomi found that the time-averaged value of the first Lyapunov exponent calculated from a human voice changed according to the speaker's psychosomatic condition [2]. The

ENRI (Electronic Navigation Research Institute) has then been studying ways of measuring human performance by analyzing a person's voice since 1998.

Figure 11 is the result of research and development of the workload measurement method according to the voice analysis method mentioned above. The CEM value calculated from normal reading voice decreased, when operators were mentally exhausted. In other words, the CEM value indicates the degree of arousal level. However, even when the arousal level is high, if it is necessary to perform more important work than speaking or calling, the CEM value calculated from the utterance as a secondary task decreases, since the processing capacity of the human brain is finite.

Since there is no situation in which the ATC controller becomes sleepy in every simulation scenario, it is thought that the decrease of the CEM values depends on the increase of thinking work. The Student's t-test was used for comparison between the "0% case" and "30% case" groups, and the "30% case" and "80% case" groups. A probability value $p < 0.05$ was considered statistically significant. As the results of analysis of the communication voice of ATC controllers, Figure 11 shows that the ATC workload increases under the condition that the aircraft with CPDLC capability and conventional aircraft without the capability are intermingled. It is then confirmed that the ATC workload is not simply dependent on the length of communication time.

Acknowledgements

The software application for calculating the CEM values from human voices has been granted the following patents in the U.S.A.: US 6,876,964 (Oct. 19, 2005), US 7,321,842 (Jan. 22, 2008) and US 7,363,226 (Apr. 22, 2008). The voice signal processing software that was used to voice analysis can be provided under certain contracts to researchers who want to evaluate the function of the software.

References

[1] http://www.icao.int/safety/SafetyManagement/Pages/SARPs.aspx.
[2] K. Shiomi and S. Hirose, *"Fatigue and Drowsiness Predictor for Pilots and Air Traffic Controllers"*, Proc. of 45th Annual ATCA Conference, Oct. 2000, U.S.
[3] K. Shiomi and et. al., *"Experimental Results of Measuring Human Fatigue by Utilizing Uttered Voice Processing"*, Proc. of IEEE-SMC 2008, P557, Oct. 2008, Singapore.
[4] K. Shiomi, *"Voice Processing Technique for Human Cerebral Activity Measurement"*, Proc. of IEEE-SMC 2008, P660, Oct. 2008, Singapore.
[5] K. Shiomi, *"Chaotic Voice Analysis Method for Human Performance Monitoring"*, Proc. of ESREL2009, Sep. 2009, Prague, Czech Republic.

Quantification of Bayesian Belief Net Relationships for HRA from Operational Event Analyses

Luca Podofillini, Lusine Mkrtchyan, Vinh N. Dang
Paul Scherrer Institute, Villigen PSI, Switzerland

Abstract: Bayesian Belief Nets represent factor relationships in the form of conditional probability distributions (CPDs). The transparency of the CPD assessment is an important element for the acceptability of BBNs. This is especially the case when expert judgment is dominant in CDP assessment, which is often the case in risk analysis and in particular HRA. Unfortunately, research and applications on BBNs have frequently focused on their modeling potential as opposed to the process of building BBNs. This paper deals with this process and examines it for a BBN developed to quantify Errors of Commission (EOCs). The derivation of CPDs is based on introducing weighted functions among the nodes, an approach from the literature. The approach builds the CPDs automatically (ie. by an algorithm) from high-level assumptions on the effect of the factors; this contrasts with approaches in which CPDs for each child node are separately elicited. The assumptions concerning the effects of the factors were determined from operational event analyses in the database of the Commission Error Search and Assessment (CESA) quantification method (CESA-Q). The application shows the feasibility of systematically building a BBN from limited information and identifies some of the research needs related to BBN building and verification.

Keywords: Bayesian Belief Nets, Human Reliability Analysis, Expert Judgment, Errors of Commission.

1. INTRODUCTION

Bayesian Belief Nets (BBNs), a mathematical framework to model probabilistic causal relationships [1], are increasingly raising interest in the Human Reliability Analysis (HRA) field. One reason is their natural ability to represent the joint effect of numerous factors that are possibly correlated and interacting. Another is that they can be built by aggregating heterogeneous sources of information: data and expert judgment of different forms [2]. The applications of BBNs for HRA have addressed different issues. A number of studies have exploited their multi-level modeling to integrate the quantitative treatment of management and organizational factors in HRA, eg. [3-5]. Other contributions proposed BBN versions of existing HRA models, such as SPAR-H [6] and CREAM [7], allowing to introduce additional modeling features, such as interdependent performance shaping factors. Further approaches to integrate cognitive models, field data and expert judgment for the development of a BBN-based HRA model are presented in [2, 8].

With few, notable exceptions [2, 8], the BBNs developed for HRA (and for many other applications in risk analysis) are developed solely from expert judgment. Indeed, their graphical structure and quantification engine are naturally suited to represent expert knowledge about factors and their influences. The most delicate part of the BBN development process is the quantification of the model relationships. In BBNs, these take the form of Conditional Probability Distributions, CPDs. Especially when resorting to expert judgment, care should be taken to avoid different types of biases – as discussed in [9]. Another issue relates to the large number of relationships to be elicited, which can indeed be impractical and potentially lead to inconsistencies [9]. Additionally, the separate elicitation of all relationships may lead to the loss of view of general model properties, e.g. the functional relationships of the factors over their entire range of variability, the overall importance of factors, and group influences. This can be overcome by resorting to algorithms to populate the CPDs: expert judgment is limited to the determination of selected relationships (selected CPDs) and/or to the definition of general tendencies in the factor influences; then, the algorithm populates the CPDs on the basis of the expert input. The application of such algorithms to HRA has been limited. Some examples are [10, 11]; however, the application of the algorithm required the important assumption of independence in the factor influences – a condition that, for HRA models, is often difficult to satisfy.

This paper presents the application of the approach from [12], in which the CPDs are generated based on associating functional relationships between the values of the influencing nodes (parent nodes) and the probability of the influenced nodes (child nodes). The functions allow modeling dominance effects on the parameter (with maximum or minimum values dominating), and therefore allow modeling some degree of dependence among the factors. The functions and their parameters, which will determine the CPDs, are assessed based on the general tendency of the effect of the factors, with no need for direct elicitation of all CPDs. An important difference of the present paper, compared to the original approach of [12], is that the functions and their parameters are determined based on information from a database of experienced EOC events in which the influencing factors have previously been identified by means of expert judgment. The implications of this difference will be discussed shortly in the conclusions of this paper.

The approach is applied for the development of a model for the quantification of Errors of Commission (EOCs), aggravating operator actions in post-initiator accident scenarios. This is an area of HRA where strongly interacting factors are expected to influence the human error probabilities and where practically no quantitative data is available. The EOC quantification model underlies CESA-Q [13], the quantification module of the Commission Error Search and Assessment (CESA) method, a method developed at the Paul Scherrer Institute [14, 15]. In the original version of CESA-Q [13], additional judgment by the analyst is required after the factors are assessed. The need to decrease the element of expert judgment in the application of CESA-Q motivates the adoption of a model-based EOC quantification approach. With a model-based approach, the analyst is only required to assess the input factors of the model; the model, which is the BBN, yields the corresponding error probability. The database of pre-evaluated situations is the CESA-Q database (a set of 26 operational events involving EOCs that have been analyzed and quantified in earlier work [16]).

The paper is organized as follows. Section 2 briefly introduces the CESA-Q method – the detailed method presentation is reported in [13], some recent advances in [17]. The approach to quantify the CPDs (based on [12], with the CESA-Q database providing the information to determine the functional relationships and their parameters) is presented in Section 3. Section 4 compares the predictions of the developed BBN with the results of the database analyses [13]. Of course, this does not represent a validation of the model, given that the database was used for its development. However, the comparison can serve as partial verification of the model response for "known" situations and it allows some conclusions to be drawn concerning the model response.

2. CESA-Q: A METHOD FOR QUANTIFYING ERRORS OF COMMISSION

The CESA method was developed with the focus on identification and prioritization of EOCs [14, 15]. The CESA method includes guidance for the quantitative analysis of EOCs as well as for the assessment of their risk importance; CESA-Q addresses the quantification, emphasizing decision EOCs, i.e. for which the inappropriate action is committed following a motivated decision (so the action is intentionally made, although its inappropriateness is not known).

The features of CESA-Q relevant for the present paper are as follows (refer to [13] for a complete description of the method). The EOC is analyzed in terms of plant- and scenario-specific factors that may motivate inappropriate decisions. Two groups of factors are introduced: situational factors, which identify EOC-motivating contexts (Table 1), and adjustment factors, which refine the analysis of EOCs to characterize how strong the motivating context is – the adjustment factors are: Verification Hints (VH), Verification Means (VM), Verification Difficulty (VD), Verification Effort (VE), Time Pressure (TP), Benefit Prospect (BP), Damage Potential (DP), Personal Redundancy (PR).

In CESA-Q, a distinction is made between the nominal context and multiple "worse-than-nominal" contexts. The nominal context is defined by the scenario (in the Probabilistic Safety Assessment) in which the error is modelled; the worse-than-nominal contexts refer to scenario variants that could lead to more challenging contexts than the nominal one (a scheme to search for worse-than-nominal contexts is provided in [13]). CESA-Q includes two levels of quantification. The first level treats only the nominal context. The second level addresses the search for worse-than-nominal contexts, the quantification of their likelihood, and the analysis and quantification of the error probability for each of the identified contexts. The quantification analysis can be terminated at the first level, depending on the specific EOC's risk significance. If this is done, the obtained EOC probability bounds the results that would be obtained in the more detailed, second-level analysis, which additionally quantifies the worse-than-nominal contexts.

An important element of the CESA-Q quantification is the analysis of the strength of the error forcing impact (of the nominal as well as of the "worse-than-nominal" contexts) on the basis of eight adjustment factors. The strength of the impact is characterized by the so-called reliability index, i, representing the overall belief regarding the positive or negative effects on the EOC probability (the reliability index i is defined from 0 for strongly error-forcing contexts to 5 for contexts with very low EOC probabilities). Table 2 shows the correspondence of the reliability index with the qualitative judgment on the error forcing impact of the context under analysis. The probability of committing the error is related to the reliability index (and therefore to the error-forcing impact) as: $\text{Prob}(\text{EOC}|i) = \exp(-c \cdot i)$, with the constant $c = 1.315$, obtained in [18] via a statistical analysis of operational events.

In its original form [13], the determination of the error forcing impact characterizing a specific context (ie. the corresponding value of the index i) is based on a match-and-adjust approach: it involves comparing the EOC under analysis with entries from the above-mentioned CESA database of operational events. The closest entry in the database provides the reference probability value for the new analysis. Given the limited number of entries in the database, the identification of a close match is indeed rare and guidelines for adjusting the reference are limited. The new concept recently developed for EOC quantification via CESA-Q is based on an explicit model, a BBN [17]. The quantitative relationships underlying the model are informed based on the existing CESA-Q database. The adoption of a model-based approach is expected to reduce the subjectivity in the quantification, because the applicable error probability directly follows from the factor evaluations, without need for additional judgments by the analyst.

Table 1: The CESA-Q situational factors [13]

Situational Factor	Short description
Misleading Indication or Instruction (MII)	An indication or instruction is misleading. It indicates or advices the acceptability or need of an action that is inappropriate under the condition in the scenario
Adverse Exception (AE)	The operators are involved a response strategy. The strategy includes an action that becomes or is inadequate due to an exceptional condition (e.g. an subsequent event or component failure)
Adverse Distraction (AD)	A cue (i.e. an occurrence that draws the operator's attention) arises, which has an association to an action or decision that is inappropriate. The action could be outside the scope of the nominal decision options, or the cue could be different from the key indications referred to in the procedural guidance
Risky Incentive (RI)	The operators have to follow a well-recognized safety rule. The deviation from the rule is associated with a prospect of a notable benefit (such as the prevention of economical loss or delay in plant stabilization), and the rule is precautionary (deviations do not necessarily lead to safety degradations)

Table 2: Correspondence of Error forcing impact, reliability index (i), mean probability of EOC in CESA-Q [13]

Error-Forcing Impact	Extremely high	Very high	High	Low	Very low	None
Reliability index (i)	0	1	2	3	4	5
Mean Prob(EOC | i)	1	2.7e-1	7.2e-2	1.9e-2	5.2e-3	1.4e-3

3. DEVELOPMENT OF THE BBN MODEL

3.1 BBNs

A BBN is a probabilistic graphical model whose structure consists in nodes linked by directed arcs [1]. Nodes represent random variables and arcs between nodes (linking parent nodes to child nodes) indicate causal or influential relationships. Typically, discrete states are associated to each nodes. The quantitative relationships between the nodes are represented by conditional probabilities: each outcome (state) of the child node has a conditional probability given each combination of the states of the parent nodes. The primary use of BBNs is the representation of knowledge and decision support under uncertainty; their application is established in diverse areas such as medical diagnosis and prognosis, engineering, finance, information technology, natural sciences [1]. Generally, a distinction is made between BBN use in data-rich applications (e.g. some medical diagnosis and financial applications) and rare-event applications (typically, risk). In the former, both the BBN structure and the quantitative relationships are learned from the data: the general use of BBNs in data-rich applications is to identify the important factors, their relationships (correlations and causal relationships) and their quantitative influence on the variables of interest. If small data sets are available, the typical approach is to construct the BBN structure with expert judgment and use the available data for quantification of the relationships. In most of the applications dealing with rare events, only expert judgment is available; in these cases, BBNs are used to represents the expert knowledge about factors and their influences.

3.2. BBN structure and node definition

As presented in Section 2, CESA-Q features eight adjustment factors used to characterize the error-forcing impact of a particular context (refer to [13] for their definition, rating scale and guidance). The ratings associated to the factors are the basis for the assessment of the EOC probability given a particular context. Naturally, the BBN has been developed by taking the adjustment factors as the model input nodes (Figure 1). Two intermediate nodes ("Verification" and "Benefit_Damage") have been introduced, modeling the presence of a single overarching group effect that can arise from multiple factors. Indeed, no verification of the inappropriateness of the action can arise because of missing verification hints, or because of the cognitively complex verification activity. The node "Benefit_Damage" models the net effect between the factor 'benefit prospect' (which represents whether the performers perceive that the action has a benefit) and the factor 'damage potential' (which represents whether the performers perceive that the action has a potential for damage).

The introduced intermediate nodes decouple the factors entering the intermediate node and the output node, the Error Forcing Impact (EFI) node: the EFI depends only on the intermediate node entering the EFI node (e.g. Verification), not on the single factors (Verification Hints, Means, Complexity, and Effort). As explained later on, the introduction of intermediate nodes largely decreases the number of conditional probabilities to be determined (note that the introduction of the intermediate nodes should, whenever appropriate, generally be sought when building BBNs, because it helps identifying and visualizing group effects and significantly simplifies the model quantification).

The states of the BBN nodes are defined in Table 3. Note the node states are different from the rating scale presented in [13]: the node states represent the revised definition of the CESA-Q factors of [17], recently developed to improve the evaluation guidance of the CESA-Q adjustment factors. The EFI node has 5 states, corresponding to the CESA-Q reliability indexes 0 to 4 (Table 3).

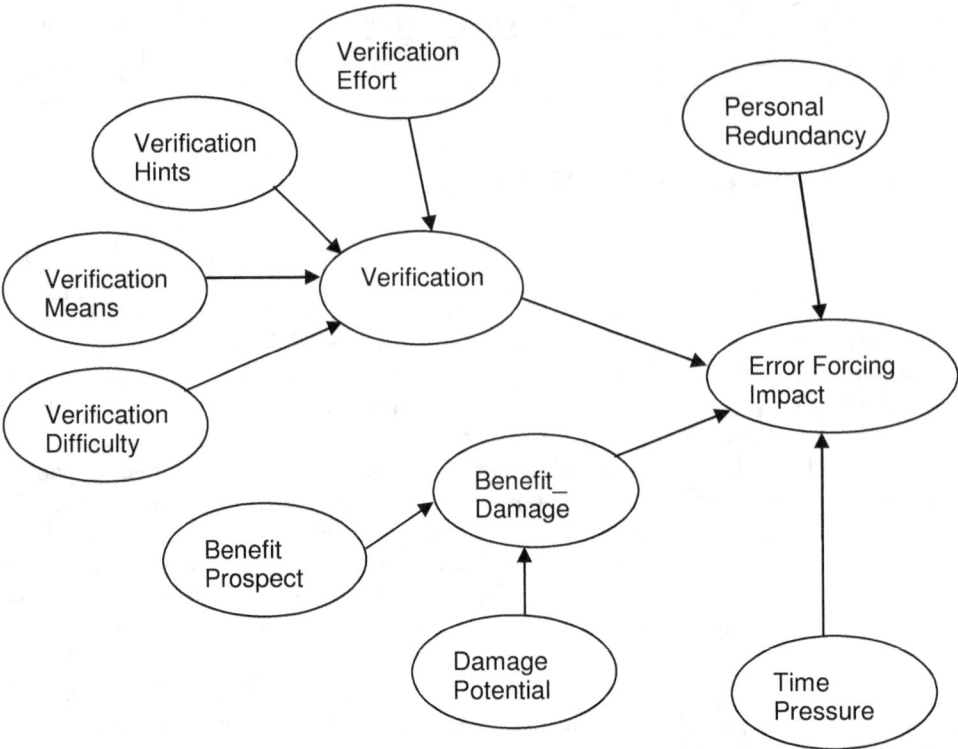

Figure 1: The BBN structure modeling influence of CESA-Q adjustment factors on the EFI (implemented in the software AgenaRisk, http://www.agenarisk.com/)

Table 3: CESA-Q adjustment factors and BBN nodes

CESA-Q factor / BBN node	States	Label in BBN
Verification Hints, Verification Means, Verification Difficulty Time Pressure	0 (error-forcing)	EF
	0.5 (moderately error-forcing)	Mod_EF
	1 (not error forcing)	NEF
Verification Effort Benefit Prospect	0 (error-forcing) and N/A	EF
	1 (not error-forcing)	NEF
Damage Potential Personal Redundancy	0 (not success-forcing)	NSF
	1 (success-forcing)	SF
Verification (intermediate node)	0 (error-forcing)	EF
	0.5 (moderately error-forcing)	Mod_EF
	1 (not error-forcing)	NEF
Benefit_Damage (intermediate node)	0 (error-forcing)	EF
	0.5 (neutral)	Neutral
	1 (success-forcing)	SF
Error forcing impact (output node)	Extremely high	Ex High
	Very high	Very high
	High	High
	Low	Low
	Very low	Very low

3.3. Quantification of the BBN relationships (the CPDs)

For each node, a CPD is associated for each combination of the states of the incoming nodes: 94 conditional probability distributions are required in total, for the developed BBN. For example, $3^3*2 = 54$ distributions quantifying the relationships between factors VH, VM, VD, VE and the intermediate factor Verification. These distributions are of the type:

Prob(Verification = "EF" | VH = "EF", VM = "EF", VD = "EF", VE = "EF")
Prob(Verification = "Mod_EF" | VH = "EF", VM = "EF", VD = "EF", VE = "EF")
Prob(Verification = "NEF" | VH = "EF", VM = "EF", VD = "EF", VE = "EF"),

of course, the above three probabilities sum to 1. To complete the table for the intermediate node Verification, distributions need to be determined for each combination of the states of the VH, VM, VD, VE nodes. Note that in case of no intermediate nodes, i.e. all eight adjustment factors directly connected with node EFI, the number of distributions to be assessed would be $3^4 * 2^4 = 1296$, thus largely complicating the quantification of the model.

For the present application, a BBN with "ranked nodes" was deemed appropriate. Such nodes represent qualitative variables that are abstractions of some underling continuous quantities, typically ranging between 0 and 1 [12]. Indeed, for example, the five states of the output EFI node discretize the underlying continuum of the error-forcing impact. An attractive feature of the "ranked nodes" BBN relates to the determination of its CPDs: the child node's probabilities are derived from a weighted (continuous) function of the parent node values (on the underlying continuous scale) [12]. Therefore, for each node, the each CPD is not elicited separately (for example each of the 54 distributions required to fill up the CPD for node "Verification"), but by choosing the appropriate function, its parameters and the factor weights.

The CPDs are determined based on the underlying doubly truncated normal distribution ("TNormal") on the continuous variables underlying the factor labels, then discretized on the range associated to each label (in the present application the 0 – 1 range is equally split among the labels, i.e. of size 0.2 for output node "EFI", centered in 0.1, 0.3, ..., 0.9). The probability density for each child is the function TNormal(μ, σ), where μ is a weighted function of the input values (on the underlying continuous scale) and σ the standard deviation, representing the degree of uncertainty on the child node value. Four weighted functions are introduced: Mean Average (Wmean), Minimum (Wmin), Maximum (Wmax), Mix of Minimum and Maximum (Wminmax). The approach is implemented in the Software AgenaRisk (http://www.agenarisk.com/).

The choice of the appropriate function depends on the effect of the value of the parent nodes on the child node. As presented in the Appendix, this can be inferred from statements elicited from experts on selected parent-child relationships and possibly other qualitative considerations. Note that, however, this choice requires a number of subjective assumptions be made, i.e. no hard rules connecting elicited information and these functions exist. The evaluations of the operational events (excerpt in Table 4) were the basis for understanding the parent node effects, along with qualitative considerations by the authors of the present paper on the relative importance and effect of the factors.

Note that two different BBN sub-models were developed, one used to represent EOC situations of types "Misleading Indications", "Adverse Exception", "Adverse Distractions" while the second sub-model covers EOC situations of type "Risky Incentive" (the present paper will present only the former). The use of two sub-models was needed to represent the substantial difference between the two groups of factor influences expected for these situational features.

Table 5 reports the data used for building the BBN with the algorithm from [12]. Several considerations entered in the determination of the specific functions and the corresponding weights. For example, function Wmax favors large values of the output (e.g. verification towards 1, i.e. error forcing, EF) in case of large values of at least one of the inputs (at least one of the inputs towards 1, i.e. EF): in other words, at least one input being error-forcing leads verification being error-forcing. The weights represent the importance of each factor. For example, for node verification, the value of "Verification Hints" has been given the highest importance: reasonably, the presence and quality of the hints are very important to the error probability. For the node "Benefit_Damage" the combined use of both functions Wmin and Wmax (Table 5) favors small values of the output (i.e. Benefit_Damage towards 0, i.e. "success forcing") in case of small values of input "Damage Potential" ("Damage Potential" towards 0, i.e. SF) and large values of the output (i.e. Benefit_Damage towards 1, i.e. "error forcing") in case of large values of input "Damage Potential" ("Damage Potential" towards 1, i.e. NSF): this allows to model on the same node "Benefit_Damage", the success forcing effect of "Damage Potential" and the error forcing effect of "Benefit Prospect". The mathematical formulation of these functions is reported in the Appendix. Generally, besides qualitative considerations on the factor importance, the function weights were tuned after several trial-and-error attempts to reproduce as closely as possible the EOC event evaluations from [13]. Coverage of these events will be returned to in the next Section 4.

Concerning the value of the standard deviation σ of the TNormal function, this represents the uncertainty in the value of the child nodes (represented by the shape of the CPDs) given the values of the parent nodes. The approach developed in [19] has been used, which aims at formally aggregating expert estimates on human error probabilities and provides the maximum confidence that can be given to each operational event evaluation. Operatively, in the development of the BBN model, the parameter σ was set to the value (Table 5) such that generally the standard deviation of the distribution of the EFI would not be higher than the limiting values provided in [19].

4. VERIFICATION OF THE DEVELOPED BBN MODEL

This section presents the response of the BBN CESA-Q sub-model on the operational events: this allows evaluating how well the model reproduces the "known" results, and with which level of confidence. Figure 2 compares the BBN predictions in terms of the reliability index i, with the reliability index assessed in [13]; Figure 3 addresses the BBN predictions in terms of the conditional EOC probability. The operational cases are shown from left to right in decreasing order of the assessed i from [13]. The figures also include the BBN predictions on extreme cases, all positive factors ("All_pos", i.e. with all factors assessed as "not error forcing" or "success forcing") and all negative factors ("All_neg", i.e. with all factors assessed as "error forcing" or "not success forcing"). The Figures show the means and the 25th and 75th percentiles of the predicted i's on the continuous variables underlying the BBN ranked node (a linear relationship is established between the BBN output variable and the reliability index, ranging from 4 to 0 as the BBN output ranges from 0 to 1, respectively).

First, the generally decreasing trend of the BBN predictions from left to right suggests that the BBN is able to represent and distinguish the increasing impact of the error forcing conditions across the events, ranging within different levels, from low impact (i around 3-4) to high impact (i around 1-0). Then, the assessments from [13] are within the 25th and 75th percentile bounds for all events with intermediate levels of error forcing impact (low, i=3 to very high, i=1). For the extreme levels very low (i=4) and extremely high (i=0), overestimation and underestimation of the error forcing impact are observed, respectively. While the underestimation of extremely high error forcing impact is certainly an issue for the use of the model in practical PSA applications, the relationships that correspond to these (and similar) combinations of input factors can be easily modified (manually) to represent the higher impact. A brief discussion of these results is presented in the next section.

Table 4: Excerpt of the CESA-Q database of 26 operational events involving EOCs [13]

Case ID	Event Title	VH	VM	VD	VE	TP	BP	DP	PR	i	p(EOC|i)
AE.2	Fire and Loss of Offsite Power (Diablo Canyon 1, 1995)	1	1	1	0	1	1	1	1	2	7.2E-2
AE.4	Loss of Coolant through RCS Hot Leg (Oconee 3, 1991)	1	0.5	0.5	0	1	1	0	0	2	7.2E-2
AE.5	Loss of Coolant through RHR Discharge Isolation Valve (Wolf Creek, 1994)	0	0.5	0.5	0	1	1	0	0	0	1.0
MI.2	Loss of Coolant through Faulted Steam Generator (Ginna, 1982)	0.5	1	0.5	1	0.5	0	0	1	1	2.7E-1
MI.3	Reactor Overheating due to Degradation of Safety Injection (Ft. Calhoun, 1992)	0.5	1	0.5	1	0.5	1	0	1	2	7.2E-2
MI.4	Core Damage due to Termination of Safety Injection (TMI 2, 1979)	0	0.5	0.5	1	0	0	0	1	1	2.7E-1
AD.2	Damage of High Pressure Injection Pumps (Oconee 3, 1997)	0.5	0.5	0.5	1	0.5	1	0	1	2	7.2E-2

The eight CESA-Q adjustment factors: VH: Verification Hints, VM: Verification Means, VD: Verification Difficulty, VE: Verification Effort, TP: Time Pressure, BP: Benefit Prospect, DP: Damage Potential, PR: Personal Redundancy.

Table 5: Quantification of the BBN relationships: data for the application of the algorithm in [12] (BBN Applicable for situational features AD, AE, MI)

BBN Node	Function	Weights
Verification	WMax	$w_{VH} = 5$, $w_{VM} = 2.5$; $w_{VD} = 2.5$; $w_{VE} = 1$; $\sigma^2 = 2e\text{-}2^{(2)}$
Benefit_Damage	WMin for BP=NEF	$w_{BP} = 5$; $w_{DP} = 5$; $\sigma^2 = 2e\text{-}2$
	WMax for BP=EF	$w_{BP} = 5$; $w_{DP} = 5$; $\sigma^2 = 2e\text{-}2$
EFI	WMax	$w_V = 5$, $w_{TP} = 2.5$; $w_{B\_D} = 2.5$; $w_{PR} = 2.5$; $\sigma^2 = 2e\text{-}2$

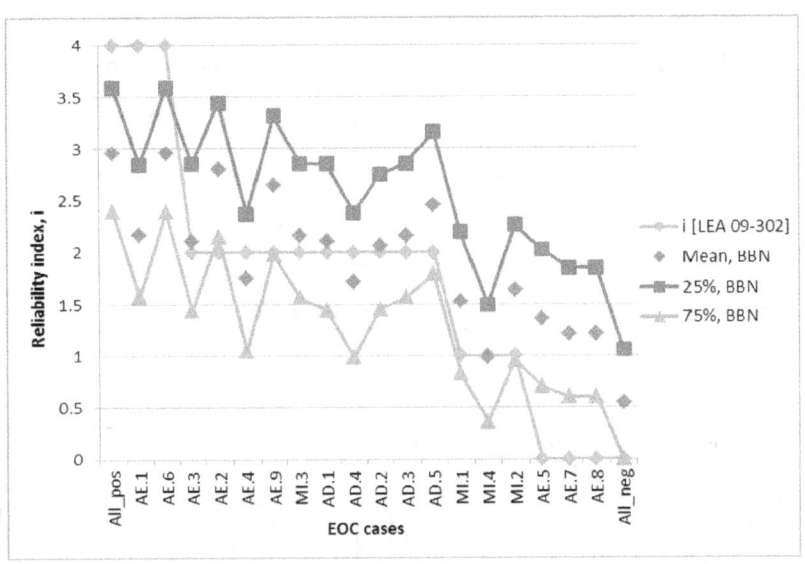

Figure 2. Comparison of the predictions (reliability index i with confidence bounds) of the BBN sub-model with the operational events from [13] (LEA 09-302 in the figure) (sub-model for "Misleading Indications": MI, "Adverse Exception": AE, "Adverse Distractions": AD).

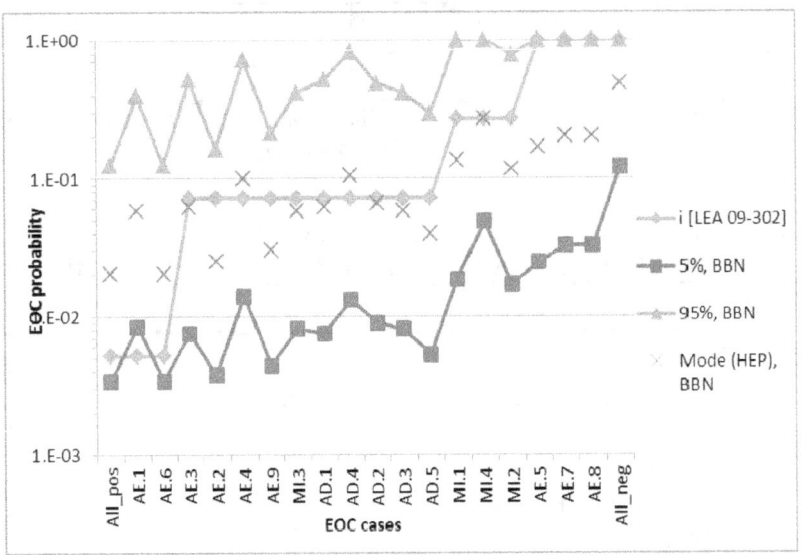

Figure 3. Comparison of the predictions (EOC probability with confidence bounds) of the BBN sub-model with the operational events from [13] (LEA 09-302 in the figure) (sub-model for "Misleading Indications": MI, "Adverse Exception": AE, "Adverse Distractions": AD).

4. DISCUSSION AND CONCLUSIONS

This paper has focused on an important aspect of the use of BBN in data-poor applications such as risk analysis and, in particular HRA: the derivation of the quantitative model relationships. When expert judgment is the main (or sole) source of information, it is important to limit the effort required to the experts. Indeed, eliciting all relationships can be impractical and may lead to inconsistencies; also, one may lose track of the overall model properties. The transparency of the elicitation process is also one important element for the acceptability of BBNs.

The paper has presented the application of an approach from the literature for the assessment of the BBN relationships, based on associating weighted functions relating the parent and the child nodes

[12]. The approach allows building the CPDs automatically (ie. by an algorithm) based on the general tendency of the effect of the factors – in contrast to the approach in which all CPDs need to be separately elicited. An attractive feature of the approach is that the general tendency can be informed by statements by experts on specific combinations of the parent node states (e.g., of the type "if factor 1 is low and factor 2 is high, then the output is high). In this paper, the tendencies used to build the CPDs are those observed in a database of pre-evaluated situations (by expert judgment).

The approach has been applied in the HRA domain to the development of a model-based (BBN-based) version of the CESA-Q method for quantifying EOCs, currently under development. The adoption of the model-based EOC quantification is generally motivated by the need to decrease the element of judgment required of analysts in the application of the original CESA-Q. In the model-based approach, the analyst is only required to assess the input factors of the model (the corresponding error probability is produced by the model). The database of pre-evaluated situations is the CESA-Q database (a set of 26 operational events including EOCs analyzed and quantified in earlier work).

The use made in the present paper of the database of pre-evaluated situations to inform the model relationships (compared to the expert statements suggested by the original BBN building approach formulation) is expected to enhance the traceability of the model development. Indeed, the database evaluations are independent on the BBN building approach and can be reviewed and accepted by experts external to those developing the models, thus providing the established foundation on which the model should build.

While the use of the independent database is expected to enhance the model development traceability, the conversion of the information from the database into the choice of appropriate weighted functions and of the values of their parameters can be quite subjective (no hard rules connecting elicited information and these functions exist). The approach used in the present paper to compare the BBN output with the CESA-Q database analyses can allow a partial verification of the model (because of the independence of the database analyses with the model development process) and provide some confidence on the chosen functions and their parameters. Of course, the question remains of evaluating the model prediction outside the input combinations addressed by the database. To improve confidence on the response of the model for these combinations as well, it would be beneficial to address future research to establish guidance (or, ideally, some level of automation) in the conversion of the database information into model relationships.

The comparison of the model predictions on the operational events has shown that the BBNs are able to represent and distinguish the increasing impact of the error-forcing conditions across the events, ranging within different levels, from low to high impacts. For the extreme levels 'very low' and 'extremely high', overestimation and underestimation of the error forcing impact are observed, respectively. The relationships that correspond to these (and similar) combinations of input factors can be easily modified (manually) to represent the higher impact. However, the latter represents an additional, subjective intervention on the model, which may be avoided by using alternative functional relationships which represent the expected effect of HRA factors over their entire range.

Finally, it is worth noting that, while being promising for the HRA domain, the approach for BBN building used in the present paper is one among different alternatives. Given the mentioned importance of building BBNs while limiting the information required from the experts, it seems worthwhile to systematically investigate the attractiveness of these alternatives for their application to HRA, addressing different aspects such as the type and amount of information required from the expert, handling of uncertainties, possible limitations on the number of node states.

Acknowledgements

This work was funded by the Swiss Federal Nuclear Safety Inspectorate (ENSI), under DIS-Vertrag Nr. 82610. The views expressed in this article are solely those of the authors.

References

[1] F. V. Jensen, T. D. Nielsen. "*Bayesian Network and Decision Graphs*". Springer science, 2007, New York, NY, USA.

[2] K.M. Groth, A. Mosleh. "*Deriving causal Bayesian networks from human reliability analysis data: A methodology and example mode*". Proceedings of the Institution of Mechanical Engineers, Part O: Journal of Risk and Reliability, 226(4), pp. 361-379 (2012).

[3] Z. Mohaghegh, R. Kazemi, A. Mosleh. "*Incorporating organizational factors into Probabilistic Risk Assessment (PRA) of complex socio-technical systems: A hybrid technique formalization*". Reliability engineering and System safety, 94(5), pp: 1000-1018 (2009).

[4] J. E. Vinnem et al. "*Risk modelling of maintenance work on major process equipment on offshore petroleum installations*". Journal of Loss Prevention in the Process Industries, 25(2), pp: 274-292 (2012).

[5] P. Trucco, E. Cagno, F. Ruggeri, O. Grande. "*A Bayesian Belief Network modelling of organisational factors in risk analysis: A case study in maritime transportation*". Reliability Engineering and System Safety, 93(6), pp: 845-856 (2008).

[6] K.M. Groth, L.P. Swiler. "*Bridging the gap between HRA research and HRA practice: A Bayesian network version of SPAR-H*". Reliability Engineering and System Safety, 115, pp. 33-42 (2013).

[7] M.C. Kim, P.H. Seong, E. Hollnagel. "*A probabilistic approach for determining the control mode in CREAM*". Reliability Engineering and System Safety, 91(2), pp. 191-199 (2006).

[8] R. Sundarmurthi, C. Smidts. "*Human reliability modelling for Next Generation System Code*". Annals of Nuclear Energy, 52, pp. 137-156 (2013).

[9] R.M. Cooke. "*Experts in uncertainty*". New York: Oxford university press (1991).

[10] M. R. Martins, M. C. Maturana. "*Application of Bayesian Belief networks to the human reliability analysis of an oil tanker operation focusing on collision accidents*". Reliability Engineering and System Safety, 110, pp: 89-109 (2013).

[11] B. Cai, Y. Liu, Y. Zhang, Q. Fan, Z. Liu, X. Tian. "*A dynamic Bayesian networks modelling of human factors on offshore blowouts*". Journal of Loss Prevention in the Process Industries, 26, pp: 639-649 (2013).

[12] N. E. Fenton, M. Neil, J. G. Caballero. "*Using ranked nodes to model qualitative judgments in Bayesian networks*". IEEE Transactions on Knowledge and Data Engineering, 19(10), p: 1420–1432 (2007).

[13] B. Reer. "*Outline of a Method for Quantifying Errors of Commission*", LEA 09-302, Paul Scherrer Institut, Villigen PSI, Switzerland (2009).

[14] B. Reer, V. N. Dang, S. Hirschberg. "*The CESA method and its application in a plant-specific pilot study on errors of commission*". Reliability Engineering & System Safety, 83(2), pp: 187-205 (2004).

[15] L. Podofillini, V.N. Dang, O. Nusbaumer, D. Dres. "*A pilot study for errors of commission for a boiling water reactor using the CESA method*", Reliability Engineering & System Safety, 109, pp: 86-98 (2013).

[16] B. Reer, V. N. Dang. "*Situational Features of Errors of Commission Identified from Operating Experience*". LEA 09-303, Paul Scherrer Institut, Villigen PSI, Switzerland, Villigen PSI, Switzerland (2009).

[17] L. Podofillini, V.N. Dang. "*CESA-Q: a method for quantifying Errors of Commission Enhanced method guidance and its technical basis*", LEA 13-302, Paul Scherrer Institute, Villigen PSI, Switzerland (2013).

[18] B. Reer. "*An Approach for Ranking EOC Situations Based on Situational Factors*". LEA 09-304, Paul Scherrer Institut, Villigen PSI, Switzerland, Villigen PSI, Switzerland (2009).

[19] L. Podofillini, V.N. Dang. "*A Bayesian Approach to Treat Expert-Elicited Probabilities in Human Reliability Analysis Model Construction*". Reliability Engineering & System Safety, 117, pp. 52-64 (2013).

APPENDIX

The appendix presents some details of the approach used for derivation of the BBN CPDs. For a more comprehensive treatment, see [12]. The CPDs of the child node are derived associating a doubly truncated normal distribution ("TNormal") to the continuous variable underling the factor labels and discretizing it on the range associated to each label. Then, the probability density for each child is the function TNormal(μ, σ), where μ is a weighted function of the input values (on the underlying continuous scale) and σ the standard deviation, representing the degree of uncertainty on the child node value. Four weighted functions are introduced: Mean Average (Wmean), Minimum (Wmin), Maximum (Wmax), Mix of Minimum and Maximum (Wminmax). The decision of which function to use depends on the effect of the parent nodes values on the child node value. As presented in [12], this can be inferred from limited information, e.g. specific evaluations in correspondence of combinations of input values (typically, the cases where the nodes have their extreme states). For example, in the case of one child node Y with two parents X_1 and X_2, if the following statements are elicited from experts [12]:

- when X_1 and X_2 parent nodes are both 'very high' the distribution of Y child node is heavily skewed toward 'very high',
- when X_1 and X_2 parent nodes are both 'very low' the distribution of Y child node is heavily skewed toward 'very low',
- when X_1 is 'very low' and X_2 is 'very high' the distribution of Y is centered below 'medium',
- when X_1 is 'very high' and X_2 is 'very low' the distribution of Y is centered above 'medium',

then it is appropriate to use the weighted average function (with possibly different importance weights for the two parents). A simple weighted sum model is used to measure the contribution of each parent node to explaining the child node as a 'credibility weight'. The higher the credibility value, the higher the correlation between the parent node and the child node. The weights are derived from judgment. Mathematically, for child node Y, having $X = \{X_1, X_2, ..., X_n\}$ causal ranked nodes as parents and each X_i parent node having w_i contribution weight, the TNormal distribution with weighted mean average function will have the following form:

$$p(Y|X) = TNormal\left[\frac{\sum_{w_i=1}^{n} w_i \cdot X_i}{\sum_{w_i=1}^{n} w_i}, \sigma, 0, 1\right]$$

As mentioned before, also other weighted rank node functions can be used to derive the probability values in CPDs. The following observation, for example, will lead to the use of weighted minimum function:

- When X_1 and X_2 parent nodes are both 'very high' the distribution of Y child node is heavily skewed toward 'very high'.
- When X_1 and X_2 parent nodes are both 'very low' the distribution of Y child node is heavily skewed toward 'very low'.
- When X_1 is 'very low' and X_2 is 'very high' the distribution of Y is centred toward 'very low.
- When X_1 is 'very high' and X_2 is 'very low' the distribution of Y is centred toward 'low'.

The corresponding function will have the following form:

$$p(Y|X) = TNormal\ [Wmin, \sigma, 0, 1]$$

With Wmin:

$$Wmin = \min_{i=1,...,n} \left[\frac{w_i \cdot X_i + \sum_{j=i}^{n} X_j}{w_i + (n-1)}\right]$$

If all the weights are large, then Wmin is close to the minimum value of the inputs, and if all the weights are 1, then Wmin is the average of the parent nodes (Wmean). Mixing the influence of the weights gives result between MIN and AVERAGE. Function Wmax operates analogously.

Task Decomposition in Human Reliability Analysis

Ronald L. Boring* and Jeffrey C. Joe
Idaho National Laboratory, Idaho Falls, Idaho, USA

Abstract: In the probabilistic safety assessments (PSAs) used in the nuclear industry, human failure events (HFEs) are determined as a subset of hardware failures, namely those hardware failures that could be triggered by human action or inaction. This approach is top-down, starting with hardware faults and deducing human contributions to those faults. Elsewhere, more traditionally human factors driven approaches would tend to look at opportunities for human errors first in a task analysis and then identify which of those errors is risk significant. The intersection of top-down and bottom-up approaches to defining HFEs has not been carefully studied. Ideally, both approaches should arrive at the same set of HFEs. This question remains central as human reliability analysis (HRA) methods are generalized to new domains like oil and gas. The HFEs used in nuclear PSAs tend to be top-down—defined as a subset of the PSA—whereas the HFEs used in petroleum quantitative risk assessments (QRAs) are more likely to be bottom-up—derived from a task analysis conducted by human factors experts. The marriage of these approaches is necessary in order to ensure that HRA methods developed for top-down HFEs are also sufficient for bottom-up applications.

Keywords: HRA, human failure events, task decomposition, error of commission, latent error.

1. INTRODUCTION

Human reliability analysis (HRA) is a systematic approach to identify sources of human error and quantify their likelihood. As depicted in Boring (2009), at a high level, HRA may be seen as having three distinct phases (see Figure 1). The first phase is to identify possible human errors and their contributors. The second phase is to determine how these errors may be modelled in terms of an overall risk model such as a probabilistic safety assessment (PSA), which includes the interplay of hardware failures and human errors. Finally, these sources of human error are quantified to produce a human error probability (HEP) and accompanying uncertainty bounds. In practice, identifying and modelling may be considered qualitative aspects of HRA, while the final phase is associated with quantification.

Figure 1: Three General Phases of Human Reliability Analysis (Boring, 2009).

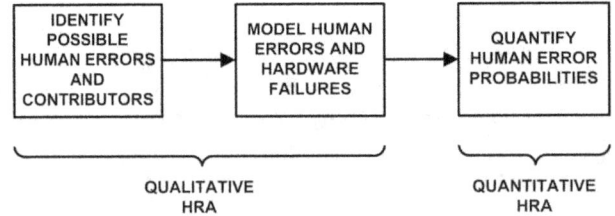

Different HRA methods have been developed to address these three phases, and some methods focus primarily on individual phases. Surprisingly, the area that remains least addressed across various HRA methods is the middle phase—modeling. This paper summarizes available guidance on modeling in HRA in support of oil and gas applications. As HRA is refined and subsequently more fully

---

* Ronald.Boring@inl.gov

incorporated into the safety and risk analyses performed in the oil and gas industry, it is imperative that all phases of HRA are thoroughly addressed.

HRA methods do not have a consistent level of task decomposition at the modeling phase. This lack of consistency can result not only in different qualitative analyses but also different HEPs. The level of task decomposition affects the dependency between tasks, which may have a further effect in driving the HEP. The issue is not that different HRA methods necessarily produce different results for the same HFE; rather, different HRA methods may decompose the HFE to different levels. Thus, the quantification of the same HFE may entail different assumptions and, to some extent, different groupings of tasks across HRA methods. In other words, because of a lack of a common task decomposition framework, HRA methods may not be using the same unit of analysis when producing the HEP.

For example, the European *Human Factors Reliability Benchmark Exercise* (Poucet, 1989), also referred to as the "Ispra Study" within the HRA community, demonstrates how central this topic is to HRA. The benchmark featured three phases of analysis to compare HRA methods. Each successive phase served to further bound the level of decomposition that defined the HFE. The first phase included identification of HFEs and their quantification by different analysis teams. Because different HFEs were identified across methods, it was difficult to compare method results directly. The second phase involved a more explicit definition of the HFEs to ensure the analysis teams quantified the same HFE. Even with a commonly defined HFE, there was considerable variability in how analysis teams modeled the HFE. Differences in task decomposition played a significant role in the differences of the HEPs for the HFEs. Some analysis teams decomposed to a finer level, resulting in lower HEPs. However, the dependencies between HFEs were not well accounted for in the analyses with finer grained task decomposition, resulting in unrealistically low HEP values in the authors' opinion. As such, a third phase was conducted, this time with an explicit decomposition of tasks and a common HRA event tree used in quantification.

The purpose of this paper is to review existing guidance on modeling human error in HRA and synthesize the disparate guidance into a simple framework that can be used in support of HRA in petroleum applications. The goal of establishing a common framework for human error modeling is to eliminate potential sources of variability in HEP quantification across methods. This paper presents initial insights derived from a literature review of applicable sources. Additional guidance will be developed and reported in the future.

2. DEFINING HUMAN ERROR

2.1 Human Error and Human Failure Events

HRA depicts a cause and effect relationship of human error. The *causes* are typically catalogued in terms of qualitative contributions to a human error, including the processes that shaped that error and the failure mechanisms. The processes—cognitive, environmental, or situational—that affect human error are typically referred to as performance shaping factors (PSFs). The resultant *effect* is the manifestation of human error—often called the failure mode. This failure mode is treated quantitatively and has an associated failure probability, the HEP.

The term *human error* is often considered pejorative, as in suggesting that the human is in him- or herself the cause of the failure mode (Dekker, 2006). This belies the current accepted understanding that human error is the product of the context in which the human operates. In other words, it is not the human as the ultimate cause of the error but rather the failure mechanisms that put the human in a situation in which the error is likely to occur. The colloquial term, human error, is further challenged in that a human error may manifest but have little or no risk consequence. Human errors may be recovered or may simply not have a direct effect on event outcomes. Such risk insignificant occurrences are typically screened out of the HRA model.

Thus, to denote a risk significant human error, the term *human failure event* (HFE) has been posited. According to the American Society of Mechanical Engineers (ASME), a human failure event is "a basic event that represents a failure or unavailability of a component, system, or function that is caused by human inaction, or an inappropriate action" (2009). The HFE is therefore the basic unit of analysis used in PSA to account for HRA. While an HFE may be incorporated as a simple node in a fault tree or a branch in an event tree, the documentation supporting the HFE represents an auditable holding house for qualitative insights used during the quantification process. These insights may be simple to detailed, depending on the analysis needs and the level of task decomposition.

In PSAs used in the nuclear industry, as per the ASME definition, HFEs are determined as a subset of hardware failures, namely those hardware failures that could be triggered by human action or inaction. This approach is top-down, starting with hardware faults and deducing human contributions to those faults. Elsewhere, there is a bottom-up approach. More traditionally human factors driven approaches would tend to look at opportunities for human errors first in a task analysis and then model them in terms of potential for affecting safety outcomes. The order of identifying vs. modeling HFEs as shown in Figure 1 may be seen as changing depending on the approach. A top-down approach would tend to model the opportunity for HFEs and only then identify the sources of human error. In contrast, a bottom-up approach would first identify sources of human error and then model them in the PSA.

The intersection of top-down and bottom-up approaches to defining HFEs has not been carefully studied. Ideally, both approaches should arrive at the same set of HFEs. This question is crucial, however, because the HFEs used in nuclear PSAs tend to be top-down—defined as a subset of the PSA hardware faults—whereas the HFEs used in petroleum QRAs are more likely to be bottom-up—derived from a task analysis conducted by human factors experts. The marriage of these approaches is necessary in order to ensure that HRA methods developed for top-down HFEs are also sufficient for bottom-up applications. Figure 2 depicts the top-down and bottom-up approaches to defining HFEs. As can be seen, it is possible that both approaches arrive at the same solution. However, the solution set for the top-down and bottom-up approaches should be seen in terms of two circles in a Venn diagram. The problem is not that the HFEs may indeed overlap; the problem is that these HFEs may not always be identical.

Figure 2: Two approaches to defining human failure events.

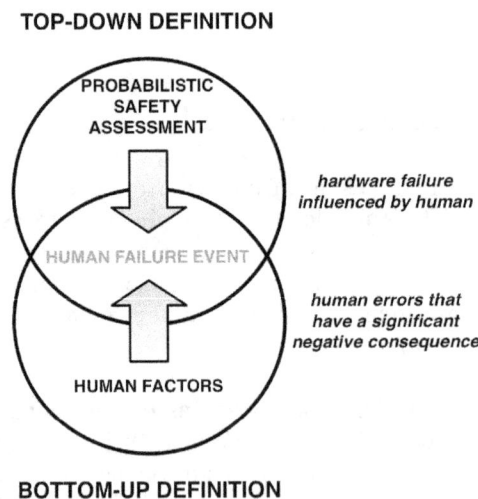

Additionally, some HFEs used in a petroleum context are derived from barrier analysis and are prospective in nature, designed to identify how the defense in depth of a system may be increased to ensure the safety of a system to be built. This approach may emphasize the evolving timescale of barrier effectiveness, whereas most conventional PSAs represent a static snapshot of an HFE. The barrier analysis approach is rarely used in contemporary PSAs for the nuclear industry. Additional guidance will be necessary to link the human factors processes for identifying vulnerabilities with the PSA fault modeling in HRA (Boring and Bye, 2008).

2.2 Limitations of the Top-Down Approach

As depicted in Figure 2, there are areas covered in the bottom-up approach that are not necessarily covered by the top-down approach (and vice versa). In this section, we discuss two noted shortcomings of the traditional top-down approach to defining HFEs—namely, errors of commission and latent errors, neither of which is adequately accounted for in traditional PSAs. We argue that the bottom-up approach provides better opportunity to incorporate these commonly omitted types of human error.

As noted, the top-down approach to defining HFEs begins by modeling those hardware systems that can fail and whose failure can be influenced by human actions or inactions. For example, if a particular electrical bus is a risk significant vulnerability to the overall system safety, the risk analyst would identify the failure of the bus as the starting point. He or she would next determine if the system is controlled by human operators. If yes, and if the human action is a significant subset of the overall risk of the bus failure, an HFE is modeled. The risk analyst must then determine what types of human errors are possible. This is often accomplished by referencing operating procedures and identifying which steps could be performed incorrectly. It is easier to identify a failure to execute particular required procedural steps than it is to postulate all the possible deviation paths the operator could follow that aren't encompassed by the procedure. In other words, the steps omitted (i.e., errors of omission) are more readily modeled than extra steps performed beyond the procedures (i.e., errors of commission). Thus, the top-down approach has exhibited far greater success in including relevant errors of omission than in anticipating possible errors of commission.

Already in one of the key early HRA textbooks, Gertman and Blackman (1994) elaborated on how the HRA methods of that time did not account for errors of commission very well, particularly ones that are more cognitive in nature. That is, while the earliest HRA method, THERP (Swain & Guttmann, 1983), provided failure rates for manual control actions (e.g., simple, skill-based tasks which can include errors of commission that can be quantified), there was and is still no widely accepted approach that can account for errors of commission that fall outside of slips and lapses (and the PSFs that influence these error types). To model errors of commission that are more cognitively based (i.e., not skill-based manual control actions), Gertman and Blackman state that the practice at that time was to quantify errors of commission using simplified commission models (e.g., selection errors), or to use screening values to estimate a crew's probability of successfully diagnosing an event (e.g., SHARP – Hannaman & Spurgin, 1984; ASEP – Swain, 1987). Yet, we find that these methods still do not provide enough specific and useful guidance to help come up with an actionable approach to bridge this gap between the top-down and bottom up approaches.

Straeter, Dang, Kaufer, and Daniels (2004) argued that first generation HRA methods were not effective in characterizing, predicting, and preventing accidents such as Three Mile Island, which featured significant errors of commission. They argued further that this failure was because first generation methods did not have the necessary human performance data, a way of representing errors of commission that are not simple, skill-based manual control actions, nor a methodological framework designed to handle these kinds of errors of commission. Second generation methods were developed in part to address these shortcomings in first generation methods, but somewhat ironically, there was also increased pressure to have these second generation methods be less labor intensive in their methodology, which often meant that some of the more time consuming aspects of HRA were truncated or eliminated altogether (e.g., no explicit guidance in many newer methods on how to define HFEs). The result is that many second generation HRA methods still struggle with how to handle errors of commission that are more cognitively based.

Work by Reer, Dang, and Hirschberg (2004); Reer and Dang (2007); and Podofillini and Dang (2012) on the Commission Errors Search and Assessment (CESA) method has been an important and proactive step in helping address a part of this errors of commission issue in HRA. These researchers have developed CESA as a systematic approach to, "Identify potentially risk-significant EOCs [errors of commission], given an existing PSA [probabilistic safety analysis]." (Reer, Dang, & Hirschberg, 2004; pg. 189). One point that these researchers emphasize is that, given their focus on identifying errors of commission that are easier to quantify, CESA provides a straightforward and streamlined approach to determining which errors of commission are important to consider in PSA. Specifically, CESA focuses on *active* errors of commission that operators make while following control room procedures, rather on the more nebulous *latent* errors of commission managers and designers could make in decision-making, which are often more difficult to quantify because they are determined by multiple distal factors that are further influenced by transient contextual circumstances.

We believe there is an important distinction to make between *active* errors of commission and *latent* errors of commission. Reason (1990) defines active and latent errors as follows:

> In considering the human contribution to systems disasters, it is important to distinguish two kinds of error: active errors, whose effects are felt almost immediately, and latent errors, whose adverse consequences may lie dormant within a system for a along time, only becoming evident when they combine with other factors to breach the system's defences....In general, active errors are associated with the performance of 'front-line' operators of a complex system: pilots, air traffic controllers, ships' officers, control room crews and the like. Latent errors, on the other hand, are most likely to be spawned by those whose activities are removed in both time and space from the direct control interface: designers, high-level decision makers, construction workers, managers and maintenance personnel. (pg. 173)

Woods, Johannesen, Cook, and Sarter (1994) and Hollnagel (1998) also identified this very strong correlation between active errors and errors that occur at the "sharp end" of the work production processes, and latent errors and errors occurring at the "blunt end" of the process. More importantly, with respect to this paper, Reason (1990) and others have repeatedly made the point that latent errors—both latent errors of commission and latent errors of omission—are more risk-significant than active errors.

The analysis of significant operating events at commercial nuclear power plants (e.g., NUREG/CR-6753, 2001) further supports the conclusions of Reason (1990), Gertman and Blackman (1994), and Woods, Johannesen, Cook, and Sarter (1994). NUREG/CR-6753 used data from the U.S. Nuclear Regulatory Commission's (NRC) Accident Sequence Precursor (ASP) Program and the Human Performance Events Database (HPED) to identify safety significant events in which human performance contributed to changes in risk. The sensitivity analyses performed using these data showed that human performance contributed significantly to analyzed events. In particular, two hundred and seventy human errors were identified in the events reviewed, and multiple human errors were involved in every event. More importantly, latent errors (i.e., failures to correct known problems and errors committed during design and maintenance activities) were present four times more often than were active errors. These results confirm the assertion that latent errors contribute significantly to risk-significant events.

The *Deepwater Horizon* accident is one notable example from the petrochemical industry on how a top-down approach to identifying HFEs can miss important human errors. We analyzed this event in greater detail for this paper for two reasons. First, it is an event that occurred in the petrochemical industry, and therefore has more relevance for this paper than other events that have occurred in the nuclear industry and elsewhere (e.g., transportation). Second, other significant events, such as Three Mile Island and Chernobyl have been analyzed extensively relative to the *Deepwater Horizon* accident.

While there were many contributing factors that were responsible for the *Deepwater Horizon* accident, analyses of the events leading up to the accident clearly show that latent errors of commission played a significant role. That is, while multiple safety systems on the *Deepwater Horizon* offshore oil-drilling rig actively failed on 20 April 2010, they failed in large part because of latent errors of commission. There were numerous errors in human decision making whereby information on the system's state that were leading indicators of impending problems were misinterpreted, discounted, or ignored, and the decision to proceed with operations was made. These decision-making errors were at both the operations (i.e., tactical) and management (i.e., strategic) level, resulting in the largest oil spill in U.S. history.

With respect to operations related decision-making errors, the well that had been drilled into the Macondo Prospect oil field had a blowout preventer (BOP) on its wellhead, but according to various news reports and event investigation reports, operations continued even though there were components on the BOP that were known to be damaged. Namely, the annular (i.e., a rubber donut-shaped gasket that closes around the drilling pipe to seal the well) and the control pods, which contain the instrumentation and control systems for the BOP, were damaged and/or not fully functional prior to the event. In the case of the annular, it had become "stripped" a month before the accident. That is, it was intentionally closed around the pipe, and the pipe was moved up and down to strip over 40 joint tools attached to it. When the stripping was completed, according to the Deepwater Horizon Study Group (2011) report, witnesses reported seeing pieces of rubber in the drilling fluid that had risen to the top of the well. According to the National Research Council (2011) report, the annular was apparently, "Untested for integrity afterwards. Annulars are often unable to seal properly after stripping" (pg. 52). With respect to the control pods, there was a primary and backup control pod installed on the BOP, but according to the American television newsmagazine program, *60 minutes* (CBS News, 2011), management and operations were aware that one of the control pods was not fully functional, meaning there were known potential issues with its reliability if it were to be called upon to actuate important safety functions in the event of an emergency.

With respect to management related decision-making errors, according to Bronstein and Drash (2010), on the day *Deepwater Horizon* event occurred, drilling operations were behind schedule by approximately five weeks. To hasten progress in finishing the well, as three concrete plugs were being placed in the column of the well, management from British Petroleum decided to use seawater to keep the pressure from the oil well under control instead of "drilling mud." Others, including the drilling rig's chief driller, challenged this decision. It was also reported by Bronstein and Drash that workers on the oil drilling platform had a tacit understanding that they could get fired or face other chilling effects for bringing up safety concerns that would delay progress on drilling operations, implying that there were problems with the organization's safety culture.

Clearly, there were both tactical and strategic errors of commission that contributed to the *Deepwater Horizon* accident. Whittingham (2004) presents another way of characterizing errors of commission that differs from the tactical and strategic delineation used above. Whittingham proposed that errors of commission with significant cognitive aspects often have errors: 1) at the organizational and management level, 2) in the design, and 3) in maintenance activities. Using Whittingham's characterization scheme reveals additional insights into the nature of the errors of commission committed prior to this event.

Namely, there were organizational and management errors in that management made decisions to put profit over safety. Design errors also contributed to this event in that the BOP did not have a well designed component called the blind shear ram (BSR). The BSR is made up of two metal blades on opposing sides of the pipe in the bore hole that are designed to come together in a scissor like motion to cut the pipe, effectively preventing the well from leaking. According to the National Academy of Engineering's final report (2011), the BSR on Deepwater Horizon's BOP had a known design flaw that would affect its ability to perform its designed safety function. In particular, engineering analyses of the BSR prior to the accident had documented the fact that it would have trouble shearing a pressurized pipe, particularly if the pipe was not perfectly aligned with cutting surfaces of the BSR.

As previously discussed, maintenance errors also contributed to this event in that the control pods on the BOP were known to be not fully functional prior to the accident and that the annular had experienced stripping in the month preceding the accident, and yet drilling operations were not halted to repair these safety systems.

The point of this more detailed analysis of the *Deepwater Horizon* accident is to show the ways in which latent errors of commission, and in particular errors of commission related to decision-making and other cognitively intensive activities, can lead to catastrophe. Furthermore, it is not apparent that a traditional top down approach, whereby active errors at the sharp end derived from the identification of the most risk significant hardware failures are converted into HFEs, would have identified these tactical, strategic, organizational, design, and maintenance errors that occurred at the blunt end of the work processes. Either the top down approach needs to identify how HFEs based on latent errors can credibly affect system safety, or additional analyses using other approaches need to be included so that these blunt end errors can be identified and converted into meaningful HFEs that can be effectively incorporated into QRA.

3. EXISTING METHOD GUIDANCE ON MODELING HUMAN FAILURE EVENTS

Switching directions now, in this section we briefly review a number of available methods, guidance documents, and standards for HRA to derive potential rules for decomposing tasks to define HFEs. The methods review is centered on U.S. approaches, since these have been the sources widely used by analysts and documented in nuclear applications. Additional insights may be derived by careful study of non-U.S. HRA methods.

3.1 U.S. HRA Methods

THERP. The task analysis model in the Technique for Human Error Rate Prediction (THERP) is described in Chapter 4 of NUREG/CR-1278 (Swain and Guttman, 1983). It uses a goal-task breakdown of human activities to answer what are the goals of the human in terms of their interface with equipment such as controls. Task analysis classifies human activities into dynamic (involving interpretation and decision-making) and step-by-step (continuous or on-going) tasks. These tasks are included in the HRA event trees as branches. Since the tasks are modeled at the level of each step in a sequence of actions, the task decomposition may be considered quite detailed. These subtasks can be combined to represent an overall human action, and THERP provides clear guidance on aggregating subtasks during quantification.

Importantly, THERP provides a dependency model—which calculates how the relationship between subtasks should be treated in mathematical terms when aggregating the HEP. In other words, related tasks should not be double-counted when computing the likelihood of error. The dependency model in THERP has been adopted by almost every subsequent HRA model. The contemporary application of dependency is, however, considerably different from the original use in THERP (Whaley et al., 2012). In the original THERP application, dependency was used to account for subtasks that were closely related, typically in terms of using the same crew, occurring close in time, with little new contextual information. Dependency in THERP modeled intra-task relations, not inter-task relations. In fact, the point at which no relationship between tasks existed was considered the point at which the task was fully defined and constituted a complete HFE. Ironically, current use of dependency is almost exclusively for inter-task relations *between* HFEs. This is a widespread misapplication of the original THERP guidance and has the potential to result in different HFEs.

ASEP. The Accident Sequence Evaluation Process (ASEP) method (Swain, 1987) came about as a simplification of THERP. It does not include a unique process to model HFEs but instead defers to THERP and to PSA judgment about relevant tasks to analyzed. In contrast to THERP, there is a stronger emphasis on the need not to analyze every task, particularly for screening analyses. Thus, the clear definition for an HFE provided in THERP was loosened by the time ASEP was released.

SPAR-H. The Standard Plant Analysis Risk-Human Reliability Analysis (SPAR-H) method (Gertman et al., 2005) is a simplified HRA approach based in part on THERP (Boring and Blackman, 2007). SPAR-H provides no explicit guidance on task decomposition or defining or modeling the HFE beyond considering action and diagnosis tasks separately. SPAR-H defers to the IEEE 1082 and ASME PSA standards (discussed below) for discussion on how to model HFEs (i.e., decompose the tasks) for inclusion in the PSA. SPAR-H assumes the HFE is predefined, and the method therefore does not devote extensive time to telling the analyst how to formulate the HFE.

ATHEANA. A Technique for Human Error Analysis (ATHEANA; US Nuclear Regulatory Commission, 2000; see also Forester et al., 2007) provides nine overall steps, several of which are related to identifying HFEs. Using the ATHEANA approach, it is possible to determine if the modeled event should be considered an HFE or an unsafe action (UA). The delimiter is based on the consequence in terms of contribution to core damage—an UA is akin to a human error that is not risk significant. The ATHEANA method also provides guidance on determining errors of commission.

In practice, the final ATHEANA step—incorporation into PSA—is not as clearly articulated as the other steps, leading to some problems using ATHEANA to define HFEs. ATHEANA takes a holistic approach to HRA, and its task decomposition may be seen at the scenario or overall unsafe action level. ATHEANA considers unsafe acts, but the specific aggregation of these into the HFE remains underspecified. Importantly, ATHEANA considers deviations from nominal scenarios. These represent possibly unsafe conditions at the plant caused by operator action or inaction. The likelihood of these nominal scenarios is considered in the quantification of the overall HFE. Additional guidance (NUREG-1880) states that the human reliability analysts should work with the PSA team to model the HFE consistent with the PSA. This latter guidance points to a lack of clear guidance on modeling the HFE at a level consistent with the PSA. Since the SPAR-H method (Gertman et al., 2005) points analysts to the ATHEANA method specifically to identify and model human errors as needed, there is a troubling disconnect between both ATHEANA and SPAR-H and the practicable HFE in a PSA.

CBDT. The Cause-Based Decision Tree (CBDT) method (EPRI, 1992), widely used in industry, uses a decision tree approach to arrive at the quantification of HFEs based on key pieces of information (decision points) about operator performance. The method uses the SHARP1 framework for task decomposition as described in the next section.

3.2 Standards and Guidance Documents in HRA

SHARP1. The Systematic Human Action Reliability Procedure Revision 1 (SHARP1; EPRI, 1992) is an extension of the original SHARP process used for integrating HRA into the PSA process. The first stage of the SHARP1 process is the identification of the HFEs that are quantified in a subsequent stage. This first stage outlines five steps to arrive at the HFE:

1. Define the human interactions with the system that are potentially of interest. These are typically those that could leave some part or function of the plant unavailable. The procedure recommends identifying both errors of omission and commission as they might impact the plant.
2. Screen these human interactions to reduce the scope of the analysis to those that are most important.
3. Break down subtasks according to procedures to identify those subtasks that may have an impact on the plant. The emphasis here is to identify any tasks that may leave specific parts or functions of the plant unavailable, even if only temporarily as part of routine plant operations.
4. Perform an impact assessment to determine what effect the human subtasks have on plant equipment and plant state.
5. Integrate the human interactions as HFEs into the plant PSA model.

The approach falls short at defining an adequate way to decompose the overall event in terms of analysis. For example, if applying the five steps, particularly Step 3, it would not be clear whether to use a method to quantify the subtasks, groupings of subtasks, or the overall human interaction with the system, which could encompass hundreds of subtasks. This lack of decomposition can result in a myriad of HEPs, as many methods are not sensitive to subtask vs. task level analysis.

IEEE-1082 (1997). This standard, currently under revision, advocates a "stepwise" incorporation of human actions into the PSA model. The process begins with a complete but not unnecessarily detailed inclusion of human actions. These actions are considered in terms of risk-significance, such that only human actions that truly drive core damage frequency should be considered. A screening analysis narrows the number of human actions that are considered in the PSA. Some actions may be revisited at a later time when additional detail is added to the PSA model. The HFEs that are risk significant are modeled in sufficient detail to allow quantification. The IEEE-1082 standard is a very high level document. As such, it better addresses screening human errors for risk significance than actually defining those errors as HFEs.

ASME/ANS RA-Sa (2009 Revision). The ASME PSA standard explicates a number of important points to consider in HRA but does not provide specific recommendations on modeling HFEs beyond providing a formal definition of HFEs (see Section 2.1). The standard requires documentation of the identification, characterization, and quantification of pre-initiator, post-initiator, and recovery human actions, but it does not advocate a particular approach or recommend the appropriate level of decomposition. According to the standard, HFEs must be defined and included for each human activity that is not screened out and must be defined to reflect the resulting unavailability of a component, train, system, or function that is modeled in the PSA. The standard does provide guidance that several human activities may be grouped into a single HFE if the impact of the activities is similar. Three levels of HFEs are defined, differentiated by those HFEs that do not perform a task analysis, those that have a high-level task analysis (e.g., human impact at the train level), and those that have a detailed task analysis (e.g., human impact on individual components).

Good Practices for HRA (NUREG-1792). As with the standards mentioned above, the *Good Practices* link the HFE to the specific hardware failure that results from the human action or inaction. The level of modeling (i.e., level of decomposition) should reflect the amount of plant hardware that is affected. Thus, the HFE may be defined at the component, train, system, or function level. Human actions may be grouped at a higher level as appropriate. For example, if multiple human actions affect multiple components in the train, the HFE should be modeled at the train level. If, however, quantification differs considerably between the component and train level of modeling, the more conservatively bounding HFE definition should be used. If grouping multiple actions masks the potential for considering subsequent dependencies, the actions should be modeled as individual HFEs. This guidance is helpful in establishing the boundaries between HFEs in an event evolution, although it fundamentally reflects the top-down definition of the HFE.

3.3 Other Considerations

Dynamic modeling. In developing human performance simulation models, the issue of task decomposition resurfaces. Simulation models like ADS-IDAC (Chang and Mosleh, 2007) feature the ability to model human performance at the very detailed subtask level, such as decision points and simple manual actions. While quantification is possible at the subtask level, the models do not provide guidance for combining subtask HEPs into the HFE level appropriate for a PSA. Such combinatorial quantification must consider dependencies between subtasks, but there is the possibility to inflate HEP values if the aggregation algorithm does not properly consider the small subtasks that the simulation models use (Boring, 2007). Additionally, dynamic modeling reveals the need for PSF latency—namely, that PSFs must not be considered discretely without the lingering effect from one time point to another. For example, stress cannot simply be turned off because the underlying cause of that stress has disappeared. This insight suggests that PSFs may need to play a role in defining the HFEs, or at least the boundaries between HFEs.

HERA/SACADA. The Human Event Repository and Analysis (HERA) database system (Hallbert et al., 2006) and its descendent, the Scenario Authoring, Characterization, and Debriefing Application

(SACADA) database (Chang et al., in press), provides guidance on how to decompose events into subevents suitable for a detailed understanding of human performance. An event may be any human action or inaction that negatively affects plant safety. A human-related event may be further broken into subevents, according to the following criteria:

- Are separate people involved across the span of the event?
- Does an action within the event have a different goal than other actions?
- Does an action involve different equipment?
- Does an action have different consequences than the overall event being modeled?

HERA and SACADA subevents may be identical to HFEs in an HRA or PSA, but they are often more detailed in nature, because the purpose of the database system is to capture as much information as possible about human performance. As such, these subevents are not screened in terms of risk significance the way HFEs are in the HRA and PSA. Still, the fundamental guidance provided on decomposing events is useful in defining HFEs, and it provides a more bottom-up approach than is typical.

4. CONCLUSIONS

Defining an HFE for use in novel HRA applications still remains somewhat elusive. Although general guidance exists for the top-down approach, there remains a large element of skill of the craft in actually decomposing groups of subtasks into an HFE suitable for inclusion in the PSA. While approaches exist for bottom-up definitions, these still do not adequately address topics such as latent errors or errors of commission. Nonetheless, several candidate principles of HFE modeling have emerged from our review in this paper:

- Until clear guidance is available to identify commonalities and differences between the top-down and bottom-up approaches, it is desirable to employ a combination of both approaches to define the HFE.
- When adopting the top-down approach, the definition of the HFE should start broad, identifying those human actions and inactions that may trigger the unavailability of components, systems, or functions.
- These broad HFEs should be screened to determine the risk significant activities. The risk significant activities are the primary HFEs that are modeled in greater detail in the HRA.
- Task analysis of these risk significant activities may reveal additional sources of failures that may not be anticipated in the initial definition of the HFE. This represents the bottom-up approach. The definition of the HFE and screening should be an iterative process to arrive at a complete and relevant model of the human contribution to the overall system risk.
- Bottom-up approaches should consider errors of commission and latent errors in crafting the HFEs.
- Subtasks may reasonably be grouped into a single HFE provided that they are logically related; they do not represent different tasks, personnel, or equipment; and they do not mask dependencies that need to be accounted for.
- The earliest HRA methods used a simple equipment-level task decomposition. This is the level of flipping a switch. As interfaces have progressed in complexity, the interaction of the human with the equipment may represent a much higher level of decomposition that includes more cognitive or diagnostic activities. It is insufficient to define HFEs in terms of simple tasks—it must include a significant cognitive component as well.

These principles will be refined and developed into comprehensive guidance for defining HFEs in the petroleum context. Ultimately, one key goal of is to bridge the gap in existing HRA guidance and application to the petroleum domain. Current practice follows a somewhat vague top-down approach of using predefined HFEs from the PSA. As HRA is refined for oil and gas applications, it will need to include a clear bottom-up approach compatible with QRA.

Disclaimer

INL is a multi-program laboratory operated by Battelle Energy Alliance LLC, for the United States Department of Energy under Contract DE-AC07-05ID14517. This work has been carried out as part of The Research Council of Norway project number 220824/E30 "Analysis of human actions as barriers in major accidents in the petroleum industry, applicability of human reliability analysis methods (Petro-HRA)". Financial and other support from The Research Council of Norway, Statoil ASA and DNV are gratefully acknowledged. This paper represents the opinion of the authors, and does not necessarily reflect any position or policy of the above mentioned organizations. Neither the United States Government, nor any agency thereof, nor any of their employees makes any warranty, express or implied, or assumes any legal liability or responsibility for the accuracy, completeness, or usefulness of any information, apparatus, product, or process disclosed, or represents that its use would not infringe privately-owned rights.

References

American Society of Mechanical Engineers. (2009). Addenda to ASME/ANS RA-S–2008, Standard for Level 1/Large Early Release Frequency Probabilistic Risk Assessment for Nuclear Power Plant Applications, ASME/ANS RA-Sa-2009. New York: American Society of Mechanical Engineers.

Boring, R.L. (2007). Dynamic human reliability analysis: Benefits and challenges of simulating human performance. In T. Aven & J.E. Vinnem (Eds.), *Risk, Reliability and Societal Safety, Volume 2: Thematic Topics. Proceedings of the European Safety and Reliability Conference (ESREL 2007)* (pp. 1043-1049). London: Taylor & Francis.

Boring, R.L. (2009). Human reliability analysis in cognitive engineering. *Frontiers of Engineering: Reports on Leading-Edge Engineering from the 2008 Symposium* (pp. 103-110). Washington, DC: National Academy of Engineering.

Boring, R.L., & Bye, A. (2008). Bridging human factors and human reliability analysis. *Proceedings of the 52nd Annual Meeting of the Human Factors and Ergonomics Society*, 733-737.

Bronstein, S. & Drash, W. (2010, June 9). Rig survivors: BP ordered shortcut on day of blast. Available from: http://www.cnn.com/2010/US/06/08/oil.rig.warning.signs/index.html

CBS News (2010, May 16). Blowout: The Deepwater Horizon disaster [Television series episode]. *60 minutes*. New York, NY.

Center for Chemical Process Safety (CCPS) (2008). *Guidelines for hazard evaluation procedures (3rd Edition)*. American Institute of Chemical Engineering/AIChE, New York, NY.

Chang, Y.J., Bley, D., Criscione, L., Kirwan, B., Mosleh, A., Madary, T., Nowell, R., Richards, R., Roth, E.M., Sieben, S., and Zoulis, A. (In press). The SACADA database for human reliability and human performance. *Reliability Engineering and System Safety*.

Chang, Y.H.J., and Mosleh, A. (2007). Cognitive modeling and dynamic probabilistic simulation of operating crew response to complex system accidents, part 1: Overview of the IDAC model. *Reliability Engineering and System Safety, 92*, 997-1013.

Deepwater Horizon Study Group. (2011). *Final report on the investigation of the Macondo well blowout. Center for Catastrophic Risk Management*, University of California at Berkeley, Berkeley, CA.

Dekker, S. W. A. (2006). *The Field Guide to Understanding Human Error*. Aldershot, UK: Ashgate Publishing Co.

EPRI. (1992). *An Approach to the Analysis of Operator Actions in Probabilistic Risk Assessment, TR-100259*. Palo Alto: Electric Power Research Institute.

EPRI. (1992). *SHARP1—A Revised Systematic Human Action Reliability Procedure, EPRI TR-101711*. Palo Alto: Electric Power Research Institute.

Forester, J., Kolaczkowski, A., Cooper, S., Bley, D., and Lois, E. (2007). *ATHEANA User's Guide, NUREG-1880*. Washington, DC: US Nuclear Regulatory Commission.

Gertman, D. I., & Blackman, H. S. (1994). Human reliability and safety analysis data handbook. John Wiley & Sons, New York, NY.

Gertman, D., Blackman, H., Marble, J., Byers, J., Smith, C., and O'Reilly, P. (2005). *The SPAR-H Human Reliability Analysis Method, NUREG/CR-6883*. Washington, DC: US Nuclear Regulatory Commission.

Gertman, D., Hallbert, B., Parrish, M., Sattison, M., Brownson, D., & Tortorelli, J. (2001). *Review of findings for human error contribution to risk in operating events, NUREG/CR-6753*. Washington, DC: US Nuclear Regulatory Commission.

Gertman, D., Hallbert, B., & Prawdzik, D. (2002). *Human performance characterization in the Reactor Oversight Process, NUREG/CR-6775*. Washington, DC: US Nuclear Regulatory Commission.

Hallbert, B., Boring, R., Gertman, D., Dudenhoeffer, D., Whaley, A., Marble, J., & Joe, J. (2006). *Human event repository and analysis (HERA) system, overview, NUREG/CR-6903, Volume 1*. Washington, DC: US Nuclear Regulatory Commission.

Hollnagel, E. (1998). *Cognitive reliability and error analysis method (CREAM)*. Oxford: Elsevier.

IEEE. (1997). *Guide for Incorporating Human Action Reliability Analysis for Nuclear Power Generating Stations, IEEE-1082*. New York: Institute of Electrical and Electronics Engineers.

Kolaczkowski, A., Forester, J., Lois, E., and Cooper, S. (2005). *Good Practices for Implementing Human Reliability Analysis, Final Report, NUREG-1792*. Washington, DC: US Nuclear Regulatory Commission.

National Research Council. (2011). Macondo Well Deepwater Horizon Blowout: Lessons for Improving Offshore Drilling Safety. Washington, DC: The National Academies Press.

Podofillini, L., & Dang, V. (2012). Progress on Errors of Commission: an Outlook Based on Plant-Specific Results. *Proceedings of the 11th International Conference on Probabilistic Safety Assessment and Management*, 16B-Th5-1, Helsinki, Finland.

Poucet, A. (1989). *Human Factors Reliability Benchmark Exercise, Synthesis Report, EUR 12222 EN*. Luxembourg: Office for Official Publications of the European Communities.

Reason, J. (1990). *Human Error*. Cambridge: Cambridge University Press.

Reer, B., & Dang, V. (2007). *The Commission Errors Search and Assessment (CESA) Method. PSI Report No. 07-03*. Baden, Switzerland: Paul Scherrer Institute.

Reer, B., Dang, V., & Hirschberg, S. (2004). The CESA Method and its Application in a Plant-Specific Pilot Study on Errors of Commission. *Reliability Engineering and System Safety, 83*(2), 187-205.

Straeter, O., Dang, V., Kaufer, B., & Daniels, A. (2004). On the way to assess errors of commission. *Reliability Engineering & System Safety, 83*(2), 129-138.

Swain, A.D. (1987) *Accident Sequence Evaluation Program Human Reliability Analysis Procedure, NUREG/CR-4772*. Washington, DC: US Nuclear Regulatory Commission.

Swain, A.D., & Guttmann, H.E. (1983) *Handbook of Human Reliability Analysis with Emphasis on Nuclear Power Plant Applications, NUREG/CR-1278*. Washington, DC: US Nuclear Regulatory Commission.

US Nuclear Regulatory Commission. (2000) *Technical Basis and Implementation Guidelines for A Technique for Human Event Analysis (ATHEANA), NUREG-1624, Rev. 1*. Washington, DC: US Nuclear Regulatory Commission.

Whaley, A.M., Kelly, D.L., and Boring, R.L. (2012). Guidance on dependence assessment in SPAR-H. *Joint Probabilistic Safety Assessment and Management and European Safety and Reliability Conference*, 27-Mo4-4.

Woods, D., Johannesen, L., Cook, R., & Sarter, N. (1994). *Behind human error: Cognitive systems, computers and hindsight, CSERIAC SOAR 94-01*. Columbus: Ohio State University.

A Comparison of Two Cognition-driven Human Reliability Analysis Processes - CREAM and IDHEAS

Kejin Chen[*a], Zhizhong Li[a], Yongping Qiu[b], and Jiandong He[b]
[a] Department of Industrial Engineering, Tsinghua University, Beijing, 100084, P. R. China
[b] Shanghai Nuclear Engineering Research & Design Institute, Shanghai, P. R. China

Abstract: Years of technology development has witnessed the increasing reliability and robustness of instruments in modern complex systems, while humans, still constitute the major incidents contributor. This article proposes a new taxonomy of various HRA methods based on how the basic probability is determined. Next focusing on the cognition-driven HRA methods, the article summarizes the general quantification model in cognitive-driven category. CREAM and IDHEAS, two representative HRA methods, are compared in terms of their analysis processes against the general qualification process. A simpler two-phase response model for new HRA is suggested and discussed.

Keywords: HRA taxonomy, cognition-driven, analysis process, CREAM, IDHEAS.

1. INTRODUCTION

Years of technology development has witnessed the increasing reliability and robustness of instruments in modern complex systems, while humans, as the most flexible while the least scrutable part of the systems [1], still constitute the major incidents contributor. HRA (Human reliability Analysis) is therefore introduced to render a description of the human contribution to risk and identify ways to reduce it by using systems engineering and behavioral science methods. Proposed firstly in 1950s, HRA gained lots of focus after the Three Mile Island Accident (TMI).

HRA is part of PRA (Probability Risk/Safety Analysis, PRA/PSA). Hollnagel [2] stated that in this PSA-cum-HRA framework, HRA has been constrained by the simplification of event trees. More and more researchers in HRA tend to believe that HRA needs combination of various disciplines like human factors, social psychology, behavior science and organization management, and so on [3,4,5,6]. Boring [7] stated that HRA can provide a comprehensive description about the contribution of human errors to safety in both qualitative and quantitative fashions. However, a comprehensive description seems impossible considering that a human itself is a highly complex system. Simplification on human behaviors has to be made and priority of HRA should be given to major behaviors that are vital to system safety. And the purpose of HRA should be the screening and evaluation of risky potentially behaviors, in lieu of analysis of every behavior.

In the past decades, various HRA methods were proposed to analyze, predict and reduce human errors in nuclear power plants, and in other process industries, for example, THERP (Technique for Human Error Rate Prediction) [8], HCR (Human Cognitive Reliability) [9], CREAM (Cognitive Reliability and Error Analysis Method) [2], SPAR-H (Accident Sequence Precursor Standardized Plant Analysis Risk Model, ASP/SPAR) [10], and so on. However, current taxonomies for HRA methods remain unclear and controversial. One objective of this article is to propose a new HRA taxonomy.

Along with increasing research on human reliability and other relevant fields like cognitive science, neural science, increasing HRA methods from 1990s transfer their focus from task to context and cognition that support tasks. Not only focusing on the behavior outcomes, recent HRA methods stress potential cognitive mechanisms and causes underlying human errors from the perspective of psychology, cognition, and neuroscience. Emphasis on the influence of context and cognition factors on the one hand make a much more reasonable and persuasive qualitative analysis possible, while it on the other hand provides a relatively vivid structure to facilitate quantitative analysis. However, the

---
[*] *chenkj10@mails.tsinghua.edu.cn*

qualitative analysis is still open to subjective bias due to the inevitable use of task analysis and the involvement of experts. And the quantitative techniques in different HRA methods are still far from convincible, suffering inadequacy of data source, bias of expert estimation, immeasurability of performance shaping factors, and so on. This article aims to compare the analysis process of two cognition-driven methods: CREAM, a cognition-driven method, and IDHEAS, a state-of-the-art HRA method freshly proposed.

2. HRA TAXONOMY

Everdij and Blom [11] summarized 726 techniques, methods, database and models related to safety. Among them, 175 techniques or models are used for risk analysis, 171 for human performance analysis, which can be adopted in nuclear power, healthcare, aviation, and other safety-critical industries. Bell and Holroyd [12] reviewed 17 methods out of 35 HRA in detail, but some developing methods like IDAC and IDHEAS are not included. Lyons et al. [13] summarized 35 HRA methods adopted in medical industries rather than in nuclear power plants. As for so many methods, researchers tried to develop taxonomies according to some common characteristics [2, 4, 14]. It's widely accepted that HRA can be classified into two categories: The first generation and the second generation. However, differences between two generations methods remain indistinct. CREAM is a typical representative of the second generation HRA. Its developer, Hollnagel, stated that two differences can be used to differentiate two generations: 1) the second generation HRA methods should extend the dichotomy of events or behaviors, and further analyse the error mechanism. 2) Influence of conditions on performance should be considered explicitly in HRA. The first difference reflects that the second generation HRA tries to explore the internal error mechanism, based on the cognitive process of human behaviors. And the second one reflects that the operation conditions do influence operators' performance in some researchers' view. In CREAM, Hollnagel addressed the importance of analysis of cognitive processes in HRA, and he developed a way to correspond tasks to different cognitive functions. Actually those first generation HRA methods did as well consider the influence of operation conditions by including performance shaping factors (PSFs) related to operation conditions. For this reason, it's the focus on the analysis of cognitive processes rather than the influence of operation conditions that differentiate CREAM from those HRAs developed earlier than it. Besides, the influence of context on performance is also considered as another distinction between the first and second generation HRA [14]. ATHEANA (A Technique for Human Error Analysis) is a HRA method that emphasizes such influence [15]. Chronological sequence is as well a criterion to classify the first and second generation HRA, but controversies remain. If all HRAs later than ATHEANA are classified into the second generation, then how can we classify SPAR-H? It was published in 2004, later than ATHEANA, while it is a succinct version of THERP without considering cognitive processes analysis. Classification according to chronological sequence fails to describe any significant characteristics of HRA methods although this taxonomy is simple to use.

Spurgin [14] categorized HRAs according to how the basic HEP (human error probability) is calculated: task-defined, time defined and context-defined methods. In his taxonomy, SPAR-H is treated as context-defined methods. HEP (human error probability) of SPAR-H nevertheless is pre-defined by its developer and context factors are merely used to adjust the HEP, therefore it seems more reasonable to put SPAR-H into task-defined category. In addition, since available time to complete a task constitutes one of task attributes, it may be better to integrate the time-defined category into task-defined one. The taxonomy by Spurgin is inspiring though. A new taxonomy based on how the HEP is determined is proposed in this article: task-driven HRA, cognition-driven HRA, and context-driven HRA. Illustration of three categories is showed in Table 1.

Table 1 HRA Taxonomy

Taxonomy Category	Representative Methods
Task-driven	THERP, ASEP, HCR, HEART, NARA, SPAR-H
Context-driven	SLIM, SLIM-MAUD, ATHEANA, HDT, IDAC
Cognition-driven	CREAM, IDHEAS

3. COGNITION-DRIVEN HRA METHODS: CREAM vs. IDHEAS

Whatever category a HRA method is in, the quantification process for one HFE (human failure event) consists of three general sub-processes: 1) derive the basic HEP for each sub-task; 2) adjust the basic HEP with adjusting coefficients; 3) Integrate sub-tasks and derive the HEP with dependency model [16]. What differs various HRA methods is how the basic HEP is determined in our taxonomy. For the cognition-driven category, unlike other two categories, a cognitive function analysis is conducted to decompose the HFE before the basic HEP is determined. To summarize therefore, cognition-driven HRA methods can be generally made up with four steps: 1) Conduct cognitive functions analysis; 2) derive basic probability; 3) derive adjusting coefficients; 4) calculate the final probability with dependency model. And HRA methods in cognition-driven categories vary in their approaches to realize each step.

3.1. CREAM

CREAM stresses the significance of cognitive factors in HRA, which differentiates it from its previous HRA methods. There are two versions of CREAM: Basic CREAM and extended CREAM. The basic CREAM is used to analyze operators' control modes which further determine the probability interval of human error. To obtain specific error probability, the extended CREAM is required, which can be decomposed into four sub-steps. 1) Describe the cognitive activities of target tasks; 2) identify possible cognitive failure types for each cognitive activity; 3) assess the effects of CPCs (if the basic CREAM has not been adopted); 4) calculate human error probability. There are 15 different cognitive activities in CREAM, and each cognitive activity can be mapped to one or two cognitive functions. 4 cognitive functions are identified (observation, interpretation, planning, and execution). Each cognitive function can be further mapped to several cognitive failure types. There are 13 cognitive failure types in total. After analysts identify cognitive activities and corresponding cognitive functions and cognitive failure types, they can then derive the basic HEP (CFP_0) of each sub-task. Then formula (1) and (2) are used to derive the error probability of HFE.

$$CFP_i = CFP_0 \times \text{Weight(CPCs)} \quad (1)$$
$$CFP_{Total} = \sum_i (CFP_i \times K) \quad (2)$$

CFP_i denotes the error probability of ith cognitive activity. The weight of CPCs can be derived by the relation between CPCs levels and four cognitive functions, and K is from expert estimation.

3.2. IDHEAS

The Integrated Decision-tree Human Event Analysis System (IDHEAS) is the latest HRA method developed under the support of the U.S. Nuclear Regulatory Commission (USNRC) [17, 18]. To address the limitations in existing HRA methods, researchers conducted elaborate review in psychology, cognition, team performance, and other related fields, which has been summarized in NUREG-2114 [19]. There are two important concepts in IDHEAS: macrocognition and proximate cause. Macrocognition is a high-level description on what humans do with their brains [17]. The macrocognition model in IDHEAS is composed of five functions: detecting/noticing, sensemaking/understanding, decision making, action, and team coordination. The concept of proximate cause is developed to describe the cause of the failure of a macrocognition function [19]. Based on these two concepts, IDHEAS provides guidance to define HFE led by crew response tree (CRT), and further constructs a cause-based quantification model [20].

Crew Failure Mode (CFM) and Decision Trees (DTs) are two major elements of IDHEAS [20]. 14 CFMs are defined and categorized into three phases of response model (plant status assessment, response planning, and execution). After a CFM is identified, a decision tree (DT) is developed for each CFM, and branching point represents the one of Performance Influencing Factors (PIFs) that is most relevant to the CFM. And analysts can select PIFs from Groth and Molesh [21]. Based upon the CFMs and DTs, the HEP in scenario S can therefore be calculated by formula (3):

$$\text{HEP (HFE|S)} = \sum_{CRT\ sequence} \sum_{CFM} Prob(CFM|CRT\ sequence, S) \quad (3)$$

3.3. Comparison between CREAM and IDHEAS

IDHEAS so far is developing and its quantification model is incomplete to some extent. For example, dependency model, DTs for internal events, and data at the end point of DTs remain unsolved [20, 22]. In another words, IDHEAS currently is not mature enough to be put into practice. Therefore this comparison would focus on their analysis processes as figure 1 shows, rather than the quantification results.

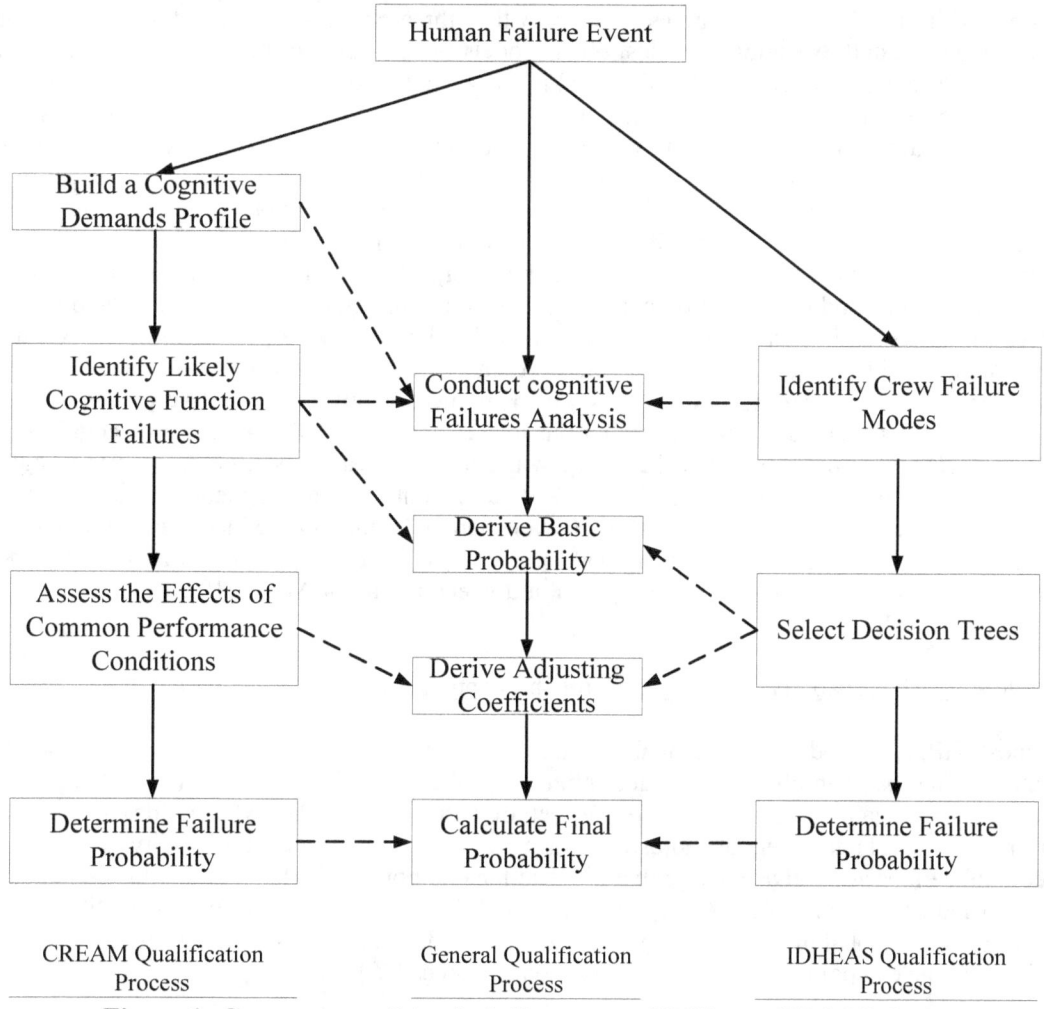

Figure 1: Comparison of Analysis Processes of Different HRA Methods

Conduct cognitive failures analysis

Cognitive failure analysis makes cognition-driven HRA methods distinct from the others. CREAM and IDHEAS differ a lot in how to analyze cognitive failures. CREAM requires analysts to identify the cognitive activities of a HFE before they identify the likely cognitive failures via four contextual control model functions (COCOM functions) [23], while in IDHEAS, analysts can identify the likely cognitive failures directly without considering cognitive demands explicitly. Both methods categorize various cognitive failures in response model which describes how operators respond plant status cognitively. Since the presentation of cognitive functions has already facilitated the identification of cognitive failures to a great extent, the 15 cognitive activities in CREAM seems no more than a white elephant that merely manifests how CREAM stresses the cognitive demands instead of facilitating the quantification. In addition, noted by Hollnagel himself, the list of critical cognitive activities is limited by its source so that it cannot be proved complete or even correct [2]. When it comes to the cognitive failures, analysts using CREAM could identify the most credible failures while in IDHEAS, analysts have to identify all relevant ones. Emphasis on the predominant failures helps CREAM analysts circumvent the dependency issue among cognitive failures, while IDHEAS developers have to face this challenging issue if they insist on considering all relevant failures (crew failure modes).

It's suggested in this article that identifying cognitive failure modes directly by response model in IDHEAS is more reasonable than CREAM. However, how to develop the response model needs further discussion. Hollnagel [2] proposed a 4-phase model: observation, interpretation, planning, and execution, while in IDHEAS, researchers put up with a three-phase model: plant status assessment, response planning, and execution. Besides, other models were proposed. For example, in the famous Information Processing Model of Decision Making [24], 5 phases are specified: sense, perception, diagnosis, choice, and execution. Current controversy lies in how to classify the processes before execution. Actually it is difficult to draw a clear line between different cognitive processes. Take "plant status assessment" and "response planning" as an example in IDHEAS. The essence of plant status assessment is information acquisition which still requires the involvement of response planning, since operator has to make "decision within decision" [24], that is, s/he has to decide where and when to collect the information, whether to stop collecting information when s/he finds an important piece of information. This kind of interaction between plant status assessment and response planning makes the selection of CFMs in each category depend on each other. Moreover, to integrate HRA with the context factors, influences of dominant factors on both cognitive and action performance need more empirical research in the process industry. However, immeasurability of cognitive process leaves no choice for most researchers in human factors but to measure merely the outcome of cognition, like diagnosis accuracy or diagnosis time [25, 26], while a few researchers tried to measure cognitive process costly using devices like eye trackers, NMR equipment (nuclear magnetic resonance) [27, 28]. To fractionize the cognitive process makes it even harder to measure. Since there is a vivid and uncontested difference between diagnosis and execution, it is suggested in this article therefore to adopt two phase response model, just diagnosis and execution as SPAR-H [10], to circumvent both issues mentioned above.

Calculate sub-task probability (Derive basic probability and adjusting coefficients)

Like most HRA methods, basic probability and adjusting coefficients in CREAM are assessed separately: The basic probability is determined by the cognitive failures while the adjusting coefficients are obtained by assessing the effect of CPCs. And CREAM provides quantification equation (formula 1) to integrate adjusting coefficients with the basic probability. Instead, from limited publications about IDHEAS, probabilities at the end points of DTs have already been adjusted by context factors, namely, IDHEAS provides probabilities under various scenarios. Unlike CREAM to provide basic probabilities from nowhere, IDHEAS developers show their attempts to assign probability at the end points of DTs from various data sources [20].

Calculate final probability with dependency model

The final step is to calculate the probability of the whole HFE by combining the sub-component of the HFE. In another word, dependency model constitutes the crux of the combination issue. Two types of dependency are identified: dependency between sub-tasks within one HFE, and dependency between separate HFEs [29]. Although the selection of dependency model highly influences the results [29, 30], most dependency models in the current HRA practice merely focus on the dependency within one HFE. In CREAM, Hollnagel suggested to solve the combination by considering the event sequence, improved and summarized later by He et al. [31]. In contrast, another type of dependency in IDHEAS between CFMs relevant to one subtask arises. No matter the dependency between two HFEs or between subtasks in one HFE, events or sub-tasks can be sorted out by time and space, making dependency model based on time or space possible, while CFMs relevant to one sub-task cannot be sorted chronologically. Since the dependency between CFMs is new, current dependency models may not fit for IDHEAS, at least not for its CFMs combination process. A good dependency model is still expected in IDHEAS.

4. CONCLUSION

This article proposes a new taxonomy (task-driven, context-driven, and cognition-driven category) of various HRA methods based on how the basic probability is determined. Next focusing on the cognition-driven HRA methods, the article summarizes the general quantification model in cognitive-driven category. CREAM and IDHEAS, two representative HRA methods, are compared. Considering the limited publications and incomplete quantification model about IDHEA, the comparison was based on the analysis process for quantification of both methods, rather than on their quantification results of a case study. A simpler two-phase response (Diagnosis/Execution) model is suggested and discussed due to the controversial definition on and immearuability of cognitive sub-processes.

Acknowledgements

This study was supported by the National Natural Science Foundation of China (Project No. 70931003).

References

[1] R. B. Fuld. "The fiction of function allocation", Ergonomics in Design, 1, pp. 20-24, (1993).
[2] E. Hollnagel. "Cognitive Reliability and Error Analysis Method", Elsevier, 1998, Oxford.
[3] K. P. LaSala. "Human Performance Reliability: A Historical Perspective", IEEE Transaction on Reliability, 47 (3), pp. 365~371, (1998).
[4] P. Pyy. "Human Reliability Analysis Methods for Probabilistic Risk Assessment", [2014-02-24]. http://www.vtt.fi/inf/pdf/publications/2000/P422.pdf.
[5] S. Hirschberg. "CSNI Technical Opinion Papers (No.4): Human Reliability Analysis in Probabilistic Safety Assessment for Nuclear Power Plants", 2004, [2014-02-24]. https://www.oecd-nea.org/nsd/reports/2004/nea5068-HRA.pdf
[6] X. H. He and X. R. Huang. "Human reliability analysis in industrial systems: Theory, methods and practise", Tsinghua Press, 2007, Beijing.
[7] R. L Boring. "Dynamic Human Reliability Analysis: Benefits and Challenges of Simulating Human Performance (INL/CON-07-12773)", 2007, [2014-02-24]. http://www.inl.gov/technicalpublications/Documents/3735688.pdf
[8] A. D. Swain and H. E. Guttmannn. "Handbook of Human Reliability Analysis with Emphasis on Nuclear Power Plant Applications (NUREG/CR-1278)", USNRC, 1983, Washington D.C.
[9] EPRI. "Operator Reliability Experiments Using Power Plant Simulators (Final Report), EPRI NP-6937-L", EPRI, 1991, Palo Alto.
[10] D. Gertman, H. Blackman, J. Marble, J. Byers and C. Smith. "The SPAR-H Human Reliability Analysis Method (NUREG/CR-6883)", USNRC, 2004, Washington D.C.
[11] M. H. C. Everdij and H. A. P. Blom. "Safety Methods Database (Version 0.9)", 2010, [2014-02-24]. http://www.nlr.nl/documents/flyers/SATdb.pdf
[12] J. Bell and J. Holroyd. "Review of Human Reliability Assessment Methods. Health and Safety Laboratory for the Health and Safety Executive 2009, Report RR 679", [2014-02-24] http://www.hse.gov.uk/research/rrpdf/rr679.pdf
[13] M. Lyons, M. Woloshynowych, S. Adams and C. Vincent. "Error Reduction in Medicine", Imperial College 2005, London.

[14] A. J. Spurgin. "Human Reliability Assessment: Theory and Practice", CRC Press, 2010, Boca Raton.
[15] S. E. Cooper, A. M. Ramey-Smith, J. Wreathall, G. W. Parry, D. C. Bley, W. J. Luckas, J. H. Taylor and M. T. Barriere. "A Technique for Human Error Analysis (ATHEANA), NUREG/CR-6350". USNRC, 1996, Washington D.C.
[16] S.J. Lee, J. Kim and S.C. Jang. "Development of a human reliability analysis tool for conventional and advanced main control rooms in nuclear power plant". Paper presented at the meeting of NPIC & HMIT 2012.
[17] A. M. Whaley, S. M. L. Henderickson, R. L. Boring, J. C. Joe, K. L. L. Blanc and J. Xing. "Bridging human reliability analysis and psychology, part 1: The psychological literature review for IDHEAS method (INL/CON-12-24798)", 2012, [2014-02-24], http://www.inl.gov/technicalpublications/documents/5517275.pdf.
[18] A. M. Whaley, S. M. L. Henderickson, R. L. Boring and J. Xing. "Bridging human reliability analysis and psychology, part 2: A cognitive framework to support HRA (INL/CON-12-24797)", 2012, [2014-02-24], http://www.inl.gov/technicalpublications/Documents/5517274.pdf.
[19] A. M. Whaley, J. Xing, S. M. L. Henderickson, R. L. Boring, J. C. Joe and K. L. L. Blanc. "Building a psychological foundation for human reliability analysis (NUREG-2114)". USNRC, 2012, Washington D.C.
[20] H. Liao, K. M. Groth and S. Steven-Adams. "Leveraging existing human performance data for quantifying the IDHEAS HRA method". Safety, Reliability and Risk Analysis: Beyond the Horizon-Steenbergen et al. (Eds), pp. 563-569, (2014).
[21] K. M. Groth and A. Molesh. "A data-informed PIF hierarchy for model-based Human Reliability Analysis", Reliability Engineering and System Safety, 108, pp. 154-174, (2012).
[22] J. Xing. "SRM-061020 on HRA methods: Progress and status in 2013", 2013, [2014-02-24], http://pbadupws.nrc.gov/docs/ML1400/ML14009A355.pdf.
[23] E. Hollnagel. "Human reliability analysis: Context and control", Academic Press, 1993, London.
[24] C. D. Wickens, and J. G. Hollands. Engineering Psychology and Human Performance (3rd Edition). NJ: Prentice Hall, (2000).
[25] G. A. Jamieson. 2003. "Ecological Interface Design for Petrochemical Process Control: Integrated Task- and System-Based Approaches." (CEL_02_01). Cognitive Engineering Lab: University of Toronto.
[26] K. J. Vicente, N. Moray, J. D. Lee, J. Rasmussen, B. G. Jones, R. Brock, and R. DJEMIL. "Evaluation of A Ranking Cycle Display for Nuclear Power Plant Monitoring and Diagnosis," Human Factors, 38, pp. 506-521., (1996).
[27] K. J. Chen, and Z. Z. Li. Eveluation of human-system interfaces with different information organization using an eye tracker. The 5th International Conference, HCI International 2013, Las Vegas, NV, USA, July 21-26, 2013.
[28] J. H. Goldberg, and X. P. Kotval.. Eye movement-based evaluation of the computer interface. In: S. K. Kumar (ed.), Advances in Occupational Ergonomics and Safety, 529–532. Amsterdam: ISO Press.(1998)
[29] M. Cepin. "DEPEND-HRA—A method for consideration of dependency in human reliability analysis", Reliability Engineering and System Safety, 93, 1452 – 1460, (2008).
[30] M. Cepin. "Comparison of Methods for Dependency Determination between Human Failure Events within Human Reliability Analysis." 2008, [2014-02-24], http://www.hindawi.com/journals/stni/2008/987165/.
[31] X. H. He, Y. Wang, Z. P. Shen and X. R. Huang. "A simplified CREAM prospective quantification process and its application", Reliability Engineering and System Safety, 93, pp. 298-306, (2008).

Human Reliability in Spacecraft Development:
Assessing and Mitigating Human Error in Electronics Assembly

[a]Obibobi K. Ndu,[*] Space Mission Assurance Group, Johns Hopkins University Applied Physics Laboratory, Laurel, MD, USA

[b]Dr. Monifa Vaughn-Cooke, Department of Mechanical Engineering, Reliability Engineering Program University of Maryland, College Park, MD USA

1.0 Introduction

Integral to the proper functioning and reliability of any spacecraft are the proper design, fabrication, assembly, and integration of its electrical and electronic systems. A critical look at the composition of such spacecraft systems reveals a preponderance of circuits built on printed wiring assemblies (PWA). Considering the highly complex process of spacecraft development and the stringent reliability and performance requirements imposed on operational spacecraft, it is apparent that human reliability during the development process is a factor in the quantification of overall spacecraft system reliability.

This paper presents the development and application of Human Reliability Analysis (HRA) methods to specifically analyze the human error associated with the task of applying a polymeric coat onto a printed wiring assembly, a process also known as conformal coating. The polymeric coat serves to protect electronic circuitry against moisture, chemical contaminants and corrosion, extremes of temperature, and dust particles. The conformal coating is typically specified to protect against the particular space environment to which the spacecraft will be subjected. Subsequently, improper application potentially leads to loss of electro-electrical functionality.

Yet a closer look at the development process shows the high degree of interface and contact between the human fabricator and the system in development however, traditional system reliability analysis does not address the issue of human error introduced during the manufacturing and integration phase. It is assumed that quality assurance and quality control standards and processes will eliminate these workmanship-related defects. Acknowledging that no system or process is perfectly capable of arresting quality escapes in manufacturing, fabrication, and integration, how then can one account for human error in the absence of more modern and precise machine controlled processes?

In the context of polymeric application, we will need to establish the boundaries of human error by defining the scope and then decide on the methods and tools relevant to analysis of human error. Human error in this process shall be defined as an action, or omission of action, by a technician, which detracts from reaching a specific target end state; in this case, the perfect application of the appropriate polymeric coat on a PWA.

2.0 HRA Overview

HRA[1] is the process of modeling the likelihood and consequence of human error and the subsequent impact on the reliability of a system. These methods were divided into first and second generation methods, discussed in subsequent sections.

2.1 First Generation Methods

Early methods of HRA, often referred to as the First Generation Methods (FGMs) developed in the 1980s, were based on modeling the human operator as a component within the system,

---

[*] Obibobi.ndu@jhuapl.edu

whereby the failure of the "human component" and the effects of its consequence on the system could be traced through a fault tree [2]. FGMs treated human error in a similar fashion as component failures in fault tree analysis. Error probabilities could then be assigned to these human errors. An attribute of FGMs is the decomposition of errors in to two basic types; errors of omission, a case where an operator fails to respond to an event, and errors of commission, a case where a human performed an unintended action.

FGMs in HRA include; Technique for Human Error Rate Prediction (THERP)[1], Success Likelihood Index Method (SLIM)[3], Standardized Plant Analysis Risk-Human Reliability Analysis (SPAR-H)[4], Human Cognitive Reliability Method (HCR)[5], and Human Error Assessment and Reduction Technique (HEART)[6].

2.2 Second Generation Methods

FGMs proved insufficient for characterizing the effects cognitive and human behavioral processes. Another limitation of the first generation methods is the inability to account for dependence of human errors on the dynamic evolution of incidents [7].
The solutions to these limitations represented a breakthrough and resulted in the development of second generation methods [7]. Some of these methods include; A Technique for Human Event Analysis (ATHEANA)[8], Assessment Method for the Performance of Safety Operation ("Méthode d'Evaluation de la Réalisation des Missions Opérateurs pour la Sûreté ") MERMOS [9], and Cognitive Reliability and Error Analysis Model (CREAM)[10]. These methods were based on four main components: 1) a cognitive model of human behavior; 2) a taxonomy or classification; 3) a database; and 4) a formal application method [7].

2.3 Application of HRA in Space Industry

HRA methods have been applied to the NASA and the Space Industry. Most recently, probabilistic risk assessment (PRA) performed on the Space Shuttle included a Space Shuttle HRA. Also the International Space Station PRA[11] included an HRA. The NASA Shuttle PRA used the Technique for Human Error Rate Prediction (THERP) as a screening tool and evaluated pre-initiating events (Shuttle ground processing errors), initiating events (crew errors), and post initiating events (crew errors) using CREAM. The International Space Station program, chose to identify human errors in their accident scenarios rather than explicitly quantifying the contribution of human error to risk [12].

3.0 Method Development

HRA utilizes a set of tools to estimate the probability of human error in the context of a PRA. An HRA methodology must include a procedure for generating qualitative and quantitative results. It must also be based on a causal model of human response rooted in cognitive and behavioral sciences. Finally, it must be detailed enough to support data collection, and empirical and theoretical validation.

The method presented in this paper is based on a task analysis, which identifies and lists potential unsafe acts. Performance Shaping Factors (PSFs) that contribute to the unsafe acts are also identified and cross-linked with the unsafe acts. Existing HRA methods were then evaluated for use by comparing each method's responsiveness to a set of assessment questions.

3.1 Task Analysis

Several task analysis methods were considered, such Hierarchical Task Analysis, Cognitive Task Analysis, and Procedural Task Analysis (PTA). A Procedural Task Analysis (PTA), based on NASA Workmanship Standards[13], was selected due to its relevance in addressing the cognitive and physical actions required of the human to successfully complete the primary task A description of the human-system interface is used to provide the requisite contextual basis to guide the proposed HRA.

The PTA was performed using a series of task flow diagrams. The first step in the PTA was to decompose the primary task into a four secondary tasks that could still be further discretized depending on the desired fidelity of the HRA. The benefit of this approach is that the essential framework for the HRA can then be applied to all levels of tasks in the polymeric application process.

The four secondary tasks are each assigned a unique identifier for ease of reference and for place keeping. They are listed as follows; Surface Preparation, Chemical Preparation, Chemical Application, and Curing and Demasking. Each secondary task is further decomposed into a set of discrete task steps. These discrete task steps are also assigned unique identifiers that link them back to their parent secondary task. These task flow diagrams aided in the identification of purely cognitive task steps and physical task steps.

3.2 Unsafe Acts

After completion of the PTA, another decision tool was introduced – a comprehensive list of Unsafe Acts (UA), grouped according to the pertinent secondary task, Table 3-1. Each UA was assigned a unique identifier for reference and place keeping. The completed PTA facilitated identification of the UAs by allowing an assessment of what could go wrong at each step in the four task flow diagrams. Recognizing that each task flow step fit into one or two categories of human action; cognitive or physical, it is then possible to evaluate potential behavioral theories and models from these two broader human factors areas that would apply to the each UA.

Table 3-1 Sample Unsafe Acts List

Identifier	Unsafe Acts for Surface Preparation Task
1	Inadequate surface cleaning
2	Improper execution of Ionic Contamination Test
7	Incorrect application of masking
8	Application of masking in areas not specified by engineering specifications

3.3 Performance Shaping Factors

PSFs are used in HRA to characterize the dimensions – cognitive, social, emotional, and physical – of human response. They aid in understanding why human error occurs and are classified as social, personal, organizational, and or technological. A natural consequence of developing the list of UAs is the ability to document a set of PSFs and link the UAs to the top-level categories of PSFs; social, personal, organizational, and technological.

3.4 Method Evaluation

The evaluation of existing HRA methods for this application was performed by addressing a set of 11 questions, which allowed for a cross-method comparison of the essential elements of any HRA. The evaluation allows for identification of deficiencies in any single method that can be compensated for by adopting a particular aspect of another method.

Table 3-2 HRA Method Evaluation Questions

3.4.1.1.1.1.1 1. Are generic or context/operator-specific tasks required?	3.4.1.1.1.1.2 7. Are task and PSF dependencies necessary?
2. Are generic or context/operator-specific PSFs required?	8. Is consideration of error recovery a necessary component?
3. Is a screening method required?	9. Do uncertainty bounds need to be estimated?
4. What type of HEP source is appropriate (analysis or method)?	10. What knowledge level is required for HRA implementation?
5. Is current data available (type, source)?	11. Is a software implementation tool available?
6. Has this method been validated for the context in question?	

The 11 questions were structured to highlight the suitability of each of the essential elements of an HRA. Figure 3-1 illustrates assessment of HRA methods against Question 9. The elements addressed were: task decomposition; number of PSFs; human factors coverage; source of Human Error Probability (HEP); error mode-specific HEPs; treatment of task/error dependencies and recovery; uncertainty bounds estimation; required knowledge level for use; industry applicability or experience base; and software implementation availability. Comparing the method-specific task decomposition – typically presented as a set of generic tasks – with the polymeric application task analysis ensures that each method is vetted for suitability. Also, the link between task, unsafe act, and PSF, combined with the cross-method comparison to aid in the method selection.

Figure 3-1 HRA Method Assessment Example

Method	Screening	Primary Source for HEP Estimates		HEPs for Specific Error Modes	Explicit Treatment of		Uncertainty Bounds Estimation
		Number provided by method	Number produced by analyst		Task/Error Dependencies	Recovery (includes actions with feedback)	
THERP	√	√		Detailed and many	√	√	√
NARA		√		None specified			
SPAR-H		√		Diagnosis & Action	√	√	√
ASEP	√	√		Diagnosis & Action	√	√	√
SLIM			√	None specified			
HEART		√		None specified			
ATHEANA			√	Expert judgment	√	√	√
CREAM	√	√		13 error modes			√

The use of the decision tools resulted in the selection of the combination of CREAM and the HEART.
Eric Hollnagel developed CREAM in 1998 after an analysis of HRA existing methods and based on the Contextual Control Model [14]. The method is applicable to retrospective analysis as well as to performance prediction. It is based on a distinction between competence and control, utilizing a classification scheme that separates causes and manifestations, also referred to as genotypes and phenotypes respectively [15].

CREAM method identifies 9 Common Performance Conditions CPCs, which are individually assessed for an Expected Effect on Performance Reliability (EEPR) based on a possible CPC state. CPCs are assumed to exist in various possible states depending on the particular CPC. Similarly, the EEPRs are assumed to have potential impacts on an operator's performance ranging from; Improved, Not Significant, and Reduced. Each EEPR is associated with a CPC State. A description of EEPR and CPC is available in the literature [10]. The pertinent Control Mode identifies the differing levels of control that an operator has in a given context and the characteristics which highlight the occurrence of distinct conditions [2]. The control modes are available in the literature. CREAM applies a set of CPCs to a particular setting in order to establish the applicable control mode. The applicable control mode is then indicative of the expected level of reliability in the given setting. This is possible because each control mode is assigned a predetermined reliability interval as shown in the Table 3-3 below.

Table 3-3 CREAM Control Modes and HEP [10]

Control Mode	Reliability Interval (probability of failure)
Strategic	$0.5E\text{-}5 < p < 1.0E\text{-}2$
Tactical	$1.0E\text{-}3 < p < 1.0E\text{-}1$
Opportunistic	$1.0E\text{-}2 < p < 0.5E\text{-}0$
Scrambled	$1.0E\text{-}1 < p < 1.0E\text{-}0$

CREAM in its current state does not provide for explicit treatment of Error Dependencies nor Error Recovery (Reduction), this is made evident in the method evaluation by addressing Question 8 for the HRA methods in the same manner as Question 9 (figure 3-1); given these two limitations one can adopt the error reduction techniques proffered by the HEART method to address the latter. However, task or error dependency methods as presented in other HRA methods are largely subjective and do not offer a viable solution. These dependency factors are similar to common cause factors in system reliability analysis, however they have not been validated for cross-context implementation in HRA.

HEART is an HRA method based on the premise that human reliability is dependent upon the nature of the task to be performed. The method also supposes that this level of reliability will be consistently achieved within uncertainty limits given perfect conditions. Given these two premises, the method also assumes that in the absence of perfect conditions, human reliability degrades as a function of the applicability of Error Producing Conditions (EPCs)[6].

To facilitate combining both methods in order to eliminate the error reduction deficiency in the CREAM, a comparison of the 9 CPCs used in CREAM and the 38 EPCs [6] used in HEART was conducted. This PSF comparison provided the relationship between both methods and served to link the error reduction techniques given for the HEART EPCs to the CREAM CPCs.

Combining aspects of CREAM and HEART results in Cognitive Reliability and Error Analysis Method with Error Reduction Techniques (CREAM+RT), adopted based on a contextual task analysis and satisfactory responsiveness to the essential characteristics of a complete HRA method. This new composite method, CREAM+RT, still lacks a method for addressing error dependency, however, the introduction of this component could potentially introduce another layer of analyst subjectivity to the method thereby decreasing the confidence level in the method.

A key motivator for the selection of the CREAM as the foundational method is rooted in the method's ability to address basic types of human functions – cognitive, physical, and social. The method discretizes human function into four areas: observation, interpretation, planning, and execution. Each of these can be used to describe the individual steps outlined in the PTA.

3.5 CREAM+RT Overview

CREAM+RT is a composite HRA method that is based entirely on the SGM CREAM albeit with a slight modification which allows for incorporating reduction techniques derived from the HEART. This modification, as discussed in the preceding section, is the inclusion of the reduction techniques presented in the HEART method by evaluation of similar PSFs of both methods.

A detailed description of the task to be analyzed is developed to allow decomposition into subtask. This is usually performed as a task analysis. The subtasks can be matched to one of the method-specified cognitive activities. CREAM specifies 15 cognitive activities available in the literature [12]. The next step is the identification of the applicable cognitive activity for each subtask identified in the task analysis. The third step is to identify the associated human function for each subtask. As earlier stated, CREAM prescribes for human functions; observation, interpretation, planning, and execution.

In the next quantitative step, the basic human error probability (BHEP) for each subtask is determined. This is achieved by determining failure modes that result from human functions and then, associating them with a BHEP and CREAM-specified uncertainty bounds.
Following their initial quantification, adjustments due to CPC effects are made to the BHEP of the subtasks. CREAM specifies adjustment factors based on the CPC states [12]. The final step is to calculate the task HEP based on the adjusted BHEP of the subtasks. Utilizing the reduction techniques adapted from HEART, mitigation and control strategies can then be proposed in order to buy down the risk or error probability identified through the CREAM process.

The advantages of the proposed CREAM+RT include the following: allows for direct quantification of HEP, allows for contextual tailoring that explicitly fits the situation under assessment, results are readily adaptable to overall system reliability and safety analysis, allows for retrospective and predictive analysis, provides a concise, structured, and highly repeatable process, provides a set of error reduction techniques, and allows for assessment of the impact of error reduction techniques

The limitations of the proposed CREAM+RT method include the following: it is resource intensive, it may be time intensive depending on the level of analysis; and it requires a level of expertise in the field of human factors. The time intensiveness can be mitigated by the repeatability of the process, hence once a suitable framework is established for a large HRA effort, the process becomes more streamlined.

3.6 HRA Method Classification

Any analysis method must refer to a consistent classification scheme that is relevant for the domain under investigation [10]. Furthermore, the classification scheme must refer to a set of supporting theoretical principles; these principles are collectively referred to as the model. The classification scheme employed in defining the categories of effects and causes should be clearly traceable to the applicable model. The CREAM+RT method classification scheme is identical to CREAM classification and based on delineation between causes or genotypes and manifestations or phenotypes.

Genotypes are divided into three categories (individual, technological, organization). The first, **individual**, contains those causes that have a link to behavior such as personality and emotional state. The second category, **technological**, contains factors that are related to the human-system interface and interaction. The third category, **organizational**, includes those that are dictated by the organization such as local environment [2]. These three genotypes are fully described in the literature[16]:

Phenotypes are manifestations that result due to operator actions or omissions of actions. There are eight basic error modes or phenotypes that are divided into four sub-groups. The sub groups and error modes are also available in the literature[16].

The advantages of this classification scheme include the ability to predict and then describe how an error would occur. It also allows one to define the links between the genotypes and the phenotypes pertinent to the analysis. The classification scheme allows coverage of the three aspects of human function ensuring an exhaustive look at the task and potential sources of error.

3.7 Model Theory Development

Cognitive theories guide the CREAM+RT model. The foundational method of CREAM+RT, CREAM, is based on the Contextual Control Model (COCOM) [14]. The COCOM is discussed in below, however the basic concept posits that the degree of control an operator holds determines the reliability of their performance [16] as a consequence operator control is directly proportional to reliability of their performance

The COCOM is a model of human behavior that advocates the study of how a person's ability to maintain control of a situation enables effective control of a process or system on which they are working. This model is a deviation from the traditional study of human cognition which tends to focus on the cognition of the individual [17].

The PTA results show the criticality of cognitive ability to several key steps in the process. Additionally the UAs identified via the PTA direct the model selection towards a model that is largely based on cognitive theories. These cognitive theory principles range from perceptual principles, to principles of detection and understanding. A survey of the UAs listed for this task reveals several instances of correlation between an act and a cognitive theory. An example is the unsafe act "Underestimation of amount of Precipitate in Part A". This UA is clearly contrary to the first Perceptual Principle; "Avoid judging the level of a variable (e.g. loudness, color, size) which contains more than 5 to 7 possible levels.

The theoretical model of CREAM+RT also incorporates other elements of human behavioral modeling. These include information processing, type of response, human capacity and tendency, social and organizational influences.

The relevance of these additional dimensions of human behavior modeling to the CREAM+RT model is demonstrated here in relation to the UAs. Looking specifically at a single UA and its consequence, the applicability of social and behavioral models to the human error is shown.

Investigation of the failure of an electronic board in the power distribution unit of a spacecraft during testing revealed that the failure mechanism was a short circuit of the electronic board. The root cause has been identified as conductive debris on the printed circuit board (PCB) during the conformal coating process. The specific human error that led to this failure has been identified as: Improper surface preparation of the PCB before spraying of the conformal coating material.

To identify the appropriate theory for evaluating this error, an assumption is made that all steps in the conformal coating process where executed. A resultant concern is performance of each step. It is noteworthy that in the polymeric process, which includes staking and bonding of components, conformal coating is the last step prior to delivery for thermal bake-out and integration. Schedule pressure may become a valid PSF, which would directly impact the thoroughness of each executed step.

The Theory of Planned Behavior (TPB) posits that *behavior may not always be under volitional control.* TPB also proposes that behavioral control represents the perceived ease or difficulty of performing the behavior. The theory further states that behavioral control is impacted by knowledge of relevant skills, experience, emotions, and external circumstances. The TPB is based on the Theory of Reasoned Action (TRA), which states that intent to perform is a critical determinant of behavior. TRA further states that intention is influenced by attitude towards the action, behavioral expectations of the individual's social network, and motivation to comply with others' wishes.

Evaluating the applicability of these theories to the human error study at hand requires some understanding of the social structure in the organization. Technicians who rank much lower in the organization than design engineers perform PWA tasks in accordance with engineering design specifications and industry workmanship standards. The design engineers are usually under management pressure to deliver products on schedule and on budget. These pressures are communicated to technicians.

The following is a one-to-one mapping of the elements of the applicable theories to the Human Error incident.

Table 3-4 Theory Element Manifestation

Theory Element	Possible Manifestation
TRB: behavioral control represents the perceived ease or difficulty of performing the behavior	Technician is very conversant with process of spraying-cleaning a board and may or may not regard it as trivial
TRB: behavioral control is impacted by knowledge of relevant skills, experience, emotions, and external circumstances	Technician is under pressure to complete the conformal coating
TRA: intention is influenced by attitude towards the action	Technician is performing a cleaning action and assumes that debris cannot be generated but removed
TRA: intention is influenced by behavioral expectations of the individual's social network	Technician is expected to perform the task without questioning engineering decisions
TRA: intention is influenced by motivation to comply with others' wishes.	The technician ranks lower than the engineer in the organizational hierarchy and relies on engineering documents for guidance

From the aforementioned it is apparent that a theoretical model that addresses cognitive, social, and behavioral effects on human performance is desired. Understanding the attributes, specifically the CPCs and cognitive activities, antecedents and consequences, and error modes, of the foundational CREAM on which CREAM+RT is based allows a wider appreciation of the encompassing nature of its underlying model.

3.8 Data Collection

Any analysis is only as reliable as the data on which it is based. However before embarking on the pursuit of reliable data, we must first define data as pertains to the polymeric application process. There are two main categories of data that are potentially applicable in this context– objective data and subjective data. Objective data is a measurable representation of facts while subjective data contains a level of personal interpretation of the facts. The pertinent data in this case includes the number of instances of human error during conformal coating resulting in either a test or operational anomaly. This type of data is clearly objective and will be truncated to exclude errors and defects found during the post-process inspection of the PWA as these errors are remediated by reworking the PWA. Subjective data is also required but more so in establishing the contextual details of the conformal coating process. The subjective data supports the development of an accurate task analysis and the crosslinking of unsafe acts to personal performance shaping factors.

A list of data necessary for such analysis is provided below:

- Number of electronics failures attributed to human error during conformal coating
- Root cause information on actual error that caused the board failure
- Instantiation of the various root causes (a count of occurrence of an unsafe act that resulted in a failure)
- Number of conformal coated PWA boards
- Number of conformal coated PWA per electronics box – box level
- Time duration of performing the conformal coating task
- Contextual information

Given the data required to quantify the HEP in conformal coating, it is apparent that a record of anomalous events is required and would prove a satisfactory resource for the analysis. Such data support retrospective statistical analysis of such anomalous events and can serve as the validation of the predictive abilities of the HRA method used.

Spacecraft development processes and standards require collection of anomaly and problem failure data starting from the acceptance and qualification testing phase [18]. Most space system development efforts require a centralized closed-loop tracking system tracking system for documentation of anomalies, problems, and test failures. This centralized database system is essential for root cause documentation in addition to failure investigation. The data collection process for performing the proposed analysis will entail review of the spacecraft developer's anomaly and problem database and identifying each instance of component failure attributed to human error during conformal coating.

3.9 Data Analysis

This section documents the CREAM+RT analysis performed on the conformal coating process. An overview of the HEP calculation steps has been provided in Section 3.1.

From the PTA, the four secondary tasks are further decomposed in to discrete subtask steps. The secondary task and the number of associated subtask step are listed below.

1. Surface Preparation – 12 subtask steps
2. Chemical Preparation – 19 subtask steps
3. Chemical Application – 17 subtask steps
4. Curing and De-masking – 5 subtask steps

There are 53 total subtasks identified for the process.

3.9.1 HEP Calculation using CREAM+RT

Below is the list of all the subtasks in the Surface Preparation Secondary Task with their attendant identifier. Note that only error-inducing tasks are steps are included. Steps such as "return to engineering" are not considered since they do not induce error. The preceding numbers are included as the subtask identifiers.

1. Surface Preparation
    1.1. Is conformal coating with Arathane 5750 A/B specified?
    1.2. Clean the surface with Isopropyl Alcohol (IPA)
    1.3. Record Time of Cleaning
    1.4. Perform Ionic Contamination Test per ICT process
    1.5. Did surface pass the ICT?
    1.6. Record Oven Time-In (OTI)
    1.7. Bake-out PWA per Bake-out Process
    1.8. Record Oven Time-Out (OTO)
    1.9. Is present time within 8 hours of OTO?
    1.10. Apply masking in areas specified in engineering drawing
    1.11. Is present time within 8hrs of OTO?
    1.12. Apply Arathane 5750 per Chemical Application Process

The next step is to identify the type of cognitive activity associated with each subtask from the 15 activities specified by the CREAM+RT method. The following table, Table 3-5, is an extract of a matrix that links the steps to the cognitive task.

Table 3-5 Subtask-to-Activity Linking

Cognitive Activity	Description	Subtask #
Coordinate	Bring system states and/or control configurations into the specific relation required to carry out a task or task step. Allocate or select resources in preparation for a task/job, calibrate equipment, etc.	1.1
Compare	Examine the qualities of two or more entities (measurements) with the aim of discovering similarities or differences. The comparison may require calculation.	1.9 1.11

The next step is to identify the type of human function associated with each subtask and the HEPs. The error modes, based on CREAM error modes, within each of the four types of human function are as follows:
- Observation (O1 – Wrong Object observed, O2 – Wrong Identification, O3 – Observation not made)
- Interpretation (I1 – Faulty Diagnosis, 12 – Decision Error, 13 – Delayed Interpretation),
- Planning (P1 – Priority Error, P2 – Inadequate Plan)
- Execution (E1 – Action of Wrong Type, E2 – Action at Wrong Time, E3 – Action on Wrong Object, E4 – Action of Sequence, E5 – Miss Action).

The UAs associated with the Surface Preparation Task are:
1. Inadequate surface cleaning
2. Improper execution of Ionic Contamination Test (ICT)
3. Misreading of ICT Data
4. Misinterpretation of Engineering Bake out Data (data includes temperature and time for the PWA)
5. Error in recording Oven Time In (OTI) and/ or Oven Time Out (OTO)
6. Error in assessing length of time since OTO
7. Incorrect application of masking

8. Application of masking in areas not specified by engineering specifications

CREAM+RT specified error modes are assigned to the unsafe acts as follows:

1. Inadequate surface cleaning with IPA → E5
2. Improper execution of Ionic Contamination Test (ICT) → E5
3. Misreading of ICT Data → O2
4. Misinterpretation of Engineering Bake out Data (data includes temperature and time for the PWA) → O2
5. Error in recording Oven Time In (OTI) and/ or Oven Time Out (OTO) → E1
6. Error in assessing length of time since OTO → I1
7. Incorrect application of masking → E5
8. Application of masking in areas not specified by engineering specifications → E3

The "X"s in Table 3-6 correspond to the specific error mode ascribed to the UAs associated with the task. For example, there are three separate E5 errors.

Table 3-6 Error-Specific HEP Determination Matrix[12]

Type of Functional Failure BHEP Type of HSI Activity	Observation			Interpretation			Planning		Execution				
	O1	O2	O3	I1	I2	I3	P1	P2	E1	E2	E3	E4	E5
	1E-3	3E-3	3E-3	2E-1	1E-2	1E-2	1E-2	1E-2	3E-3	3E-3	5E-4	3E-3	3E-2
Coordinate													
Communicate													
Compare													
Diagnose					X								
Evaluate													
Execute											X		X X X
Identify		X X											
Maintain													
Monitor													
Observe													
Plan													
Record									X				
Regulate													
Scan													
Verify													

The next step is to adjust the basic HEPS with the CPC coefficients. For simplicity, we will assume that all CPC states are optimal. The CPC states should in reality be determined by collected contextual and objective data.

The most likely HEP is the final task HEP, calculated from the following equation:

$$Final\ HEP = Prob(Most\ Likely\ Error\ Mode | Activity\ Type) \times \prod_{i-1}^{9} Adjustment\ Coefficient\ of\ CPCs$$

The calculated Final HEP is 0.0201 and is representative of performance reliability in the Strategic Control Mode. It is important to note, however, that the HEP calculation assumed the most positive CPC states.

4.0 Conclusion

The contextual control model informs the risk assessment process given that it is the basis for the HRA method. Increasing the operators' proficiency in the four CREAM+RT cognitive functions, but specifically the execution function, will improve operator performance reliability. The results support the fact that a large contribution of human error in this process can be attributed to the higher level Personal PSF that is related physical and cognitive ability.

Given availability of resources, a suitable risk mitigation strategy would address all three unsafe acts. However, a higher fidelity analysis may be required to fully discriminate amongst the three and select the most likely contributor. This higher fidelity analysis would necessitate collection of more contextual and personal data.

The proposed HRA method leverages the attributes of existing HRA techniques and is extremely implementable given the existence of policies, procedures, and organizational resources currently in place. These policies and procedures -- such as the NASA Workmanship Standards, IPC J-Standards, NASA Handbook for Program Managers and Program Management of Problems, Nonconformance, and Anomalies, organizational plans, procedures, tools and databases for implementation of NASA and industry standards -- all can serve to facilitate implementation of this method.

To implement this method, spacecraft development organizations would need to establish a framework to periodically analyze the anomaly, problem, and failure reporting system data. The data would directly inform the HRA and alert program management of risk areas associated with the human element in the development process. Spacecraft requirements include system reliability analysis; these requirements could be expanded to require, at a minimum, a preliminary HRA based on the CREAM+RT. Analysis results that are indicative of unacceptable HEPs (based on the control modes) would trigger risk mitigation activities and in light of cost benefit analyses, stakeholder acceptance would be anticipated.

A limitation of the CREAM+RT is that it does not account for task or error dependency, although dependency is implicitly accounted for in the treatment of the genotypes and phenotypes. As was stated earlier, this dependency modeling is structurally similar to common cause modeling in system reliability analysis. Common cause failure modes are identified and then modeled explicitly for their contribution to overall system failure; as a consequence, component-specific failure modes are then adjusted to account for the common cause failures. Such a practice has been proven and validated for system and component reliability, however in the realm of HRA, it only serves to introduce additional analyst subjectivity. Decisions on how to assign weighting or adjustment factors on error-specific HEPs for cross-context tasks by analysts would devolve into arbitrary guesses since no validated method exists. It is in light of this that CREAM+RT is presented without error dependency. Investigating improvements on this deficiency could benefit from using the common cause failure modeling as a starting point.

The far-reaching impacts of human error during the fabrication phase of spacecraft require that a structured and methodical yet easily implementable approach be adopted for identifying and arresting these consequences. It is in consideration of this circumstance that the CREAM+RT HRA method is proposed. The *a priori* and *a posteriori* knowledge of cause and effect as related to human error would significantly improve system reliability for the high cost yet indispensable technological marvels known as spacecraft.

5.0 References

[1] A. D. Swain and H. E. Guttman, "Handbook of Human Reliability Analysis," 1983.

[2] S. French, T. Bedford, S. J. T. Pollard, and E. Soane, "Human reliability analysis: A critique and review for managers," *Saf. Sci.*, vol. 49, no. 6, pp. 753–763, Jul. 2011.

[3] D. E. Embrey, "SLIM-MAUD: A Computer-Based Technique for Human Reliability Assessment," *International Journal of Quality & Reliability Management*, vol. 3, no. 1. pp. 5–12, 1986.

[4] D. Gertman, H. Blackman, and J. Marble, "NUREG/CR-6883: The SPAR-H Human Reliability Analysis Method," 2005.

[5] G. W. Hannaman, A. J. Spurgin, and Y. D. Lukic, "NUS-4531, Rev 3: Human cognitive reliability model for PRA analysis," 1984.

[6] J. Williams, "A data-based method for assessing and reducing human error to improve operational performance," *Hum. Factors Power Plants*, pp. 436–450, 1988.

[7] P. C. Cacciabue, *Guide to Applying Human Factors Methods: Human Error and Accident Management in Safety-Critical Systems*. Springer, 2004.

[8] S. E. Cooper, A. M. (NRC) Ramey-Smith, and J. Wreathall, "NUREG/CR-6350: A Technique for Human Error Analysis (ATHEANA)," 1996.

[9] P. Meyer, P. Le Bot, and H. Pesme, "MERMOS : An extended second generation HRA method," *2007 IEEE 8th Hum. Factors Power Plants HPRCT 13th Annu. Meet.*, pp. 276–283, Aug. 2007.

[10] E. Hollnagel, *Cognitive reliability and error analysis method: CREAM*. Elsevier Science, 1998.

[11] C. A. Smith, "Probabilistic Risk Assessment for the International Space Station," *Jt. ESA-NASA Space-Flight Saf. Conf. Ed. by B. Battrick C. Preyssi. Eur. Sp. Agency, ESA SP-486, 2002. ISBN 92-9092-785-2., p.319*, 2002.

[12] F. Chander, Y. Chang, and A. Mosleh, "Human reliability analysis methods: Selection guidance for NASA," 2006.

[13] NASA HQ, "NASA-STD-8739.1A Workmanship Standard for Polymeric Application on Electronic Assemblies." 2008.

[14] E. Hollnagel, *Human reliability analysis: context and control*. Academic Press, 1993.

[15] J. Bell and J. Holroyd, "HSE RR679 Research Report: Review of human reliability assessment methods Review of human reliability assessment methods," 2009.

[16] Http://www.skybrary.aero/, "Cognitive_Reliability_and_Error_Analysis_Method_(CREAM) @ www.skybrary.aero." [Online]. Available: http://www.skybrary.aero/index.php/Cognitive_Reliability_and_Error_Analysis_Method_(CREAM).

[17] Y. Waern, *Cooperative Process Management: Cognition And Information Technology: Cognition And Information Technology*. Taylor & Francis, 2004.

[18] NASA HQ, "NASA-HDBK 8739.18 Procedural Handbook for NASA Program and Project Management of Problems, Nonconformances, and Anomalies," 2008.

Use of Bayesian Network to Support Risk-Based Analysis of LNG Carrier Loading Operation

Arthur Henrique de Andrade Melani[a], Dennis Wilfredo Roldán Silva[a], Gilberto Francisco Martha Souza[a]*

[a] University of São Paulo, São Paulo, Brazil

Abstract: This paper presents a methodology for risk analysis of LNG carriers operations aiming at defining the most critical pieces of equipments as for avoiding LNG leakage during loading and unloading operations. The pieces of equipment considered critical for loading and unloading operations are identified and the Cause-Consequence diagram is built. The probability of occurrence of each event listed in the diagram is calculated based on Bayesian network method. The consequences associated with those scenarios are estimated based on literature review. Based on the calculated risk profile some maintenance and operational recommendations are presented aiming at reducing the probability of occurrence of the critical failure scenarios.

Keywords: LNG, Cause-Consequence diagram, Risk analysis, Bayesian networks.

1. INTRODUCTION

The increasing worldwide demand for Liquefied Natural Gas (LNG) has corroborated its importance as a component of the world's supply of energy. Once the great consumers (concentrated in Asia, Europe and North America) of natural gas are not the great worldwide producers (concentrated in the Middle East, Africa and Central America) the need for transportation of that hydrocarbon has still increased [1].

Natural gas can be transported in its liquid form by large LNG carriers between liquefaction plants at exporting countries (where LNG is loaded into the tank ship) and regasification plants at importing countries (where LNG is unloaded from the tank ship).

As well as any other industrial activity, the natural gas industry is not free from accidents, which can cause serious consequences to the integrity of people and properties. For this reason, it is necessary to develop studies to determine what are the possible causes and scenarios of these faults specifically in the area of LNG transportation [2].

Considering the high quantity of equipments involved in the LNG carriers loading and unloading operations and also the large volume of flammable liquid that is transferred during these proceedings, the use of risk analysis techniques available in the literature are recommended to avoid accidents during such procedures.

Based on the Formal Safety Assessment (FSA) guidelines proposed for petroleum industry [3], this paper presents a risk-based method to analyze the failure scenarios and associated consequences that may happen during the cargo handling operation of a LNG carrier. The paper determines the possible leakage causes and classifies their consequences. The risk involved in loading and unloading operations is described considering the probability of occurrence of each failure mode and the consequences of the leakage based on a risk matrix method. Recommendations to avoid the occurrence of LNG leakage are discussed.

2. METHODOLOGY

The method used for LNG carrier operation analysis is a risk-based approach based on Formal Safety Assessment (FSA) guidelines proposed for petroleum industry [3]. The first step is to identify and to select the pieces of equipment and components which are part of the loading and unloading system.

* gfmsouza@usp.br

The functional tree is developed to explain the functional relation between the pieces of equipment aiming at the reliable operation of the system.

The second step is called hazard identification, which applies the Preliminary Hazard Analysis, and is used to identify the hazards associated with loading and unloading operations, related to the failure of pieces of equipment. The Cause-Consequence Diagram is used to identify the failure scenarios associated with a given hazard occurrence and the control and alarm systems used as barriers to avoid failure propagation. The probability of each of the events listed in the cause-consequence diagram is calculated based on Bayesian network method. That method is an alternative for the use of Fault Tree Analysis to define the probability of occurrence of a specific event.

The third step is the development of the risk analysis in order to obtain a quantitative value of risk which allows the classification of the risk associated with a given hazard as low, acceptable and unacceptable, in accordance with a risk matrix. Finally, the recommendations to reduce the occurrence of the events that may cause a given failure scenario can be proposed. In Figure 1 a flowchart is used to illustrate each step of the proposed method.

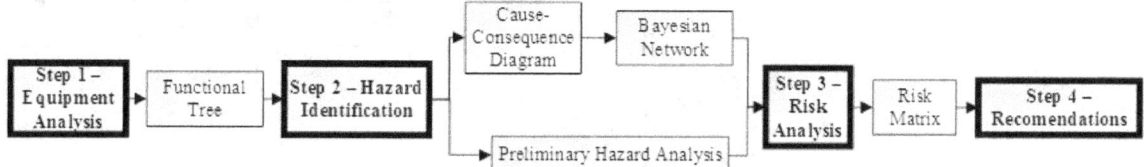

Figure 1 - Flowchart of the Proposed Method

3. APPLICATION OF THE METHOD

The present study analyzes the loading and unloading operations of a Mark III LNG carrier with four storage tanks and cargo capacity of 138,000 m3. The loading and unloading system is composed by subsystems, as for example: pumping, storage, distribution, relief system and the manifold. The main components used in the cargo handling system are shown in Fig 2.

During loading operation, LNG is loaded through the manifold and it is carried through two secondary pipelines to the liquid header line, which distributes it to each tank. Loading is completed when all tanks are loaded with 98,5% of its full capacity. After that, LNG is drained from the valves and pipelines and sent to a cargo tank, avoiding the presence of methane in the inactive lines.

The LNG is unloaded with the use of one main cargo pump for each tank which is submerged inside the respective tank. The main cargo pumps discharge the LNG to the main liquid header and then this fluid is transferred to the terminal through the manifold connections. Each tank is not fully discharged leaving a volume of LNG corresponding to a level of about 0,1m. On completion of discharge, the loading arms and pipelines are purged and drained into one cargo tank and the loading arms are then also disconnected

The loading and unloading system has a liquid header line that have two relief valves which function is to transfer the LNG relief to the cargo tanks when the liquid pressure is higher than 10 bar (relief valves set-up pressure). Usually, the pressure inside the pipelines is 1 bar. In the loading and unloading piping system are installed relief valves to avoid the raise in liquid pressure.

The storage system consists of four insulated cargo tanks, separated from each other by transverse cofferdams, and from the outer hull of the vessel by wing and double bottom ballast tanks. The insulation covers the entire primary barrier which purpose is to maintain the cryogenic temperature and to prevent the generation of the boil-off gas. According to [3], the LNG carriers have a secondary barrier that is used to contain the LNG in the case of primary barrier failure and to avoid the contact of the ship's structure with the low temperatures of the cryogenic substance.

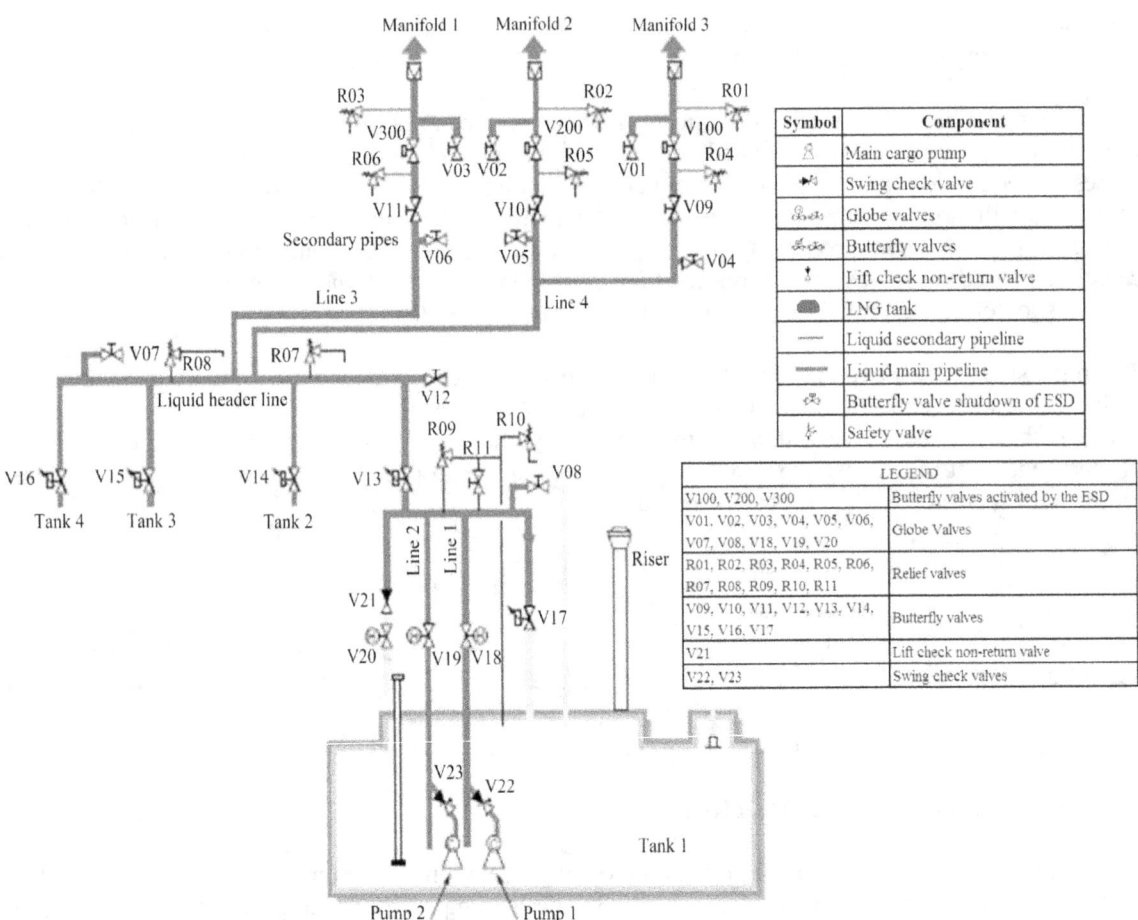

Figure 2 - LNG Circuit for Loading and Unloading Operations [14]

3.1. Functional Tree

The functional tree (Figure 3) is used to describe a system, determining its functions and the contribution of each of its components to the system performance. The cargo handling system is divided in five subsystems: pumping, storage, distribution, manifold and relief. Those subsystems are divided in components each of one performing a specific function linked with subsystem main function. A failure in a component at the bottom of the tree affects the performance of the subsystem above it, causing a possible interruption in loading or unloading operations, including LNG leakage.

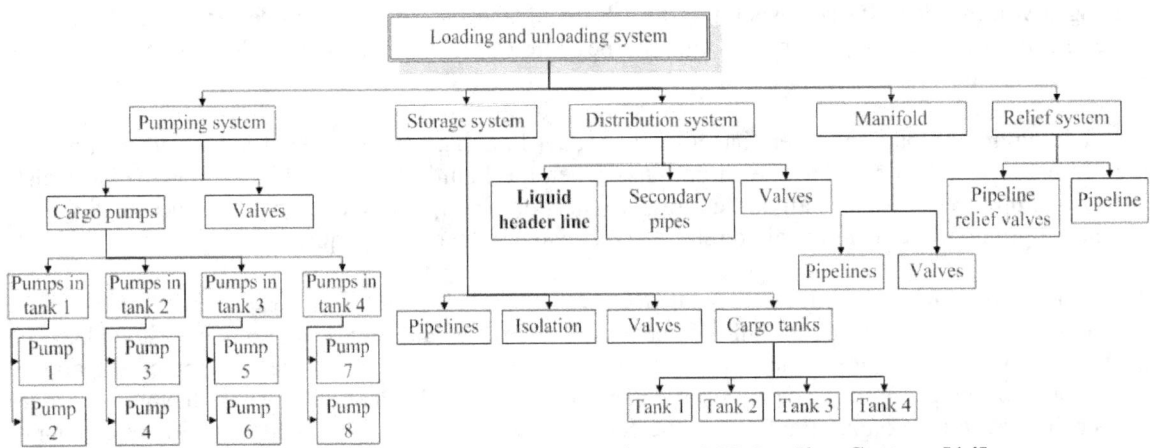

Figure 3 - Functional Tree of the Loading and Unloading System [14]

3.2. Preliminary Hazard Analysis

The Preliminary Hazard Analysis (PHA) is used to identify those accidental events that will be subject to the further risk analysis. The PHA technique was chosen to be applied here because it can be used in any period of the equipment lifecycle, including design and operation.

The present analysis studies the hazard LNG leakage during the loading and unloading operations. The causes of occurrence, consequences and safeguards associated with that hazard are identified and analyzed to develop the PHA table. This analysis is shown in the Table 1. The causes of LNG leakage considered in the PHA associated with the valves are structural deficiency, external leakage (process medium), and valve leakage in closed position. In the case of the cryogenic pumps the failure mode considered in the analysis is fail to stop on demand. In pipelines the main failure modes are presence of a through thickness crack, partial and total pipe cross section rupture.

3.3. Cause-Consequence Analysis

The Cause-Consequence Analysis is used to identify and evaluate the sequence of events that can happen given a initiating event. The analysis aims at determining if the initiating event can induce an accident or if the accident is avoided by the protection barriers of the system. In this paper, the analysis begins with the failure of the components of the cargo handling system and is centered in the occurrence of LNG leakage. Considering LNG leakage in the loading/unloading system, a series of events can happen, as shown in the Cause-Consequence Diagram presented in Figure 4.

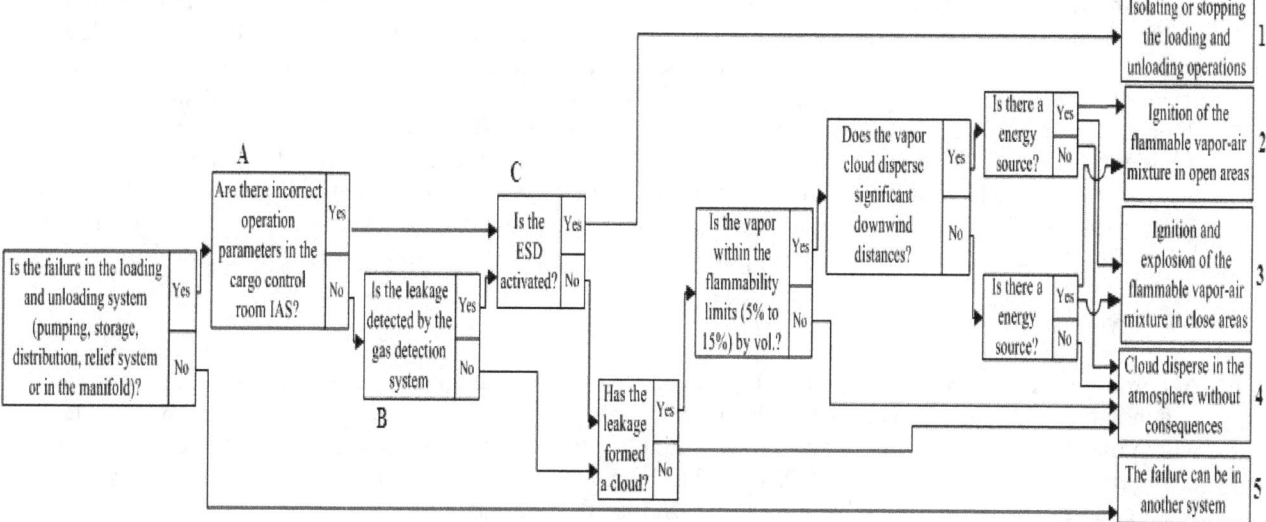

Figure 4 - Cause-Consequence Diagram

The safety barriers are designed to stop the loading/unloading operation and to avoid the continuous LNG leakage. These barriers ideally should not fail because any failure can cause major consequences.

The first barrier is the Cargo Control Room, which remotely controls and monitors the cargo handling operations. All major valves such as the manifold valves, also called ESD Manifold valves, and individual tank loading and discharge valves, are remotely operated from the IAS, so that all normal cargo operations can be carried out from the cargo control room.

The second barrier is the Gas Detection System, which detects the presence of gas, especially in spaces where gas is not normally expected to be presented. Various sensors monitors both hazardous and non-hazardous gas zones. If gas is detected, alarms are activated, indicating the occurrence of a leakage. The entire cargo piping system and cargo tanks are considered gas hazardous zones.

The third barrier is the Emergency Shutdown System (ESD). During the loading and unloading operations in case of LNG leakage, the emergency shutdown (ESD) system can be used to isolate the leaking pipe section and to stop the primary pumps and to close the ESD valves to avoid a large liquid spill. The ESD system is automatically activated in response to hazard detectors (gas and fire detectors), process alarms (pressure loss in pipe) or an operator pushing an ESD button, as defined by [4]. This system acts in response to a liquid release, interrupting the duration of the release and so affecting the consequences associated with that leakage. In case of LNG leakage, the ESD system can automatically isolate the cargo handling system or stop the process by shutting down the primary pumps and/or closing the ship-side valves located in the manifold V100, V200, V300 showed in Figure 2.

The possible scenarios for the LNG leakage are listed in the sequence, according to Figure 4 and described in Table 1.

The first scenario is the failure in the cargo handling system but the IAS works not causing important consequences once the ESD operated. In this case the failed line must be isolated, controlling the leakage, and the operation must be stopped. The second scenario occurs when after the LNG leakage the IAS does not detect the variation of the main process parameters like pressure, temperature, or flow. If a variation of these parameters is not detected by the IAS the ship has a gas detection system which works and consequently the ESD is activated, causing the stop of the loading or unloading operation.

The third scenario takes place if no one of those safety systems activates the ESD. In the place where leakage occurs a pool can be formed with a vapor cloud which concentration can be between the low and upper flammability limit but in the absence of an ignition source the vapor disperse into the atmosphere without causing effects to the ship or to the terminal. The downwind distance that flammable vapors might reach is a function of the volume of LNG spilled, the rate of the spill, and the weather conditions. The last scenario has the same sequence of the third scenario but the difference is that the vapor cloud is ignited by an energy source from the ship or from the terminal. The result is an ignition of the flammable vapor-air mixture in open areas and an ignition with explosion in close areas. The flame will burn back to the vapor source possibly causing a pool fire, according to [5].

3.4. Bayesian Network

According to [6, 7], a Bayesian Network (BN) consists of a directed acyclic graph in which each node is annotated with quantitative probability information. Each node corresponds to a random variable, which may be discrete or continuous. A set of directed links or arrows connects pairs of nodes. If there is an arrow from node X to node Y, X is said to be a parent of Y. To each node Y with parents X_1, ..., X_n, a conditional probability table $P(Y|X_1, ..., X_n)$ is attached, quantifying the effect of the parents on the node.

BNs have become a widely used formalism for representing probabilistic systems and have been applied in a variety of areas, especially in Artificial Intelligence. In dependability and risk analysis, however, other techniques, like Fault Tree Analysis (FTA), are yet more employed for evaluations of safety-critical systems. But the modelling flexibility of the BN formalism can accommodate various kinds of statistical dependencies that cannot be included in the FTA, for example, obtaining a more precise result [8].

In this paper, BNs are built to obtain the reliability of the barriers presented in the Cause-Consequence Diagram. Although three barriers are described, only the BN from the ESD system (third barrier), is shown (Figure 5). Databases [9] and [10] are used to define the reliability of the different components of the barriers. Table 2 shows the calculated reliabilities of the barriers, for one year of operation.

Table 1 - Preliminary Hazard Analysis

N°	Hazard	Cause	Consequence	Safeguards
1	LNG leakage	Failure in the connection of the loading arms with the ship's manifold.	Structural damage of the ship's structure due to the LNG leakage. There is the possibility of vapor cloud formation. Stopping the loading or unloading process. Activation of the emergency system.	Drip tray is installed in the manifold areas in order to collect any spillage and drains it overboard. The ship has a monitoring system that monitors and indicates which are the internal conditions of the circuit of LNG and an alarm system that indicates the occurrence of natural gas leakage allowing the interruption of the transfer process.
2	LNG leakage	Structural deficiency in the valves.	Structural damage of the ship's structure due to the LNG leakage. There is the possibility of vapor cloud formation. Stopping the loading or unloading process. Activation of the emergency system.	The ship has a monitoring system that monitors and indicates which are the internal conditions of the loading and unloading system and an alarm system that indicates the occurrence of natural gas leakage allowing the interruption of the transfer process.
3	LNG leakage	Valve leakage in closed position allowing that LNG can circulate in other systems such as the relief system, the emergency system or the spray system.	Entry of LNG into the spray system. Entry of LNG into the relief system. Stopping LNG transfer process.	There are valves to contain the LNG preventing the flow of LNG to a line where it should not be as for example the emergency pipelines or the pipes that are used for the spray operation. The ship has a monitoring system that monitors and indicates which are the internal conditions of the loading and unloading system and an alarm system that indicates the occurrence of natural gas leakage allowing the interruption of the transfer process.
4	LNG leakage	Crack or rupture in the liquid header line, in the secondary pipelines or in the relief pipelines.	Damage in the ship's structure. Possibility of vapor cloud formation. Freezing in the surrounding areas. Possibility of entry of atmospheric air into the LNG system and breaking the inert environment. Stopping the LNG transfer process.	The ship has an emergency system that stops the loading and unloading process in case of a leakage. The ship has a monitoring system that monitors and indicates which are the internal conditions of the loading and unloading system and an alarm system that indicates the occurrence of natural gas leakage allowing the interruption of the transfer process.
5	LNG leakage	Crack or rupture in the primary cargo tanks.	Damage in the ship's structure. Possibility of vapor cloud formation. Freezing in the surrounding areas. Possibility of entry of atmospheric air into the LNG system and breaking the inert environment. Stopping the LNG transfer process.	The ship has a monitoring system that monitors and indicates which are the internal conditions of the loading and unloading system and an alarm system that indicates the occurrence of natural gas leakage allowing the interruption of the transfer process. There is a secondary tank that has the function of containment the LNG in case of any leakage from the primary tank. There is an emergency system that stops the loading and unloading process.
6	LNG leakage	Failure of the alarm level inside the LNG storage tanks that will cause the overfilling of one or more storage tanks.	Stopping of the loading or unloading transfer process. Damage to the ship's structure. Freezing in the surrounding areas. Possibility of vapor cloud formation.	There are three levels of alarms that paralyze the cargo pumps. The ship has a monitoring system that monitors and indicates which are the internal conditions of the loading and unloading system and an alarm system that indicates the occurrence of the natural gas leakage and allow the suspension of the transfer process.
7	LNG leakage	Pressure increase within the loading and unloading system due to high output discharge pressure in the cargo pumps.	Rupture or crack in the LNG circuit due to high pressure inside it. Stopping of the LNG transfer process. Possibility of vapor cloud formation.	The ship has a monitoring system that monitors and indicates which are the internal conditions of the loading and unloading system and an alarm system that indicates the occurrence of the natural gas leakage allow the suspension of the transfer process.

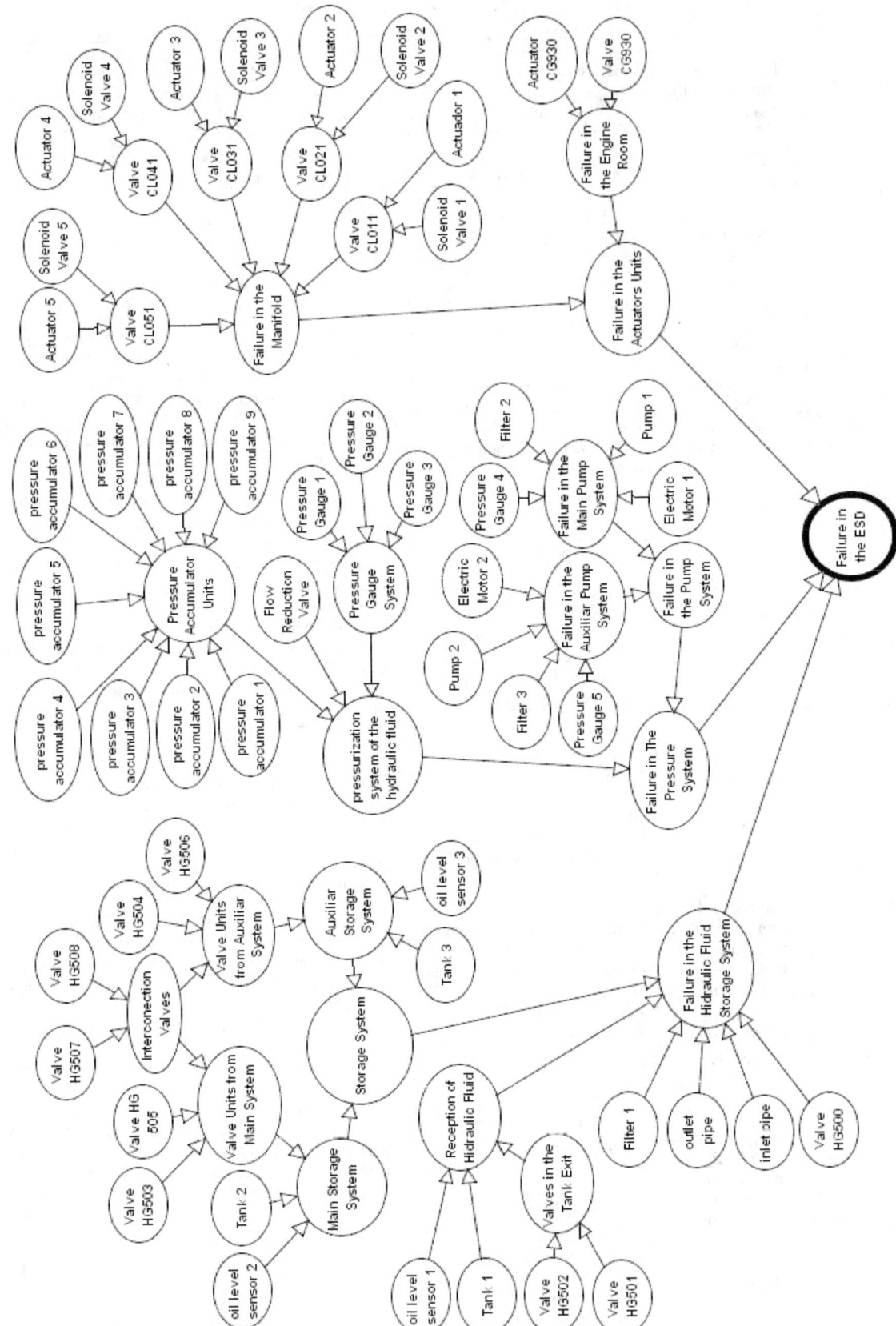

Figure 5 - Bayesian Network of the ESD System

Table 2 - Calculated Reliabilities of the Barriers

Barrier	Reliability (for one year)
Cargo Control Room (CCR)	R = 0,99
Gas Detection System (GDS)	R = 0,93
Emergency Shutdown System (ESD)	R = 0,83

In order to compute the probability of the scenario 1, "Isolating or stopping loading and unloading operations", in the Cause-Consequence Diagram, it is necessary take into account all the possibilities that lead to this scenario given a leakage. In this case, there are two possibilities:

- Cargo Control Room detects a incorrect operational parameter (A) and the ESD is activated (C);
- Cargo Control Room doesn't detect a incorrect operational parameter (A'), the Gas Detection System detects a leakage (B) and the ESD is activated (C).

The probability of occurrence of the first scenario is, then, given by:

$$P(Sc\ 1) = A*C + A'*B*C = 0,99*0,83 + 0,01*0,93*0,83 = 0,829 \quad (1)$$

The high probability of occurrence of the first scenario shows that the three barriers have an important role in preventing accidents given an LNG leakage in the loading/unloading system. The evaluation of the probability of occurrence of scenarios 2 to 5 are not discussed in the present paper once it involves cloud dispersion analysis which is not focus of the paper. Nevertheless, the barriers reliability has influence on those scenarios development. Clearly the loading/unloading operations can be considered safe due to the presence of reliable barriers. The ESD system presents the lowest reliability among the safety barriers once it presents a great number of components that must be working to keep system functionality.

3.5. Risk Analysis

Risk is defined as the evaluation of the probability of occurrence of an event and consequences associated with the occurrence of this event. The risk must be calculated based on the failure analysis of each component that can cause LNG leakage.

To classify the probability of occurrence of an event as well as its consequences, the technical standard N2784, which corresponds to the Petrobrás risk classification [11], is used. The Risk Matrix used is also extracted from [11].

Table 3 shows the frequency categories used in this study. The probability of occurrence is defined by the number of occurrence of an event in one year. To calculate the probability of occurrence, databases [9] and [10] are used to define the failure rate of the different components of the loading and unloading system.

The probability of occurrence of a given number of events in a period of time is determined by the use of the Poisson's distribution, once all components reliability is modeled with exponential distribution. The following equation shows the expression for the calculation of that probability:

$$P(n) = \frac{(\lambda t)^n \cdot e^{-\lambda t}}{n!} \quad (2)$$

The number of occurrence for all cases during the operational time is n=1 which represents one event occurrence during that time. The failure rate selected in the databases was the upper value which represents the most conservative approach for this analysis. To allow the frequency category classification according to Table 3, the operational period considered for the analysis is 8760 hours (one year). This period refers to the effective operation of the cargo handling system and not to the calendar year.

Table 4 shows the consequence categories used in this study. The consequence is measured as the impact that the LNG leakage can cause, such as injures of people (crew or third parts) or material damages (in the LNG carriers or in the terminal). The environmental impact is not analyzed in this paper.

Table 5 shows the frequency and consequence categories of certain components. The list of components in Table 5 was extracted from the Preliminary Hazard Analysis (Table 1). The probability of occurrence was calculated using the equation above.

Table 3 - Frequency Categories

Category	Denomination	Range (occurrence/year)	Description
A	Extremely remote	Less than 1 in 10^6 years	Conceptually possible but extremely unlikely to occur during the lifetime of the facility. There is no reference to historical occurrence
B	Remote	Between 1 in 10^4 years and 1 in 10^6 years	Not expected to occur during the lifetime of the facility, even though this may have occurred somewhere in the world
C	Less probable	Between 1 in 10^2 years and 1 in 10^4 years	Likely to occur once during the lifetime of the facility
D	Probable	Between 1 in a year and 1 in 10^2 years	Expected to occur a few times during the lifetime of the facility
E	Frequent	Over to 1 in a year	Expected to occur many times during the lifetime of the facility

Table 4 - Consequence Categories

Consequence category		Description	
		Personal safety	Safety of the facility
I	Negligible	Do not cause injuries or deaths of employees or third parts; and/or neighbour community; the maximum consequences are cases of first aid or minor medical treatment	No damage or minor damage to equipment or in the facility
II	Marginal	Minor injuries in employees and/or in third parts	Slight damage in the equipments or in the facility (damages are controllable and/or low-cost repair)
III	Critical	Minor lesions in third parts. Lesions of moderate severity in employees, contractors and/or people from outside the facility (remote probability of death of employees and/or other people)	Severe damage in the equipment or in the facilities
IV	Catastrophic	Causes death or serious injuries to one or more people (employees, contractors and/or third parties)	Irreparable damage in the equipment or in the facilities (repair is slow or impossible)

Having Tables 3, 4 and 5, it is now possible to analyse the risk of the loading/unloading system using a Risk Matrix. According to [12], the risk matrix approach, combining the likelihood of occurrence of an event and the consequence, defines risk as a pair located in a given matrix. Risk matrices have been used extensively for screening of risks in many industries. The risk matrix used in the present study is presented in Table 6, according to [11]. Risk increases in the direction of the upper-right side of the matrix and the category changes from NC (non critical) through M(moderate) and C (critical).

For each event listed in Table 5 a risk analysis is performed considering the probability of failure and the consequences of failure, according to the scenarios developed in the Cause-Consequence diagram. In case of the failure mode in the pump corresponding to 'fail to stop on demand' and in case of butterfly valves activated by the ESD failure mode 'fail to close on demand' the risk is classified as C (Critical) once if those components fail, there can be a large leakage because the loading and unloading operations are not interrupted by ESD.

On the other hand, the risk value associated with the majority of components is classified as M (Moderate). In case of total rupture of the pipelines cross section or primary tank barrier failure, the risk is considered M (moderate) once the probability category for those events is extremely remote (according to Table 5), although the occurrence of those events can cause serious consequences due to the enormous quantity of LNG leaked if the ESD system fails.

Table 5 - Probability of Occurrence and Consequence for Failures

Component	N°	Failure Modes	Failure Rate [failure/hour]	P (n=1) Probability	Probability Category	Consequence Category
Centrifugal pump	1	Fail to stop on demand	1.55E-06	0.01340243	D	IV
Lift non-return valve	2	All modes	1.46E-06	0.01263428	D	II
Swing check valve	3	All modes	1.46E-06	0.01263428	D	II
Globe valve	4	External leakage-Process medium	1.94E-06	0.01671740	D	II
	5	Structural deficiency	3.90E-07	0.00340677	C	II
	6	Valve leakage in close position	9.70E-07	0.00843012	C	II
Butterfly valve	7	Structural deficiency	3.90E-07	0.00340677	C	II
	8	External leakage-Process medium	4.31E-06	0.03637678	D	II
	9	Valve leakage in close position	7.40E-07	0.00644423	C	II
Cargo tank (primary barrier)	10	Catastrophic	5.70E-12	0.00000005	A	IV
	11	Minor failure	1.14E-10	0.00000100	A	II
Secondary and relief pipelines (400 mm)	12	Crack of 4 mm in pipelines between 300 and 499 mm	9.13E-09	0.00008000	B	II
	13	Rupture 1/3 pipeline diameter in pipelines between 300 and 499 mm	2.28E-09	0.00002000	B	IV
	14	Guillotine in pipelines between 300 and 499 mm	7.99E-10	0.00000700	B	IV
Liquid header line (600 mm)	15	Crack of 4 mm pipelines between 500 and 1000 mm	7.99E-09	0.00007000	B	II
	16	Rupture 1/3 pipeline diameter in pipelines between 500 and 1000 mm	1.14E-09	0.00001000	B	IV
	17	Guillotine in pipelines between 500 and 1000 mm	4.56E-10	0.00000400	B	IV
Butterfly valve activated by the ESD	18	Fail to close on demand	5.42E-05	0.29550184	D	IV
	19	Structural deficiency	5.16E-06	0.04322743	D	II
	20	External leakage-Process medium	6.47E-05	0.32155361	D	II
Relief Valve	21	Structural deficiency	8.50E-07	0.00739505	C	II
	22	External leakage-Process medium	3.00E-06	0.02561267	D	II
	23	Valve leakage in close position	2.62E-06	0.02244309	D	II
Manifold	24	All modes	2.27E-05	0.16283750	D	II

Table 6 - Risk Matrix

			Consequence			
			Negligible	Marginal	Critical	Catastrophic
			I	II	III	IV
Frequency	Frequent	E	M	M	C	C
	Probable	D	NC	M(2;3;4;8; 19;20;22;23; 24)	C	C (1;18)
	Less Probable	C	NC	M(5;6;7;9; 21)	M	C
	Remote	B	NC	NC(12;15)	M	M(13;14;16; 17)
	Extremely remote	A	NC	NC(11)	NC	M (10)

The failure of the tank primary barrier is not so critical because it has the secondary barrier which function is to collect the LNG in case of primary barrier failure. Table 6 shows the risk of each one of the failure modes presented on Table 5 (observing their numeration). The analysis indicates 3 failure modes that are considered NC, the majority (19 failure modes) are considered M and finally 2 failure modes correspond to Critical category which are number 1 and number 18 ('fail to stop on demand' of the pump and 'fail to close on demand' of the butterfly valves activated by the ESD respectively).

3.6. Recommendations

The maintenance procedures and the operational recommendations can be used as contingency measures aiming at reducing the probability of failures of the components listed in the Cause-Consequence diagram.

The ESD system presents the lowest reliability among the safety barriers once it presents a great number of components that must be working to keep system functionality. It is mainly constituted by sensors that are considered electronic devices as for reliability analysis. The pieces of equipment, such as valves, that are controlled by ESD also have actuators that are usually hydraulic powered. As for reliability analysis the duration of the useful life of electronic components is very long and the failure rate is considered constant. The maintenance activities associated with sensors are typically corrective aiming at restoring the functionality of the device after the loss of function or performance. The hydraulic actuators reliability can be modeled as aging components with monotonically increasing failure rate during operational life. For those components preventive inspection and maintenance can reduce reliability deterioration. To check the sensors and actuators operational condition, at the beginning of loading/unloading operation, the ESD system is tested, including the actuation of valves and cargo pumps of LNG carrier and terminal. Also, after the cool down operation, the valves are actuated in order to verify any detrimental effect of the cargo low temperature on the valves performance. Nevertheless, due to its random failure nature, the sensors can fail during loading/unloading operations, without presenting previous performance deterioration. Those failures can affect ESD system reliability as proposed in the failure scenario presented in the cause-consequence diagram.

For the valves controlled by ESD the maintenance recommendations are preventive inspection and time based substitution of components subjected to wear. A periodically tested and repaired component can have its failure rate modeled as constant, provided that the maintenance activities cause no deterioration of the valve. Nevertheless, there is still a chance that a valve can fail during loading/unloading operations.

For structural components such as piping system and cargo tank primary barrier the Linear Elastic Fracture Mechanics concepts can be used to calculate the number of load cycles to cause the propagation of a crack until it becomes a through thickness crack, causing LNG leakage [13]. The crack propagation in pipelines is associated with the cyclic loads induced by temperature change during cooling down operations. In the primary barrier the main load is the weight of the cargo. Structural inspection during ship life and pressure test of piping system before loading/unloading operations will reduce the probability of structural failure. Special care must be taken during the execution or repair welds aiming at not introducing more defects in the structural part.

4. CONCLUSIONS

The proposed method of risk analysis allows understanding the events that cause the LNG leakage and the consequences of those events during cargo handling operations. The method helps to determine the critical components, which failures lead to a high level risk.

The use of Bayesian Networks to help the quantitative analysis of the Cause-Consequence Diagram proved to be efficient. BNs are very appropriate to represent complex dependencies between components. Unlikely FTA, however, BN does not allow an easy study of the system just by analyzing its configuration, been necessary to know the Conditional Probability Table from each node.

Although the paper identified some failure scenarios that could cause critical consequences in case of LNG leakage, it also stressed the safety measures adopted by LNG transportation industry to prevent an accident. The sophisticated safety systems include gas detection and low temperature monitoring, heat and fire detection and cargo-related emergency shutdown devices. All processes involved in LNG handling are certified by classification societies to ensure international standard of safety.

Nevertheless, the paper also shows the possibility of improving operational safety based on developing a reliability database specific for equipment used in LNG carriers and terminals. That database would support precise reliability estimate that would improve risk analysis and design of this type of ship. It would also support the improvement of maintenance procedures developed for this type of ships.

Acknowledgements

The authors thank for the financial support of Financiadora de Estudos e Projetos (FINEP) and Conselho Nacional de Desenvolvimento Científico e Tecnológico (CNPq).

References

[1] G. F. M. Souza, E. M. P. Hidalgo, D. W. R. Silva, M. R. Martins. *"Probabilistic Risk Analysis of a LNG Carrier Loading Pipeline"*, Proceedings of the 31st International Conference on Ocean, Offshore and Arctic Engineering, OMAE 2012, Rio de Janeiro, Brazil.

[2] E. M. P. Hidalgo, D. W. R. Silva, G. F. M. Souza. *"Application of Markov Chain to Determine the Electric Energy Supply System Reliability for the Cargo Control System of LNG Carriers"*, Proceedings of the 32nd International Conference on Ocean, Offshore and Arctic Engineering, OMAE 2013, Nantes, France.

[3] International Maritime Organization. *"Formal Safety Assessment (FSA) - Liquefied Natural Gas (LNG) Carriers. Detail of the Formal Safety Assessment"*. MSC 83/INF 3, IMO 2007.

[4] S. R. Cheng, B. Lin, B. M. Hsu, M. H. Shu. *"Fault-tree analysis for liquefied natural gas terminal emergency shutdown system"*, International Journal of Expert Systems with Applications, Vol 36, pp. 11918 – 11924, 2009.

[5] American Bureau of Shipping. *"Consequence Assessment Method for Incidents Involving Releases from Liquefied Natural Gas Carriers"*, Rep# GEMS 1288209, Federal Energy Regulatory Commission, ABS 2004.

[6] S. Russell, P. Norvig. *"Artificial Intelligence: A Modern Approach"*, Third Edition, Prentice Hall, 2010, New Jersey.

[7] F. V. Jensen, T. D. Nielsen. *"Bayesian Networks and Decision Graphs"*, Springer Science, 2007, New York.

[8] A. Bobbio, L. Portinale, M. Minichino, E. Ciancamerla. *"Improving the Analysis of Dependable Systems by Mapping Fault Trees Into Bayesian Networks"*, Reliability Engineering and System Safety, Vol 71, pp. 249-260, 2001.

[9] DNV- Det Norske Veritas. *"OREDA – Offshore Reliability Data Handbook"*, 4th Edition, 2002, Norway.

[10] HSE - Health and Safety Executive. *"Failure Rate and Event Data for use within Land Use Planning Risk Assessments 2012,"* Available at http://www.hse.gov.uk/landuseplanning/failure-rates.pdf. Accessed in 2014, Jan, 20th.

[11] Petrobrás. *"N2784 - Confiabilidade e Análise de Riscos"*, 2005, Rio de Janeiro.

[12] E. Skramstad, S. U. Musaeus, S. Melbo. *"Use of Risk Analysis for Emergency Planning of LNG Carriers"*, Available at http://www.ivt.ntnu.no/ept/fag/tep4215/innhold/LNG%20Conferences/2000/Data/Papers/Skramstad.pdf . Accessed in 2014, Jan, 20th.

[13] J.J. Wilson. *"An Introduction to the Marine Transportation of Bulk LNG and the Design of LNG Carriers"*, International Journal of Cryogenics, Vol 14, pp 115-120, 1974.

[14] G. F. M. Souza, D. W. R. Silva *"Risk-Based Analysis of LNG Carriers Loading and Unloading Operations"* Proceedings of the 22th International Offshore and Polar Engineering Conference, ISOPE 2005, Rhodes, Greece.

Probabilistic Analysis of Geological Properties to Support Equipment Selection for a Deepwater Subsea Oil Project

Christopher J. Jablonowski[a], Edward E. Shumilak[a], Kenneth F. Tyler[a], Arash Haghshenas[b]

[a]Shell Exploration and Production Company, Houston, TX, U.S.A.
[b]Boots & Coots Services LLC, Houston, TX, U.S.A.

Abstract: This paper describes the method and results of a probabilistic risk analysis that was used to provide a quantitative basis for a complex and high-stakes design decision for a deepwater subsea oil project. The analysis specified probabilistic simulations of geologic properties based on information from a small number of exploration and appraisal wells. Each iteration of the simulated data was then fed into a deterministic engineering model to simulate various operational scenarios. Conventional probabilistic sampling and a more efficient experimental design approach were both employed. The key results are cumulative density functions for critical operational variables that drive design decisions.

Keywords: Risk Analysis, Probabilistic Analysis, Experimental Design, Oil and Gas.

1. INTRODUCTION

In almost all oil and gas projects, incomplete information about the geologic properties of the asset (e.g. rock and fluid properties) leads to uncertainties in derivative computations that are used for design decisions. Options are available to collect additional information by drilling additional exploration and appraisal wells, by completing additional modeling and analysis, etc., but this information typically comes at a significant cost in time and/or expenditures. Therefore, at some point in the project maturation process, the cost of additional information destroys project value, and decisions of all kinds must be made giving consideration to the residual uncertainty. There are numerous decision analysis frameworks and quantitative methods that can be applied depending on the decision setting, for example, stochastic and/or deterministic optimization. There is a rich literature that demonstrates the application of these concepts and others for oil and gas problems [3,4,5,6,8,9,11,12,13,15,17, 18, 19,22,25,27].

In the deepwater oil and gas industry, wells routinely cost over $100 million each, and this expenditure only adds one additional data point for analysis. For the project examined in this study, the exploration and appraisal drilling phase is complete, and no additional geologic information was going to be obtained prior to most of the major design decisions for the project. Therefore, the design decisions are based on assumptions about the probability density functions (PDFs) of the geologic properties.

The design decision examined in this study was the specification of the pressure rating of the wellheads for a deepwater subsea oil project, the "Project." Specifically, the design decision was whether the wellheads should be specified for 15,000 psi or 20,000 psi. The 15,000 psi equipment provides less operational flexibility under certain geologic and operational outcomes and could cause a loss of reserves, or at least a delay in production. Specifying 20,000 psi equipment would eliminate almost all of the risks. While the 15,000 psi equipment is readily available, the 20,000 psi equipment does not exist and would impose a three to four year delay in the project to allow the equipment to be designed, tested, and certified. From a risk analysis perspective, the question is "What is the likelihood of a loss or delay if 15,000 psi equipment is specified, given the current assumptions (PDFs) regarding the geologic uncertainty?"

2. WELL ENGINEERING BASICS

2.1. Wellbore, Wellhead, and Access

A typical wellbore schematic is provided in Figure 1. In a conventional oil well, steel casings are cemented in place as the well is drilled deeper and deeper until the oil reservoir is penetrated. At the top of the wellbore is a wellhead that is appropriately pressure-rated to contain the maximum reservoir pressure, and to enable monitoring and control of fluids. In a subsea setting, the wellhead sits close to the seafloor. Figure 2 depicts a typical subsea wellhead. After a well is drilled and the wellhead installed, the well is put on production and flows back to a gathering facility through a subsea flow line.

During the life of a well, it is probable that some form of intervention will be required. Interventions are required to repair damage, to re-complete the well in a different reservoir, to plug a depleted reservoir, and for other reasons. During an intervention, the wellhead is accessed by a floating drilling rig or similarly capable vessel, as depicted in Figure 3. During an intervention, it is possible that a process called "bullheading" will be required. In a bullheading operation, a high-density fluid is pumped down the well, displacing the fluid in the well back into the reservoir. After a high-density fluid is in the well, intervention operations can proceed in an efficient and safe manner.

Figure 1. Generic Wellbore Schematic
(figure courtesy of the EPA)

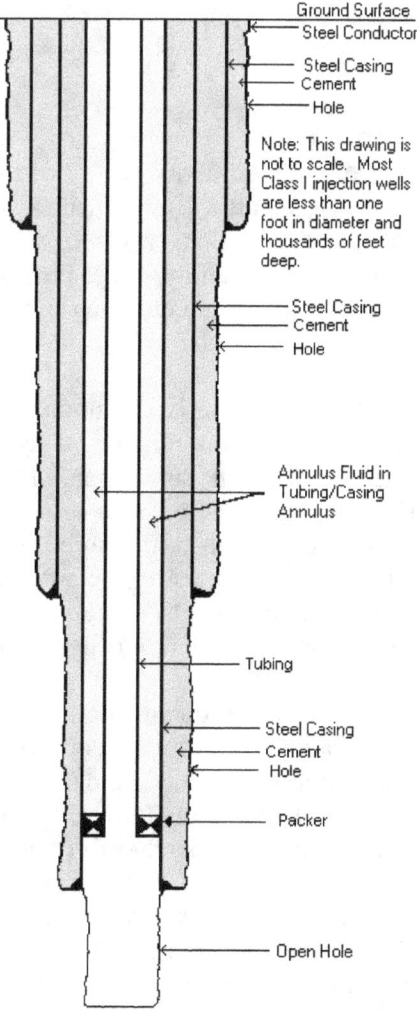

Figure 2. Subsea Wellhead (typical)
(figure courtesy of FMC)

Figure 3. Accessing Subsea Wellheads and Equipment
(figure courtesy of Oceaneering)

2.2. Specification of System Model

When a well is shut-in, the pressure at the wellhead builds up to the reservoir pressure less the hydrostatic gradient of the fluid in the well, and this is called the shut-in pressure. If the well is to be bullheaded, the shut-in pressure must be increased to overcome the pressure losses in the system, most notably the pressure loss incurred when pumping into a permeable reservoir (per Darcy's Law), or in some cases the pressure required to fracture the formation. This pressure is called the bullhead pressure. The expected shut-in and bullhead pressures are important inputs into the specification of the pressure rating of the wellhead.

To estimate the expected bullhead pressure, a common first step is to conduct a simple hydrostatic analysis. If the resulting estimate of bullhead pressure leads to an obvious and economic choice, then a more detailed analysis probably is not warranted. However, if the resulting estimate of bullhead pressure is close to the cross-over point between a lower and higher pressure rating of the wellhead, then a more detailed analysis is warranted, especially if the incremental cost of the higher-rated wellhead is significant. In the extreme case where the higher-rated wellhead does not exist, the analysis of bullhead pressure may be central to the economic viability of the project.

A more detailed analysis of bullhead pressures requires a shift from a simple static hydrostatic analysis to a more complex dynamic analysis. Modeling a dynamic bullheading operation is not a trivial exercise. The information requirements are significant: reservoir rock properties, fluid properties (reservoir fluids and bullhead fluids), reservoir pressures, geothermal gradients, mechanical properties of the hydraulic flow path, bullhead rates, completion efficiency, and the depletion plan.

A physics-based deterministic model of the system was specified for the Project that explicitly accounts for all of these inputs. The system model serves as the computational core of the subsequent risk analysis.

3. WORKFLOW AND RESULTS

The system model is deterministic and solves one case at a time. But as described above, many of the geologic variables are defined only as PDFs. Therefore, a workflow was specified that uses the deterministic system model in a probabilistic manner. Two approaches were employed. In the following descriptions, a *scenario* is defined as the collection of the PDFs of the uncertain variables. Because there may be uncertainty in the properties of the PDFs, it may be desirable to investigate different scenarios. A *sample* is defined as one random observation from each of the uncertain input PDFs for a given scenario. An *iteration* is one run of the system model using one sample. A *simulation* is the collection of multiple iterations for one scenario.

Full Probabilistic. A scenario is defined and a simulation is run. Because of the large number of uncertain variables, a somewhat large simulation size of 2000 iterations was used. The combination of system model complexity and sample size entails significant resource requirements for each simulation. This resource requirement increases linearly with the number of scenarios. After the results from a simulation are available, it is possible to specify regression models that relate variables of interest, e.g. shut-in pressures and bullhead pressures, to the uncertain variables. The resulting models can be used as fast surrogates for the system model for future probabilistic analysis or other analytical needs. The surrogates can also be used to make point predictions and associated probability statements. This approach has been employed in various oil and gas settings and is well-documented in the literature [1,2,10,14,20,21,23,24].

Experimental Design. In contrast to the large simulation size used in the full probabilistic analysis, one can specify a reduced number of iterations for each simulation. That is, the samples are not random, but rather are *designed* to explore the range of uncertainty in the variables. Again, after the results from a simulation are available, it is possible to specify regression models that relate variables of interest to the uncertain variables, and to use the resulting model as a fast surrogate for the system model. This approach is also known to the oil and gas literature [7,16,26,28,29,30,31].

The experimental design, if properly constructed, should yield a regression model of similar explanatory power as that from the full probabilistic analysis. So why do it? First, it was desired to demonstrate that the experimental design approach produces such an equivalent result. In the future on this project, it may be necessary to update the system model and/or run many different scenarios, and the experimental design will be significantly more efficient than reproducing the full probabilistic analysis. Also, for other projects in the future, it is desired to use experimental design only, and this comparison can be referenced to demonstrate their equivalence.

3.1. Probabilistic Analysis

The full probabilistic analysis workflow is depicted in Figure 4. Its major steps include sampling, computations using the system model, analysis of the cumulative distribution, estimation of the surrogate equation, and finally use of the surrogate equation in place of the system model.

Figure 4: Probabilistic Workflow

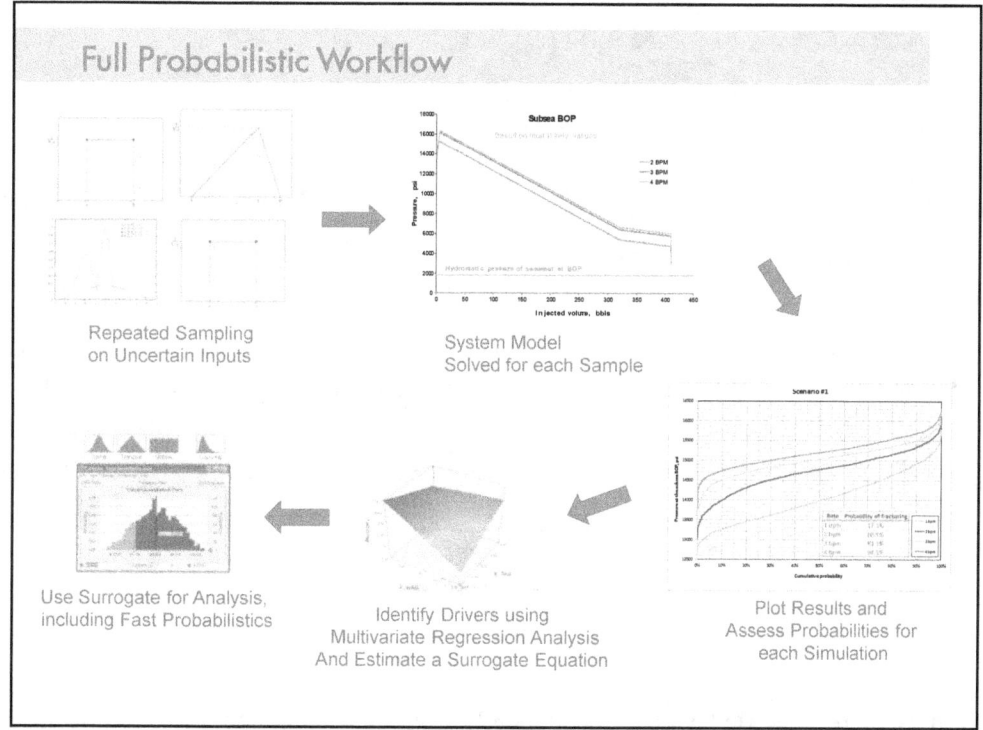

A probabilistic model was specified for three scenarios representing different reservoir pressure regimes (different PDFs): initial conditions, 6 months, and 1 year after initiation of production. This is intended to generate information regarding the rate at which the risk is reduced. In each of these scenarios, a simulation was run for each of four bullhead flow rates: 1, 2, 3, and 4 bpm. Each simulation consisted of 2,000 iterations. This setup results in 3 x 4 x 2,000 = 24,000 model runs or observations.

The shut-in pressure never exceeds 13,200 psi and thus does not impact the wellhead pressure specification decision except as an input into the bullhead pressure computation. The cumulative distribution functions for the initial condition scenario and each of the four bull-head rates are depicted in Figure 5. As can be observed in Figure 5, the probability that the internal pressure will exceed 15,000 psi during a 1 bpm bullhead operation initiated at initial reservoir conditions (worst case) is about 8%. After 6 months of production and pressure decline, this same probability decreases to less than 1%, and after 1 year it approaches 0%. As expected, the probabilities of exceeding 15,000 psi increase as a function of flow rate. At 2, 3, and 4 bpm bullhead rates the probabilities of exceeding 15,000 psi are 30%, 50%, and 70%, respectively. If the 15,000 psi wellhead pressure rating is specified, then prevailing conditions will dictate the maximum bullhead rate. These results indicate that it is very likely that a 1-2 bpm rate will be attainable without exceeding the wellhead pressure rating.

The next step is to specify regression models that relate variables of interest, e.g. shut-in pressures and bullhead pressures, to the uncertain variables. The resulting models can be used as fast surrogates for the system model for future probabilistic analysis or other analytical needs. For example, if it is desired to analyze a new scenario for one or more of the uncertain variables, the surrogate can be used to generate a probabilistic simulation in minutes with the caveat that the range of the revised PDFs are not dissimilar from the original scenario.

Figure 5: Cumulative Distribution of Maximum Bullhead Pressure for Scenario 1 (Initial Reservoir Pressure)

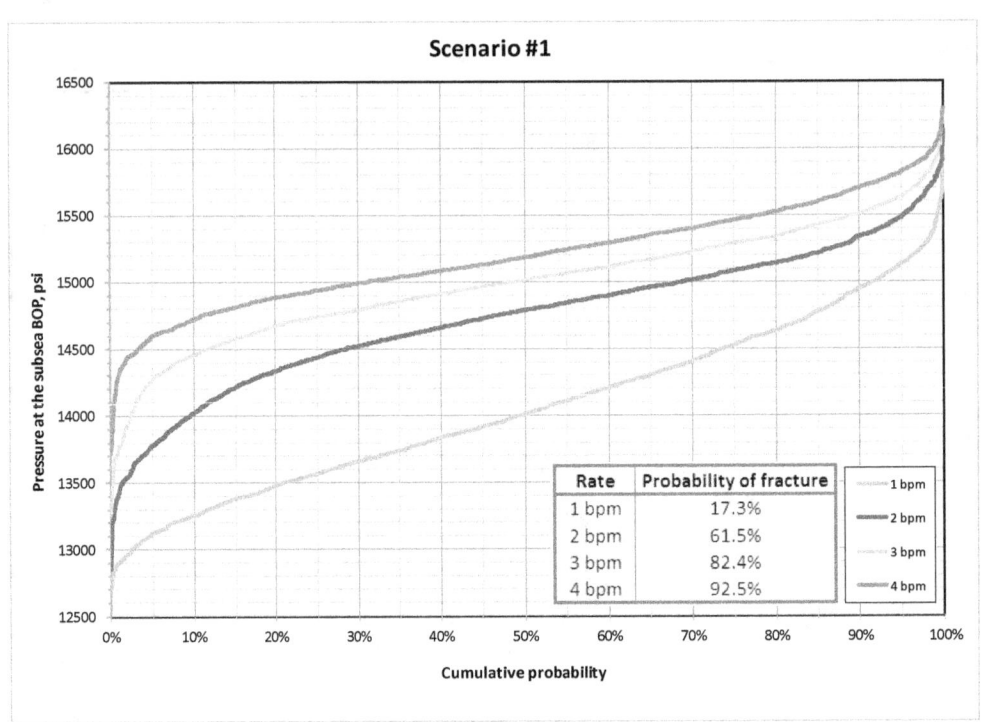

Regression models were specified where the dependent variable was specified as the maximum bullhead pressure, and the independent variables were defined as the uncertain variables. Because the system model is physics-based and deterministic, it is known that there will be two distinct cases. One case is governed by the Darcy equation where the fluid is radially displaced into the pore space of the reservoir. A second case, where the Darcy differential pressures are large, is governed by the formation fracture gradient where fluid is displaced into the fracture. Note, the Darcy differential is defined as follows: Darcy differential (psi) $\approx \frac{203328 q B_o \mu_o}{kh}\left(ln\frac{r_w}{r_r} + s\right)$.

3.1.1. Surrogate Model Using the Full Probabilistic Results: The "No-Fracture" Model

In this regression model, the Darcy differential is small and thus no fracture occurs. Initial analysis showed that results could be pooled across all three (i) pore pressure scenarios and all four (q) bullhead rates, and the regression was specified as $y_{iq} = c + x_{iq}\beta + \varepsilon_{iq}$. Based on knowledge of the design of the system model, the independent variables are defined using the uncertain variables: $x_1 = \frac{B_o \mu_{oil} q}{kh}$, $x_2 = \frac{B_o \mu_{oil} q Skin}{kh}$, shut-in pressure, oil compressibility, q, and q^2. The linear and quadratic q terms are to account for friction losses. The radius terms are constant for all observations and can be ignored. The results of this regression are provided in Table 1, and a plot of the regression model predictions (x-axis) versus the system model output (y-axis) is depicted in Figure 6.

Table 1: The "No Fracture" Surrogate Model (full probabilistic)

```
. reg   maxsspresspsi x1 x2 shutinpresssspsi avgcomppsi q qsq if frac10==0

      Source |       SS       df       MS              Number of obs =   13457
-------------+------------------------------            F(  6, 13450) =       .
       Model |  3.1840e+10        6  5.3067e+09         Prob > F      =  0.0000
    Residual |   4657456.3    13450   346.279279        R-squared     =  0.9999
-------------+------------------------------            Adj R-squared =  0.9999
       Total |  3.1845e+10    13456   2366590.35        Root MSE      =  18.609

------------------------------------------------------------------------------
maxsspress~i |      Coef.   Std. Err.      t    P>|t|     [95% Conf. Interval]
-------------+----------------------------------------------------------------
          x1 |    1520749    1074.29   1415.58   0.000     1518643     1522855
          x2 |     196632   45.71866   4300.91   0.000    196542.4    196721.6
shutinpres~i |   .9995036   .0003444   2902.04   0.000    .9988285    1.000179
  avgcomppsi |  -1.11e+08    8891966    -12.49   0.000   -1.28e+08   -9.36e+07
           q |   11.51706   .8594087     13.40   0.000    9.832496    13.20162
         qsq |   17.01699   .1727986     98.48   0.000    16.67828     17.3557
       _cons |   400.5776   35.76365     11.20   0.000    330.4758    470.6793
------------------------------------------------------------------------------
```

Figure 6: The "No Fracture" Surrogate Model Predictions versus the System Model (full probabilistic)

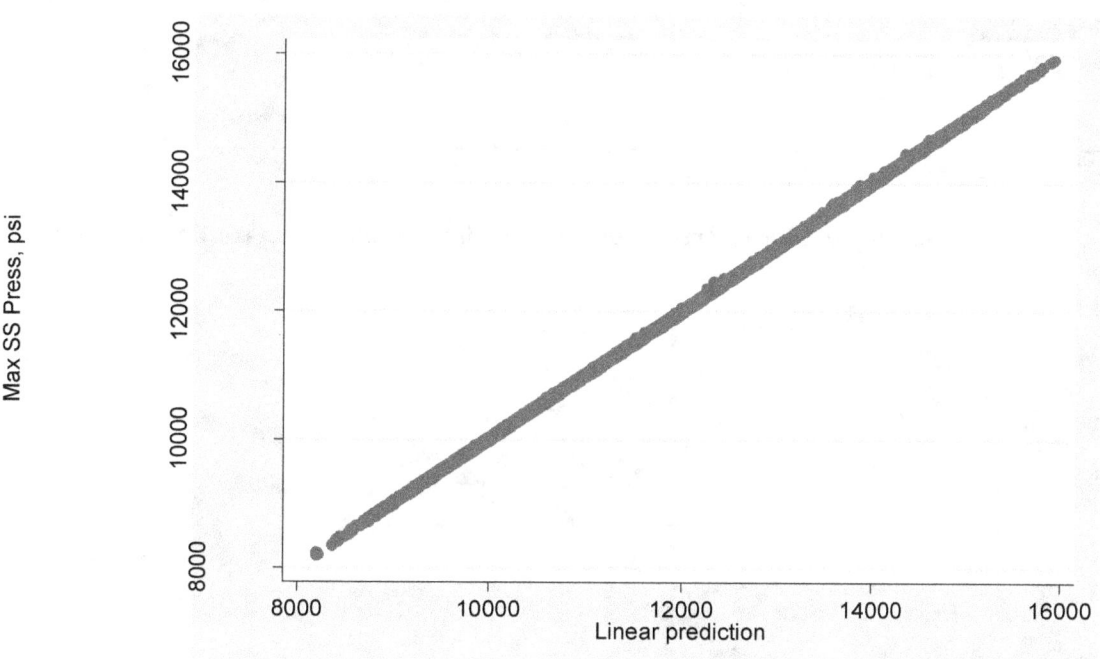

3.1.2. Surrogate Model Using Full Probabilistic Results: The "Fracture" Model

In this regression model, the Darcy differential is large and fracture occurs. Again, initial analysis showed that results could be pooled across all three (i) pore pressure scenarios and all four (q) bullhead rates. The independent variables are defined using the uncertain variables: shut-in pressure, fracture pressure minus reservoir pressure, oil compressibility, q, and q^2. The results of this regression are provided in Table 2, and a plot of the regression model predictions (x-axis) versus the system model output (y-axis) is depicted in Figure 7.

Table 2: The "Fracture" Surrogate Model (full probabilistic)

```
. reg  maxsspresspsi shutinpresssspsi  deltap avgcomppsi q qsq if frac10==1

      Source |       SS       df       MS              Number of obs =   10543
-------------+------------------------------            F(  5, 10537) =       .
       Model |  2.7271e+09     5   545429627           Prob > F      =  0.0000
    Residual |  2527230.05 10537   239.843414          R-squared     =  0.9991
-------------+------------------------------            Adj R-squared =  0.9991
       Total |  2.7297e+09 10542   258933.349          Root MSE      =  15.487

------------------------------------------------------------------------------
 maxsspress~i |      Coef.   Std. Err.      t    P>|t|     [95% Conf. Interval]
-------------+----------------------------------------------------------------
shutinpres~i |   .9996581   .0005427  1841.98   0.000     .9985942    1.000722
      deltap |   .9692905   .0004745  2042.90   0.000     .9683604    .9702205
  avgcomppsi |  -1.41e+08    7790680   -18.15   0.000    -1.57e+08   -1.26e+08
           q |   14.91068   1.096346    13.60   0.000     12.76163    17.05972
         qsq |   16.79008   .1905019    88.14   0.000     16.41666     17.1635
       _cons |   488.0527   33.86037    14.41   0.000        421.68    554.4254
------------------------------------------------------------------------------
```

Figure 6: The "Fracture" Surrogate Model Predictions versus the System Model (full probabilistic)

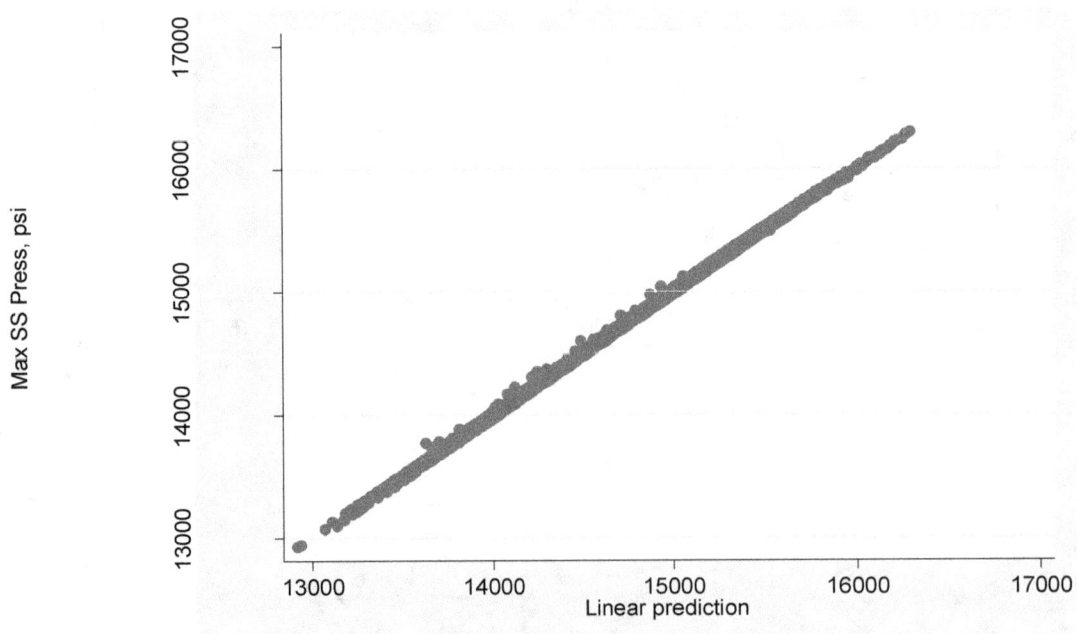

Both the "No Fracture" and "Fracture" surrogate models accurately replicate the system model and are judged to be acceptable surrogates.

3.2. Experimental Design

The experimental design workflow is depicted in Figure 7. It is identical to the probabilistic workflow except for the first step. Instead of repeated sampling, a smaller set of samples is specified for simulation in the system model. That is, the samples are not random, but rather are designed to explore the range of uncertainty in the variables. The smaller simulation size reduces the time required for the computations, and specification of the PDFs for the uncertain variables is not required. The results of the simulation are used to estimate regression models that relate variables of interest to the uncertain variables, and to use the resulting model as a fast surrogate for the system model. Of course, when the surrogate equation is used to conduct a probabilistic simulation, the uncertain variables would need to be fully specified, and these results could be used to create the desired cumulative distribution plots for maximum bullhead pressure as depicted in Figure 5.

Experimental design was used to specify 72 samples for each of the four bullhead rates. Whereas the probabilistic model results in 24,000 observations, the experimental design only requires 72 x 4 = 288 observations (the full range of reservoir pressure can be sampled rather than sampling the three distinct regimes as was done is Section 3.1.). If the surrogate equation from the experimental design is judged to be sufficiently accurate when compared to the surrogate from the full probabilistic model, the full probabilistic model does not need to be repeated in the future.

Figure 7: Design of Experiments Workflow

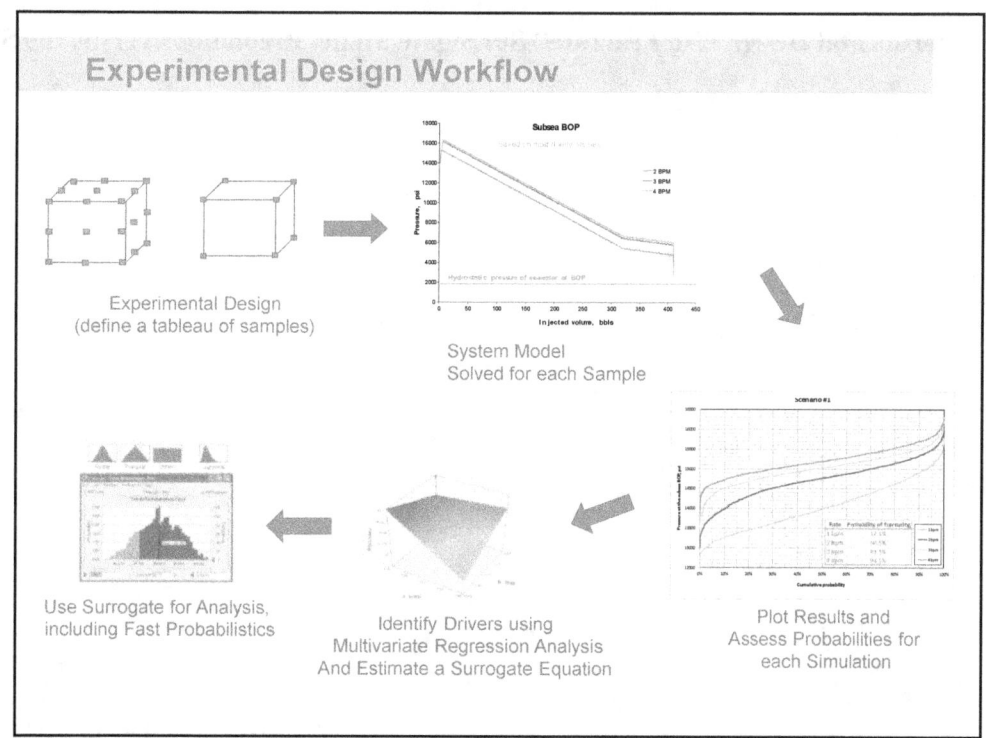

3.2.1. Surrogate Model Using Experimental Design Results: The "No-Fracture" Model

The identical specification of the "No Fracture" model from Section 3.1.1. was specified and estimated on the appropriate subset of the 288 experimental design observations. The results of this regression are provided in

Table 3, and a plot of model predictions versus the system model are depicted in Figure 8.

Table 3: The Experimental Design "No Fracture" Surrogate Model

```
. reg  maxpress x1 x2 sitp avgcomppsi q qsq if  simplefrac==0

      Source |       SS       df       MS              Number of obs =     146
-------------+------------------------------           F(  6,   139) = 9575.98
       Model |  942783887     6    157130648           Prob > F      =  0.0000
    Residual |  2280827.92  139    16408.8339          R-squared     =  0.9976
-------------+------------------------------           Adj R-squared =  0.9975
       Total |  945064715   145    6517687.69          Root MSE      =   128.1

------------------------------------------------------------------------------
    maxpress |      Coef.   Std. Err.      t    P>|t|     [95% Conf. Interval]
-------------+----------------------------------------------------------------
          x1 |    1252225   18128.68    69.07   0.000     1216381     1288068
          x2 |   162545.8   2587.263    62.83   0.000     157430.3    167661.3
 shutinpres~i|   .9771403   .0041214   237.09   0.000     .9689915    .9852892
  avgcomppsi |  -2.00e+08   3.27e+08    -0.61   0.541    -8.47e+08    4.46e+08
           q |   8.622247   53.36682     0.16   0.872    -96.89344    114.1379
         qsq |    14.2422   10.67501     1.33   0.184    -6.864193    35.34859
       _cons |   910.5656    1164.52     0.78   0.436    -1391.898     3213.03
------------------------------------------------------------------------------
```

Figure 8: The Experimental Design "No Fracture" Surrogate Model Predictions versus the System Model

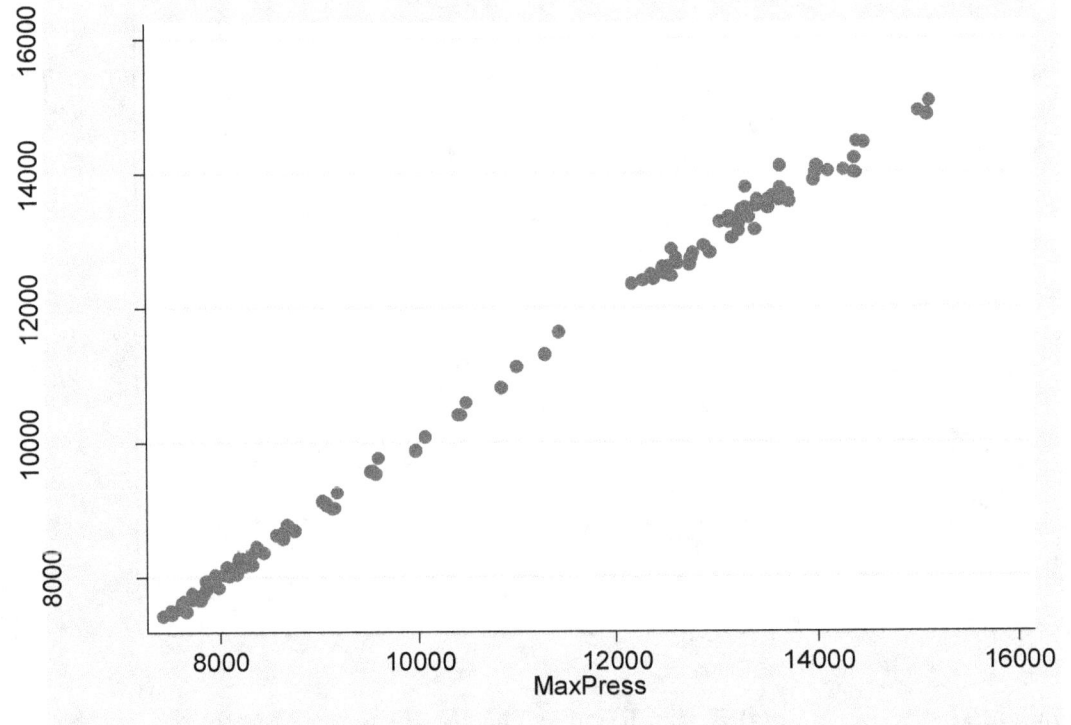

3.2.2. Surrogate Model Using Experimental Design Results: The "Fracture" Model

The regression process was repeated for the appropriate "Fracture" subset of the 288 experimental design observations. The results of this regression are provided in Table 4, and a plot of model predictions versus the system model are depicted in Figure 9.

Table 4: The Experimental Design "Fracture" Surrogate Model

```
. reg  maxpress sitp deltap avgcomppsi qsq if  simplefrac==1

      Source |       SS       df       MS              Number of obs =     142
-------------+------------------------------             F(  4,   137) = 1462.83
       Model |   109718984     4   27429745.9           Prob > F      =  0.0000
    Residual |   2568899.96  137   18751.0946           R-squared     =  0.9771
-------------+------------------------------             Adj R-squared =  0.9765
       Total |   112287884   141   796367.969           Root MSE      =  136.93

------------------------------------------------------------------------------
    maxpress |      Coef.   Std. Err.      t    P>|t|     [95% Conf. Interval]
-------------+----------------------------------------------------------------
        sitp |   .8861487   .0205793    43.06   0.000     .8454544    .9268429
      deltap |   .8370611   .0290006    28.86   0.000     .7797144    .8944079
  avgcomppsi |  -1.70e+08   3.68e+08    -0.46   0.646    -8.98e+08    5.59e+08
         qsq |   16.72607   2.052675     8.15   0.000     12.66705    20.78509
       _cons |   2129.825   1351.013     1.58   0.117    -541.7104    4801.361
------------------------------------------------------------------------------
```

Figure 9: The Experimental Design "Fracture" Surrogate Model Predictions versus the System Model

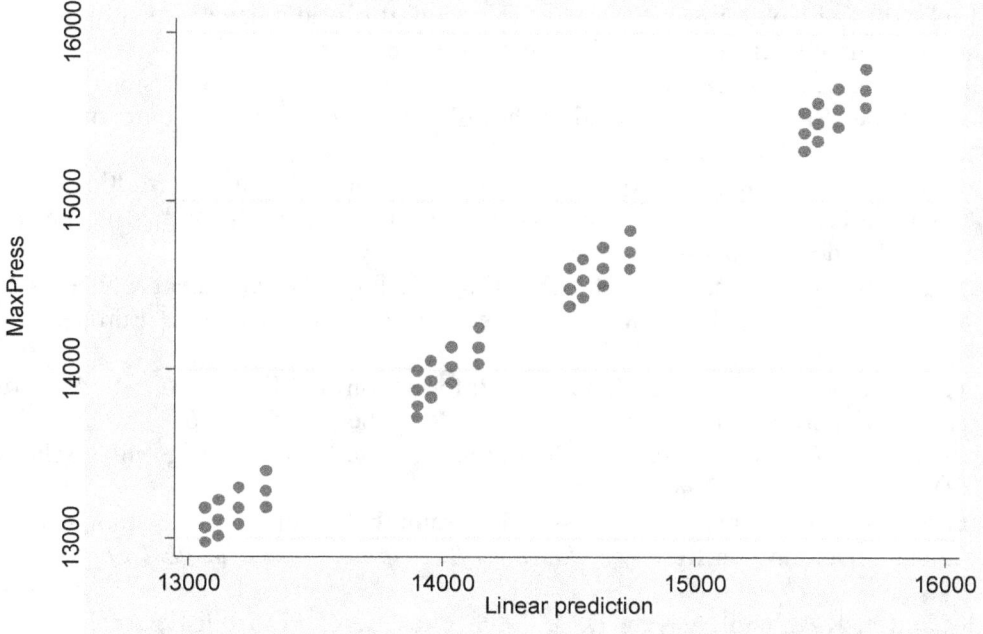

The surrogate models that are based on the experimental design yield very good fits to the system model results. There is more variance in the prediction when compared to the full probabilistic surrogate because of the smaller number of observations in the experimental design. However, for this decision-setting, these small differences are not decision-relevant, and the experimental design approach is judged to be adequate for analyzing different scenarios in the future.

4. NOMENCLATURE

bpm = barrels per minute

B_o = formation volume factor (rb/stb)
h = reservoir thickness (feet)
k = permeability (md)
µ = viscosity (cp)
q = flow rate (bpm)
r_r = radius of drainage (feet)
r_w = radius of well (feet)
s = skin factor

5. REFERENCES

[1] Adams, A.J., Gibson, C., Smith, R. 2009. Probabilistic Well Time Estimation Revisited. Paper SPE 119287 presented at the SPE/IADC Drilling Conference and Exhibition, Amsterdam, 17–19 March. doi: 10.2118/119287-MS.

[2] Akins, W.M., Abell, M.P., and Diggins, E.M. 2005. Enhancing Drilling Risk and Performance Management Through the Use of Probabilistic Time and Cost Estimating. Paper SPE 92340 presented at the SPE/IADC Drilling Conference, Amsterdam, 23–25 February. doi: 10.2118/92340-MS.

[3] Barnes, R.J., Linke, P. and Kokossis, A. 2002. Optimization of oil-field development production capacity. European Symposium on Computer Aided Process Engineering, **12**: 631.

[4] Begg, S.H., Bratvold, R.B. and Campbell, J.M. 2001. Improving Investment Decisions Using a Stochastic Integrated Asset Model. Paper SPE-71414 presented at the SPE Annual Technical Conference and Exhibition, New Orleans, Louisiana, 30 September-3 October.

[5] Cullick, A.S., Heath, D., Narayanan, K., April, J. and Kelly, J. 2003. Optimizing Multiple-field Scheduling and Production Strategy with Reduced Risk. Paper SPE 84239 presented at the SPE Annual Technical Conference and Exhibition, Denver, Colorado, October 4-6.

[6] Cullick, A.S., Cude, R. and Tarman, M. 2007. Optimizing Field Development Concepts for Complex Offshore Production Systems. Paper SPE 108562 presented at the SPE Offshore Europe, Aberdeen, 4-7 September.

[7] Damsleth E., Hage A., Volden R. 1992. Maximum Information at Minimum Cost: A North Sea Field Development Study with an Experimental Design. *J. Pet Tech* (December): 1350-1356.

[8] Ettehad, A., Jablonowski, C.J. and Lake, L.W. 2010. Gas Storage Facility Design under Uncertainty. *SPE Proj Fac & Const* **5** (3): 155-165. SPE-123987-PA.

[9] Goel, V. and Grossmann, I.E. 2004. A Stochastic Programming Approach to Planning of Offshore Gas Field Developments under Uncertainty in Reserves. *Journal of Computers and Chemical Engineering* **28**: 1409–1429.

[10] Hariharan, P.R., Judge, R.A., and Nguyen, D.M. 2006. The Use of Probabilistic Analysis for Estimating Drilling Time and Costs While Evaluating Economic Benefits of New Technologies. Paper SPE 98695 presented at the IADC/SPE Drilling Conference in Miami, Florida, USA, 21–23 February. doi: 10.2118/98695-MS.

[11] Haugen, K. 1996. A Stochastic Dynamic Programming Model for Scheduling of Offshore Petroleum Fields with Resource Uncertainty. *European Journal of Operational Research* **88**: 88-100.

[12] Kabir, C.S., Chawathe, A., Jenkins, S.D., Olayomi, A.J., Aigbe, C. and Faparusi, D.B. 2004. Developing New Fields Using Probabilistic Reservoir Forecasting. *SPE Reservoir Evaluation & Engineering*, 7(1): 15-23. SPE-87643-PA.

[13] Kabir, C.S., Gorell, S.B., Portillo, M.E. and Cullick, A.S. 2007. Decision Making With Uncertainty While Developing Multiple Gas/Condensate Reservoirs: Well Count and Pipeline Optimization. *SPE Res Eval & Eng* **10** (3): 251-259. SPE-95528-PA.

[14] Kitchel, B.G., Moore, S.O., Banks, W.H., and Borland, B.M. 1997. Probabilistic Drilling-Cost Estimating. *SPE Comp App* **12** (4): 121–125. SPE-35990-PA. doi: 10.2118/35990-PA.

[15] Kosmidis, V.D., Perkins, J. D. and Pistikopoulos, E.N. 2002. A Mixed Integer Optimization Strategy for Integrated Gas/Oil Production. European Symposium on Computer Aided Process Engineering, **12**: 697.

[16] Li, B., Friedmann, F. 2005. Novel Multiple Resolutions Design of Experiment/Response Surface Methodology for Uncertainty Analysis of Reservoir Simulation Forecasts. Paper SPE 92853 presented at the SPE Reservoir Simulation Symposium, Houston, TX, 31 January-2 February.

[17] Lund, M.W. 2000. Valuing Flexibility in Offshore Petroleum Projects. *Annals of Operations Research* **99** (1-4): 325-349.

[18] McFarland, J.W., Lasdon L. and Loose V. 1984. Development Planning and Management of Petroleum Reservoirs Using Tank Models and Nonlinear Programming. *Operations Research* **32** (2): 270-289.

[19] Murray, J.E. and Edgar, T.F. 1978. Optimal Scheduling of Production and Compression in Gas Fields. *J Pet Technol* **30** (1): 109-116. SPE-6033-PA.

[20] Murtha, J. 1997. Monte Carlo Simulation: Its Status and Future. Distinguished Author Series, *J Pet Technol* **49** (4): 361–370. SPE-37932-MS. doi: 10.2118/37932-MS.

[21] Noerager, J.A., Norge, E., White, J.P., Floetra, A., and Dawson, R. 1987. Drilling Time Predictions From Statistical Analysis. Paper SPE 16164 presented at the SPE/IADC Drilling Conference, New Orleans, 15–18 March. doi: 10.2118/16164-MS.

[22] Ortiz-Gomez, A., Rico-Ramirez, V. and Hernandez-Castro, S. 2002. Mixed-integer Multi-period Model for the Planning of Oil-field Production. *Computers and Chemical Engineering* **26** (4–5): 703.

[23] Peterson, S.K., Murtha, J.A., and Roberts, R.W. 1995. Drilling Performance Predictions: Case Studies Illustrating the Use of Risk Analysis. Paper SPE 29364 presented at the SPE/IADC Drilling Conference, Amsterdam, 28 February –2 March. doi: 10.2118/29364-MS.

[24] Peterson, S.K., Murtha, J.A., and Schneider, F.F. 1993. Risk Analysis and Monte Carlo Simulation Applied to the Generation of Drilling AFE Estimates. Paper SPE 26339 presented at the SPE Annual Technical Conference, Houston, 3–6 October. doi: 10.2118/26339-MS.

[25] Purwar, S., Jablonowski, C., Nguyen, Q. "Development Optimization Using Reservoir Response Surfaces: Methods for Integrating Facility and Operational Options," *Natural Resources Research*, 20 (1): 1-9. March, 2011.

[26] Schiozer, D.J., Ligero, E.L., Maschio, C., Risso, F.V.A. 2008. Risk Assessment of Petroleum Fields—Use of Numerical Simulation and Proxy Models. *Petroleum Science and Technology*, 26 (10): 1247-1266.

[27] Van den Heever, S., Grossmann, I.E., Vasantharajan, S. and Edwards, K.L. 2001. A Lagrangean Decomposition Heuristic for the Design and Planning of Offshore Hydrocarbon Field Infrastructure with Complex Economic Objectives. *I&EC Research* **40** (13): 2857.

[28] Vanegas Prada, J.W., Cunha, L.B. 2008. Assessment of Optimal Operating Conditions in a SAGD Project by Design of Experiments and Response Surface Methodology. *Petroleum Science and Technology*, 26 (17): 2095-2107.

[29] White, C.D., Willis, B.J., Narayanan, K., Dutton, S.P. 2001. Identifying and Estimating Significant Geologic Parameters with Experimental Design. *SPE Journal*, 6 (3): 311-324.

[30] White, C.D., Royer S.A. 2003. Experimental Design as a Framework for Reservoir Studies. Paper SPE 79676 presented at the SPE Reservoir Simulation Symposium, Houston, Texas, 3-5 February.

[31] Yamali, N., Nguyen, Q.P., Srinivasan, S. 2007. Optimum Control of Unwanted Water Production in Stratified Gas Reservoirs. Paper SPE 106640 presented at the SPE Production and Operations Symposium, Oklahoma City, Oklahoma, 31 March-3 April.

Gas Detection for Offshore Application

Peter Okoh[a]

[a] Norwegian University of Science and Technology, Trondheim, Norway

---

Abstract: Release of hazardous and flammable gas is a significant contributor to risk in the offshore oil and gas industry and various types of automatic systems for rapid detection of gas are therefore installed to accentuate the elimination or reduction of the dangerous releases. There are different types of gases which may be released and gas may be released in different environments and under different conditions. Several principles for detecting gas are therefore applied and a variety of types of gas detectors are in use. However, a significant percentage of gas releases remain undetected by the dedicated detectors and hence unaccounted for and uncontrolled.

The objectives of this paper are: (1) to present a state-of-the art overview of gas detection in relation to offshore applications, (2) to present an overview of requirements for gas detection in the Norwegian offshore industry, and (3) to do a comparative study of performance standards for gas detection worldwide. The paper builds on a review of literature, standards and guidelines in relation to gas detection offshore.

Keywords: Flammable, Toxic, Gas, Detection

---

1. INTRODUCTION

In the offshore industry, dangerous gases are naturally occurring or man-made in petroleum operations. Release of hazardous and flammable gas is a significant contributor to risk in the oil and gas industry. The ignition of flammable gas clouds or vapors can lead to major fire and explosion with highly devastating consequences as was the case in the Piper Alpha disaster [7]. Similarly, toxic gas release can lead to multiple fatalities over a wide area as was the case in the Bhopal gas tragedy [7], although this happened in an onshore location.

Gas detection is a crucial topic in the process industries, e.g. the Norwegian offshore industry where focus has been on safety barriers supported by the policy of reporting hydrocarbon leaks [20]. The process industry learns from related incidents/accidents in addition to being proactive to predict what can go wrong and how to control it. Various types of automatic systems for rapid detection of gas are therefore installed to control the risk. In fact, the safety of the offshore industry depends on the efficiency and effectiveness of the gas detection systems.

Gas detection has, however, experienced mixed results in the industry. Although it has achieved more success than failure, the failure statistics is significant. According to the report of a research conducted by [19], about 44% of all gas releases, or 38% of major gas releases were undetected by the gas detectors deployed. The offshore environment is being characterized by a complex mix of open and enclosed areas, low and high areas, diversity of hazardous gases and conditions for release, potential gas traps, and varieties of operational and environmental conditions that may influence the unreliability of gas detection. Besides, offshore installations have different challenging gas detection needs that require specific solutions; e.g. some facilities require that detectors identify gases at the lowest possible level (in ppm or LEL range), whereas other facilities are exposed to compounds or other gases that can undermine the detectability of the target gases (i.e. a problem of cross-sensitivity or non-specificity). Furthermore, complexities in ventilation patterns and in the size and composition of modules often make it extremely difficult to site gas detectors; this implies that availability of detectors is no guarantee for detection [19]. Other gas detection related problems include incorrect selection of gas detectors, deficiencies in design, installation, calibration and maintenance of gas detectors as well as users' lack of knowledge of the limitations of a given detection principle.

Several authors and companies have made or are making efforts to improve gas detection technologies. Work on enhancing the effectiveness of single technology has been carried out by several authors, e.g. [18] and [8] etc. The optimal placement of sensors under uncertainty has been studied by [12]. Furthermore, the development of versatile single technology or integrated technologies for multiple gas detection has been or is being explored by several others, e.g. [16], [5, 6] and [11].

The main objective of this paper is to present a state-of-the art overview on gas detection in relation to offshore applications. The rest of the paper is structured as follows. First, the types of releases that may occur and the conditions under which they may occur are described. This is followed by a description of relevant principles for detecting gas, and relevant Norwegian regulatory requirements and standard. A comparative study of gas detection standards worldwide is then presented, followed by conclusion and recommendations.

2. TYPES OF GASEOUS RELEASES, THEIR CHARACTERISTICS AND POTENTIAL SOURCES

The different types of gases which may be released in the offshore petroleum industry are described in the following:

- Hydrocarbon (HC) gas release: This is a flammable release that may occur at atmospheric conditions or under pressure from containment systems [13]. Such releases can occur in containment systems subject to the following failure mechanisms: corrosion, erosion, wear, manufacturing defects, operational loading, well pressure etc. Other possible causes include human factors in the form of normal operational releases, operators error and third party damage [13]. The release can occur at the topside where the process equipment are located or subsea in the form of blowouts from wells and leaks from subsea pipelines and isolation valves etc. Specific areas where releases are likely such as the rig floor, the vicinity of the test separator and the choke manifold require permanent rather than portable gas detection system [9]. It is also common to release methane (CH_4), a lower hydrocarbon, from combustion to generate electricity and to power compressors and pumps as well as from flaring of excess gas for safety and during well testing [15].

- Hydrogen Sulphide (H_2S) release: This is an extremely toxic release that usually occurs as a contaminant in produced gases. It occurs naturally together with natural gases from wells. During well testing, it is advisable to monitor the area to check the presence of Hydrogen Sulphide (H_2S) concentrations and that it is safe for working, since even in relatively low concentrations this release can readily lead to fatality [9]. The first significant presence of H_2S is readily noticeable from samples taken downstream of the choke manifold and at the gas outlet from the separator [9]. Furthermore, H_2S usually collect at the lowest points on rigs such as the cellar deck area (offshore) and on land rigs since it is heavier than air [9].

- Carbon Dioxide (CO_2) release: This is a release that becomes dangerous usually in relatively high levels in confined spaces. The release usually results from combustion of fossil fuels to generate electricity and to power compressors and pumps, as well as from flaring of excess gas for safety and during well testing [15].

- Carbon Monoxide (CO) release: This is a highly toxic release. It usually results from combustion of fossil fuel to generate electricity and to power compressors and pumps, as well as from flaring of excess gas for safety [15].

3. PRINCIPLES FOR DETECTING GAS

Offshore gas detection system is necessary to warn about the presence of hazardous and flammable gases in unacceptable concentrations within a given ambiance in order to prevent major accidents. Several principles for detecting gas exist to cover the different types of gases which may be released under different environments and conditions.

The types of gas detection technologies applicable to the offshore petroleum industry, a brief description of their principles of operation as well as their safety-related applications are shown in Table 1.

Table 1: Operational principles and safety-related applications of gas detectors [1, 3, 10]

Operational principles of gas detectors	Description of principles	Applicable gases	Safety-related applications
Catalytic	Uses a catalytic bead to oxidize combustible gas; a Wheatstone bridge converts the resulting change in resistance into a corresponding sensor signal.	All combustible gases (non-selectively)	Flammable gas detection
Electrochemical	Uses an electrochemical reaction to generate a current proportional to the gas concentration.	Many toxic gases, environmental pollutants, combustion products and oxygen.	Toxic gas detection
Solid state	Measures the change in resistance of a metal oxide in response to the presence of a gas; the change in resistance translates into a concentration reading.	HCs, CO, O_3, H_2S, organic vapors, etc.	Flammable gas detection
Thermal conductivity	Measures the gas' ability to transmit heat by comparing it with a reference gas (usually air). The change in electrical resistance as a result of the heat transmission is proportional to the gas concentration.	Binary gas mixtures (often a known gas in air); combustible and toxic gases	Flammable and toxic gas detection
Photoacoustic Infrared (IR)	Uses a gas ability to absorb IR radiation and generating an audible pressure pulse whose magnitude indicates the gas concentration present.	Many IR absorbing gases; combustible gas, toxic gas	Flammable and toxic gas detection
Infrared (Absorptive)	Applies absorption spectroscopy such that a specific gas absorbs a specific wavelength in the infrared (IR) spectrum, and the gas concentration is proportional to the amount of IR light absorbed.	Many mid-IR absorbing gases, e.g. CO_2, CO, CH_4, NO etc.	Flammable and toxic gas detection
Ultrasonic (or acoustic)	Uses ultrasonic sensors to detect leak based on the sound generated by escaping gas at ultrasonic frequencies.	All types of gases whether combustible, toxic or inert.	Flammable and toxic gas detection
IR gas cloud imaging	Applies an absorption imaging technique whereby the image of an area illuminated by infrared radiation is captured by an infrared camera.	Gases that absorb IR radiation at the wavelength of the IR radiation, e.g. hydrocarbon gases.	Flammable gas detection

The coverage of gas detection is a crucial factor to consider in addition to the vulnerabilities of the detection technology that can be exploited by certain operational and environmental conditions. To this end, the various principles for gas detection are classified based on coverage and each of the detection principles was described further by its application area, strengths and weaknesses as shown in Table 2.

Table 2: Further gas detection characteristics [10, 11, 14]

Coverage	Detection principles	Application areas	Strengths	Weaknesses
Point detection	Catalytic	Point sources - potential leakage points (e.g. pumps, compressors, major packing, seal or gasket vulnerable points, etc).	Robust, easily installable and operable, simple to calibrate, long lifetime with a low life-cycle cost, detectability of a variety of gases, wide range of operating temperature, easily calibrated to gases undetectable by infrared absorption, e.g. hydrogen.	Passive detection (not fail-safe), gas must diffuse into catalytic bead so as to be detected, contaminants can poison or deactivate catalyst, the only means of identifying loss of sensitivity due to catalyst's poison is by testing with appropriate gas regularly, requires oxygen for detection, sensor performance may become degraded from prolonged exposure to high concentrations of ignitable gas.
	Electrochemical	Same as above	Speedy response, high accuracy, versatility (detects a wide range of toxic gases), low power consumption.	Less effective at low ambient temperatures ($\leq -40°C$), cannot withstand dry environment (<15% RH) over several months, operates in a narrow pressure range (1 ± 0.1 atm)
	Solid state	Same as above	Robust, versatile (detects a wide range of gases), wide range of operating temperature, resistant to corrosive and low-humidity environment, long operating life (2-10 years)	Usually not selective, although some new improvements have overcome this limitation, high power consumption, operation is not fail-safe.
	Thermal conductivity	Same as above	Wide measuring range	Non-specific (cross-sensitive), unsuitable for gases with thermal conductivities (Tc) close to one. Gases with Tc < 1 are more difficult to measure. Output signal not always linear.
	Photoacoustic Infrared (IR)	Same as above	High sensitivity, linear output, simple to use, not subject to poisoning, long-term stability	Not suitable for hydrogen detection
	Fixed-point IR (absorptive)	Same as above	Immune to poisoning by contaminants, fail-safe operation, absence of routine calibration, can operate in the absence of oxygen or in enriched oxygen, can operate in continuous presence of gas	Gas must pass by the sampling path so as to be detected, the gas to be detected must be infrared active (e.g. a hydrocarbon), gases that do not absorb IR energy cannot be detected, highly humid and dusty environments can increase the maintenance cost of IR detector, routine calibration to a different gas is impractical, a relatively large amount of gas is required for response testing, ambient temperature limit of detector use is 70°C, not suitable for multiple gas applications, the IR source is not replaceable in the field, but in the factory.
Open path (line or perimeter) detection	Open-path IR (absorptive)	Boundaries with public areas and between fire areas or equipment, along rows of items and perimeters	Same as above and long line coverage.	Same as above
Area detection	Ultrasonic (or acoustic)	General process areas, loading/offloading facilities, gas turbines, flow stations, tank farms etc.	Very high detection rate of pressurized gas leaks, versatility (detects pressurized leaks irrespective of gas type), unaffected by ambient conditions (fog, heavy rain and others), minimal maintenance, absence of consumable parts, robust, fail-safe, insensitive to gas dilution and changing wind direction, wide area coverage, gas must not be at the device for detection.	Unsuitable for low pressure leaks, under certain conditions influenced by artificial or natural ultrasonic sources, requires estimation of background noise levels before installation, cannot determine concentration of gas, cannot pin-point leak source.
	IR gas cloud imaging	Large gas clouds monitoring in unmanned platforms, pipelines	Wide field of view and detection coverage, no gas calibration is required in the field, highly immune to spurious alarm sources, simultaneous detection of multiple gases.	Detectability of gases is poor when the contrast with the background is poor, heavy fog and rain reduces detection range, suitable only for large leaks - not a small leak detector.

4. REQUIREMENTS FROM NORWEGIAN REGULATIONS AND STANDARDS

Gas detection in the Norwegian offshore petroleum industry is being regulated by some standards and regulations briefly described in Tables 3 and 4.

Table 3: Requirements from Norwegian regulations

Provisions Relevant to Gas Detection	Norwegian Guidelines	Related Standards
Safety barriers: Safety functions being regarded as barriers against hazards and accidents	Sections 3 and 8 of the Facilities and 4 and 5 of the Management Regulations in PSA Guidelines	
Design of safety functions: Requirements for design of safety functions	Section 8 of the Facilities Regulations in PSA Guidelines, OLF 070 Guideline	NS-EN ISO 13702, NORSOK S-001 and IEC 61508
Design of fire and gas system: Requirements for design of fire and gas detection systems	Section 8 of the Facilities Regulations in PSA Guidelines, OLF 070 Guideline	NS-EN ISO 13702 with Appendix B.6, NORSOK S-001 Chapters 12 and 13
Disconnection: When it becomes necessary to disconnect safety functions, the requirements shall be applied.	Sections 8 of the Facilities and 26 of the Activities Regulations in PSA Guidelines	
Performance requirements: Performance requirements shall be established for all safety barriers on an installation	Section 1 and 2 of the management regulations in PSA Guidelines, OLF 070 Guideline	IEC 61508
Availability: The requirement for available status shall be fulfilled.	Section 8 of the Facilities Regulations in PSA Guidelines	NORSOK I-002, Chapter 4
Independence: The fire and gas detection system shall come in addition to systems for management and control and other safety systems.	Section 32 of the Facilities Regulations in PSA Guidelines	
Interface: The fire and gas detection system may have an interface with other systems as long as it cannot be adversely affected as a consequence of system failures, failures or isolated incidents in these systems.	Section 32 of the Facilities Regulations in PSA Guidelines	
Limiting consequences: Relevant safety functions shall be activated when there is a demand on the detection system	Section 32 of the Facilities Regulations in PSA Guidelines	
Not Permanently Manned Facilities: They should also have a dedicated gas detection function for the area around and on the helicopter deck	Section 32 of the Facilities Regulations in PSA Guidelines	
Visual perception of detection: Detection of gas should be shown by means of a light signal that is visible at a safe distance from the facility.	Section 32 of the Facilities Regulations in PSA Guidelines	
Gas detection for mobile units: For mobile facilities that are not production facilities, and that are registered in a national ships' register	Section 32 of the Facilities Regulations in PSA Guidelines	DNV-OS-D301 Chapter 2, Section 4, subsection D

The role of gas detection as stipulated in NORSOK S-001 (subsection 12.1) shall encompass the continuous monitoring of flammable or toxic gases. The standard focused primarily on hydrocarbon (HC) gas detection (including H_2 as relevant), H_2S gas detection, CO_2 gas detection and CO gas detection wherein it sets alarm limits for each of these. For hydrocarbon gas detection (including H_2 as relevant), i.e. flammable gas detection, the alarm limits (both low and high) are fixed in relation to the types of detectors in use whether point detectors or IR open path detectors. It is possible to use a single alarm limit

Table 4: Requirements from Norsok S-001 Standard

Provisions	NORSOK S-001 Requirements	References
Role	Continuous monitoring of flammable or toxic gases.	subsection 12.1
Interfaces	Link between gas detection system and ESD, BD system, ISC, ventilation, PA and alarms system and fire fighting systems.	subsection 12.2
Required utilities	Uninterrupted Power Supply (UPS) and instrument air supply (if aspiration system is applied) are required in gas detection system.	Subsection 12.3
Detection design coverage	Speedy and reliable detection before gas cloud reaches critical concentration/size.	subsection 12.4, subsubsection 12.4.1
Leak detection	All potential flammable gas leak points shall have flammable gas detection.	subsection 12.4, subsubsection 12.4.2
	Herein, the smallest gas cloud with the least unacceptable consequence shall be the basis for confirmed gas detection.	Same as above
	In naturally ventilated area, a smaller leak rate for warning (alarm) is enough and is typically 0.1 kg/s.	Same as above
	In mechanically ventilated areas, detection of smaller leaks shall be subject to expert judgment.	Same as above
	Deploying detectors shall be based on an assessment of gas leak scenarios in relation to potential leakage source and rate, dispersion, density, equipment arrangement, ventilation and the probability of small leak detection therein.	Same as above
	The basis for selection and placement of detection in each area shall be documented.	Same as above
	Open path detectors are preferred where the layout enables good coverage by them.	Same as above
	Detection principle to apply shall be subject to considerations for environmental conditions and availability of protection for detectors.	Same as above
	Catalytic detectors shall not be used unless other detectors do not perform as required.	Same as above
Detection location	Sufficient detectors shall be located by natural passageways along flow direction, in different levels in an area or module, in potential gas traps and in the air inlets of heat sources and accessible without scaffolding.	subsection 12.4, subsubsection 12.4.3
Detection characteristics and calibration	The detector characteristics and calibration shall guarantee good estimation for gas concentration (point detectors), gas amount (open path detectors) or leakage rate (acoustic detectors)	subsection 12.4, subsubsection 12.4.4
Detection actions and voting	The detection system shall activate all actions according to the Fire and Explosion Strategy (FES).	subsection 12.4, subsubsection 12.4.5
Detection levels	Detectors used shall give alarms as soon as possible and within the recommended alarm limits/settings. Detection, failure of further action on demand and system defect shall be reflected in central control room (CCR) as alarms. Use alarm limits and outputs for annunciation as stipulated by standard.	subsection 12.4, subsubsection 12.4.6
Detection response time	Maximum response time of detection shall be defined so as to ensure fulfillment of total reaction time for each safety function. Apply recommended response times unless reduction is needed	subsection 12.4, subsubsection 12.4.7
Detection logic solver	Logic solver compliance with the intended use and safety integrity requirement shall be demonstrated.	subsubsections 12.4.8 and 9.4.6 and IEC61508
Fire and gas independence	The fire and gas detection system shall operate as an independent system.	subsection 12.4, subsubsection 12.4.9
Survivability requirements	The gas detection system shall not be dependent on local instrument rooms with location less safe than the central control room.	subsection 12.5
	Equipment critical to effectuation of system actions shall be protected against mechanical damage and accidental loads until all actions from the detection system have been activated.	subsection 12.5

for hydrocarbon gas detection, but this must be the low alarm limits. The alarm limits for area detection systems (e.g. ultrasonic/acoustic detectors) are left to the operators to decide and adjust on the basis of the background noise peculiar to their operating environment. However, guidelines on the use of IR gas cloud imaging, a type of area detection, has yet to be treated by the standard. This is probably due to its being a new technique that has yet to be applied extensively in the Norwegian industry. Furthermore, the alarm limits for toxic gases are defined in the standard based on the effect of toxic gas in relation to concentration or exposure time and these vary for H_2S gas detection, CO_2 gas detection and CO gas detection.

5. COMPARATIVE STUDY OF PERFORMANCE STANDARDS FOR GAS DETECTION WORLDWIDE

The performance standards for gas detection do specify the performance levels to which gas detectors should be tested and operated, and several variations of these exist across the geographical regions of the world. The variations are probably as a result of diversity of regulatory agencies. Some of the standards available in different countries have little differences, whereas the differences between some are significant. However, they all have a common goal which is the prevention of accidents.

In North America, as regards flammable gas detection performance specifications, FM 6310/6320 (used mainly in the US) is similar to C22.2 152 (used mainly in Canada) and both of them are closely related to ANSI/ISA 12.13.01-2000 [4]. As regards offshore toxic gas detection, the ANSI/ISA 92.00.01 is widely used worldwide [4] and emphasizes on repeatability, step-response and recovery as part of requirements for toxic gas detection performance tests with the worst case accessory attached [2]. The ANSI/ISA 12.13.04 recommends instrument measurements in LEL-m (lower explosion limit meters) or ppm-m (parts per million meters) for flammable gas open-path detection [2]. It also recommends several rigorous tests covering solar immunity, simulated fog/mist and water vapor, partial obscuration of optics, long range operation with 95% obscuration of optics, vibration and temperature extremes and long term stability, either while under stress or before and after stress [2]. The ANSI/ISA 92.00.04 also demands measurement in ppm-m (only) of the toxic gas in the optical beam of the open-path toxic gas detector as well as a misalignment test [2].

In Europe, the national standards are becoming harmonized with the European standards and the IEC standards. For example, as regards both point and open-path flammable gas detection performance specifications, the IEC 60079-29 series have been adopted by many European countries [4]. The same also applies for IEC 45544 series which are dedicated to toxic gas detection [4]. The IEC/EN 60079-29 series recommends that a detector for flammable gas should be used where the accumulation of a combustible air-gas mixture can pose a hazard to life and assets. Furthermore, such a detector is required to sound alarms, show visual warnings or initiate mitigative actions. In addition, IEC/EN 60079-29 and IEC/EN 45544 series advise on considering the effects of variations in temperature and humidity of the gas marked for detection.

In Norway, NORSOK standards which are developed by the Norwegian petroleum industry are widely in use. They are a range of standards intended to serve as references or bases upon which relevant Norwegian regulatory bodies can prescribe statutory requirements and evaluate their compliance. In addition, NORSOK standards serve as replacements for oil company specifications and they normally make necessary additional provisions to recognized international standards in order to address some needs peculiar to the Norwegian petroleum industry [17]. The NORSOK standard that treats gas detection is NORSOK S-001 (Technical safety) which has been described to some extent earlier. NORSOK S-001 is an all-in-one standard generally covering point, open-path and area detection of both flammable and toxic gases. This is unlike the other standards that are separated such that each covers not more than one of the following aspects: flammable-gas point detection (IEC/EN 60079-29-1 and ANSI/ISA 12.13.01), toxic gas point detection (IEC/EN 45544 series and ANSI/ISA 92.00.01), flammable gas open-path detection (IEC/EN 60079-29-4 and ANSI/ISA 12.13.04), toxic gas open-path detection (IEC/EN 45544 series and ANSI/ISA 92.00.04) and area detection of flammable or toxic gas.

A table briefly juxtaposing performance specifications for toxic-gas point detection across Norway, Europe and America is shown in Table 5.

Table 5: A brief comparison of performance standards for toxic-gas point detection

Toxic gas detection specifications	NORSOK S-001	IEC/EN 45544 series	ISA 92.00.01 to 92.06.01, FM 6341, NFPA 70
Gas concentrations		0, 20%, 50%, 90% of full scale	10 to 100 ppm H2S
Temperature range		-10 to 40°C	14 to 122°F (−10 to 50°C)
Relative humidity range		20% RH, 50% RH, 90% RH	15 to 90%
Response time	T90 < 2 seconds	T50 < 60 seconds, T90 < 2.5 minutes	T20 < 10 seconds, T50 < 30 seconds
General alarm limits	Maximum is 10 x 10-6 /20 x 10-6 (low/high for H2S), maximum is 5000 x 10-6 /15000 x 10-6 (low/high for CO2), maximum is 30 x 10-6 /200 x 10-6 (low/high for CO)	70 Db(A) at 0.3 meters from apparatus	
Accuracy/Linearity		0.3% (for 0.5 STGC to 10 STGC) to 0.5% (for 0.1 STGC to 0.5% STGC)	10% of applied gas concentration or 3 ppm

6. CONCLUSION AND RECOMMENDATIONS

According to recent statistics, a significant percentage (about 44%) of gas releases remain undetected in spite of the application of the detection technologies in use [19]. The main objective of the paper has been to present a state-of-the-art knowledge of gas detection for offshore application.

This paper has given more insights into the various aspects of applicable gas detection and will be useful to students and practitioners in offshore petroleum related fields. The paper has reviewed literature, standards and guidelines in relation to gas detection in the offshore oil and gas industry. It has covered the description of the various gaseous releases, the applicable detection technologies and their pros and cons as well as standards and guidelines being applied in the offshore industry in Norway and worldwide. In addition, a comparative study of performance requirements across international boundaries has been done.

Based on the aforementioned, it can be inferred that no single detection technology is a complete solution to offshore gas detection. There is the need to link various technologies together in order to achieve complete coverage and enhanced redundancy. In this way, detection layers of protection (barriers) will be established and made independent. This will enhance the prevention of major accidents characterized by fire, explosion and toxic release. Besides, the associated flammable and hazardous gases, of which exposure is inevitable, need continuous monitoring since the processes generating them are continuous.

Furthermore, it has been seen that no single detection standard across the world areas can be regarded as "'the standard of everything about gas detection offshore'". Hence, there is the need for continuous improvement as regards the harmonization of standards.

References

[1] Analog Devices. ADI Gas Detector Solution Based on NDIR and PID, Project Code: APM-Gas-Detection (EC)-2012. Technical report, Analog Devices, 2012.

[2] Shankar Baliga. ISA Performance Standards for Gas Detection. Technical report, General Monitors, Lake Forest, 2012. URL http://www.isa.org/~oranc/DOC/Meetings/2012/11_nov/isa_performance_standards_for_gas_detection.pdf.

[3] RW Bogue. Technology Roadmap: Optoelectronic gas sensors in the Petrochemicals, Gas and Water industries. Technical report, OptoCem.Net, 2006. URL www.optocem.net.

[4] Bill Crosley and Simon Pate. Survey of Standards Related to Gas Detectors. Technical report, Minneapolis, 2011. URL http://safety.det-tronics.com.

[5] Dräger. Introduction to Gas Detection Systems. Technical report, Dräger Safety AG & Co. KGaA, 2008. URL http://www.draeger.net/media/10/03/05/10030546/introduction_gds_fl_9046421_en.pdf.

[6] Dräger. Multi-gas detection, 2013. URL http://www.draeger.com/sites/enus_us/Pages/Fire/X-zone-5000.aspx?navID=427.

[7] Trevor Kletz. *Learning from accidents*. Gulf Professional Publishing, Oxford, third edition, 2001.

[8] S.W. Legg, C. Wang, a.J. Benavides-Serrano, and C.D. Laird. Optimal gas detector placement under uncertainty considering Conditional-Value-at-Risk. *Journal of Loss Prevention in the Process Industries*, 26(3):410–417, May 2013. ISSN 09504230. doi: 10.1016/j.jlp.2012.06.006. URL http://linkinghub.elsevier.com/retrieve/pii/S0950423012000952.

[9] Stuart McAleese. Safety Procedures. In *Operational aspects of oil and gas well testing*, chapter 4, pages 19–43. Elsevier, Amsterdam, 1st ed edition, 2000.

[10] MSA. *MSA Gas Detection Handbook*. Mine Safety Appliances Company (MSA), Pittsburg, 5th ed. edition, 2007.

[11] Edward Naranjo and Gregory A Neethling. Integrated Solutions for Fixed Gas Detection. Technical report, General Monitors, 2010. URL http://www.gassonic.com/fileadmin/user_upload/documents/GM-Gassonic_Whitepaper_Final.pdf.

[12] P W Nebiker and R E Pleisch. Photoacoustic gas detection for fire warning. *Fire Safety Journal*, 37:429–436, 2002.

[13] Dennis Nolan. Characteristics of Hydrocarbon Releases, Fires, and Explosions. In *Handbook of Fire and Explosion Protection Engineering Principles*, chapter 5, pages 49–69. Elsevier Inc., 2011. ISBN 9781437778571. doi: 10.1016/B978-1-4377-7857-1.00005-7.

[14] Dennis Nolan. Fire and Gas Detection and Alarm Systems. In *Handbook of Fire and Explosion Protection Engineering Principles*, chapter 17, pages 181–201. Elsevier Inc., 2011. ISBN 9781437778571. doi: 10.1016/B978-1-4377-7857-1.00017-3.

[15] Oil & Gas UK. Atmospheric emissions, 2009. URL http://www.oilandgasuk.co.uk/knowledgecentre/atmospheric_emissions.cfm.

[16] R. Rubio, J. Santander, J. Fonollosa, L. Fonseca, I. Gràcia, C. Cané, M. Moreno, and S. Marco. Exploration of the metrological performance of a gas detector based on an array of unspecific infrared filters. *Sensors and Actuators B: Chemical*, 116(1-2):183–191, July 2006. ISSN 09254005. doi: 10.1016/j.snb.2006.03.018. URL http://linkinghub.elsevier.com/retrieve/pii/S0925400506001821.

[17] Standards Norway. Technical Safety. Technical report, Standards Norway, Lysaker, 2008. URL www.standard.no/petroleum.

[18] S Stueflotten, T Christenser, S Iversen, J Hellvik, K Almå s, T Wien, and A Graav. An Infrared Fibre Optic Gas Detection System. In *Proc. SPIE. 0514, 2nd Intl Conf on Optical Fiber Sensors: OFS'84 87*, pages 87–90. SPIE, 1984. doi: 10.1117/12.945064.

[19] AM Thyer. Offshore Ignition Probability Arguments, Report Number HSL/2005/50. Technical report, Health and Safety Laboratory, Buxton, 2005.

[20] J.E. Vinnem, R. Bye, B.A. Gran, T. Kongsvik, O.M. Nyheim, E.H. Okstad, J. Seljelid, and J. Vatn. Risk modelling of maintenance work on major process equipment on offshore petroleum installations. *Journal of Loss Prevention in the Process Industries*, 25(2):274–292, March 2012. ISSN 09504230. doi: 10.1016/j.jlp.2011.11.001.

BOP Risk Model development and applications

Xuhong He[a], Johan Sörman[a], Inge A. Alme[b], and Scotty Roper[c]
[a] Lloyd's Register Consulting, Stockholm, Sweden
[b] Lloyd's Register Consulting, Kjeller, Norway
[c] Lloyd's Register Drilling Integrity Services Inc., Houston, USA

Abstract: Deepwater drilling operations typically involve a critical safety system called BOP (Blow Out Preventer), which is latched onto a wellhead and situated on the seabed. It is the final and ultimate line of defense in protecting life and the environment throughout drilling operations. It is thus important to make sure the BOP will function when it is required. When a failure is detected in a certain system or component on the submerged BOP, the industry's typical response is to analyze the possible consequences and perform a risk assessment in order to define risk levels. If the risk level is increased above the certain level, the drilling has to be stopped and BOP needs to be pulled to the surface to fix the problem. To stop the drilling and pull the BOP to the surface for manual inspection is a very costly and timely operation.

To support the BOP pull or no pull decision, there is a need of a risk-informed model which quickly defines the change in operational risks based on the BOP status. This model must be transparent, verifiable and with the subjectivity removed. The BOP Risk Model is realized using traditional risk- and reliability modeling methods. The risk model is made available for drilling rigs staff using a risk monitor interface that can be used for visualizing operational risks.

The paper describes the model development process, involving: (1) identify key BOP functions; (2) establish block diagrams for each function; (3) FMEA; (4) establish fault trees based on the logic block diagrams and FMEA; (5) Integrate the fault tree model into the risk monitor. A case study will be given using the BOP Risk Model for decision making.

Keywords: Blowout Preventer (BOP), Risk Model, Drilling, Oil & Gas.

1. INTRODUCTION

A Blowout Preventer (BOP) is a critical part of the safety of an offshore drilling, and is often the final line of defense for protecting life and environment. In deepwater and ultra-deepwater drilling operations, a BOP is required to be latched onto a wellhead, situated on the seabed. When a failure is detected in a certain system or component on the submerged BOP, a great challenge for the industry is to decide what to do with the failure [1].

Stopping drilling and pulling the BOP to the surface for manual inspection and reparation is a very costly and time-consuming operation. Some failures are critical and if they happen, the drilling operation must be stopped. Some of the potential failures are not critical to the safety of a drilling rig. There are also levels of redundancy available in the BOP.

The industry is therefore in need of a BOP Risk Model to quickly define changes in the operational risks reflected by the available redundancy for the BOP functions. This model must be fully transparent, verifiable and with the subjectivity removed [2].

The BOP Risk Model is realized using traditional risk- and reliability modeling methods in RiskSpectrum® PSA software. The risk model is then made available for drilling rigs staff using RiskSpectrum® RiskWatcher that can be used for visualizing operational risks.

The paper describes the BOP Risk Model developm ent process, involving: (1) identify key BOP functions; (2) establish block diagrams for each function; (3) FMEA; (4) establish fault trees based on the logic block diagrams and FMEA; (5) Integrate the fault tree model into the risk m onitor. A case study will be given using the BOP Risk Model for decision making.

2. BOP RISK MODEL DEVELOPMENT

2.1. Identify key BOP functions

A BOP is a quite com plicated system consisting of a num ber of subsystems and redundancy in the design. The following figure shows a typical example of BOP functions, and the generic failure modes for these functions.

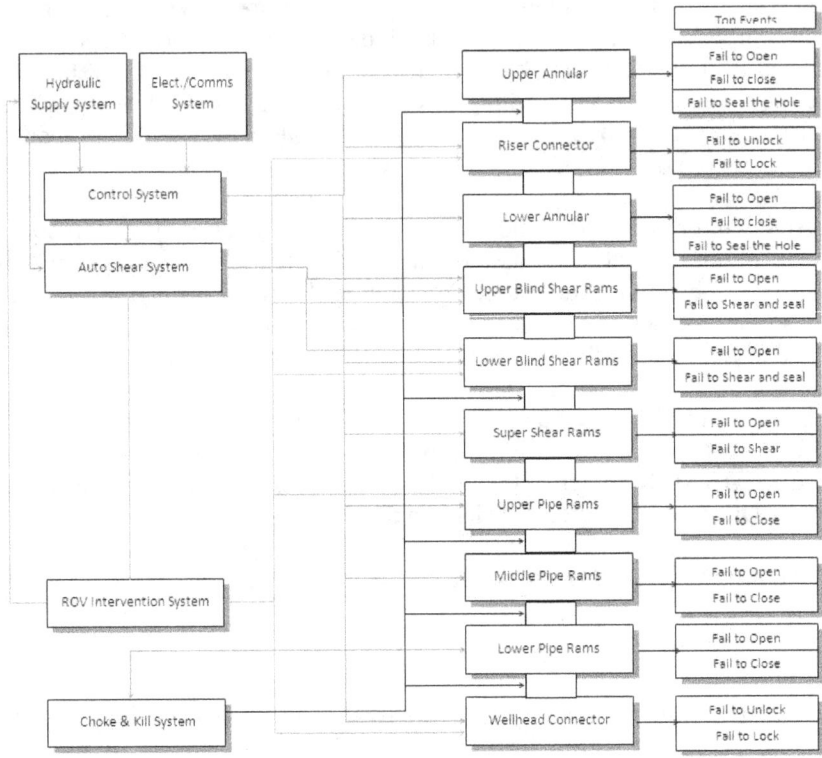

Figure 1. Overall BOP functions (example)

2.2. Establish block diagrams

After all criti cal functions have been identified, logic block diagram s will be built to describe the functions and the needed components and their logics. This is further broken down to each component all the way down until every main, minor, and subcomponents needed for the function are identified

An example of a block diagram for one of the functions is shown in the following figure.

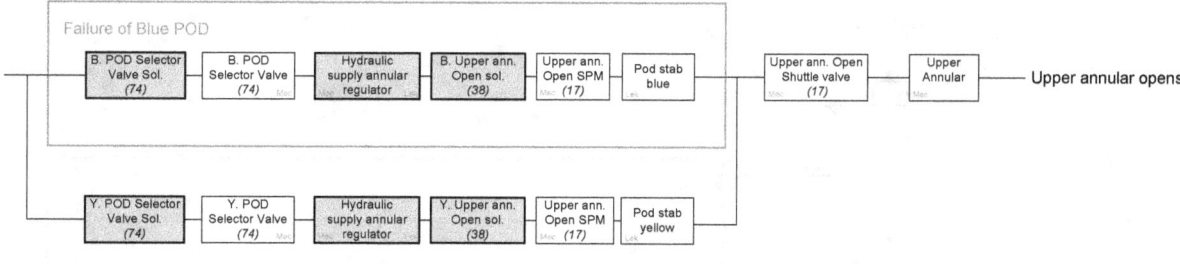

Figure 2. Example of logic block diagram showing components involved in one main function.

2.3. FMEA

The Failure Mode and Effects Analysis (FMEA) method is a tool for analysis of risk, through identification of failure modes and corresponding effects (risks) to personnel, equipment and production.

The main intention of the analysis is to identify possible failures of the system, and the consequences of the failures, which are then used as inputs to the BOP Risk Model.

The FMEA is performed by dividing the relevant BOP stack into subsystems and main components, and looking at the functions these subsystems and main components needed to perform in order for the BOP stack to work as intended. It is then assessed how these functions could fail (failure modes) and the effects of these failures. Redundancy in components is also analyzed.

2.4. Fault trees

The outputs from logic block diagram and FMEA are used as inputs to build the fault trees. A fault tree is a logical diagram that illustrates the coherence between an undesirable top event in a system and the causes for this event.

An example of fault tree is given in Figure 3. The complete fault tree model is built in RiskSpectrum® PSA software.

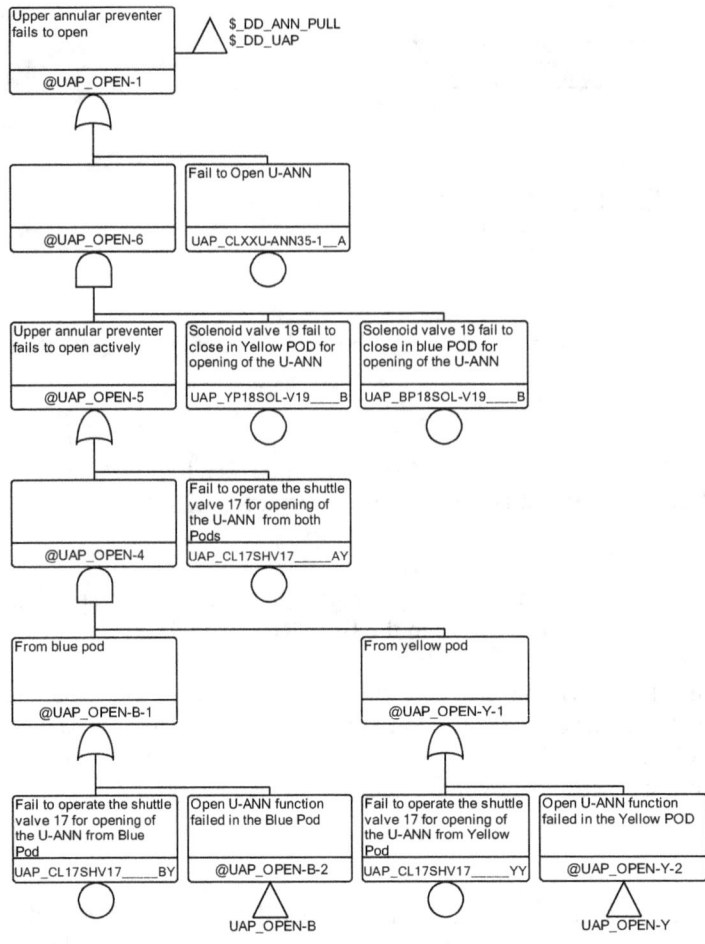

Figure 3. An example fault tree in the BOP model

2.5. Realize the BOP Risk Model in RiskWatcher software.

In addition to the fault trees, the system/subsystem/component information is also added to the BOP Risk Model.

The model is finally implemented in RiskWatcher, a risk monitor software in RiskSpectrum family. This program is unique in that the operator interface is easily navigable and does not require experience in risk analysis or fault tree modelling.

The BOP risk levels can be assessed easily, by comparing the remaining available redundancy of the BOP capabilities with the minimum requirements in the company policy, industry standards and the regulatory regulations. The minimum requirements are identified and verified by the experienced BOP experts. These requirements are the basis for the "Pull" or "No Pull" suggestion from the BOP Risk Model.

For the drilling rigs operated in the Gulf of Mexico, the following regulatory and industry requirements are relevant:

- CFR 250
- API 16D
- API Standard 53

The applicable requirements will need to be investigated for the drilling rigs operated in the other areas, e.g. NORSOK requirements for Norwegian Continental Shelf.

Four colours are introduced in the RiskWatcher software to indicate the different risk level based on the applicable requirements and the 'real-time' BOP status:

- RED   Red on the top level means that critical functions cannot be operated and that the BOP is under the minimum requirements. Critical functionality less than 100 %
- ORANGE   Orange means loss of redundancy. Detailed risk assessment of the failure taking into account the actual risk of the drilling operation must be performed.
- YELLOW   Yellow means that at least one component in the BOP has failed, maintenance is needed at the next available opportunity
- GREEN   Green means all fine, no problems. This must be the colour when the stack is deployed and after landing

An example of the defence-in-depth structure can be seen in the pictures below.

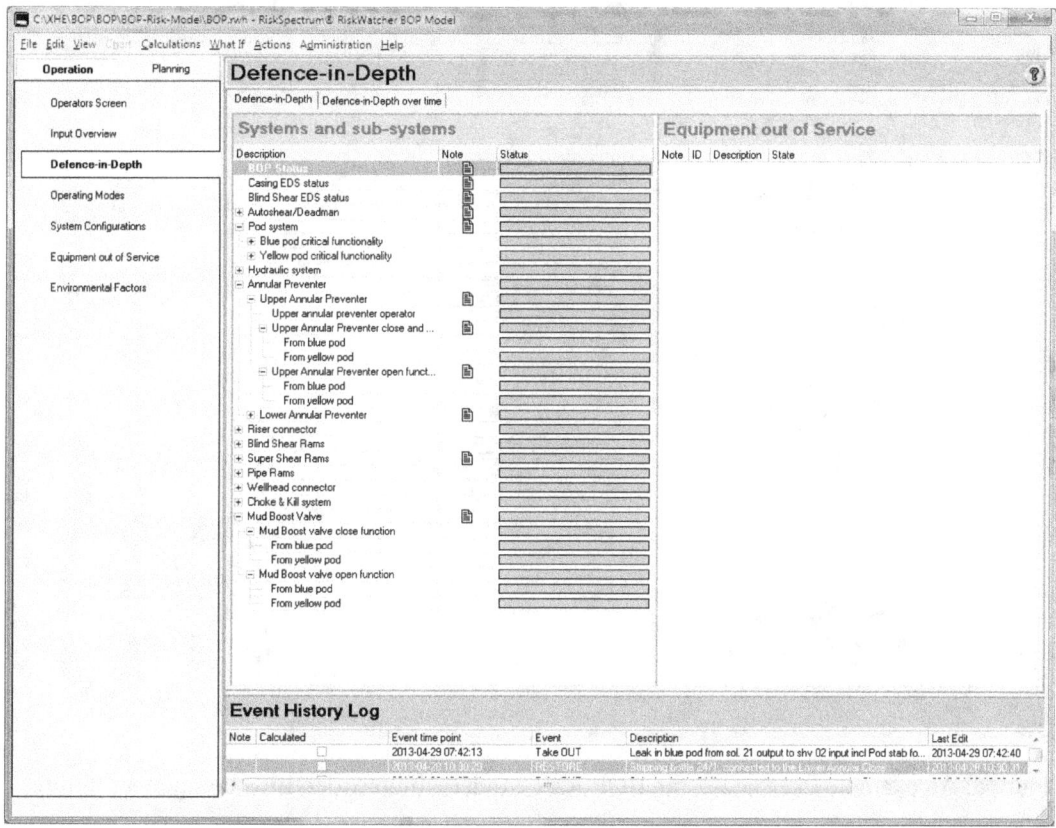

Figure 4. An example interface in the BOP Risk Model

As can be seen from the picture, in addition to the overall status of the BOP, the status of each of the functions and sub-functions is represented by a status bar.

3. A CASE STUDY

An offshore drilling rig developed several simultaneous leaks. Most of the leaks by themselves were minor issues; however the sum total made the personnel uneasy and need to decide on continuing with operations, pull the LMRP, or to pull the entire BOP stack.

Over the next few days, a number of teleconferences were organized between the drilling manager, his staff and shore-based experts to discuss issues and to clearly identify the root causes. The following failures were identified:

- Leak on mud boost valve open function from blue pod
- Leak on the lower annular open function
- Leak on surge-stripping accumulator bottle on upper annular which rendered the upper Annular inoperative

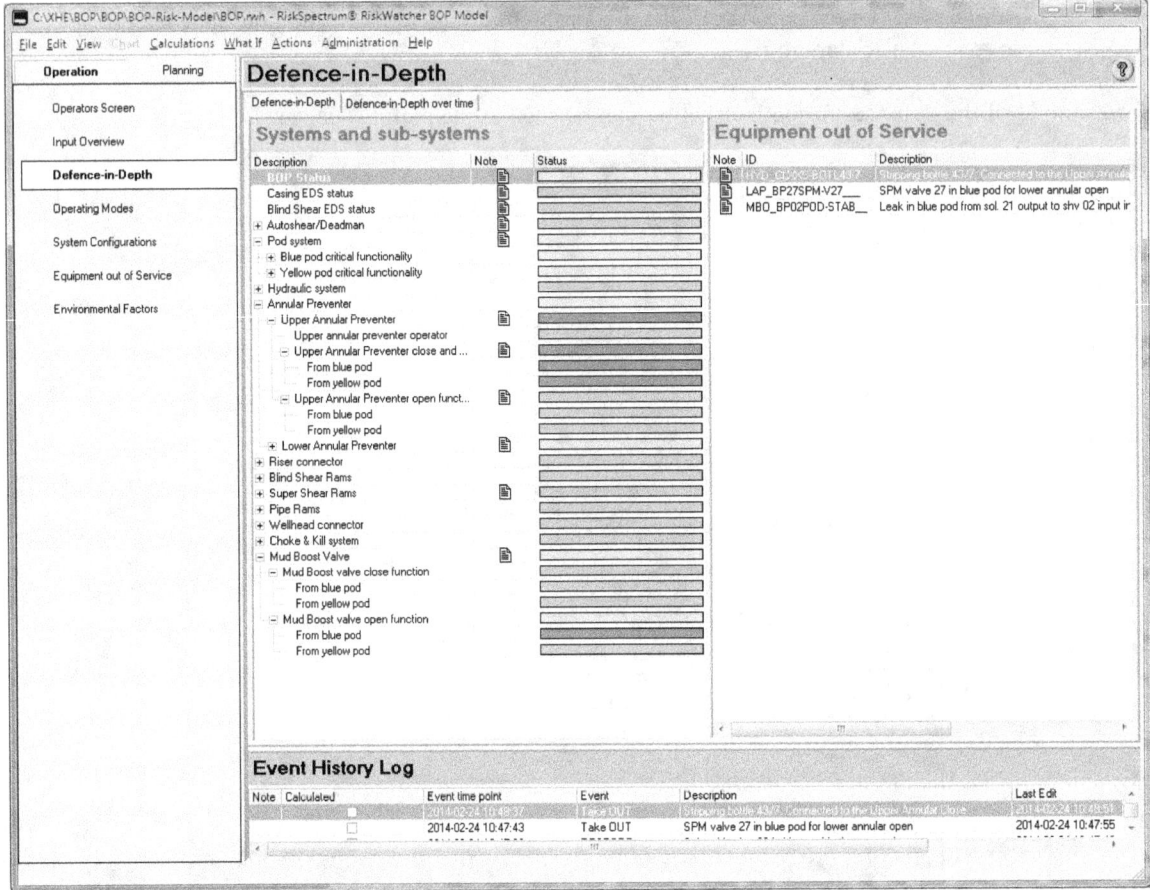

Figure 5. An example interface in the BOP Risk Model with the identified component failures

The overall BOP status for this particular BOP in this particular case is yellow from the BOP Risk Model and drilling is able to continue without pulling BOP. The reasons that the BOP was able to avert pulling include:

- The leak in the lower annular open function could be controlled by closing a valve.
- Lower annular can close
- Lower annular can open by itself (rubber memory) as long as close pressure is vented
- Mud boost valve function is not critical for safety

4. WEB VERSION BOP RISK MODEL

A web version RiskWatcher software for BOP Risk Model is currently under development. The web version software can open the model through the web browser. Any users with authorized user name and password can open the model within the company intranet.

All the model information, the event history including equipment out of service, restoration, etc. will be saved in the server and available for all the users. Multiple users are able to open the same model and check the up-to-date BOP status.

5. RELIABILITY OF THE BOP CRITICAL FUNCTIONS

BOP Risk Model is different from the traditional BOP monitor. The traditional BOP monitor could be an advanced feature of BOP control system where the real time BOP equipment status could be monitored. While BOP Risk Model is built on the logic block diagram and fault tree model and it is able to help decision maker to decide the current BOP risk level to make BOP pull or no pull decision.

BOP Risk Model can also be used to evaluate the reliability of the BOP critical functions in the future. In some countries/regions, there are specific safety integrity level (SIL) requirements for BOP critical functions. An example is the requirement in the OLF 070:

- As a minimum the SIL for isolation using the annulus function should be SIL 2 and the minimum SIL for closing the blind/shear RAM should be SIL 2

SIL2 means the probability of failure on demand is between 10^{-3} and 10^{-2}.

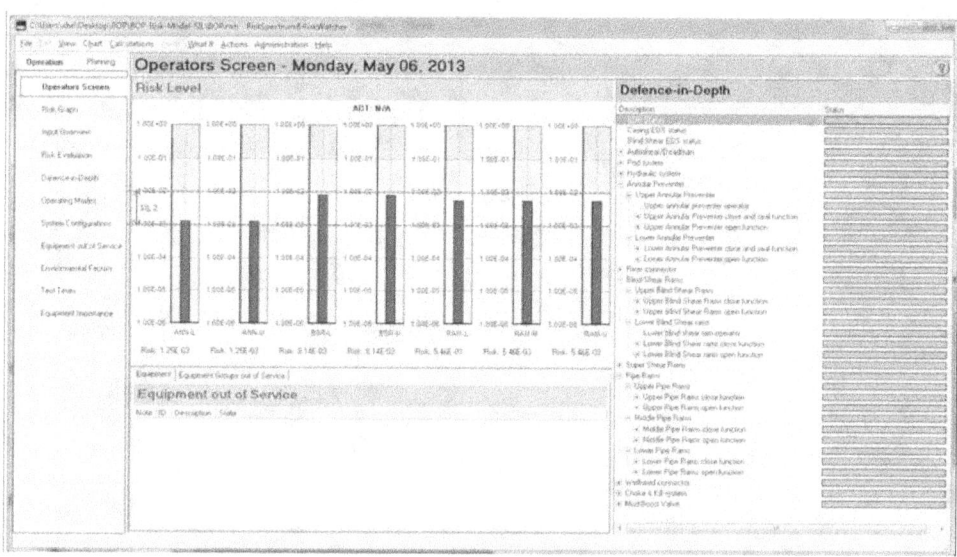

Figure 6. Reliability of the BOP critical functions (example)

6. DISCUSSIONS AND CONCLUSIONS

BOP Risk Model projects have been successfully implemented in a number of offshore drilling rigs. Positive feedbacks are received from the regulatory body. The model provides a good common platform for different parties including operator, drilling company, regulatory body to look at the BOP risk level. It helps to get accurate logic decision and fasts the decision making process.

Additional advanced features are under evaluation based on feedbacks and the project experiences. One big challenge is the fault identification in the BOP system. This may be solved using the available functions in the existing BOP control system as well as the other diagnosis technologies.

References

[1] Jeff Sattler, WEST Engineering Services. *"Pull Your BOP Stack-Or Not? A Systematic Method to Making This Multi-Million Dollar Decision"*, SPE/IADC 119762. SPEIIADC Drilling Conference and Exhibition held In Amsterdam, The Netherlands, 17-19 March 2009.

[2] I. A. Alme, X. He, T. B. Fylking, et, al. *"BOP risk and reliability model to give critical decision support for offshore drilling operations"*, 11th International Probabilistic Safety Assessment and Management Conference and The Annual European Safety and Reliability Conference, Helsinki, Finland, 25–29 June 2012.

Determination of the Design Load for Structural Safety Assessment against Gas Explosion in Offshore Topside

Migyeong Kim[a], Gyusung Kim[a,*], Jongjin Jung[a] and Wooseung Sim[a]
[a] Advanced Technology Institute, Hyundai Heavy Industries, Ulsan, Republic of Korea

Abstract: The possibility of gas explosion accidents always exists at offshore facilities of the oil and gas industry. In the design of those structures, the structural safety assessment against explosion loads is necessary to prevent loss of lives or catastrophic failure of structures. One of the essential parts in the structural assessment is to determine design explosion loads. Nowadays, the explosion loads are calculated by probabilistic approach rather than deterministic approach. In the recommended probabilistic load calculation procedure of the offshore explosion accidents, for instance, the NORSOK standard, the design load shall be established based on the relation between the predicted overpressure and its duration from numerous scenarios, and also, the exceedance frequency of the loads to the risk acceptance criteria. In most offshore projects, however, the conservative approach has been used, which derives only the design overpressure from the probabilistic load scenarios, or considers overpressure and duration independently because there is no efficient application method for the suggested procedure. In this paper, the practical method to determine the design explosion load, especially considering both overpressure and duration, are presented by using the response surface model with the joint probability distribution and compared with the present industrial practices.

Keywords: Explosion, Design load, Risk, Probabilistic approach

1. INTRODUCTION

Explosion accidents always exist at offshore drilling or production facilities in the oil and gas industry due to the characteristics of containing hydrocarbons, and they result in serious impact on personnel, environmental and property. Therefore, the safety assessment of offshore facilities against explosion should be performed on design phase to prevent loss of lives or catastrophic failure of structures and it is preceded by defining design accidental load.

The state of the art in determining design explosion loads is the probabilistic approach rather than the deterministic approach usually based on the worst-case scenario because the explosion accident is a complex phenomenon derived by a number of random variables which act along a chain of events connecting the potential hazards to consequence of possible explosion accidents [1]. The design variables considered on calculating explosion loads are sizes and directions of oil or gas leakage, locations of ignition, congestion and confinement in a layout, geometrical shapes of the impacted structures, and so on. The distribution of explosion load is defined based on numerous explosion scenarios using sophisticated analysis method, like CFD(Computational Fluid Dynamics). However, simulating all of possible scenario is very expensive for both man power and computer time. Moreover, the transient characteristic of explosion load leads another difficulty of determining explosion loads being expressed by not a singular item but two components; overpressure and duration. Each value of these load components has a great impact on structural safety because the response of a structure under explosion loads has inconsistent tendency showing some hollows and humps in dynamic region when one component of explosion load is changed as shown in Figure 1 [2]. Therefore, the selection of adequate duration corresponding to the defined overpressure is an important matter on determination of the design explosion load. In the present probabilistic methods of offshore explosion analyses, however, overpressure and duration(or impulse) are usually not treated as a combined term because of the missing probabilistic relation ignored when design explosion loads are determined with given criteria.

In this paper, a practical method to determine the design explosion load, especially, considering both of overpressure and impulse(or duration), is suggested using the response surface model with the joint probability distribution of them. So, the design explosion loads are determined as a set of overpressure

---

* Corresponding author: gyusung@hhi.co.kr, gyuskim@gmail.com

and impulse. In this new method, exceedance surface method, the structural response is considered in load determination procedure and acceptance criterion is selected as the load exceedance frequency. For these studies, the explosion scenario data from CFD simulations for the topside structure were used.

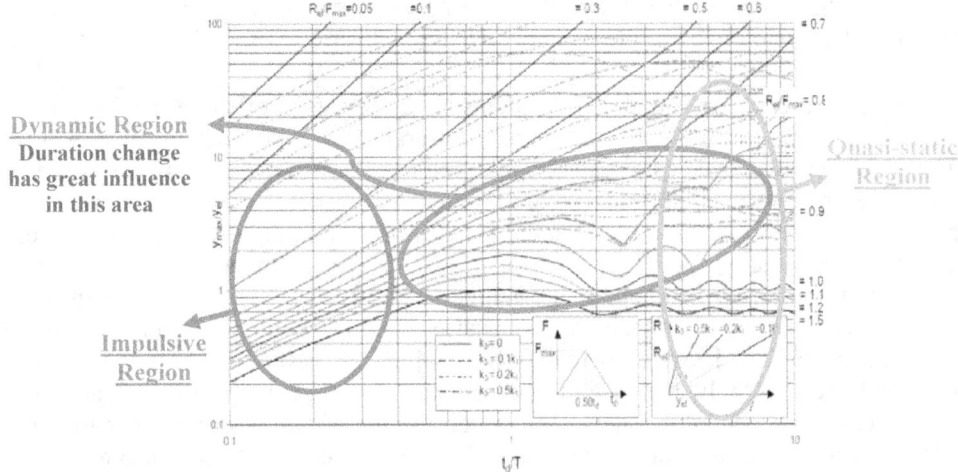

Figure 1: Dynamic response of a system under explosion load [2]

2. PROBABILISTIC ANALYSIS OF EXPLOSION LOAD

As mentioned above, explosion accidents at offshore installation are widely distributed in terms of their consequences and have high uncertainties due to the associated many random variables. Nowadays, the probabilistic approach combined with CFD analysis is practically used to reflect these complex characteristics of explosion loads.

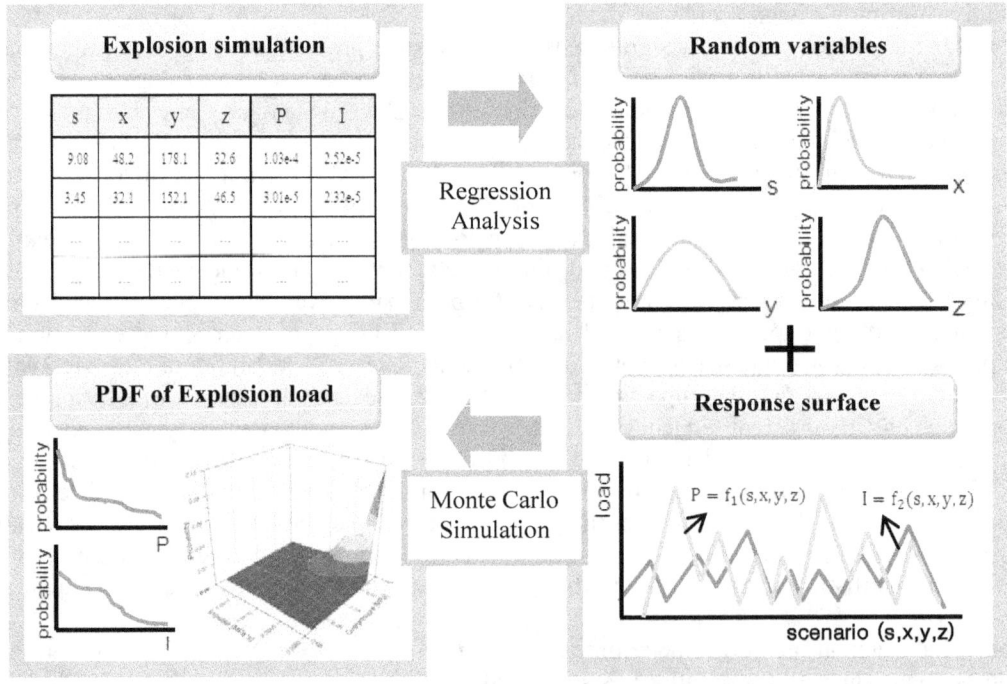

Figure 2: Process for determining explosion design load

In this paper, existing probabilistic approach is improved in following aspects. For each analysis step, all of the factors which affect the explosion loads are treated as random variables and final results are given as the joint probability distribution of explosion loads which can be applied for the risk based determination of design explosion loads. The modified method for the probabilistic definition of explosion loads is shown schematically in Figure 2.

2.1 Analysis model

The topside structure of a FPSO is selected as the analysis model [3]. Figure 3 shows the analyzed FPSO and the layout of its topside module. The module of separation train is considered for the explosion accidents. All leakage scenarios are selected in this area and the design explosion load is determined.

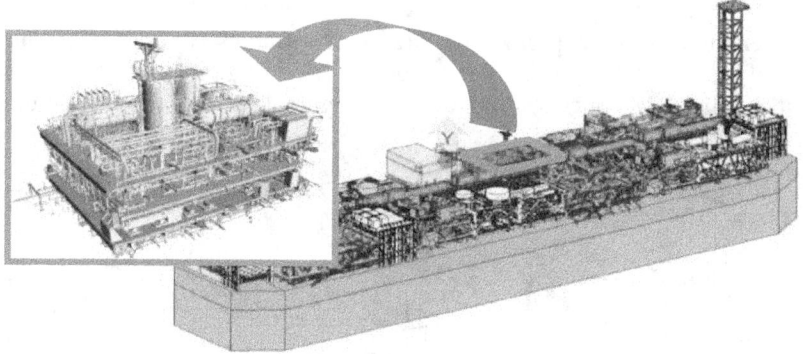

Figure 3: Module arrangement and separation module of analyzed FPSO

2.2 Leakage scenario & dispersion analysis

Explosion only occurs when enough amount of flammable gas mixture is cumulated and ignited by some internal or external sources. Therefore, the variables related to gas cloud are the most important factors for the frequency and consequence of the explosion load. To verify the location and size of the flammable gas clouds when leakage is occurred, dispersion analysis can be performed using CFD software, like FLACS [4]. With the leakage of gas, environmental conditions also affect formation of gas cloud and these random variables for the dispersion analysis are given below.

Table 1: Random variables for dispersion analysis

Random variable	Distribution type	parameters
Wind direction	Normal distribution	$\mu=180.9$, $\sigma=50.57$
Wind speed	Weibull distribution	$\alpha=4.84$, $\beta=2.563$
Leak rate	Weibull distribution (3para.)	$\alpha=29.39$, $\beta=1.454$, $\lambda=1.0$
Leak duration	Weibull distribution (3para.)	$\alpha=76.76$, $\beta=1.712$, $\lambda=0.362$
Leak direction	Uniform distribution	$C=1.6667$
Leak position X	Normal distribution	$\mu=0.437$, $\sigma=0.222$
Leak position Y	Normal distribution	$\mu=0.442$, $\sigma=0.198$
Leak position Z	Uniform distribution	$C=1$

For dispersion analysis, total 50 scenarios are defined via LHS(Latin Hypercube Sampling) method [5]. LHS is a sampling technique to extract the pre-selected number of samples from the individual areas which are defined by dividing probabilistic distribution function of each variable to have the same probability of occurrence. For all leakage scenarios, the result of dispersion analysis can be represented by 8 values as below.

- Gas cloud volume
- Gas concentration.
- Gas cloud position in the X dir.
- Gas cloud position in the Y dir.
- Gas cloud position in the Z dir.
- Gas cloud size in the X dir.
- Gas cloud size in the Y dir.
- Gas cloud size in the Z dir

2.3 Explosion analysis

2.3.1 Variables for explosion analysis

The results of dispersion analysis expressed by 8 random variables are applied as input for explosion analysis. To simplify the analysis process, non-regular shape of flammable gas is converted to cube form which has the highest gas concentration level [4]. Figure 4 shows this concept graphically.

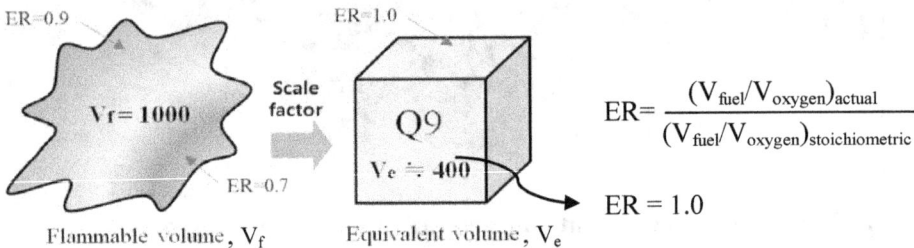

Figure 4: Conversion of non-uniform flammable gas to equivalent volume

With this assumption, pre-mentioned 8 variables can be reduced to 4 random variables in terms of gas cloud size and location in the X, Y, and Z direction. Figure 5 shows the probabilistic distribution of reduced random variables fitted as normal distribution, gumbel distribution, log-normal distribution, and normal distribution respectively.

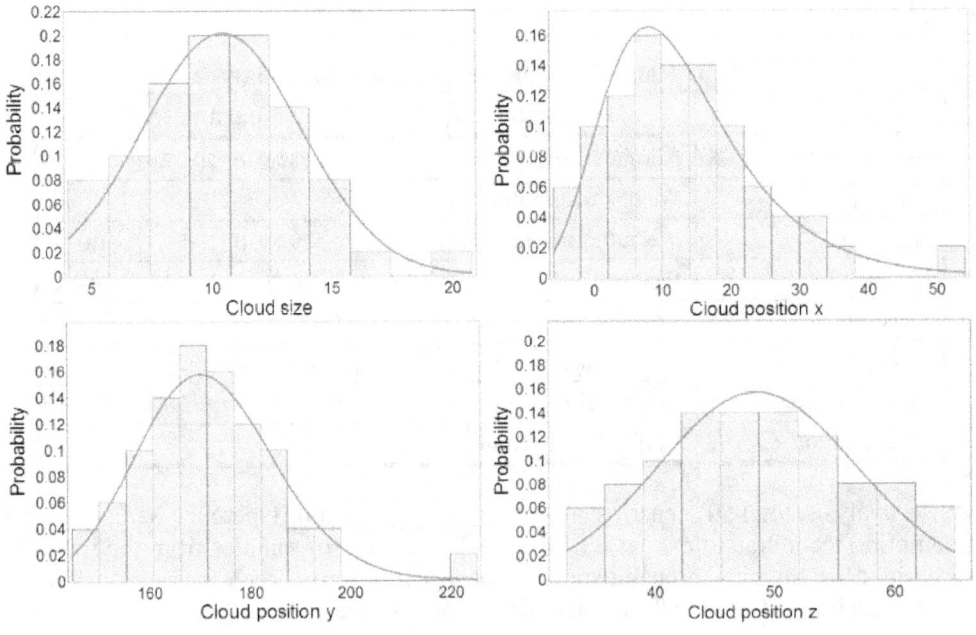

Figure 5: Probabilistic density function of random variables

When a large number of variables act compositely, ignoring the correlation between two or more variables can leads the misunderstanding to the effect of each variable on induced phenomenon. So, in this paper, Pearson product-moment correlation coefficients are calculated to confirm the degree of connection between two variables.

Table 2: The correlation coefficient of two variables

Variable		Cloud size	Cloud position		
			X dir.	Y dir.	Z dir.
Cloud size		1.0			
Cloud position	X dir.	0.0602	1.0		
	Y dir.	0.0491	-0.1400	1.0	
	Z dir.	-0.1160	-0.0851	-0.0560	1.0

As shown in Table 2, most of the absolute values for correlation coefficients are less than 0.1 and it means no linear relation exists between two variables. Only two results marked as red show weak linear relation but it is small enough to exclude its influence. Additionally, the scatter diagrams of two variables are plotted to check the case of non-linear relation and no strong correlation is found from these graphical results. So, in this paper, each random variable is treated as an independent value.

2.3.2 The results of explosion analysis

Table 3 shows some results of 50 explosion analyses which are conducted for EFEF JIP [3]. For these explosion analyses, 4 random variables defined in previous section are used and FLACS software is selected as the solver. All of the calculations performed using explosion analysis results in following sections are based on these data.

Table 3: Analysis results of explosion simulation with FLACS

Scenario No.	Cloud size [m]	Cloud position [m]			Pressure [MPa]	Impulse [MPa*s]
		X dir.	Y dir.	Z dir.		
1	9.18	-0.689	181.3	48.2	3.01 E-04	5.49 E-05
2	7.40	9.08	225	52.0	1.030 E-04	1.820 E-05
3	4.64	6.98	172.5	54.1	1.190 E-04	2.35 E-05
⋮	⋮	⋮	⋮	⋮	⋮	⋮
48	12.15	13.68	175.9	51.0	8.41 E-04	9.97 E-05
49	9.85	15.42	183.3	46.3	1.640 E-02	1.010 E-03
50	4.86	-1.694	182.3	54.6	9.21 E-05	1.070 E-05

2.4 Mathematical model of explosion load

The inputs of explosion analysis are 4 random variables as referred in Section 2.3.1. So, in this scheme, the mathematical model of explosion loads also can be established in terms of 4 variables by a regression analysis and it is defined as a response surface of explosion load. Figure 6 shows the real explosion analysis results and various mathematical models generated.

In Figure 6, the estimated explosion load by higher order mathematical model shows good agreement with the CFD analysis results. However, it doesn't mean that the complicated mathematical model has high accuracy because the adequate function of mathematical model for each variable is always different due to its physical characteristics.

In this paper, 2nd order polynomial function which has no combined term is selected for the each explosion response surface. Basic form of selected load response surface is expressed as equation (1).

$$\text{'Overpressure' or 'Impulse'} = a_0 + \sum a_i x_i + \sum b_i x_i^2 + \varepsilon \qquad (1)$$

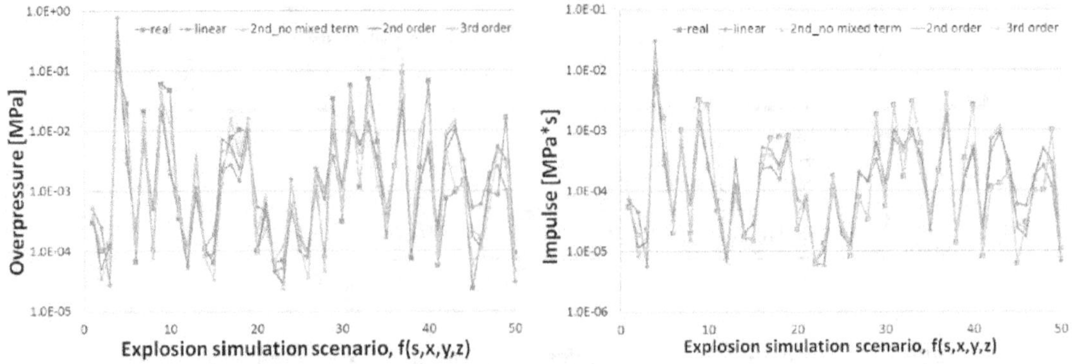

Figure 6: Mathematical models for the overpressure and impulse

2.5 Probabilistic distribution of explosion load

2.5.1 Probabilistic density function of each explosion load

For the accurate determination of probabilistic distribution for explosion loads, simulating all probable explosion scenarios with a sophisticated analysis method is the best way, if it is possible. However, this approach is not effective due to the time consumption induced by the large number of scenarios or long simulation time spent by the analysis tool itself. To reduce the calculation cost, in this paper, Monte Carlo simulation using response surface method is introduced to determine the probabilistic density function of overpressure and impulse. Total 1 million leak scenarios are selected considering the distribution of 4 random variables and the consequence of explosion loads are calculated adopting the pre-defined load response surface in Section 2.4. Figure 7 show the determined probabilistic density function of each explosion loads fitted as log-normal distribution function.

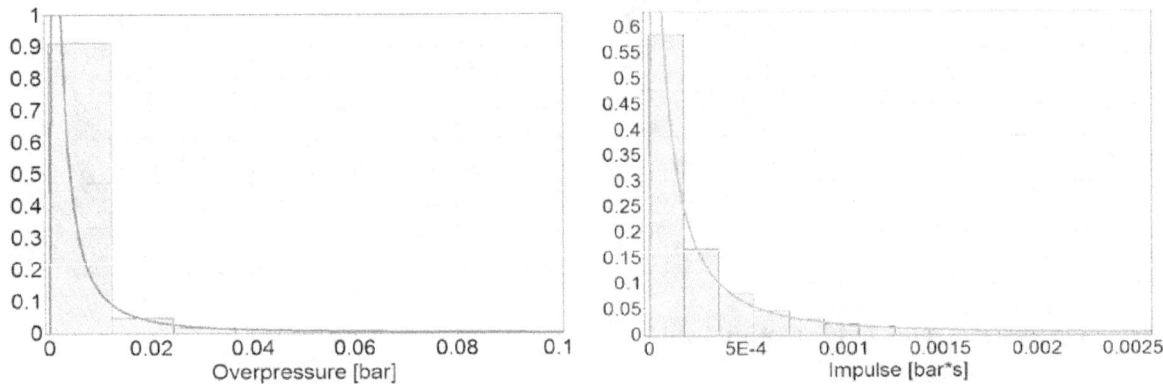

Figure 7: Approximation of the overpressure & impulse mathematical model

2.5.2 Joint probability of overpressure & Impulse

Scatter diagram for pre-selected 50 scenarios in Figure 8 is showing a bivariate distribution almost entirely supported on its diagonal line. It means that there is high linear relation between two

components of explosion load, overpressure & impulse. So, treating these two loads as a combined term instead of independent expressions is more realistic way in defining the explosion load.
Considering the correlation between overpressure and impulse, Joint probability of two load components are found using each load probabilistic density function. The calculated joint probability of explosion loads is graphically shown above Figure 9.

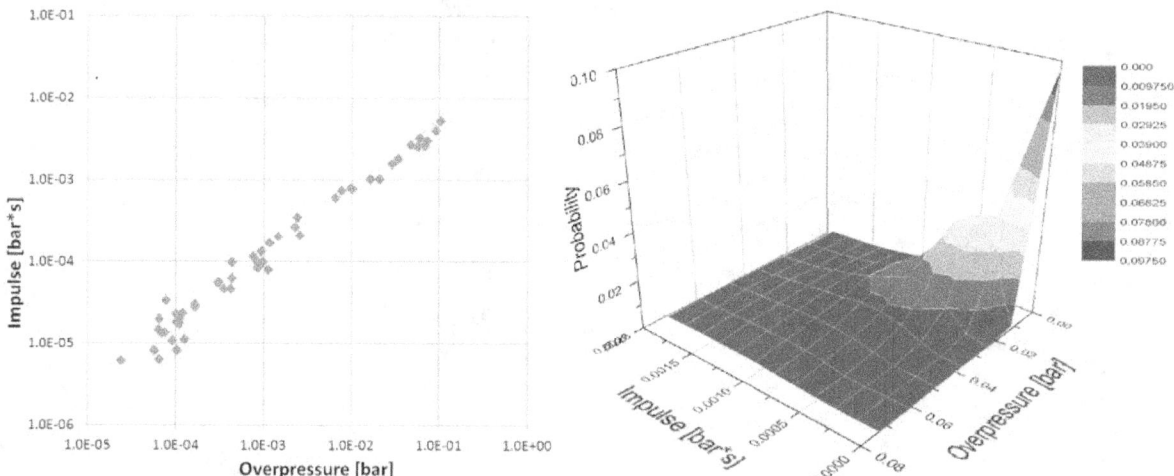

Figure 8: Scatter plot with explosion loads   Figure 9: Joint probability density function of explosion load

3. DETERMINATION OF DESIGN EXPLOSION LOAD

As mentioned in previous sections, overpressure and impulse are highly influenced by each other. Furthermore, the response of a structure under explosion loads also shows different tendency as one component of explosion load is changed. So, the adequate selection of duration (or impulse) corresponding to defined overpressure is an important matter on determination of the design explosion load. However, in existing methods, determining the combined form of a design load is not clear because it usually depends on the experience of engineers. So, in this paper, a practical way to determine combination of design explosion loads are suggested applying acceptance criteria of load frequency, and details of the method are specified in following sub sections.

3.1 Existing methods (Dual curve method, P-t curve method)

There are two kinds of risk based method practically used for determination of design load. First one is the dual exceedance curve method which adopts one curve for each load component as shown in Figure 10. In this method, overpressure and impulse are listed in descending order and the value of which cumulative frequency corresponds to 1.0×10^{-4}/yr is selected as the design load [6]. However, in this procedure, the correlation of two load component is not considered due to the independency of load curves. Another method is the P-t curve method to complement the shortcoming of dual exceedance curve method. P-t curve defines the explosion duration (or impulse) as a function of overpressure so, it can select the proper duration which is corresponding to the design overpressure defined by exceedance frequency curve. Although P-t curve is considering the relation between two load components, it is still not enough to reflect the whole probabilistic characteristic of explosion loads because the variation of duration is ignored when it is generated. Therefore, the selected design duration can be a value for specific scenario having no probabilistic meaning.
Additional method to improve this uncertainty of P-t curve is suggested applying the exceedance frequency surface method described in following section. It supplements the P-t curve defining the relation of overpressure and impulse with given acceptance load frequency criterion.

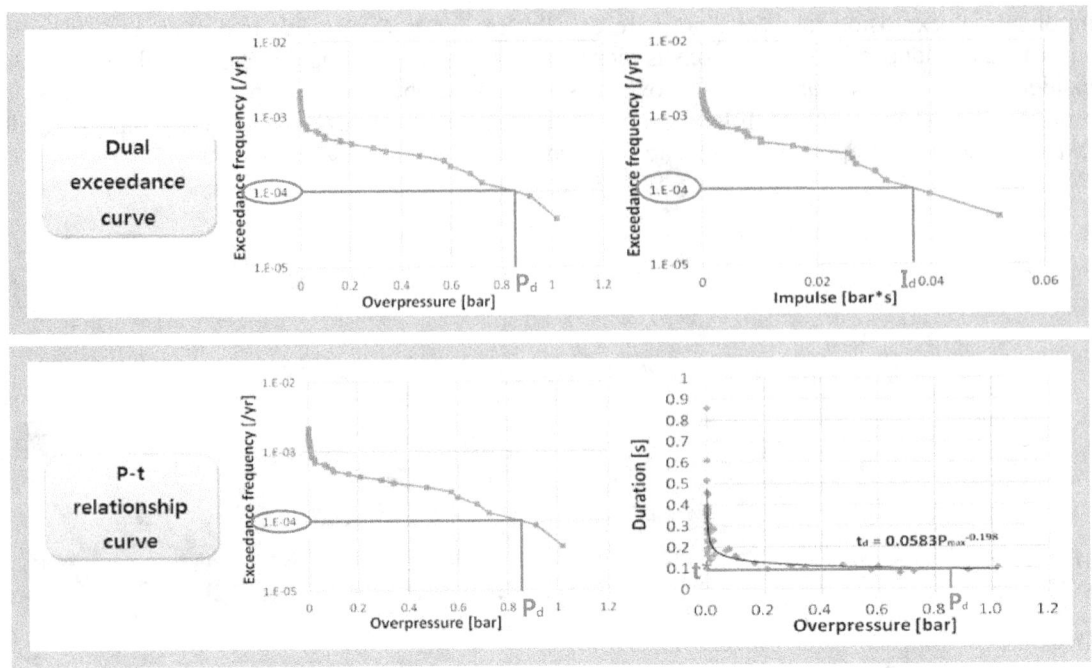

Figure 10: Determination of dimensioning explosion load with exceedance curves

3.2 New method (Exceedance frequency surface method)

Although dual exceedance curve and P-t curve method are dealing with both component of an explosion load, combination of selected values can be different from the probabilistic expression of an explosion load because the correlation between overpressure and impulse is not considered in detail. So, in this paper, above methods are modified to improve this aspect as mentioned in previous section. From the joint probability density function defined in Section 2.5.2, the exceedance probability surface is generated and it is changed to the leakage frequency term applying the HSE data which includes the information of leak amount and its corresponding frequency [7].

It is assumed the ignition probabilities of all possible scenarios are the same value to simplify the calculation in this paper. To consider the realistic explosion phenomenon, the distribution of ignition probability should be defined in terms of the variable which affects the value of it, like cloud size, location, time, etc. This ignition probability is combined with leak frequency as expressed in equation (2) to calculate the explosion frequency, and exceedance explosion frequency surface is generated with these data.

$$\text{Explosion frequency} = \text{Leak frequency} \times \text{Ignition probability} \qquad (2)$$

Applying the general risk acceptance criteria as the exceedance frequency, 1.0×10^{-4}/yr, the sets of overpressure and impulse can be determined as a form of a curve that is shown below Figure 12. This Overpressure-Impulse(P-I) design load curve means the relation of two load components at the given risk level. To compare the result of this approach with dual exceedance curve and P-t curve method, the determined design loads by those methods are also plotted.

As shown in Figure 12, determined design loads by dual exceedance curve and P-t curve are not located on the P-I curve although these methods are employing the same acceptance criteria. The correlation of overpressure and impulse can be considered as a main reason for these differences.

For the determination of a design explosion load with the exceedance surface method, additional criteria should be introduced to select a critical combination of overpressure and impulse among the numerous sets on the curve. As shown in Figure 12, one component of the load converges to a specific value when the other component of the load is getting larger. So, the limit point of each load component should be selected as a reference point to be checked with response criterion because this combination can results in one of the most serious responses among the load sets which is located on an asymptotic curve. To choose the proper limit value of each load component, the generated P-I curve is limited as a restricted region by adopting the design overpressure and impulse determined by dual exceedance curve or P-t curve method.

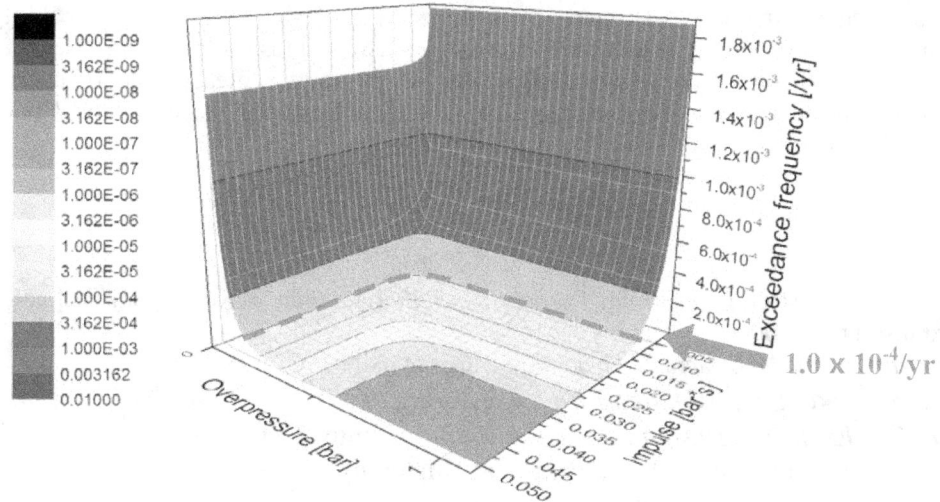

Figure 11: Exceedance explosion frequency surface

With the 2 load reference points corresponding to maximum value of each load component, final one is defined considering the natural frequency of a target structure, T, and duration of the load, t_d, to reflect the inconsistent characteristics of response in dynamic loading region. Finally, design explosion load is selected as the combination of load component which results in the maximum response among the 3 load points.

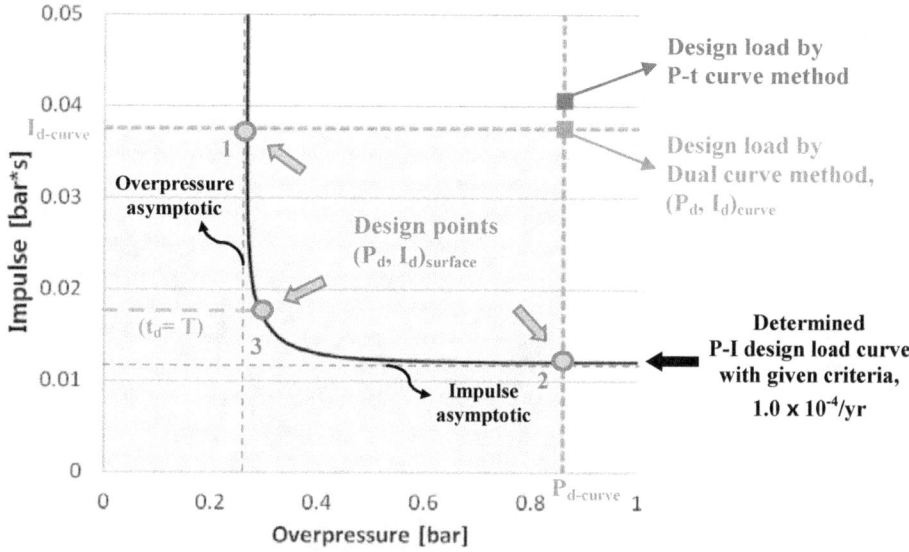

Figure 12: P-I design load curve from exceedance surface

4. CONCLUSIONS

This paper presents the procedure for probabilistic determination of design explosion loads at an offshore facility. All design factors which affect explosion events are expressed as random variables and used to determine the probabilistic density function of explosion loads. When applying load response surface, Monte Carlo simulation is performed for the purpose of reducing calculation time of CFD analysis. Based on the probabilistic density functions, new method to determine the design explosion loads is suggested. In this method, exceedance frequency curves are generated and the design explosion load is selected as the corresponding value to the risk acceptance criteria. Additionally, exceedance frequency surface is generated considering the joint probability of overpressure and impulse to include the correlation effect of two explosion load components. With this exceedance surface, therefore, the sets of design loads which result in great impact on a structure are determined. By the new method above, more exact design loads can be determined than the existing method which uses exceedance curves only.

References

[1] NORSOK Z-013, *Risk and Emergency Preparedness Analysis, Rev. 2,* Norwegian Technology Standard Institution, 2001
[2] J. Biggs, *Introduction on Structural Dynamics*, McGraw-Hill, 1964
[3] JK Paik, J. Czujko, *Explosion and Fire Engineering of FPSOs (phase II) – Definition of Fire and Gas Explosion Design Loads,* Research Institute of Ship and Offshore Structural Design Innovation, Pusan National University, 2010
[4] FLACS, GexCon AS (www.gexcon.com), Norway
[5] J. Czujko, *Design of Offshore Facilities to Resist Gas Explosion Hazard – Engineering Handbook*, Corrocean ASA, 2001.
[6] Oil & Gas UK, *Fire and Explosion Guidelines*, Issue 1, 2007
[7] HSE, *Hydrocarbon Release System*, http://www.hse.gov.uk/hcr3

Propagating Uncertainty in Phenomenological Analysis into Probabilistic Safety Analysis

A. El-Shanawany[a,b*]

[a] Imperial College London, London, United Kingdom
[b] Corporate Risk Associates, London, United Kingdom

Abstract: The operation of nuclear power plants is supported by numerous analyses, both computational and experimental. Probabilistic risk analysis models attempt to quantify the risk of power plants, and implicitly use the supporting analyses during this process. The way in which these analyses are used in risk models is usually conservative, but could instead be represented as an uncertainty distribution. The conservatisms are often hidden, but affect every aspect of risk models; for example in the definition of success criteria. This paper uses operator reliability as an example to quantitatively demonstrate how conservative interpretations of supporting analyses can affect risk model predictions.

The influence of human factors is recognised to be crucially important to risk models for nuclear power plants. Human error probability quantification is a key aspect in determining the relative risk importance of human actions in the context of a holistic probabilistic safety analysis model. However, there are large degrees of uncertainty in numerous aspects of human factors analysis and in the resulting quantification, many of which can be traced back to supporting transient analyses, such as thermal hydraulic and neutronic analyses. Risk models have historically used conservative judgements resulting from these analyses as an input into human reliability assessment. This paper presents a method for incorporating uncertainty distributions arising from phenomenological analyses into human reliability quantification. The method is illustrated using uncertainty in the timescale available to the operator for performing specified actions. This paper shows how to include uncertainty distributions over the time available to the operator and provides updated quantitative analysis. An illustrative example of operator initiated long term hold down of reactivity is presented.

Keywords: PRA, Uncertainty, Success Criteria, Risk, Human Reliability.

1. INTRODUCTION

Previous studies have considered the effect of success criteria uncertainties (using auxiliary feedwater pumps as an example) on the risk model results [1, 2]. In summary, it was found that model uncertainty, for these case studies, was order of magnitudes larger than parameter (statistical) uncertainty. This type of result suggests there is an unmet need to properly characterise the model uncertainty in the results of risk models. This paper seeks to contribute to the issue of model uncertainty by considering the plant based source of uncertainty. Nuclear power plants (NPP) have numerous supporting nuclear analysis codes; the nuclear analysis codes cover a wide spectrum of knowledge domains. Assessing uncertainty is a key part of the scientific method, and computational advances have allowed quantitative uncertainty estimates to be routinely calculated in a number of contexts. However, frequently these estimates are only used within a small knowledge domain and the uncertainty information is often passed on to other experts in a summarised form, for example as a conservative estimate. This type of information reduction certainly features in the construction of Probabilistic Risk Analysis (PRA) models, which uses conservative success criteria estimates. In this paper, the effect of uncertainty arising from phenomenological analyses on human reliability estimates is considered.

---

[*] Author email: ashanawany@c-risk-a.co.uk

In this paper the interaction of success criteria with operator actions is considered. Operator actions are typically assessed to be significant contributors to the overall estimated risk of operating nuclear power stations. The effect of operator actions on the predicted plant risk is typically a significant fraction of the total risk, although the precise figure is highly dependent on the specifics of the plant design and operation. The task of estimating human reliability, hence, has a very significant bearing on the results and insights of probabilistic risk models, and is a valuable example to use.

Quantitative human reliability assessment has numerous open questions associated with it and is an active area of research. Many methods have been developed over a number of years to quantitatively estimate human reliability. For example, some early methods include Technique for Human Error Rate Prediction (THERP), Accident Sequence Evaluation Programme (ASEP), and Human Error Assessment and Reduction Technique (HEART) [3, 4, 5]. A 2009 review [6] by the UK Health and Safety Executive (HSE), identified 72 potential methods. Some of these methods are public domain, while others are proprietary methods. This paper is written without reference to any specific method, but it does draw upon ideas that have been developed by the human reliability quantification methods. In particular the concept of factors which modify human reliability is considered in this paper.

It is widely accepted that there are numerous variables that can affect the performance of operators. These factors can be coarsely split into internal station factors and external station factors. The internal station factors include issues such as familiarity with a task, the time available to complete a task, and clarity of feedback. External factors include all those aspects of life which will impinge on an operator's state of mind, for example their personal life outside of work. Generally human reliability quantification methods consider only the internal to station factors. A description of the factors which can affect human reliability is provided in Reference 7. Quantitative techniques have been developed to estimate the impact of these factors on human reliability. However, uncertainty arising from these factors has not previously been considered quantitatively. Instead the implicit ethos adopted by human error quantification techniques has been the same as the ethos traditionally used throughout probabilistic safety analysis; that is to make conservative judgements whenever a judgement needs to be made. However, the tools used to estimate the impact of a given factor can usually be extended to incorporate the effect of uncertainty. For example the effect of available time on operator reliability has been characterised by several methods, and can be extended to consider uncertainty over the time available. This paper demonstrates how uncertainty over the time available for an operator to perform an action can be incorporated into risk models, and links the uncertainty in the time available to the underlying plant physics.

2. NITROGEN EXAMPLE

The large range of factors which affect human performance can each have several different sources of uncertainty. In this paper the problem is restricted to the time available to the operator to perform a specified task, and the uncertainty in the time available which is attributable to the underlying phenomenological modelling. For example, following a trip at a gas cooled reactor, the primary shutdown mechanism is to fully insert the control rods, thereby making the reactor subcritical and terminating the fission reaction. However, if some control rods fail to insert then nitrogen gas can be used to provide an additional safety margin for shutdown and to ensure the long term hold down of reactor reactivity [8]. The key parameter determining the timescale available for nitrogen shutdown, assuming a fixed number of rods fail to insert, is the Xenon-135 transient. Xenon-135 is a strong neutron absorber (~2.6E+06 barns) and reaches a steady state in an at power reactor [9]. Xenon-135 formation and decay is primarily dictated by the processes shown in Figure 1 below.

Figure 1: The major Xenon formation and decay processes

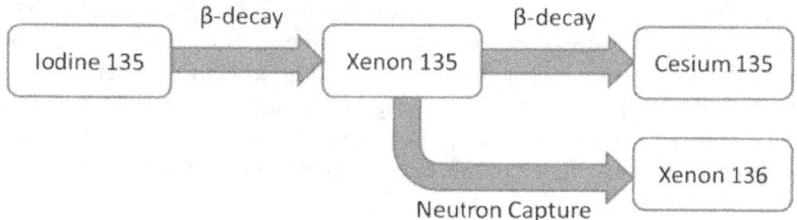

During at power operation Xe-135 is continually "burnt off" due to the high neutron flux. However, once the fission process is terminated the neutron flux falls to a low value, and the neutron capture method of decay shown in Figure 1 ceases to be a significant decay mechanism of Xe-135. This causes the Xe-135 levels to rise initially as I-135 is converted into Xe-135 faster than Xe-135 decays into Ce-135. However, I-135 is primarily produced from at power decay processes, and once the reactor is shutdown, the production of I-135 falls to a low value. Following shutdown the decay rate of I-135 is greater than the production rate, causing a steady decline in the level of I-135. As the level of I-135 falls, the production rate of Xe-135 falls. Eventually the production rate of Xe-135 falls below that of the decay rate, and the level of Xe-135 falls asymptotically to zero. Once the Xe-135 level begins to fall this effectively represents a reactivity insertion into the reactor. This phenomenon is well understood and described in many texts, for example Reference 9.

The rationale behind nitrogen injection is to insert extra negative reactivity before the Xe-135 transient causes an insertion of positive reactivity, thus maintaining the shutdown margin. However, the precise form of the Xe-135 transient depends heavily on the operating history of the reactor, although it will follow the general form described above. For any given reactor state, the Xe-135 transient can be accurately predicted and can be used to infer the time available to start nitrogen injection in order to maintain the safety margin in the event that primary shutdown is not fully achieved. The possible scenarios can be bounded by a "worst case" scenario which is used to define the minimum time that could elapse before positive reactivity insertion due to Xe-135 decay could occur. This minimum time is then, conservatively, used to define the time available to the operators to commence nitrogen injection. This conservatism is largely hidden from view during a typical procedure for quantitative human reliability estimation. The framework for considering time pressure as a factor in the estimation of human reliability has existed since the first generation HRA methods [6]. The existence of such methods facilitates including uncertainty over the time available for an action into the estimate of human reliability. The method used to incorporate time pressure can be complex, depending on the overall human reliability method used, but this will be substantially simplified for the purposes of this paper. The next section gives an outline of a simple theoretical way of incorporating time factors into HRA.

3. TIME FACTORS

There are numerous ways in which time could be incorporated into HRA estimates, and different human reliability quantification methods take different approaches. Most modern methods acknowledge that time pressure has a quantitative impact on human reliability, and to a varying extent provide a mechanism for quantitatively estimating this effect. However, as noted above, these methods can be involved. The aim here is to demonstrate an overall method for incorporating important thermodynamic uncertainties into HRA estimates, rather than the HRA methods themselves. To that end a grossly simplified method of incorporating time factors will be used by using a multiplicative factor applied to the human reliability estimate. The multiplicative factor is a negative exponential decay model of the effect of additional available time over some baseline time for performing a task, as shown in the equation below:

$$TF = \exp(-(t_a - t_b)/s) \qquad (1)$$

Where t_a and t_b are the time available and a "baseline time" respectively. The baseline time is a nominal "normal" amount of time which should be available in order for the operator to perform the action successfully. 's' is a scaling factor that can be used to adjust for the effect of the selection of the units. Deviations from the baseline time are weighted using a negative exponential factor. In this paper the baseline time will be taken as being the conservative time limit that would be used in existing risk modelling processes. Hence in this paper t_a is always greater than or equal to t_b.

It is emphasised that the model described above does not have a basis in quantitative human reliability research, and is chosen primarily for simplicity. The curve does also satisfy the intuitive properties that increases in the available time increase reliability while increases in reliability decline as more time is made available, so that in a finite time it is not possible to achieve an arbitrarily low reliability using this model. Any alternative method for assessing the impact of time on human reliability could be substituted for this model without changing the overall method described, and without altering the overall message of the paper.

4 INCORPORATING UNCERTAINTY

Section 3 has provided a simplified method for incorporating time factors into quantitative human reliability assessment. Hence to create an uncertainty distribution **based on the effect of time uncertainty only**, we can assess the reliability at all of the possible time points. In general this is a continuous distribution, but practically we will restrict it to a discrete distribution over possible time states. This is demonstrated below in the context of the nitrogen injection example.

To incorporate uncertainty in the time factor of nitrogen injection we need to estimate the probability distribution over the time to injection. This is a difficult task. Fortunately, known phenomena and transients that can occur at nuclear power stations invariably have already been subjected to rigorous analysis. From a risk analysis perspective we can simplify the task to collating existing analyses which provide estimates on uncertainty. In this example the information needs to be discretised into "reasonable" time chunks permit analysis. Each time period is assigned a probability mass, which in this case is nominal, but in practice can be based on existing analyses. This is illustrated with a hypothetical operator action which is assigned a nominal human error probability of 1.00E-03. This defines the human error probability in the baseline time, which is set to be 3 hours in this example. Table 1 below gives a breakdown of time periods and probability masses, together with the multiplicative factor calculated using our simple model, and the revised human error probability estimate for that time period.

Table 1: Discrete Probability Distribution Over the Human Error Probability Estimate

ID	Time Available	Probability	Multiplicative Factor	Human Error Probability Estimate
1	3-6 hours	0.1	0.94	9.39E-04
2	6-12 hours	0.2	0.78	7.79E-04
3	12-18 hours	0.5	0.61	6.07E-04
4	18-24 hours	0.2	0.47	4.72E-04

In each case the midpoint of the range has been used in estimating our multiplicative factor. The greater the number of intervals used the more detailed the resulting uncertainty distribution will be. The method is just as well demonstrated using a small number of intervals as using many intervals, since it still provides the shift from a conservative assessment to a best estimate plus uncertainty case. It is in making this shift that the greatest difference is observed, rather than by increasing the number of intervals further. Beyond a 24 hour time horizon there is likely to be significant offsite support available and the problem fundamentally changes character from predicting the reliability of a single team or single operator. For this reason the simple model presented above is considered wholly inapplicable for times beyond 24 hours and times on longer scales are not considered in this paper.

Table 1 shows that the best estimate of the time available to complete the action is in the 12-18 hour bracket, rather than the conservative estimate of 3 hours. A probability distribution is formed by the intervals defined in Table 1 above, and this is assumed to be available as the result of phenomenological modelling. As noted previously, this information is already available in many instances. The next section considers the impact of incorporating the information presented in Table 1 into a simple model. Using the values from Table 1 the mean estimate of the human error probability is calculated to be 6.48E-04 rather than the conservative estimate of 1.00E-03.

5 RESULTS USING A SIMPLE MODEL

A simple example fault tree is shown below, and will be used to demonstrate the concept of including time uncertainty for human error probability.

Figure 2: Simple Fault Tree Example

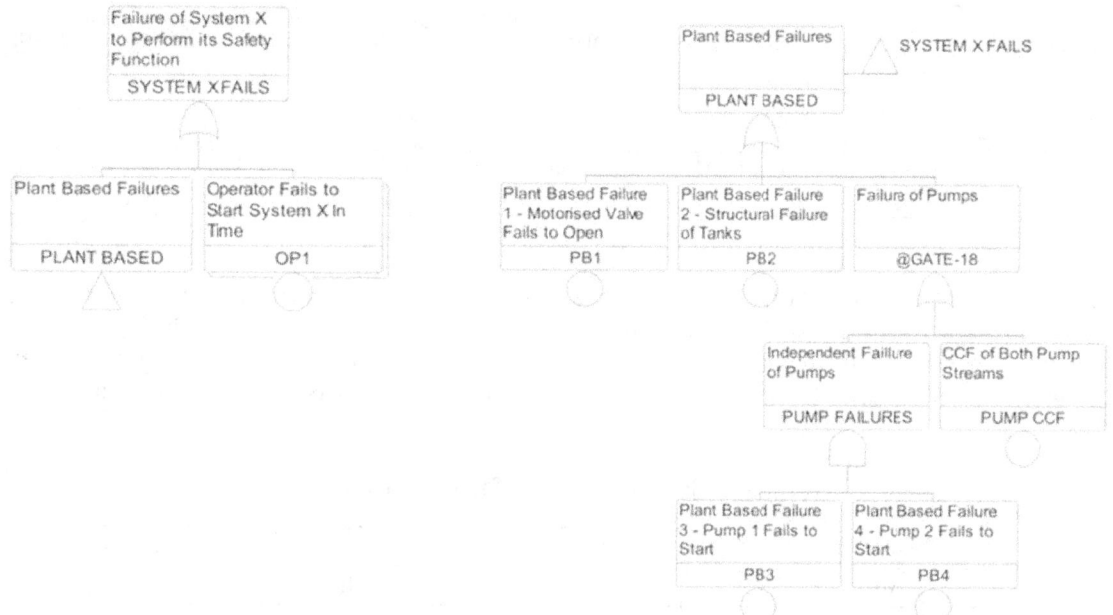

A table of basic events used in the model is provided in Table 2 below:

Table 2: Summary of Basic Events

ID	Description	Mean
OP1	Operator Fails to Start System X In Time	6.48E-04
PB1	Plant Based Failure 1 - Motorised Valve Fails to Open	2.00E-04
PB2	Plant Based Failure 2 - Structural Failure of Tanks	1.00E-05
PB3	Plant Based Failure 3 - Pump 1 Fails to Start	1.00E-03
PB4	Plant Based Failure 4 - Pump 2 Fails to Start	1.00E-03
PUMP CCF	CCF of Both Pump Streams	1.00E-04

In the base case the conservative assumption is that 3 hours are available to perform the operator action. This conservative assumption is the one which would be made in any standard probabilistic risk model, and represents the current practice in incorporating human error probabilities into risk models. The parameters used for this model are shown in Table 3 below.

Table 3: Model Failure Parameters Table

Parameter Name	Description	Mean	Uncertainty Distribution	Error Factor	Median	5th Percentile	95th Percentile
Base Case HEP	Base Case Conservative Value for Human Error Probability	1.00E-03	None	N/A	1.00E-03	1.00E-03	1.00E-03
Uncertainty HEP	Human Error Probability Inc Time Uncertainty	6.48E-04	Discrete	See Table 1	6.07E-04	9.39E-04	4.72E-04
Pump CCF	Pump CCF	1.00E-04	Lognormal	5.00	6.20E-05	1.24E-05	3.10E-04
PumpFS	Pump Fails to Start	1.00E-03	Lognormal	2.00	9.15E-04	4.58E-04	1.83E-03
Tankfail	Structural Failure of Storage Tanks	1.00E-05	Lognormal	5.00	6.20E-06	1.24E-06	3.10E-05
ValveFC	Valve Fails Closed	9.00E-04	Lognormal	2.00	8.23E-04	4.12E-04	1.65E-03

The model has been analysed using RiskSpectrum PSA v1.1.4.3, using standard analysis settings, to provide an estimate of the probability of failure on demand of the top gate shown in Figure 2. The software has been used to find the minimal cutsets of the fault tree, calculate importance metrics and estimate the uncertainty distribution of the top gate. The minimum cutsets for the base case are shown in Table 4 below:

Table 4: Minimum Cutsets for the Base Case

ID	Sequence Probability	Percentage Contribution	Event 1	Event 2
1	1.00E-03	49.8	OP2	
2	9.00E-04	44.8	PB1	
3	1.00E-04	4.98	PUMP CCF	
4	1.00E-05	0.5	PB2	
5	1.00E-06	0.05	PB3	PB4

The percentage contribution is a standard importance measure used to evaluate cutset results. Note that the percentage contribution has the property that the sum of all cutset percentage contributions is not (usually) 100. The formula used for calculating fractional contributions is copied below for reference.

$$FC_i = 1 - \frac{Q_{TOP}}{Q_{TOP}(Q_i = 0)} \qquad (2)$$

Where FC_i is the fractional contribution of the ith component, Q_{TOP} is the probability of failure on demand of the top event (or unavailability in some problem setups), and $Q_{TOP}(Q_i=0)$ is the probability of failure on demand of the top event with the probability of failure of the ith component set to zero.

Table 5 below shows summary results for the basic events in the base case:

Table 5: Basic Events Importance and Sensitivity for the Base Case

ID	Basic Event	Mean	Fractional Contribution	Risk Decrease Factor	Risk Increase Factor	Sensitivity (RIF/RDF)
1	OP2	1.00E-03	4.97E-01	1.99	4.98E+02	9.90
2	PB1	9.00E-04	4.47E-01	1.81	4.98E+02	8.41
3	PUMP CCF	1.00E-04	4.97E-02	1.05	4.98E+02	1.51
4	PB2	1.00E-05	4.97E-03	1.00	4.98E+02	1.05
5	PB3	1.00E-03	4.97E-04	1.00	1.50E+00	1.00
6	PB4	1.00E-03	4.97E-04	1.00	1.50E+00	1.00

The risk decrease factor is the ratio of the top event failure probability with the defined model parameters to the top event failure probability if that specific basic event has a zero failure probability. In simple examples, such as this one, this is easily relatable to the fractional contribution: for example OP2 contributes ~1/2 of the probability of top event failure, and the risk decrease factor is hence 2. The risk increase factor is the same ratio but with the specific basic event failure probability set to one. Sensitivity is the ratio of the risk decrease and risk increase factors.

The analysis has then been repeated but replacing the point estimate of 1.00E-03 with the discrete distribution given in Table 1 above. The minimum cutsets in this case are shown in Table 6 below:

Table 6: Minimum Cutsets for the Uncertainty Case

ID	Sequence Probability	Percentage Contribution	Event 1	Event 2
1	9.00E-04	62.1	PB1	
2	4.38E-04	30.2	OP1	
3	1.00E-04	6.9	PUMP CCF	
4	1.00E-05	0.69	PB2	
5	1.00E-06	0.07	PB3	PB4

Table 7 below shows the fractional contributions of basic events in the uncertainty case:

Table 7: Fractional Contributions of Basic Events for the Uncertainty Case

ID	Basic Event	Mean	Fractional Contribution	Risk Decrease Factor	Risk Increase Factor	Sensitivity (RIF/RDF)
1	PB1	9.00E-04	6.21E-01	2.64	6.90E+02	14.9
2	OP1	4.38E-04	3.02E-01	1.43	6.90E+02	5.11
3	PUMP CCF	1.00E-04	6.89E-02	1.07	6.90E+02	1.73
4	PB2	1.00E-05	6.89E-03	1.01	6.90E+02	1.07
5	PB4	1.00E-03	6.89E-04	1.00	1.69E+00	1.01
6	PB3	1.00E-03	6.89E-04	1.00	1.69E+00	1.01

Comparing Table 4 and Table 6 it can be seen that the importance ranking of the operator action is altered when uncertainty is included in the estimate. This type of permutation in the importance of cutsets is very significant since the analysis of cutsets is a primary method for understanding the plant risk, and providing input into risk informed decision making. In the list of fractional contributions of basic events (Table 5 and Table 7), there is a corresponding permutation in the ordering of basic event importance in the uncertainty case compared to the conservative case.

6. DISCUSSION

PSA models are filled with hidden examples of uncertainties that arise from uncertainties in the underlying analyses of physical processes that have been performed. The incorporation of uncertainties is not always as straight forward as in this case. For example some uncertainties could only be incorporated through changes to the structure of the model. For example if using a fault tree paradigm, then some uncertainties could only be incorporated by structural changes to the number of inputs to gates.

The observation arising from Section 5 is essentially a very simple one; that is that using a best estimate plus uncertainty to represent a failure parameter value can have significant effects on the risk profile of a model, and of the risk importance of cutsets. This is an obvious statement, but the insight and contribution of the paper really comes from the source of the re-assessment of the value of the failure parameter. The source is one that falls between domains of knowledge; the Relap analyst typically knows little about human factors analysis, and the human factors analyst rarely considers the details and implications of Relap analysis. The uncertainty that would be considered by quantitative human reliability methods is that associated with statistical uncertainty in the data used; and that uncertainty is only considered in the latest methods. This clearly misses the uncertainty considered in this paper, and as a result provides a misleading representation of having assessed the uncertainty; as noted by Zio and Aven [10], recourse to a quantitative evaluation method without detailed understanding of the underlying factors can easily lead to a misrepresentation of the risk results, and may place an unwarranted level of certainty about the results of the analysis

Hence the significant uncertainty considered included here would normally fall between the gaps of knowledge domains. The need for multi-disciplinary teams in general is, of course, well established, but appeals for multi-disciplinary collaboration are often vaguely justified or even presented as an end in itself, rather than serving a specific useful purpose. The type of observation made in this paper is one specific justification for increased interaction between diverse domain experts to allow high level understanding of how domains of knowledge interact. This gestalt understanding is important to nuclear safety.

7. CONCLUSION

There are many hidden conservatisms within the model structure of probabilistic risk models. The source of these conservatisms can often be traced back to a conservative interpretation of a supporting physical analysis. Often the conservative interpretation is a simplification of the actual analysis available and uncertainty is routinely calculated in many domains. This paper provides a joined up use of uncertainty for the purposes of risk assessment. This paper has presented a method to quantitatively incorporate uncertainty in human reliability due to underlying uncertainties in physical analyses. This type of uncertainty has not been quantitatively assessed before, but probabilistic models have instead relied on conservative judgements. Indeed this uncertainty remains largely hidden from view in probabilistic models, unless a joined up view of the analysis processes involved, including an understanding of the basic physical processes and how uncertainties in one sphere of knowledge can propagate through to other analyses in the course of risk analysis.

The uncertainty analysed here is also an example of how quantitative human reliability analysis can unwittingly hide uncertainties. This is true of all existing methods, even those which purport to estimate the uncertainty of operator reliability. The methods surveyed which do provide an estimate of uncertainty provide only an estimate of statistical variation between responses to controlled conditions. This is a useful factor to try to estimate but it is a misleading to portray this as a full characterisation of the uncertainty associated with the problem. Aleatory uncertainty is only the tip of the iceberg in terms of the uncertainties that contribute to human reliability. There is a significant body of work to be performed in identifying and, where possible, quantifying, latent uncertainties in human reliability assessment that have, to date, been masked by conservative judgements.

More generally the method presented here can be extended to conservatisms latent throughout probabilistic risk models. A model with numerous hidden conservatisms limits itself to make statements of the form "the risk is at least be tter than X". Iteratively re-assessing conservatisms is a vital part of transforming risk analysis of hazardous plants and the type of statement that risk models can be used to make. This work is likely to modify our understanding of the o verall risk profile of nuclear power plants, and as shown in this paper it can significantly affect our understanding of the risk importance of operator actions.

The contribution made here is one par t of ass essing model uncertainty. The assessment of model uncertainty in this sense could be expanded to a ll parts of PRA models using existing uncertainty estimates. This is similar in spirit to the uncer tainty analysed in the case st udy of success criteri a uncertainty for auxiliary feedwater pumps in Reference 2.

In addition to the uncertainty induced by time uncertainty, there is also uncertainty in the "effects model" used to represent the effect of time. That is, the correct form of equation 1 is unknown. This is another area for further work.

ACKNOWLEDGEMENTS

The author is very grateful to the EPSRC for funding and to Corporate Risk Associates for ad ditional funding and technical sup port. Thanks to Rebec ca Brewer for reviewing the paper and providin g useful feedback. Thanks to Lavinia Raganelli for helpful discussions and feedback. The author would like to thank Dr Simon Walker, Dr Charles Shepherd and Jasbir Sidhu for their continuing support.

REFERENCES

[1] J. M. Reinert and G. E. Apostolakis, *"Including model uncertainty in risk-informed decision making"*, Annals of nuclear energy 33.4, pp354-369, (2006).

[2] J. K. Knudsen and S. L. Curtis, *"Estimation of system failure probability uncertainty including model success criteria"*, Proceedings of the 6th International Conference on Probabilistic Safety Assessment and Management (PSAM 6), Vol. 1, (2002).

[3] A. D. Swain and H. E. Guttmann, *"Handbook of Human Reliability Analysis with Emphasis on Nuclear Power Plant Applications"*, Final Report, NUREG/CR- 127 8 SAND80-0200 RX, (1983).

[4] A. D. Swain, *"Accident Sequence Evaluation Program Human Reliability Analysis Procedure"*, NUREG/CR-4772, (1987).

[5] J.C. Williams, *"A Data Based Method for Assessing and Reducing Human Error to Improve Operational Performance"*, IEEE Fourth Conference on Hum an Factors and Power Plants, pp 436 – 450, (1998).

[6] Health and Safety Laboratory, *"Review of human reliability assessment methods"*, Health And Safety Executive, (2009).

[7] J.C. Higgins, J.M. O' Hara, P.M. Lewis, J.J. Persensky, J.P. Bongarra, S.E. Cooper, and G.W. Parry, *"Guidance for the Review of Changes to Human Actions"*, NUREG-1764, U.S. Nuclear Regulatory Commission Office of Nuclear Regulatory Research, (2007).

[8] E. Nonbøl, *"Description of the Advanced Gas Cooled Type of Reactor (AGR)"*, NKS/RAK2(96)TR-C2, Risø National Laboratory Roskilde, Denmark, (1996).

[9] J.R. Lamarsh, A.J. Baratt a, *"Introduction to Nuclear Reactor Theory"*, 3rd edition, Prenti ce Hall, (2001).

[10] E. Zio and T Aven, *"Industrial disasters: Extreme events, extremely rare. Some reflections on the treatment of uncertainties in the assessment of the associated risks"*, Process Safety and Environmental Protection 91, pp 31–45, (2013).

A Procedure Estimating and Smoothing Earthquake Rate in a Region with the Bayesian Approach

J.P. Wang
The Hong Kong University of Science and Technology, Kowloon, Hong Kong

Abstract: Reliable instrumentation earthquake data are considered limited compared to the long return period of earthquakes, especially the major events. As a result, earthquake rate estimating could become unrealistic based on the limited earthquake observation with classical statistics algorithms. For example, given no $M \geq 6.0$ earthquakes were recorded in the past 50 years, a best-estimate for the earthquake rate around the region should be zero, and such a zero estimate is considered unconvincing owing to the short observation period and long earthquake return periods. In this paper, a Bayesian calculation is proposed to earthquake rate estimating and smoothing given a reliable, but relatively short, earthquake catalog compiled with instrumentation data recorded since the last century. The key to this Bayesian application to engineering seismology is to utilize the observed rates in neighboring zones as the prior information, then updated with the likelihood function governed by the earthquake observation in a target zone.

Keywords: The Bayesian approach, earthquake rate estimating and smoothing

1. INTRODUCTION

Earthquake frequency or annual rate is an important parameter for earthquake analyses such as seismic hazard assessment. Understandably, the most reliable approach to estimate the parameter is based on sufficient instrumentation data, such as an earthquake catalog compiled with instrumentation data in the past 100 years. However, the data is considered a limited observation, compared to the long return period of earthquakes, especially the major events. As a result, based on the classical statistics algorithms, the estimates on earthquake frequency would become unrealistic because the observation data are limited. For example, given no $M \geq 6.0$ events recorded in the past 100 years, a best-estimate annual rate for such an event would be equal to zero, which is somehow unrealistic and not convincing.

Different from classical statistics algorithms relying on samples only, the Bayesian approach is to develop an estimate considering both the information from samples and from general prior knowledge. In the Bayesian terminology, the method is to update the prior probability density function (PDF) with the likelihood function from samples to develop a posterior PDF. Next, a Bayesian estimate can be developed according to the posterior PDF, a result of general prior information and site-specific observations from samples.

The Bayesian method is commonly used in a variety of studies, such as estimating dissolved oxygen in a river [1], evaluating the reliability of pile foundations [2], reconstructing a Synthetic Aperture Radar image [3], and quantifying the risk of offshore drilling [4]. In addition to those applications, the Bayesian approach was also used in earthquake engineering and engineering seismology, such as evaluating earthquake-induced slope failures [5], characterizing the structure's vulnerability against earthquakes [6], and estimating the parameters of an active fault [7]. As a result, it is understood that the Bayesian approach is in a general framework, and the data used as prior information and likelihood function can vary case-by-case.

This paper presents a new Bayesian application to engineering seismology, estimating and smoothing earthquake frequency in a region given a relatively short earthquake observation in a comparison to the long return period of earthquakes. In addition to the methodology, a case study was also given as a demonstration example to this Bayesian application to earthquake rate estimating and smoothing.

2. THE OVERVIEW OF THE BAYESIAN APPROACH

As mentioned previously, the Bayesian approach is to develop a best estimate with multiple sources of data. Take a discrete case for example (i.e., the prior probability density function is discrete), the algorithm of the Bayesian approach can be expressed as follows [8]:

$$P''(\theta_i) = \frac{P'(\theta_i) \times P(\varepsilon \mid \theta_i)}{\sum_{i=1}^{n} P'(\theta_i) \times P(\varepsilon \mid \theta_i)} \qquad (1)$$

where $P'(\theta_i)$ is the prior probability for an estimate θ_i in the prior probability mass function (PMF), $P(\varepsilon \mid \theta_i)$ is the likelihood function given observation ε, $P''(\theta_i)$ is the posterior probability after updating, and n is the number of estimates in the prior PMF.

An example was demonstrated in the following to help describe the Bayesian approach. First, the function shown in Fig. 1 is the prior PMF for $M \geq 6.0$ earthquakes in a region, and the probabilities for annual rate v equal to 0.01 and 0.02 per year are 30% and 70%, respectively. Second, the observation data indicate no such events in the past 50 years were observed around the region. As a result, in this case the likelihood function for $v = 0.01$ per year in this example can be calculated as follows:

$$P(\varepsilon = zero\ event\ in\ 50\ years \mid mean\ rate = 0.01\ per\ year)$$

$$= P(\varepsilon = zero\ event\ in\ 50\ years \mid mean\ rate = 0.5\ per\ 50\ years) \qquad (2)$$

$$= \frac{e^{-0.5} 0.5^0}{0!} = 0.61$$

It is worth noting that the calculation shown in Eq. 2 is on a customary presumption that earthquake occurrence follows the Poisson distribution [8], with its probability mass function defined as follows [8]:

$$P(X = x \mid v) = \frac{e^{-v} v^x}{x!} \qquad (3)$$

where v is the mean rate or the mean value of the Poissonian random variable X.

The same calculation can be applied to the likelihood function of the other estimate (i.e., $v = 0.02$ per year). Along with the two prior probabilities given in Fig. 1, therefore the prior probabilities and likelihood functions available, the two posterior probabilities for $v = 0.01$ and $v = 0.02$ can be updated with the Bayesian algorithm (i.e., Eq. 1) as follows:

$$P''(v = 0.01) = \frac{0.3 \times 0.61}{0.3 \times 0.61 + 0.7 \times 0.37} = 0.41 \qquad (4)$$

$$P''(v = 0.02) = \frac{0.7 \times 0.37}{0.3 \times 0.61 + 0.7 \times 0.37} = 0.59 \qquad (5)$$

Fig. 1 shows the posterior function for this example after the Bayesian updating. Accordingly, the Bayesian estimate (i.e., the mean value of the posterior PMF) is 0.016 per year for $M \geq 6.0$ earthquakes after the Bayesian updating, given the prior information and the limited observation in the past 50 years.

3. EARTHQUAKES WITHIN SUB-REGIONS

For developing a more refined earthquake analysis, sometimes a region is further divided into a few sub-zones in an application. However, this practice would cause a challenge in the earthquake rate estimating because earthquake observation is also divided into a smaller sample size. We made an example shown in Fig. 2 with the earthquake data around Taiwan to further explain this situation: For a pre-defined seismic source in southeastern Taiwan [9], the annual rate of $M_L \geq 6.5$ events (i.e., M_L = local magnitude) could be around 0.036 per year given a total of 4 events recorded in the past 110 years. However, after dividing the zone into nine sub-zones for some application, the rate for $M_L \geq 6.5$ earthquakes in sub-zones S1, S4, S5, S6 and S9 should be zero, because no such events were observed within those zones during the time. (Note that the earthquake data around Taiwan have been used in a few earthquake studies, including seismic hazard assessments and earthquake statistics analyses [9, 10]. More details about the seismic source model and the earthquake catalog are available in those publications.)

Figure 1: The prior and posterior PMFs in a demonstration Bayesian example

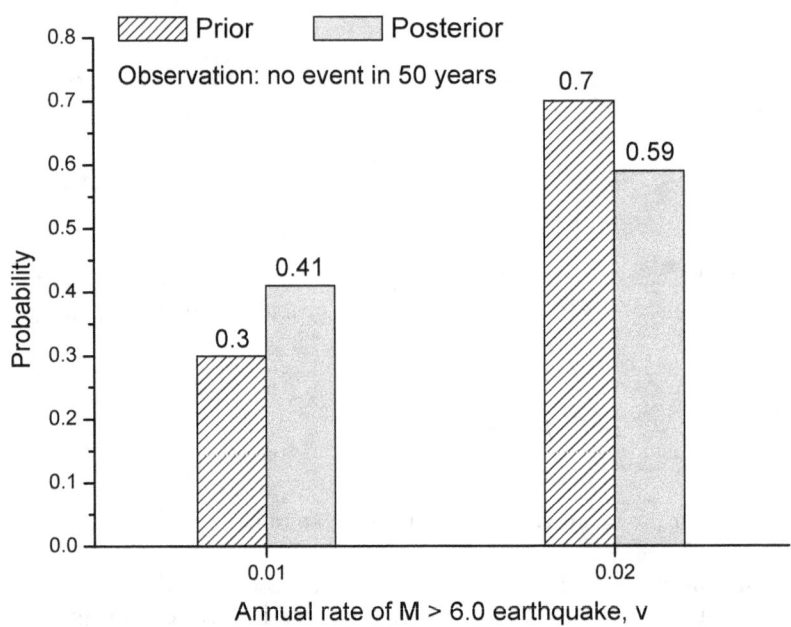

4. A BAYESIAN APPLICATION TO EARTHQUAKE RATE ESTIMATING AND SMOOTHING

As mentioned previously, the scope of this technical note is to provide a Bayesian application to estimate and smooth earthquake rates given limited data. In the following sections, we continued using the example (Fig. 2) around Taiwan to help explain and demonstrate this Bayesian calculation.

4.1. The methodology overview

Essentially, any of a Bayesian calculation is to integrate prior information with observation data to develop a Bayesian estimate. Because observation data are only available in this study, the key task in this Bayesian calculation is to develop some prior information and to integrate it with the observation. Under the circumstances, the analytical presumption in the development of the Bayesian calculation is to use the observations in neighboring areas as a source of prior information, one of the key presumptions in this study.

4.2. The definition of "neighbors"

Because the prior information is based on the observed earthquake rates in neighboring zones, the definition of "neighbors" needs clarified in the first place. In short, two areas are considered "neighbors" as long as they are in contact with each other in any form. Take Fig. 2 for example, S2, S4 and S5 are the neighbors of S1; S1, S3, S4, S5 and S6 are the neighbors of S2; and so on so forth….

Figure 2: The observed $M_L \geq 6.5$ earthquake rates within a seismic source in southeastern Taiwan based on the seismicity in the past 110 years; the value in the parenthesis is the observed earthquake rate during the time [9, 10]

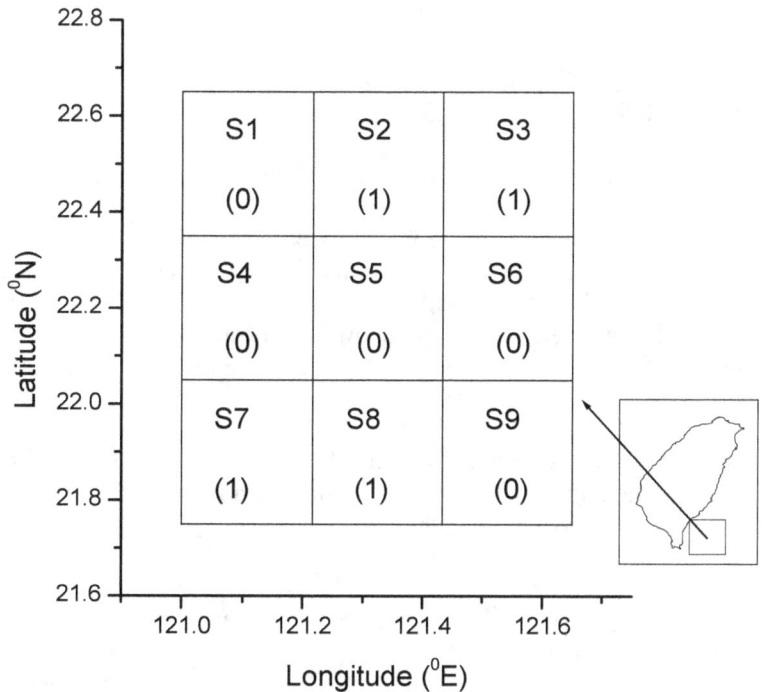

4.3. The prior probability mass function

With the observed rates (denoted as v) in a target zone and in its neighboring zones, a prior PMF about earthquake rates in the target zone could be developed. Take Fig. 2 for example, the observed rates for $M_L \geq 6.5$ earthquakes within S1, S2, S4 and S5 are 0, 1, 0, and 0 in the past 110 years. Therefore, Fig. 3a shows the prior PMF for $M_L \geq 6.5$ earthquake rates in the target zone S1 based on the information from the four zones. Accordingly, the prior probabilities are 75% and 25% for the two estimates $v = 0$ and $v = 1$, respectively. In contrast, Fig. 4a shows the prior PMF for the center zone S5, with a prior probability of 56% and 44% for $v = 0$ and $v = 1$, respectively.

4.4. The updating

With the prior and observation data available, then the Bayesian estimate can be developed. In this case study demonstration, the target zone S1 has a prior PMF like Fig. 3a, and a zero-event observation for $M_L \geq 6.5$ earthquakes in the past 110 years. As a result, considering earthquake occurrence follows a Poisson distribution, the likelihood function for the estimate $v = 0$ can be calculated as follows:

$$P(\varepsilon = zero\ event\ observation \mid v = 0)$$

$$= \frac{e^{-0} 0^0}{0!} = 1$$

(6)

The same calculation was then applied to the other estimate $v = 1$, and with the two prior probabilities and likelihood functions available, the posterior probability for $v = 0$ can be updated with the underlying Bayesian algorithm (i.e., Eq. 1) as follows:

$$P''(v=0) = \frac{P'(v=0) \times P(\varepsilon \mid v=0)}{P'(v=0) \times P(\varepsilon \mid v=0) + P'(v=1) \times P(\varepsilon \mid v=1)}$$

(7)

$$= \frac{0.75 \times 1}{0.75 \times 1 + 0.25 \times 0.37} = 0.89$$

where ε here denotes the observation that is no $M_L \geq 6.5$ earthquakes in S1 during the time.

Repeating the calculations for the other estimate $v = 1$, its posterior probabilities can be calculated as well. As a result, Fig. 3b shows the posterior PMF for $M_L \geq 6.5$ earthquakes within S1, and the Bayesian estimate on the earthquake rate is equal to 0.09 per 110 years, or 0.0008 per year, according to the posterior PMF. On the other hand, Fig. 4b shows the posterior PMF for the center zone S5. According to the posterior PMF, the Bayesian estimate on the rate of $M_L \geq 6.5$ earthquakes within S5 is 0.18 per 110 years, equivalent to 0.002 per year.

Figure 3: a) the prior probability mass function for two $M_L \geq 6.5$ earthquake rates for zone S1 (see Fig. 2), and b) the posterior function after the Bayesian updating with no $M_L \geq 6.5$ observed in the target zone S1 in the past 110 years

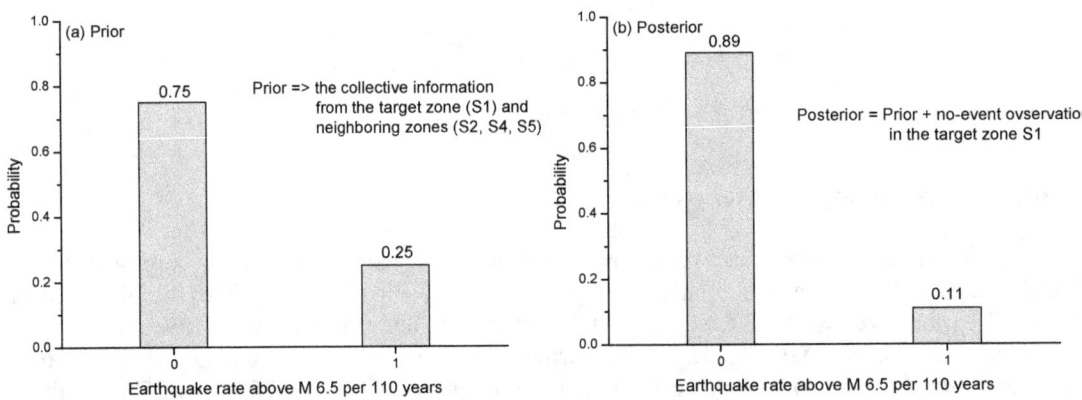

In the two Bayesian calculations, we can see although the observed earthquake rates in S1 and S5 are both equal to zero, the updated Bayesian estimates are varied because the different prior information, or different "neighbors" of the two target areas.

4.4. The earthquake rate after the Bayesian smoothing

With the updating repeated for the rest of the zones, the smoothing of the earthquake rate within a region could be achieved. For this example using the earthquake data around Taiwan, Fig. 5 shows the smoothed rates for each of the nine sub-zones with the Bayesian approach. It is worth noting that the total of the smoothed rates is equal to the observed rate, four events in total, based on the seismicity detected in the past 110 years.

From this case study, we can see that now the estimates on earthquake rates become more realistic with the Bayesian calculation, rather than a zero-event estimate with the classical statistical algorithms based on limited observations available. Therefore, the new Bayesian application to earthquake rate estimating and smoothing could be a useful option for earthquake studies, given the reliable earthquake instrumentation data are limited compared to the long return period of earthquakes, especially the major events.

Figure 4: a) the prior probability mass function for two $M_L \geq 6.5$ earthquake rates for S5, and b) the posterior function after the Bayesian updating with no $M_L \geq 6.5$ events observed during the time

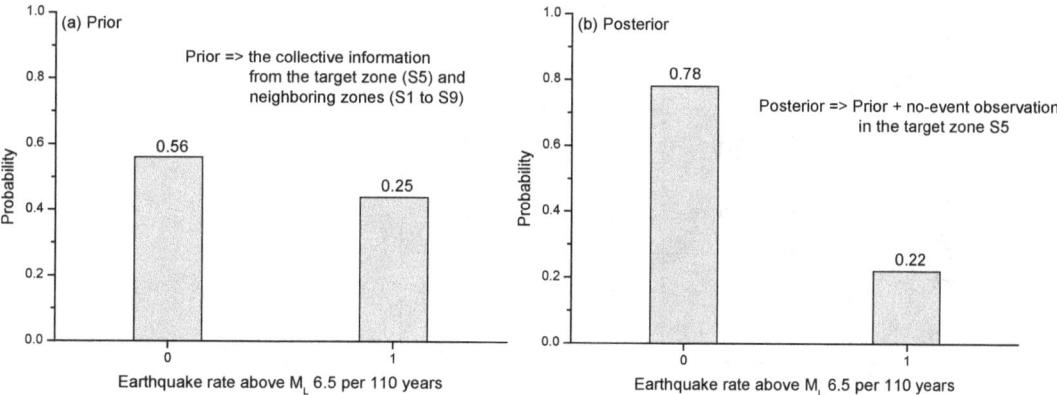

5. DISCUSSIONS

5.1. Analytical presumption behind the Bayesian calculation

An implicit analytical presumption behind the development of this Bayesian smoothing is that we weighted the observations differently for a given target zone. That is, we considered the observed rate in the target zone is more "reliable," and it should play a heavier role in the calculation than those observed rates from neighboring zones. As a result, we utilized the observed rate of the target zone in both the prior PMF and the likelihood function also governed by that observation. In other words, the observed rate of the target zone was "double counted" in this Bayesian calculation for smoothing the earthquake rate in a region.

5.2. On the proper use of the Bayesian approach

Explicitly, observation data are more reliable than prior information from judgment, experience, etc. In other words, when observation data are in a large sample size, it is not necessary to take general prior information into account for developing a best estimate, also pointed out in other Bayesian studies in geotechnical site characterizations [11, 12]. To sum up, the purpose of using the Bayesian approach was to utilize some prior information to compensate the sample-size issue when encountered. In other words, when the site was well investigated with sufficient site-specific data, the classical algorithms relying on reliable observation data should be able to develop a representative estimate, without the involvement of general prior information.

The same basics on the proper use of the Bayesian approach should be applied to this Bayesian calculation for earthquake rate estimating and smoothing. For estimating large-earthquake rates like the demonstration example, the 110-year-long instrumentation earthquake data could be too limited to estimate the rates with classical statistics algorithms, considering the long return period of the large earthquakes. By contrast, if the problem is to estimate small-earthquake rates such as $M_L \geq 3.0$ around Taiwan, the 110-year-long observation with 50,000 $M_L \geq 3.0$ events should be sufficient to develop a reliable estimate using the classical algorithms.

Figure 5: The estimates on $M_L \geq 6.5$ earthquake rates for each of the nine sub-zones within a seismic source in southeastern Taiwan with the Bayesian calculation, based on the prior information from the neighboring zones and the likelihood function governed by the earthquake observation in the target zone

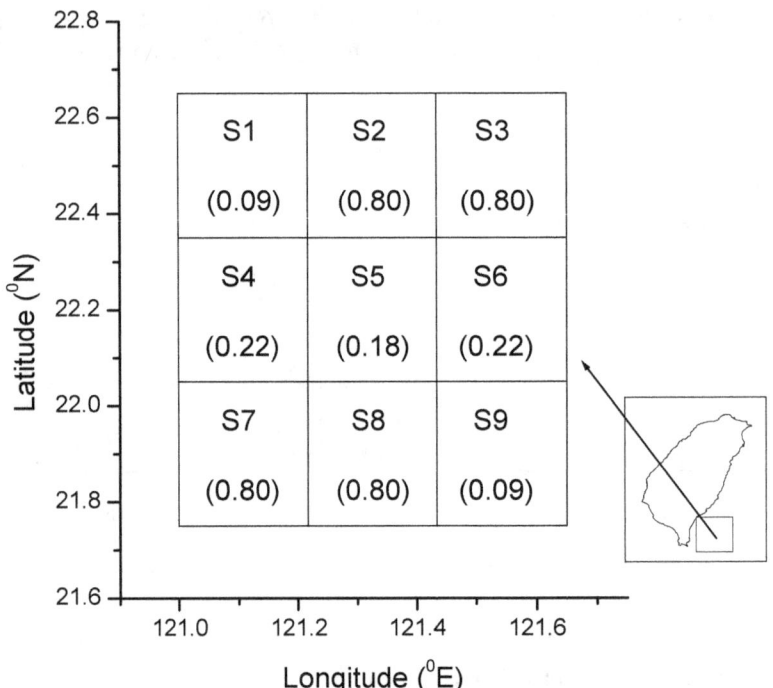

6. SUMMARY

Owing to the long return period of earthquakes, one challenge in earthquake parameter estimating is the lack of a representative sample size. As the demonstration examples using earthquake data around Taiwan, the limited observation could lead to an unrealistic "zero" estimate for earthquake occurrences, because of the relatively short period of observation compared to the long return period of earthquakes, especially the major events.

Different from classical statistics algorithms, the Bayesian approach is to utilize both prior information and observation data to develop a best estimate, and it has been practiced in many studies especially when observation data are limited. Like those Bayesian applications, this paper introduces a new Bayesian calculation for earthquake rate estimating and smoothing, given reliable, but limited, instrumentation earthquake data available.

The key to this Bayesian calculation includes the use of the observed earthquake rates in neighboring zones as the prior information, and "double counting" the earthquake observation in a target zone during the Bayesian updating. In terms of the result, the Bayesian calculation can develop a more realistic estimate on earthquake rates rather than zero, with limited instrumentation earthquake data.

Acknowledgements

The author is very thankful for the provision of the earthquake data from Prof. Yih-Min Wu of National Taiwan University.

References

[1] A. Patil and Z. Q. Deng. "*Bayesian approach to estimating margin of safety for total maximum daily load*," Journal of Environmental Management, 92, pp. 910-918, (2011).
[2] L. M. Zhang, D. Q. Li and W. H. Tang. "*Impact of routine quality assurance on reliability of bored piles*", Journal of Geotechnical and Geoenvironmental Engineering, 132, pp. 622-630, (2006).
[3] D. Vu, M. Xue, X. Tan and J. Li. "*A Bayesian approach to SAR imaging*," Digital Signal Processing, 23, pp. 852-858, (2013).
[4] N. Khakzad, F. Khan and P. Amyotte. "*Quantitative risk analysis of offshore drilling operations: A Bayesian approach*," Safety Science, 57, pp. 108-117, (2013).
[5] Y. Song, J. Gong, S. Gao, D. Wang, T. Cui and B. Wei. "*Susceptibility assessment of earthquake-induced landslides using Bayesian network: A case study in Beichuan, China*," Computers & Geosciences, 42, pp. 189-199, (2012).
[6] P. S. Koutsourelakis. "*Assessing structural vulnerability against earthquakes using multi-dimensional fragility surfaces: A Bayesian framework*," Probabilistic Engineering Mechanics, 25, pp. 49-60, (2010).
[7] M. Amighpey, B. Voosoghi and M. Motagh. "*Deformation and fault parameters of the 2005 Qeshe earthquake in Iran revisited: A Bayesian simulated annealing approach applied to the inversion of space geodetic data*", International Journal of Applied Earth Observation and Geoinformation, 26, pp. 184-192, (2014).
[8] A. H. S. Ang and W. H. Tang. "*Probability concepts in engineering; emphasis on applications to civil and environmental engineering*," John Wiley & Sons, 2007, New Jersey.
[9] C. T. Cheng, S. J. Chiou, C. T. Lee and Y. B. Tsai. "*Study on probabilistic seismic hazard maps of Taiwan after Chi-Chi earthquake*," Journal of GeoEngineering, 2, pp. 19-28, (2007).
[10] J. P. Wang, D. Huang, S. C. Chang and Y. M. Wu. "*New evidence and perspective to the Poisson process and earthquake temporal distribution from 55,000 events around Taiwan since 1900*," Natural Hazards Review ASCE, 15, pp. 38-47, (2014).
[11] Y. Wang, S. K. Au and Z. Cao. "*Bayesian approach for probabilistic characterization of sand friction angles*," Engineering Geology, 114, pp. 354-363, (2010).
[12] J. P. Wang and Y. Xu. "*Estimating the standard deviation of soil properties with limited samples through the Bayesian approach*," Bulletin of Engineering Geology and the Environment, doi: 10.1007/s10064-014-0619-5 in press (2014).

Open Conceptual Questions in the Application of Uncertainty Analysis in PRA Logic Model Quantification

Sergio Guarro[a]
[a] ASCA Inc., Redondo Beach, USA

Abstract: The final stage of quantification of Probabilistic Risk Assessment (PRA) or reliability logic models is usually carried out via processes that first obtain estimates of key component-level reliability and risk parameter values, such as the failure rate in the time domain, or the probability of failure (PoF) for a specific function or mission duration, and then propagate such values from the component to the subsystem and system levels according to the component logic arrangements reflected in system reliability and failure logic models. When applying uncertainty analysis techniques to the estimation of reliability or PoF parameters of components belonging to a given system, a conceptual problem arises as to whether the same bottom up process may be applied to the definition of the prior distributions of the parameters of interest, or whether better state-of-knowledge consistency and coherence may be achieved by a top down process that proceeds from the initial construction of a system level prior distribution for the parameter of concern. This paper examines and discusses this and related issues that arise in the application of Bayesian analyses to a system PRA or reliability assessment.

Keywords: Probabilistic risk assessment, uncertainty analysis, Bayesian prior probability.

1. INTRODUCTION

The final stage of quantification of Probabilistic Risk Assessment (PRA) or reliability logic models is typically carried out via processes that seek to first obtain estimates of key component reliability and risk parameter values, such as the component failure rate in the time domain, or the probability of failure (PoF) for a specific function or mission duration, and then propagate such values from the component to the subsystem and system levels according to the arrangement of the component logic functions as represented in the combination of logic models used to represent the system reliability or risk, which in the PRA domain most commonly consists of a combination of binary event-trees and fault-trees.

A common estimation process for component failure rates and PoF's applies Bayesian techniques, by which a *prior distribution* of the parameter value is first constructed based on what may be referred to as "soft knowledge," e.g., generic handbook indications and/or engineering judgment concerning where the range of the parameter may lie; then the best data related to the parameter is formally brought into the process via the definition of a Bayesian *likelihood function*; finally, in the last step of the estimation process the prior distribution and the likelihood function are combined, by application of Bayes' theorem, to yield a *posterior distribution*" of the parameter, from which any desired statistics, such as mean or median value, standard deviation, percentiles and *credible intervals* – the Bayesian version of confidence intervals –, can also be extracted.

While the application of this process for a single item or component is relatively straightforward, some challenging questions arise when the process is applied to a number of components that are part of a complex system and contribute to the overall reliability or failure characteristics of the system according to both functional logic arrangement and individual reliability features. Within a probabilistic framework these questions can be viewed as primarily concerning the relation between probability assessments relative to the individual components of a system or subsystem on one hand, and the system or subsystem itself on the other. However the same questions may also be addressed from the broader perspective of what forms of uncertainty representation are appropriate under the conditions of concern. This paper examines these questions and perspectives specifically in relation to the objective of establishing and maintaining the self-consistency of a PRA or reliability model quantification framework, while using knowledge and data that may be applied at different levels of system indenture, i.e., depending on user preference, at different levels in the system functional hierarchy.

2. PRA RELEVANCE OF ISSUES

The issue of consistency between Bayesian assessments for reliability or probability of failure (PoF) conducted at different levels of system indenture was initially discussed in papers published in the 90's and initially thought of as an anomaly of Bayesian probability estimation [1, 2]. The observation generating these discussion was that the application of the same evidence, if alternatively carried out at basic component or whole system level, seemed to produce substantially different results. This subject was also discussed in more general terms in [3] as an issue of "perfect (or imperfect) aggregation" of Bayesian estimates.

Although the initial discussion of the aggregation or Bayesian anomaly issues dates several years back in time, it has gained new relevance in the context of PRAs executed in more recent times for launch and space vehicle applications. The initial PRA applications of the 80's and 90's were primarily for nuclear power plants, for which the bulk of the reliability data resided at the basic component levels, so that it was natural in this context to apply Bayesian estimations from the bottom up, i.e., by first constructing component-level PoF or reliability prior distributions using generic data, and then applying plant-specific evidence at the component and/or higher level, while progressing in the quantification process up the logic structure of a reliability or failure model, such as a fault tree. With many space systems, on the other hand, and more so in the particular case of launch vehicles, it is generally easier to construct system or subsystem-level priors than basic component priors, because generic knowledge of the range of reliability of any such a system is much better and more broadly established than corresponding knowledge for each of the basic components of which the system itself is composed. For example, it is relatively easy and defensible to identify the interval $\{0.01, 0.10\}$ as a reasonable range for the per mission PoF of a typical medium to heavy class U.S. Government certified launch vehicle. It is not as straightforward to identify any "prior" range of PoF or reliability for a check valve, electronic board, or other low level component of such a system.

3. UNCERTAINTY REPRESENTATION VIA PROBABILITY DISTRIBUTIONS VERSUS UNCERTAINTY INTERVALS

A topic discussed in the literature that is related to what is being discussed here concerns whether "epistemic" uncertainty, i.e. the uncertainty that exists because of lack of knowledge, can properly be represented via probability distributions, as quite commonly is done in PRA. Ref. [4] argues against this, and uses several examples to illustrate the issue. Among these is the example of a quantity AB which is the product of a component A, known to have a value somewhere between 0.2 and 0.4, and a component B, known to have a value somewhere between 0.3 and 0.5, with no additional information available as to where the "true value" of either component may lie. Ref. [4] observes that all that can be said about this situation is that the value of AB must then be somewhere between 0.06 and 0.2. It also argues that equating the lack of knowledge concerning where the values of A and B may be with the probabilistic assumption of a uniform distribution in the respective intervals is beyond the real knowledge about those quantities. Ref. [4] observes in fact that such an assumptions would lead to the probabilistic computation of a distribution for AB that is strongly peaked at an intermediate value between the two extremes, whereas the actual knowledge about AB cannot really go beyond the original assertion that it is somewhere in the interval $\{0.06, 0.2\}$.

Translating the above into a PRA context of assessing and propagating uncertainty on probability of failure (or failure rate) values, we can see the analogy with a situation in which the quantities A and B are uncertain PoF values, about which the only knowledge is that they lie in the stated intervals, for two components that represent redundant functionality for a system or subsystem. Under these conditions, the PoF of that system or subsystem is given by the product AB, and the considerations made by [4] about what may be valid or invalid to determine about the AB value apply to this analogous situation.

An additional observations about the situation discussed in [4] can be made if we consider the practice, long established in the probabilistic arena and known as Laplace's Principle of Insufficient Reason, of assuming a uniform probability distribution in a given range when nothing is known about a random variable besides the fact that it must fall somewhere within that range. As shown below, this implies a contradiction between interval representation of uncertainty in the terms argued by [4] and a Bayesian probabilistic representation of the same uncertainty.

The assertion by [4], based on interval analysis, is that all can be said about A, B and AB is that they are somewhere in the respective intervals {0.2, 0.4}, {0.3, 0.5} and {0.06, 0.2}, as represented by Figure 1. Translated in probabilistic terms by application of the above mentioned Principle of Insufficient Reason, this would lead to a representation of all three parameters as random variables uniformly distributed in the corresponding ranges, as shown in Figure 2. However, once probability distributions are defined to represent the A and B factors of a product AB, the axioms and laws of probability, together with the mathematical formulation linking the three variables, univocally define a specific form of the probability distribution of AB itself. In the case where A and B are completely independent variables uniformly distributed in the intervals specified above, the distribution of their product AB looks like what is shown in Figure 3. In summary, within a probabilistic framework, it appears contradictory to claim complete lack of knowledge of where the value of a dependent variable like AB may lie, if the same has been claimed for the values of the associate independent variables, like A and B, and vice versa.

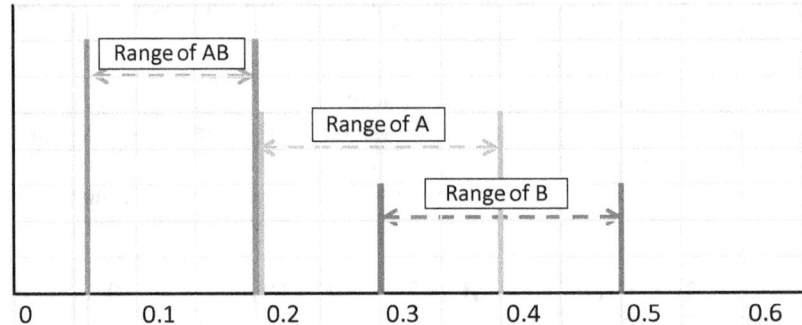

Figure 1: Interval Representation of Variables A, B, and AB

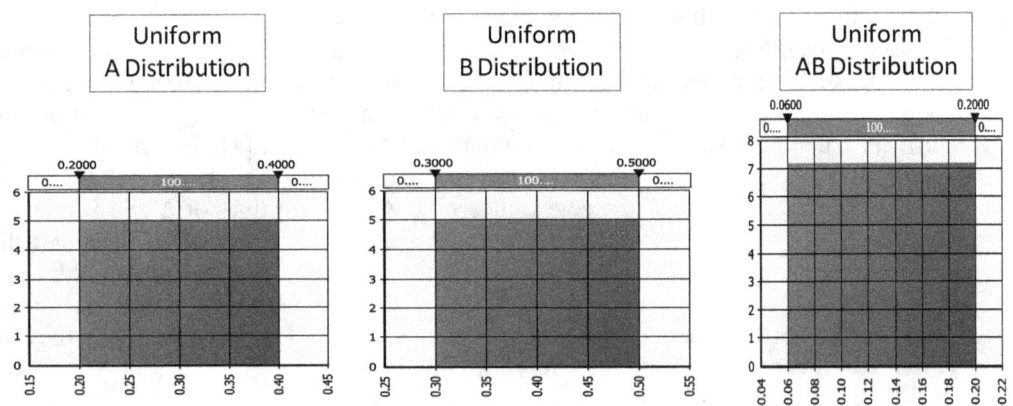

Figure 2: Principle of Insufficient Reason Probabilistic Representation of Variables A, B, AB

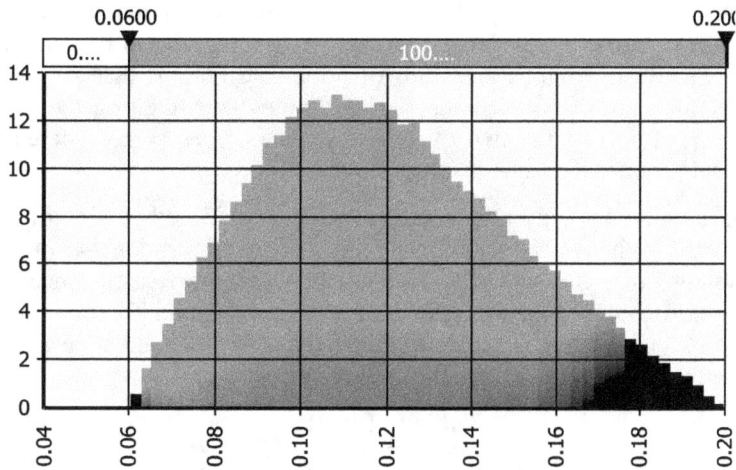

Figure 3: Probability Distribution of AB Derived as Product of Uniform A and B Distributions

For completeness, we observe that a way by which a uniform distribution can be obtained for the product AB when uniform distributions have separately been assumed for A and B does exist. This is if the value taken by one of the two terms, let us say B, is completely determined by the value assumed by the other in a certain specific fashion. For our example, this specific fashion would be that of a deterministic relation between A and B, such that, given that a random value of A occurs at a location corresponding to a fraction F of the A interval, then B is forced to a value that makes the product AB fall in a location corresponding to the same fraction F of the AB interval. This de-facto reduces the situation to one where there is only one independent variable A, and the product AB is linked to it by a relation of proportionality. This extremely correlated condition is certainly not representative of the most common uncertainty assessment conditions that PRA analysts are faced with, thus the basic terms of the issue being discussed are not affected by this observation.

3.1. Relation to the Bayesian Aggregation Issue

Even if it may not be immediately apparent, a relation exist between the issue discussed in [4] and the "Bayesian aggregation anomaly" issue.

Refs. [1] and [2] presented the anomaly subject by considering a situation where two components, which herein we shall refer to as C1 and C2, are arranged functionally in series for the successful operation of a system S. This logic arrangement happens to be in reliability terms the functional opposite of the one discussed above as a reliability analogy of the AB product example discussed in [4], but this is purely coincidental and unimportant for the purposes of the present discussion, as shall be evident from the following. For the stated situation the cited references assume a prior knowledge of the PoF values for the components C1 and C2, and for the system S. Starting from this premise, a Bayesian assessment is applied with the benefit of hypothetical binomial evidence resulting from the system S being tested a number of times, with one observed system failure occurring because component C1, but not C2, fails. Under these conditions, a Bayesian assessment is carried in two different ways, which Refs. [1] and [2] consider to be equivalent:

- In the first mode of assessment the test evidence is applied separately to the two component priors, and then the updated probabilities of failure (PoFs) are combined to provide the system PoF.

- In the second mode of assessment the test evidence is applied directly to the system prior to yield the system PoF.

Refs. [1] and [2] observe that the posterior results for PoF(S) turn out to be different in the two versions of the Bayesian assessment and consider this to constitute a probabilistic anomaly in light of the fact that the same test evidence is applied in the two cases, which would lead to the expectation of same ultimate results. In reality, as discussed in [5], the "anomaly" can be explained by the fact that the two assessments start out from assumptions of prior distributions that are not equivalent to each other; thus, although the same test evidence is applied to the two situations, the non equivalence of the assumed priors makes it so that the two overall Bayesian processes considered in are not actually equivalent and cannot produce the same results.

To better illustrate how the anomaly may occur, consider a modified version of the example initially used in Refs. [1] and [2], in which, for the same system configuration, no prior knowledge of the component and system PoFs is assumed to be available, and the test evidence is as summarized in Table I. The main modification of the example with respect to [1] and [2] consists of assuming a lower number of system tests (10 instead of 100), which has the effect of making certain aspects of the issue being discussed even more evident than when using the original case.

Table I: Test Results

	Component C1	**Component C2**	**System S**
Prior PoF	"Unknown"	"Unknown"	"Unknown"
Number of tests	10	10	10
Number of failures	1	0	1

Repeating the steps used in Refs.[1] and [2], in one version of Bayesian assessment the following Process 1 is applied:

A. Reflecting the assumed complete lack of knowledge of the PoF values before the test, uniform priors are assumed for C1 and C2 in the theoretical PoF interval [0, 1];
B. the test evidence of 1 failure in 10 trials for C1, and 0 failures in 10 trials for C2 is applied to obtain posterior PoF distributions;
C. the S system PoF is obtained from the OR-gate formula:

$$PoF(S) = 1 - [1 - PoF(C1)] [1 - PoF(C2)] \quad (1)$$

In a second version of the assessment the evidence is not applied at the component level, but directly at the system level, i.e. according to Process 2:

A. Reflecting the assumed complete lack of knowledge of the PoF value for S before the test, a uniform PoF prior is assumed for the whole system;
B. the evidence of 1 system failure in 10 trials is applied to obtain the posterior PoF for the system S.

Figure 4 shows that indeed the two processes produce different results for the posterior System PoF. However, the reason for the difference lies in the fact that the setting of priors in the two versions of the assessment is not equivalent, which in turn makes the two above processes to be themselves not altogether equivalent. In Bayesian "state-of-knowledge" terms this seems to imply that for a case like this an assessor cannot claim complete lack of knowledge of the PoF whereabouts at the same time at both the system and component levels.

To verify what was just said above about the cause of the difference in results between Process 1 and 2, a Process 3 can be applied to modify Process 2 in a manner that makes the use of the test evidence at the whole system level consistent with the Process 1 utilization of the same evidence at the component level. Process 3 would be applied as follows:

A. Reflecting the assumed lack of knowledge of the PoF values before the test, uniform priors are assumed for C1 and C2 in the theoretical PoF interval [0, 1];
B. a system prior PoF is obtained from the OR-gate formula, eqn.(1);
C. the evidence of 1 system failure in 10 trials is applied to obtain the posterior PoF for the system S.

The results of Process 3 do coincide at the S system level with those of Process 1, i.e., the system S posterior PoF produced coincides with the burgundy-shaded curve of Figure 4, as is actually shown by the direct comparison in Figure 5. It is noted in this regard that the small difference visible in Figure 5 between the Process 1 and Process 3 System POF distribution results is actually due to a distribution-fit approximation introduced to simplify the Bayesian updating process computations.

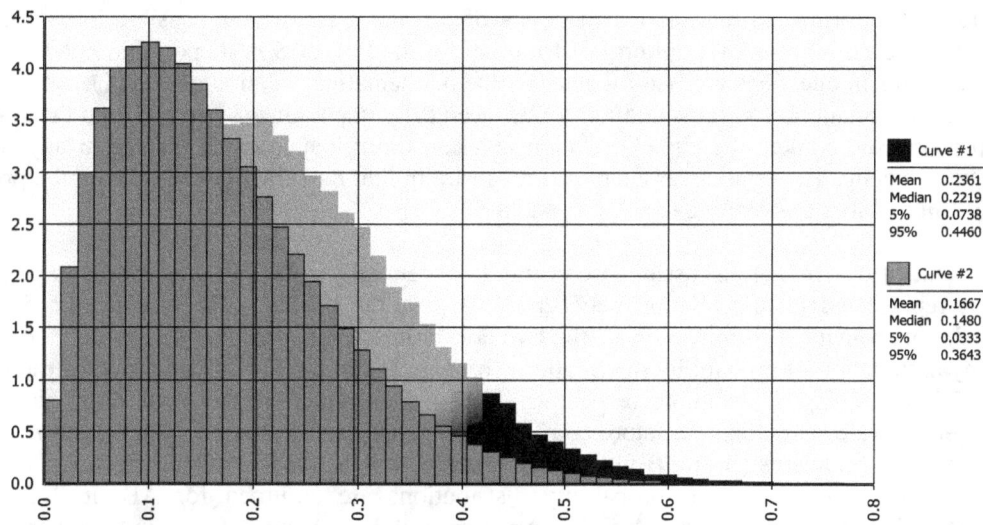

Figure 4: Posterior Probability Distributions for System S PoF per Processes 1 and 2

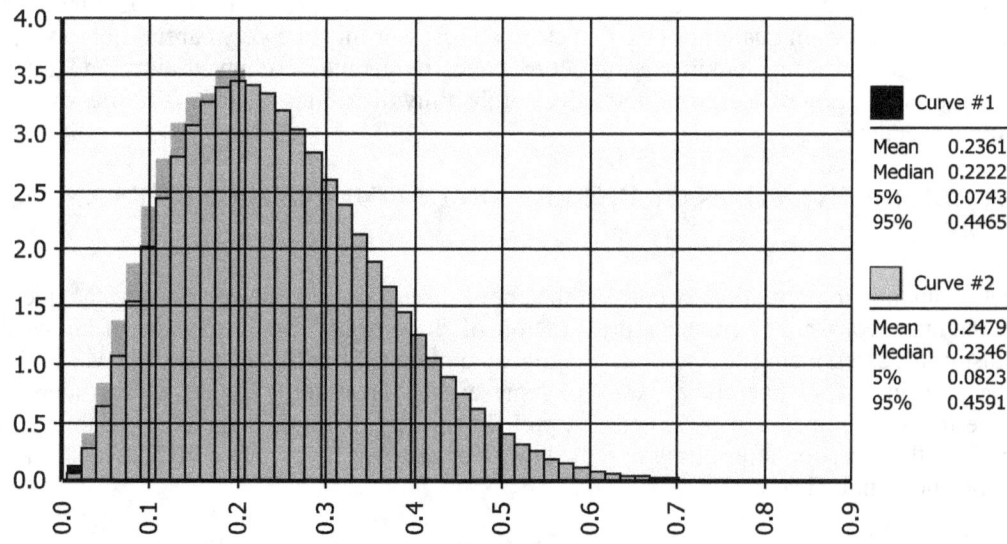

Figure 5: Posterior Probability Distributions for System S PoF per Processes 1 and 3

It is also noted that if conversely one desired, in a situation like the one in the above example, to set up a system level prior PoF and then apply component-level test evidence, to maintain the overall consistency of the estimation process it would then be necessary to decompose the S system prior PoF into component priors, in such a way as to satisfy eqn.(1). That is, the two component prior distributions, when combined by means of eqn.(1), would need to produce the initially defined prior PoF(S) distribution.

Recalling the observation made in [4] with regard to the AB product example discussed earlier in Section 3, the non-equivalence of the two sets of prior distributions assumed in Refs.[1, 2] for the two types of Bayesian assessments can be conceptually related to the difference between obtaining the AB product distribution from assumed uniform distributions of A and B, versus assuming its value to be somewhere in between its known bounds. Ref. [4] is correct in pointing out this difference, regardless of whether one also accepts the associated argument that to assume any probabilistic distribution for a parameter when only its bounds are known represents an overstepping of knowledge. In the Bayesian assessment context of the Refs. [1, 2] discussion, the part of the observation that matters is that the assumption of uniform distributions for the PoFs of C1 and C2, in their possible intervals, is not equivalent to the assumption of a uniform distribution for the PoF of S in its possible interval. This is because the PoFs in questions are tied together by the mathematical relation of eqn.(1), which in turn reflects the logic relation of success and failure between the components C1 and C2 and the system S. In any probabilistic context the eqn.(1) relation is a constraint that must be obeyed at any stage of assessment. Thus in a Bayesian assessment it is a constraint that must be obeyed by both the prior and posterior PoF distributions relative to C1, C2, and S.

In the context of the Ref.[4] discussion, a constraint conceptually corresponding to that expressed by eqn.(1) is represented by the relation between A, B, and the product AB. Despite the difference between the mathematical formulations of the two constraints, their effect is conceptually equivalent, in that any assumption of probability distributions for A and B in the Ref.[4] example, or the PoFs of C1 and C2 in the Refs.[1, 2] example, respectively forces the product AB and the PoF of the system S to have themselves distributions of nature and form which are determined by what has been assumed for the composing elements (A and B in the first example, PoF(C1) and PoF(C2) in the second). The converse is also true, so that if probability distributions are assumed for AB or PoF(S), any distributions of the composing terms (A and B in one case, PoF(C1) and PoF(C2) in the other) must be such that, when combined according to the mathematical formulations for AB and PoF(S), they produce the distributions assumed for the latter entities.

The bottom line of what we have discussed above is that, in the presence of a logic-mathematical relation between uncertain quantities or parameters, an assessor must be very careful not to violate the associated constraints when making assumptions concerning ranges of uncertainty and translating these into the assumption of "state of knowledge" probability distributions. This is further discussed in the following sections.

4. PRA REPRESENTATION OF UNCERTAINTY FOR COMPONENTS AND SYSTEMS

This section attempt to more directly connect the above discussion and observations to the question of how to maintain consistency in the representation of uncertainty and prior state-of-knowledge in probabilistic risk assessment (PRA) of multi-component systems. For this purpose, we will initially set aside the question of whether other ways of representing uncertainty should be considered as an alternative to treating uncertain parameters as probabilistically distributed random variables, but will come back to this question after discussing the key issue of consistency that a PRA assessor may face even in this more limited context.

4.1. Top-down versus Bottom-up Prior Knowledge Representation in PRA

The discussion in Section 3.1 has brought to attention the constrained nature of the process of constructing probability distributions in the presence of a functional and/or mathematical relation among the parameters for which such distributions are constructed.

In the earlier classical forms of PRA application the consistency issues discussed in Section 3.1 did not come to the forefront, as these applications, primarily concerning nuclear power plants, carried out probabilistic quantification entirely from the bottom up, i.e., defining first "generic prior" probability distributions for reliability parameters at the component level, and then updating these priors with more plant-specific data available at increasingly higher levels of indenture in the overall plant PRA models. In the simple conceptual example discussed in Section 3.1, this would be equivalent to defining prior PoF distributions for the two components C1 and C2, without having to worry about what definition of system S PoF prior distribution would be consistent with the priors assumed for C1 and C2.

As previously mentioned in Section 2, the situation may be different in other types of PRA applications, such as the space system PRAs which have become more common in recent years, where "prior generic knowledge" may be better assessed at the whole system level. After this is done first, mission-specific evidence is usually applied at lower levels in order to obtain output results that provide insights into subsystem or component behaviour, e.g. for the purpose of identifying contributors to which reliability and risk-reduction measures may be applied to obtain better overall system performance. Under these conditions, and specifically when Bayesian techniques are used in the probability assessments, it becomes necessary to construct individual subsystem and component prior distributions that, taken all together, maintain consistency with the overall uncertainty distribution assumed at the whole system level.

4.2. Constructing Component Priors from a System Prior

An approach to the problem posed at the end of the above section has been discussed in [5], and it can be related back to what has been presented in Section 3.1. Using the same example of a system S constituted of two components C1 and C2, let us then assume the following:

a) the assessor's initial knowledge of possible PoF values concerns in absolute terms primarily the system S;

b) PoF results at the "posterior" level are desired for practical purposes at the component C1 and C2 level;

c) test data for the two components are available to carry out assessments at the component level;

d) before conducting any component level assessment with the available data, the assessor has no other direct knowledge or information concerning the possible values of the two component PoFs.

Under the stated conditions, and more specifically because of what stated above in a), it would be reasonable for the assessor to initially start a Bayesian PoF estimation process from the definition of a system-level prior distribution for PoF(S). However, points b) and c) make also clear that results are desired at the component level and that test data to carry out a statistical estimation are available at that level. Because of this it becomes then necessary for the assessor to recast the system-level knowledge represented by the PoF(S) prior distribution into individual prior distributions for the component PoFs. Per the discussion carried out in Section 3.1, this cannot be done loosely, but in such a way that the interdependence of the three PoFs represented by eqn.(1) of Section 3.1 is satisfied. That is, whatever form of distribution is assumed for the prior PoF(S), the two component PoFs, as distribution functions, must be such that their combination in the form of eqn.(1) yields back exactly the distribution function assumed for PoF(S).

In the above we essentially have a "decomposition" problem for a distributed parameter, i.e., a distribution function (rather than a single number) must be decomposed into sub-distribution functions that obey a given mathematical constraint, more specifically eqn.(1) for the example situation considered here. Ref.[5] discusses in detail how this problem can be set up and handled in mathematical terms for the same example situation. It must be noted that, unless additional information is available, the solution to the decomposition problem is not unique when only one constraint applies, as represented by eqn.(1) or by any other single equation dictated by a different reliability logic linking component PoFs to system PoF. Thus, for example, in the solution discussed in [5] an additional constraint is introduced by assuming that the two component PoFs have the same mean values. In the case of real-life systems with multiple components arranged in series / parallel logic combinations, the decomposition problem may become correspondingly more complex and challenging, yet it remains conceptually the same as that discussed here and in [5] as an example. Solution processes for these more challenging conditions have been carried out and documented for large scale launch vehicle PRA applications concerning planetary NASA missions [6, 7].

4.3. Practical Consideration in Uncertainty Analysis for Space System PRAs

A question that deserves consideration, given the mathematical challenges posed by the decomposition of a system-level PoF distribution into lower level priors for Bayesian estimation purposes, is the why such a process may be given preference over the more traditional one of setting priors up from the lowest level at which any estimation is carried out.

In the execution of a space launch vehicle PRA, the top-down choice of process for developing prior distributions is primarily suggested by the lack of generic component-level statistical data that reflect the high stress conditions (vibration, acoustic, pyro-shock, thermal) of a space launch. The approach suggested by existing reliability data compilations or formulations (such as MIL-HDBK-217, the more recent "217-Plus, and NPRD-95) is to apply multiplier factors to failure rate and PoF data compiled for components operating in conditions without the stress factors mentioned above. The lack of validation for these factors and the large discrepancies between the factors suggested by different sources for the various classes of components that are of interest, however, give reason to doubt the validity and accuracy of the outcomes of this approach. When applied in the past, in fact, approaches following this route typically resulted in launch vehicle reliability predictions that were unrealistically high. I.e., when compared with the launch vehicle system record at the end of a sufficiently high number of actual launches, the PoF initially predicted by following such a kind of approach would be systematically underestimated by as much as an order of magnitude or more.

Considering the above, some launch vehicle PRA practitioners have concluded that it would be more reasonable to start the prior distribution setup process from the assessment of ranges of reliability for entire launch vehicle systems, which, being based on generic but well established knowledge of actual launch mission outcomes, would be more defensible than processes based on data compilations of dubious applicability to space launch conditions. From the launch vehicle system top level one would then proceed, using considerations of relative ratios of failure rates or PoFs between different launch vehicle subsystems and major component, to the definition of prior distributions at progressively lower levels of indenture. The considerations and discussion carried out above in Sections 3.1, 4.1 and 4.2 are directly relevant to this alternative process of defining Bayesian priors.

5. ALTERNATIVE REPRESENTATION OF EPISTEMIC UNCERTAINTY

Much discussion can be found in the literature on the nature and proper treatment of epistemic and aleatory uncertainty. Refs. [4, 8-10] provide a sample of these discussions, and [10] offers an exhaustive bibliography on the subject. Epistemic uncertainty is commonly referred to as arising from lack of knowledge, while aleatory uncertainty is considered as originating from the intrinsic variability of some physical or other type of process, such as in the toss of a die. At the philosophical level it may be argued that aleatory uncertainty is also the result of lack of knowledge, as after all if one could exactly know and represent the impulse imparted to the die and all the boundary conditions for the bounces that follow, one may in theory be able to predict the outcome of each toss. However, the distinction is well established and accepted in the PRA community. In the practical terms of interest in the PRA context, epistemic uncertainty, unlike aleatory, typically refers to situations where the unknown parameter or variable is not really changing from case to case in the theoretical set of cases of interest for a specific assessment. Rather, it may have a very specific constant value, but that value is unknown within a certain range of uncertainty.

Based on the arguments that we have tried to summarize in this paper, some authors [4, 9] have argued that the probabilistic representation of epistemic, i.e. "lack of knowledge", uncertainty is inappropriate and leads to under-estimation of the true range of uncertainty when probabilistically propagated up the logic structures of hierarchically arranged logic-probabilistic models, such as those routinely used in PRA. To counter this, these authors have proposed alternative methods of epistemic uncertainty representations and propagation within a system model, such as "interval analysis" [4] and Dempster-Schafer evidence theory [9, 10]. Both Refs. [4] and [9] present examples of propagation and presentation of both aleatory and epistemic uncertainty within a given assessment, using separate frameworks and techniques for each.

Aside from other types of considerations that have been presented in the literature debate on the probabilistic versus non probabilistic representation of uncertainty, the one that appears to perhaps have greater relevance in a practical sense is the distinction between:

a) uncertain quantities with true full wide variability within a given range;
b) uncertain quantities which are believed to have an unknown fixed value (or unknown narrow range of variability) within a wide "ignorance range."

The above could be used as a working level criterion for the classification of aleatory versus epistemic uncertainty. Regardless of whether this is 100% valid at the conceptual and philosophical level, if the above distinction is al least viewed as a valid means of categorizing the large majority of uncertain parameters that appear in a typical risk assessment, then it can be argued that the application of probabilistic mathematical models and rules to represent uncertainty relative to quantities of type a) seems to be both justified and defensible. Equally justified, however, may be some of the doubts and questions cast against the use of the same approach to represent, and carry through a complex logic-mathematical model, the uncertainty relative to quantities of type b).

To see why the difference between the two situations depicted above may possibly affect the legitimacy of a purely probabilistic view for both cases, we may in fact visualize a theoretical experiment, in which it were possible to repeatedly sample variables of the two types. In such an experiment the sampled values of the type a) variables would indeed be varying across the respective wide variability ranges according to same specific form of distribution. Each of the type b) variables, however, would have essentially the same value appear over and over in each of the sampling tests, although this essentially fixed value would be somewhere within the initially identified wide range of possible values. The implication is that any dependent variable produced by the combination of type a) independent variables would indeed be distributed according to the probabilistic-combinatorial laws governing the independent variables, but any dependent variable produced by the combination of type b) independent variables would instead simply assume a fixed value, somewhere within a range that could be pre-identified by means of interval-analysis from the initially identified ranges of the independent variables. Essentially based on this observation, Ref.[4] proposes a means of combining probabilistic (for aleatory) and interval (for epistemic) representation of uncertainty. Unfortunately such a method appears to be practically viable only for relatively simple logic-probabilistic system models and not scalable to the very complex PRA models that are common in today's practical contexts.

6. CLOSING OBSERVATIONS AND COMMENTS

This paper has discussed a set of issues relative to the representation and handling of uncertainty within the practical objectives of a probabilistic risk assessment. Some of these issues have been previously been discussed in the literature, but further discussion in the PRA technical community of their conceptual and practical relation to the overall problem of effective representation of uncertainty in PRA may be beneficial.

At the present time the use of Bayesian-style state-of-knowledge probability distributions is the preferred means of uncertainty representation and handling in large scale PRAs. Aside from any conceptual preferences in regard, this type of representation offers without doubt practical advantages, as the computational framework to support it and executed it is well established within the software tools at the disposal of the PRA technical community. The discussion in this paper shows, however, that issues of consistency may arise in the setting of Bayesian prior uncertainty distributions, depending on whether the assessor seeks to a bottom-up or top-down representation of his/her prior state-of-knowledge concerning the elements of the system being analyzed. It also shows that the concerns that have been raised by some experts with regard to the use of Bayesian probability for the representation of epistemic uncertainty may still deserve further attention and discussion, as neither probabilistic formulations nor the alternative uncertainty representations proposed in the academic literature of the more recent years appear yet to provide a widely applicable and practically implementable solution to the related issues.

7. REFERENCES

[1] L. Philipson, *"Anomalies in Bayesian launch range safety analysis,"* Reliability Engineering and System Safety, vol. 49, pp 355-357, (1995).

[2] L. Philipson, *"The Failure of Bayes System Reliability Inference Based on Data with Multi-level Applicability,"* IEEE Transactions on Reliability, vol. 45(1), pp. 66-69, (1996).

[3] V. Bier, *"On the concept of perfect aggregation in Bayesian estimation,"* Reliability Engineering and System Safety, vol. 46, pp. 271-281, (1994).

[4] S. Ferson and L. Ginzburg, *"Different methods are needed to propagate ignorance and variability,"* Reliability Engineering and System Safety, vol. 54, pp. 133-144, (1996).

[5] S. Guarro and M. Yau, *"On the Nature and Practical Handling of the Bayesian Aggregation Anomaly,"* Reliability Engineering and System Safety, vol. 94 (6), pp. 1050-1056, (2009).

[6] New Horizons Mission, *"Atlas V 551 Final SAR Databook, Revision A,"* (May 2005).

[7] Mars Science Laboratory Mission, *"Safety Analysis Report Databook, Revision B,"* (February 2010).

[8] A. DerKiureghian and O. Ditlevsen, *" Aleatory or epistemic? Does it matter?,"* Special Workshop on Risk Acceptance and Risk Communication, Stanford University, (March 26-27, 2007).

[9] L. Swiler, T. Paez and R. Mayes, *" Epistemic Uncertainty Quantification Tutorial,"* Proceedings of the IMAC-XXVII, Society for Experimental Mechanics Inc., Orlando, Florida USA, (February 9-12, 2009).

[10] E. Zio and N. Pedroni, *"Literature review of methods for representing uncertainty,"* Cahiers de la Sécurité Industrielle, vol. 2013-03, Fondation pour une Culture de Sécurité Industrielle, Toulouse, France (April 2013) – (Available at http://www.FonCSI.org/fr/).

System Initiating Event Frequency Estimation using Uncertain Data

Kurt G. Vedros[*]

NuScale Power, LLC, Corvallis, Oregon, United States

Abstract: Presented is an application of Bayesian inference methods for quantifying a prior distribution used for a system initiating event with no directly applicable system information available. New safety systems in development require the use of uncertain data and estimations where the applicability must be determined by the analyst. Presented is an approach for utilizing available, germane data that both captures the appropriateness of the contributing data and quantifies and maintains the uncertainty of the data.

Keywords: Probabilistic Risk Assessment, Parameter Estimation, Initiating Events.

1. INTRODUCTION

When a new safety system to be included in a Probabilistic Risk Assessment (PRA) is developed there is rarely a direct example with operational experience from which to draw an initiating event frequency. Additionally, there may be little to no test data on the system from which to draw for updating the estimation. Using the Bayesian inference software OpenBUGS[†] [1], the system initiating event can be modeled through its component parts with their associated failure modes to develop one distribution for use as a system initiating event prior. Similar system data can also be used for another prior distribution modeling the initiating event. I show how to incorporate all data available and how to apply proportional weights to the priors to find a weighted posterior distribution. I then show how to fit this developed distribution to a probability distribution that can be used as an initiating event frequency for PRA.

2. METHODOLOGY FOR UTILIZATION OF MULTIPLE DATA SOURCES

Sometimes there are multiple data sources available for the same initiating event. In other cases, the initiating event is unique to the plant design and it becomes necessary to compile applicable data associated with contributors to the initiating event to develop the complete system initiator. In both situations, a methodology is required to incorporate all applicable data to develop an initiating event frequency distribution for use in the PRA.

Event frequencies are reported in data sources such as NUREG/CR-5750 [2] and the NUREG/CR-6928 update [3] as a mean value with an uncertainty distribution parameter. When adding events for system initiator estimation, it is important not to lose the uncertainty information by just adding the means. Likewise, when utilizing multiple data sources for the same initiator it is important to not lose the unique uncertainty surrounding each source of information by only utilizing their means then applying some generic uncertainty to the result.

There is not a standard distribution to use when quantifying uncertainty. Several distributions can be utilized for parameterization for demand based failures or for rate based failures. Therefore, it is important to utilize a method that can mathematically manipulate different distributions. One method for this is using Bayesian inference and the software package OpenBUGS. OpenBUGS utilizes Markov Chain Monte Carlo (MCMC) to converge to a Bayesian inference solution and is practical for

---

[*] Kurt Vedros email address: kvedros@nuscalepower.com
[†] OpenBUGS is free for download at http://www.openbugs.net

use with any distribution. Prior to high speed processors and MCMC programs such as OpenBUGS, the multiple integration of the denominator in a Bayesian inference calculation made the method impractical due to the amount of time required to solve.

Instructions for running OpenBUGS, including multiple ways of checking for convergence and sample scripts that utilize these methods are found in NASA/SP-2009-569 [4], where it is referred to as its non-open source, still available predecessor, WinBUGS. The following paragraphs explain the process for adding and combining initiating events in OpenBUGS. If one does not have knowledge of OpenBUGS script, the model is defined between the "{ }" and anything behind a "#" is a comment to help in understanding what that line of code does. Note that the examples were run with 1,000 non-counted samples for convergence, followed by 200,000 samples for results.

2.1. Using OpenBUGS to Add Initiating Events

When an initiating event considered for inclusion in the plant model is a combination of failure events that are quantified separately in documentation it becomes necessary to add the events. An example of this would be where a system initiator includes a piping failure with multiple failure modes such as weld failure per weld, per year and pipe rupture per foot, per year. When adding event frequencies, it is also necessary to add the distributions. In OpenBUGS it is as easy to add dissimilar distributions as it is to add like distributions.

In BUGS language distributions are designated by the letter "d" followed by an abbreviation of the distribution used. The parameters are then given in the parenthesis immediately following. For instance a Gamma distribution is denoted by x~dgamma(r, mu) and read aloud as: "x is distributed as a gamma distribution with parameters r and mu". Parameters r and mu are defined in the OpenBUGS documentation accessible within the "help" option of the program. It is recommended to review these parameterizations because OpenBUGS may not use the same parameters you may be familiar with, sometimes the parameters are manipulated differently or inverse.

A length of pipe is used as an example system. Assume that two events (pipe weld failure and pipe rupture) are independent and either event can cause the system initiating event. Adding frequency distributions within the OpenBUGS language is accomplished simply by listing the distributions then denoting another expression to add the distributions. For instance, diverse distributions (Gamma and log-normal) for two components contributing to a system frequency of a failed pipe system could be added as follows:

Script 1: Adding Distributions in OpenBUGS

```
Model {
    A ~ dgamma(r, mu)              # Failure Mode (FM) A pipe failure per length per hour
    Lambda.A <- A * length * h.y   # FM A rate per year (length of pipe, hours per year)
    Lambda.B ~ dlnorm(mu.ln,tau)   # FM B weld failures
    m <- welds * p.weld/h.y        # mean weld failures/yr
    mu.ln <- log(m) - pow(log(EF)/1.645,2)/2 # lognorm parms given mean & EF
    tau <- pow(log(EF)/1.645,-2)
    Lambda.Sum <- lambda.A + lambda.B   # rate of system leak is sum of two FM rates
}

Data
list(r = 1.0, mu =1.0E+10, length = 100, h.y = 8760, welds = 20, p.weld = 8.0E-4, EF = 10)
```

Note that the data used are not from any reference and are used solely for the purpose of illustrating the methodology.

Table 1: Results of addition of initiating event distributions

Parameter	Mean	5th Percentile	Median	95th Percentile
Lambda.A	8.8E-5	4.6E-6	6.1E-5	2.6E-4
Lambda.B	1.8E-6	6.7E-8	6.8E-7	6.9E-6
Lambda.Sum	8.9E-5	5.9E-6	6.3E-5	2.6E-4

It can be seen from the results that means can be added, but the uncertainties must be added through the distributions. Frequencies are number of occurrences over a year and while the frequency is small in this example, it is not out of the question to find a frequency above 1.0.

2.2. Combining Posteriors of Analogous Systems

When more than one data source of an initiating event frequency is appropriate for use, a methodology used to consider all inputs is the weighted average of the posterior distributions. For instance, if the system initiating event is similar in most regards to historical data from published sources, the data should be considered as input to the estimation. Another valid data point would be to piece together sub-components of the system, for instance: piping, welds, heat exchangers, etc for a unique system initiating event estimation. The analyst would then use both estimates and give a weight to them to determine the posterior when the distributions are combined and inferred.

Consider a hypothetical "System A" which is made up of piping. There is a similar, but not identical, system found in the NRC operational database NUREG/CR-5750. System A can also be modeled by adding the frequencies of its component's initiators as above or a completely different failure mechanism can be considered. For instance, consider the following data sources for "System A":

- Datum #1: NUREG/5750, similar but not identical system
- Datum #2: The added frequency distributions of two component failure modes that make up the system
- Datum #3: An published study covering a unique failure mode applicable to the entire system that would encompass the failures in Datum #2

All three data sources are valid and meaningful to the estimation of the system initiating event frequency. The methodology used to incorporate their appropriate influence on the final estimation is to give each datum a weight based on its importance to the system where all weights used in the analysis sum to 1.0. This is accomplished by utilizing the categorical distribution in OpenBUGS. The categorical distribution requires that its components add up to a value of 1.0. Use the number of categorical components (Lambdas in the example) equal to the number of data inputs with each index number equivalent to the index in the script. Then place a proportional value on the components equal to the confidence the analyst has in the applicability of the data to the design of the system. The result is an informed posterior distribution which keeps input from all of the data, but applies proportional emphasis on each element. Of course, if all input data are equally important, then the elements of the categorical distribution are listed as equal.

The way this looks in an OpenBUGS script is shown in Script 2:

Script 2: System Initiating Event Analysis with Weighted Posterior Averaging

```
Model {
    Lambda[1] ~ dgamma(r[1], mu[1])           # Datum 1
    A ~ dgamma(r[2], mu[2])                    # Datum 2 gamma dist. frequency/ryr
    Lambda.A <- A* length * h.y
    Lambda.B ~ dlnorm(mu.ln,tau)               # Datum 2 lognorm dist frequency/ryr
    m <- welds * p.weld/h.y                    # mean weld failures/yr
    mu.ln <- log(m) - pow(log(EF)/1.645,2)/2   # lognorm parm given mean & EF
    tau <- pow(log(EF)/1.645,-2)
    Lambda[2] <- Lambda.A + Lambda.B           # Datum #2
    Lambda[3] ~ dgamma(r[3], mu[3])            # Datum #3
    Lambda.avg <- Lambda[S]
    S ~ dcat(p[])                              # p=array of weights given to the data
}
```

Data
list(r=c(1.6, 1.0, 2.3), mu=c(365000, 1.0E+10, 170000), welds = 20, p.weld = 8.0E-4, p=c(0.3, 0.4, 0.3), EF=10, h.y = 8760, length = 100)

The input distributions are prior likelihoods and their output distribution is a posterior which is averaged with the categorical distribution using the weights given to Datum 1, 2, and 3 as 0.3, 0.4, and 0.3 respectively. Weights are subjective and left to engineering discretion. Documentation should support the weighting decisions. The result, once fit to a proper distribution, will be used for initial estimation of system performance. This estimation should be updated once actual performance of the system is gathered by using it as an informed prior in a Bayesian update. Note the resulting statistics and density graphs and how each of the prior's features can be seen influencing the posterior Lambda.avg.

Table 2: Results of initiating event analysis using posterior averaging

Parameter	Mean	5th Percentile	Median	95th Percentile
Lambda[1]	4.4E-6	5.6E-7	3.5E-6	1.1E-5
Lambda[2]	8.9E-5	5.9E-6	6.3E-5	2.6E-4
Lambda[3]	1.4E-5	2.8E-6	1.2E-5	3.1E-5
Lambda.avg	4.1E-5	1.2E-6	1.3E-5	1.8E-4

Figure 1: Component and posterior probability density functions

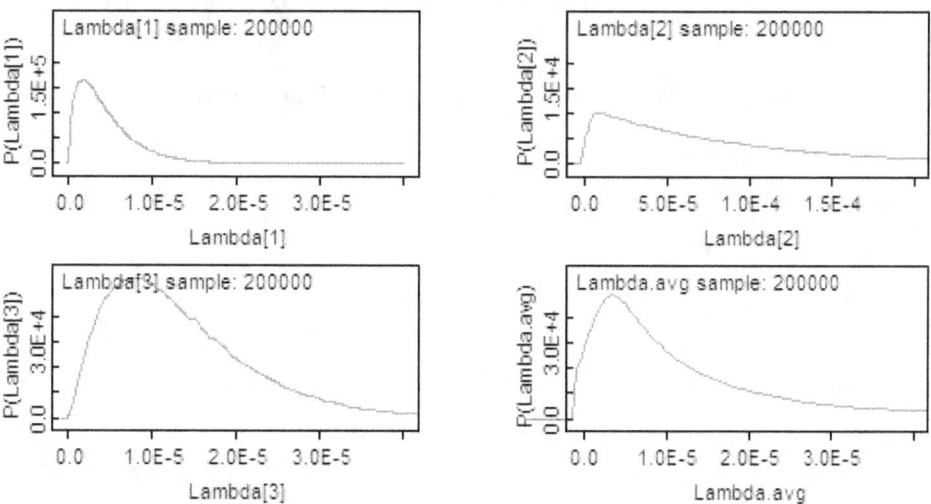

2.2. Fitting Averaged Posteriors as an Initiating Event Frequency

Whether the initiating event is calculated through the process of adding dissimilar distributions or averaging distribution posteriors, the resulting distribution is improper and not belonging to any distribution family. Visually, this can be seen by reviewing the probability density function graph for the initiating event calculated in the previous section (Lambda.avg in Figure 1). The resulting distribution is not smooth and is "pulled" in the direction of its components. In extreme cases, these posterior averaged distributions can be multi-modal and have misshapen tails. Fortunately, the posterior's parameters generated by OpenBUGS can be used to fit a proper distribution. The log-normal distribution is a good choice for this. The log-normal distribution is typically represented by a mean and an error factor. The log-normal error factor (EF) is a measure of deviation and can be determined in a few ways, all of which are a measure of the spread of the distribution akin to variance [5] [6]:

The error factor of a log-normal distribution is defined in one of three ways:

$$\text{Log-normal Error Factor: } \sqrt{\frac{95th\ Percentile}{5th\ Percentile}} \approx \frac{95th\ Percentile}{Median} \approx \frac{Median}{5th\ Percentile} \quad (1)$$

While all three formulas are accepted for determining the EF, the shape of the distribution has an effect on the EF calculation. The three formulas for determining the EF do not provide identical results unless it is close to a normal distribution. If the distribution has a long trailing tail, the ratio of the 5th to the median will be the largest EF. If the distribution has a long preceding tail, the ratio of the 95th to the median will be largest EF. The square root of the ratio of the 95th and 5th percentiles will give a result in the middle of the other two methods and this is the method chosen for the example.

Taking the results of the example initiating event posterior above for the 5th and 95th percentile, the log-normal EF is:

$$EF_{Lambda.avg} = \sqrt{\frac{1.8E-4}{1.2E-6}} = 12.2 \quad (2)$$

Therefore, the log-normal fit parameters are the mean posterior from the analysis (Lambda.avg) of 4.1E-5 and an EF calculated of 12.2.

Insert the following lines of code into Script 2 to verify the resulting parameter fit for the log-normal distribution:

Script 3: Script Additions to Compare Log-Normal Fit

```
# insert these lines under the S~dcat(p[]) line:
LogNormFit ~ dlnorm(mu.lnfit, tau.lnfit)
mu.lnfit <- log(mean) - pow(log(EFln)/1.645,2) / 2
tau.lnfit <- pow(log(EFln)/1.645,-2)

# and add to the data list:
mean = 4.1E-5, EFln = 12.2
```

A comparison of the improper posterior distribution of the analysis and the proper log-normal fit distribution is shown in Table 3 and Figure 2.

Table 3: Distribution Statistics for Resultant Distribution and Log-normal Fit

Parameter	Mean	5th Percentile	Median	95th Percentile
Lambda.avg	4.1E-5	1.2E-6	1.3E-5	1.8E-4
LogNormFit	4.1E-5	1.1E-6	1.3E-5	1.6E-4

Figure 2: Density comparison of log-normal fit to initiating event analysis posterior distribution

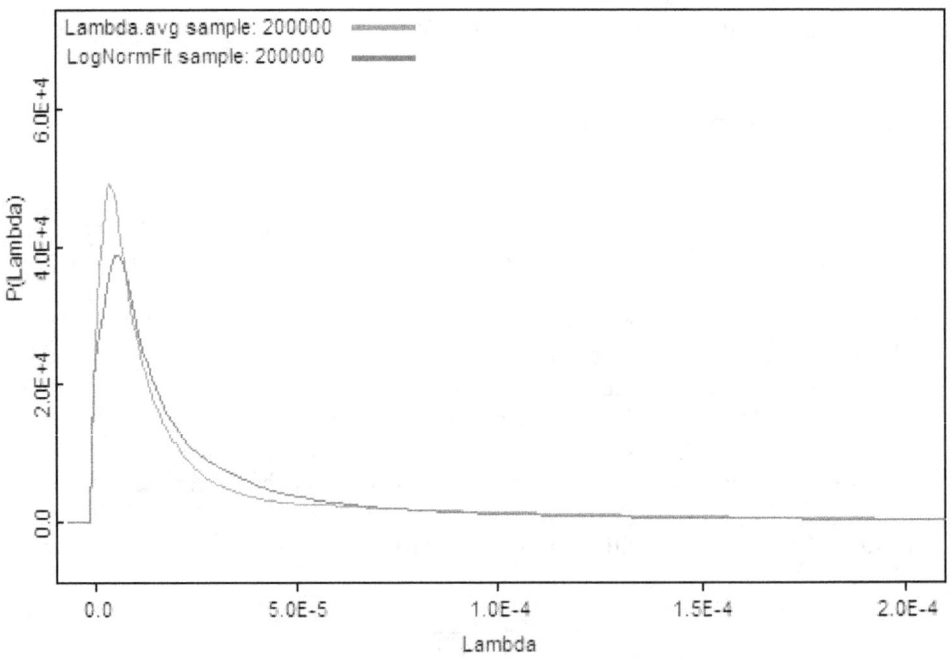

The fitted log-normal function from the results of the analysis (mean = 4.1E-5, Error Factor = 12.2) can now be used as a PRA initiating event frequency input.

3. CONCLUSION

A practical use of existing methodology using Bayesian inference software to develop a system initiating event was presented. This method illustrates a process of parameter estimation that utilizes the entire scope of the information available, preserving uncertainty and providing a mechanism to attach proper influence to each element of information. Although the exercises in the use of OpenBUGS and fitting a Log-Normal distribution from other distribution parameters are not in and of themselves new processes, the entire scheme illustrated brings together a useful technique for initiating event frequency estimation using uncertain data.

Going forward, this estimate could be used as an informed prior to update the system initiating event with operational performance in a similar manner.

References

[1] D. Lunn, D. Spiegelhalter, A. Thomas, N. Best. *"The BUGS project: Evolution, critique, and future directions"*, Statistics in Medicine, volume 28, pp. 3049-3067, (2009).
[2] U.S. Nuclear Regulatory Commission. *"Rates of Initiating Events at U.S. Nuclear Power Plants: 1987-1995"*, NUREG/CR-5750, (1999).
[3] U.S. Nuclear Regulatory Commission. *"Initiating Event Data Sheets, 2010 Update"*, Update of NUREG/CR-6928, (2012).
[4] H. Dezfuli, et al. *"Bayesian Inference for NASA Probabilistic Risk and Reliability Analysis"*, NASA/SP-2009-569, (2009).
[5] D. Kelly, C. Smith. *"Bayesian Inference for Probabilistic Risk Assessment: A Practitioner's Guidebook"*, Springer-Verlag, (2011), London.
[6] M. Rausand, A. Høyland. *"System Reliability Theory: Models and Statistical Methods"*, John Wiley & Sons, 2nd edition, (2004), Hoboken.

SUnCISTT
A Generic Code Interface for Uncertainty and Sensitivity Analysis

Matthias Behler[a], Matthias Bock[b], Florian Rowold[a], and Maik Stuke[a,*]

[a] Gesellschaft für Anlagen und Reaktorsicherheit GRS mbH, Garching n. Munich, Germany
[b] STEAG Energy Services GmbH, Essen, Germany

Abstract: The GRS development SUnCISTT (Sensitivities and Uncertainties in Criticality Inventory and Source Term Tool) is a modular, easily extensible, abstract interface program designed to perform Monte Carlo sampling based uncertainty and sensitivity analyses. In the field of criticality safety analyses it couples different criticality and depletion codes commonly used in nuclear criticality safety assessments to the well-established GRS tool for sensitivity and uncertainty analyses SUSA. SUSA provides the necessary statistical methods, whereas SUnCISTT handles the complex bookkeeping that arises in the transfer of the generated samples into valid models of a given problem for a specific code. It generates and steers the calculation of the sample input files for the used codes. The computed results are collected, evaluated, and prepared for the statistical analysis in SUSA.

In this paper we describe the underlying methods in SUnCISTT and present examples of major applications in the field of nuclear criticality safety assessment:
- Uncertainty and sensitivity analyses applied in criticality calculations.
- Monte Carlo sampling techniques in nuclear fuel depletion calculation.
- Uncertainty and sensitivity analyses of burnup credit analysis.
- Analysis of correlations between different experimental setups sharing uncertain parameters.

The examples and results are shown for SUnCISTT, coupling SUSA to different SCALE sequences from Oak Ridge National Laboratory and to OREST from GRS.

Keywords: Uncertainty and Sensitivity Analysis, Monte Carlo Sampling, Nuclear Engineering, Burnup Credit, Criticality Safety.

1. INTRODUCTION

In recent years, the increase of computational power allowed the development of probabilistic assessment strategies for the field of nuclear criticality safety. In particular, Monte Carlo sampling methods became applicable in different areas of research. These methods use repeatedly calculations of a given model with randomly varied input parameters to determine uncertainties and sensitivities of the model results.

The GRS development SUnCISTT (Sensitivities and Uncertainties in Criticality Inventory and Source Term Tool) is a modular, easily extensible abstract interface program designed to perform Monte Carlo sampling based uncertainty and sensitivity analyses for technical parameters, such as manufacturing tolerances. The methods offer a complement to traditional best-estimate analyses and can help to reduce conservatisms while still keeping the high safety standards required in the field of criticality safety calculations.

For the field of criticality safety, different criticality and depletion codes, commonly used in nuclear criticality safety assessments, have currently been coupled to the well-established GRS tool SUSA [1]. These couplings represent what is called a SUnCISTT application. Among the codes are the CSAS1, CSAS5, CSAS6 and T-NEWT sequences of the SCALE package, developed at Oak Ridge National Lab [2], the MCNP5 code developed at Los Alamos National Lab [3], and the OREST code [4]. OREST is the well-established GRS developed code for a 1 dimensional cell burnup system.

SUSA provides statistical methods for sensitivity and uncertainty analyses, whereas SUnCISTT handles the necessary bookkeeping and provides methods to translate the generated samples into valid computation models of the given problem. It generates and steers the execution of the sample input files of any used code, and collects and processes the computed results.

*) contact of corresponding author: maik.stuke@grs.de

In this paper an overview of the major capabilities of SUnCISTT will be presented. In the following chapter the SUnCISTT itself is described in more detail. We explain the core functionalities and the data flow. In chapter 3 we give a short overview of the SUnCISTT's capabilities by summarizing some analyses that have successfully been performed so far. Finally, in chapter 4 we draw a resume.

2. SUnCISTT

SUnCISTT was developed to provide a general analyses tool for uncertainty and sensitivity analysis associated with technical parameters such as manufacturing tolerances. The original goal was to determine uncertainties arising through these manufacturing tolerances in assessments related to the nuclear fuel cycle. However, due to the abstract code concept of SUnCISTT it's use is not restricted to these analyses. It can be used under both, Linux and Windows operating systems and it is, due to its object oriented programming in Python3, easy to extend in its functionalities and fields of applications. SUnCISTT is able to investigate any given mathematic model. Here we describe the couplings of SUnCISTT with the GRS code SUSA, used as pre- and postprocessor, and some specific codes for nuclear criticality and burnup calculations.

The GRS code SUSA (Software for Uncertainty and Sensitvity) is a well-known and for more than 20 years established tool for uncertainty and sensitivity analyses. For example it is also used in combination with the thermohydraulic code ATHLET [5] and with COCOSYS [6], a software designed for the simulation of severe accidents in nuclear power plants.
For the presentation at hand, SUSA serves two purposes: the generation of statistically independent samples of the uncertain input parameters of the model to investigate and to perform the uncertainty and sensitivity analysis of the computational results. Note, that for both purposes the interfaces implemented in SUnCISTT offer the opportunity to apply any other tool with similar features. The approach to couple SUnCISTT to SUSA leads to the following diagram:

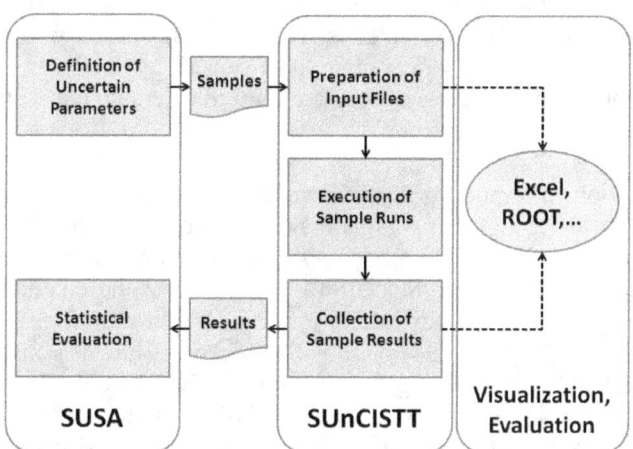

Figure 1: Possible sequence of analyses with SUnCISTT.

In the depicted sequence in Figure 1 SUSA is used for the preparation of the independent Monte Carlo samples, based on the probability density functions of the uncertain parameters. Possible dependencies between input parameters can either be considered in SUSA or using the user interface implemented in SUnCISTT. The result is an ASCII formatted list of independent samples, that is used as input to SUnCISTT.
This list is used in SUnCISTT to convert each generated sample into a valid input file for the applied codes, e.g. the CSAS5 sequence of the SCALE package. SUnCISTT steers the execution of the input files and collects the results for the subsequent statistical evaluation in SUSA. The SUnCISTT modes include mechanisms that allow their visualization, evaluation and supervision by tools like ROOT [7] or Microsoft Excel.

In the approach presented here, the statistical evaluation of the model run results is done with SUSA by calculating mean values, sample standard deviations or perform hypothesis tests. The sensitivity analysis describing the influence of the uncertainty of the input parameter on the uncertainty of the result can be performed by using several sensitivity measures, e.g. ordinary correlation or the correlation ratio.

In the following the SUnCISTT core is described in more detail.

2.1. The SUnCISTT Core

To perform a Monte Carlo sampling analysis of a given mathematical model, SUnCISTT needs information about the generated samples, the computational model to be analyzed and the specific input file requirements of the code to be executed. For the generated samples, the aforementioned ASCII list of the statistical varied input parameter has to be provided. From the input file of the nominal case, a template file is derived in which user defined keywords replace the nominal values of the uncertain parameters. The third file to be provided in the SUnCISTT mode *prepareSamples* is a configuration file that sets the information of the other files into relation. It is prepared in the *.json* format, an exchange format that can be read and written from several programming languages. The user has the opportunity to trigger auxiliary calculations that are necessary to transform the parameters of the generated samples into the parameters needed in the computational model. For example, if samples were generated for the diameter of a sphere but the model requires the radius, this conversion can be triggered from SUnCISTT. If any of the uncertain parameters requires such a conversion, SUnCISTT offers the user the possibility to provide its own Python module, defining the conversion formalism. Note, that this implementation of the user interaction allows the user to easily call any other program from its Python module during runtime. This opens a wide range of opportunities for complex analyses. Besides other configuration options, information about this Python module is also part of the configuration file.

Informations about the executable code to run the sampled input files can be easily implemented in the *runSamples* mode, due to the object oriented structure of the core. With only little extensions, SUnCISTT is capable of performing uncertainty and sensitivity analyses based on Monte Carlo sampling for any given problem, if the executable program to calculate the individual samples is using input files.

With the given information, the mode *prepareSamples* generates the desired number of individual input files with the statistical varied input parameters defined in the sample list. SUnCISTT than steers the execution of the generated files in the mode *runSamples*. With the knowledge about the structure of the provided, individual output files, SUnCISTT also collects the calculated individual result in mode *collectResults* and prepares them for further analysis. In our case files for the postprocessing in SUSA are generated as well as ROOT files or files for Microsoft Excel. For the different SUnCISTT applications this is achieved by overloading only method of the SUnCISTT core with the application specific implementation. For an overview over the whole analysis procedure see Figure 1. Two graphical sketches of the modes *prepareSamples* and *collectResults* with more details are shown in Figure 2 and Figure 3.

The mode *prepareSamples* has the capability to handle additional input variables, if needed. Often, in order to define a valid computation model, model parameters that have a functional dependence on one or more of the uncertain parameters have to be calculated, although these additional parameters are not treated as uncertain parameters themselves. For example, let's assume a material steel with three constituents: iron, nickel and chromium. If the weight percentages of nickel and chromium are defined as uncertain parameters, the contribution of iron has to be derived by subtracting them from 100%. This kind of parameter and its dependencies on uncertain parameters can be defined in the configuration file, mentioned before. SUnCISTT will calculate the final values of these additional parameters based on the user defined Python module. Then, the corresponding keywords in the input files will be replaced, just like for the uncertain parameters. The *prepareSamples* mode includes many check routines for direct feedback that help to reduce potential error sources. For example, SUnCISTT checks if the keywords defined by the user are unique and unambiguous. A bookkeeping method gives the user a quick overview about how often each keyword has been replaces in total and which keywords were replaced in each individual line of the template file. This information can be used to crosscheck the automated procedure with the users' expectation.

Once the sample input files are executed, the mode *collectResults* prepares the result files for further analysis or visualizations. This mode needs to have informations about the structure of the result files and the results of interest. In the development of new SUnCISTT applications, the main task is to implement the parsing of the result files of the codes, searching for the result parameters of interest. By default, SUnCISTT produces a result file that can be transferred to SUSA for the statistical evaluation. For further visualizations and analyses optional files for ROOT or Microsoft Excel can be generated. A graphic sketch is shown in Figure 3.

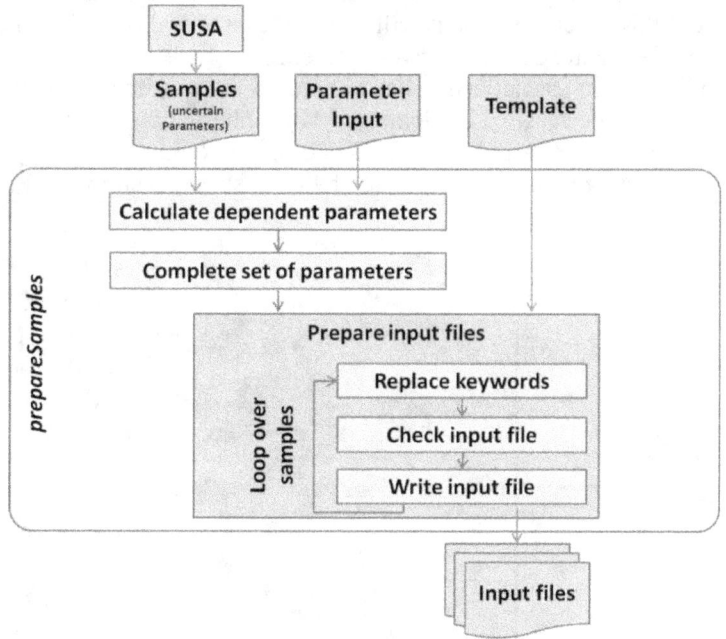

Figure 2: Graphic sketch of the SUnCISTT mode *prepareSamples*.

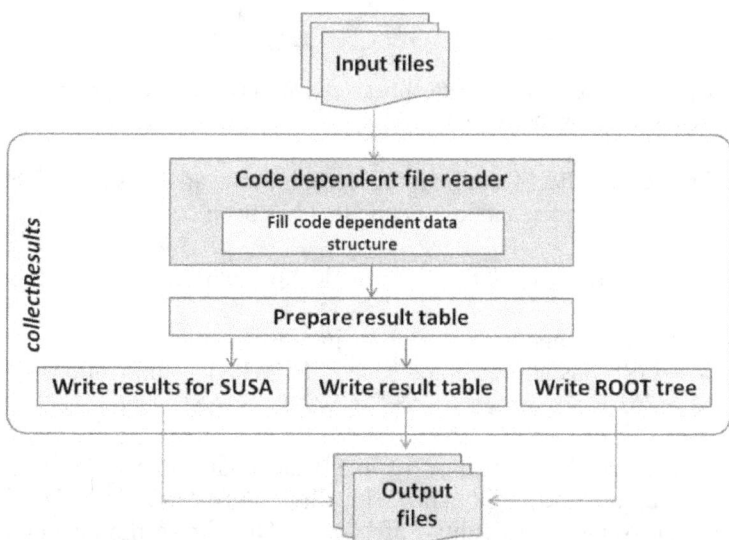

Figure 3: Graphic sketch of the SUnCISTT mode *collectResults*.

The result files obtained from the *collectResults* mode can then be used to assess the statistical quantities of interest and to evaluate the uncertainty and the sensitivity measures. The implementation presented here takes advantage of the numerous SUSA capabilities. However, with the interfaces defined in SUnCISTT, other statistical tools could easily be utilized, too.

3. ANALYSES EXAMPLES OF SUNCISTT APPLICATIONS

The acronym SUnCISTT stands for Sensitivities and Uncertainties in Criticality Inventory and Source Term Tool and as suggested by the name, it was developed to provide a general analysis tool associated with the nuclear fuel cycle. However, providing well defined interfaces to establish the coupling to specific codes, SUnCISTT can easily be adopted to perform uncertainty and sensitivity analyses in any desired field.

In the following we show some results and capabilities of the SUnCISTT in the field of nuclear criticality safety. Since there are numerous applications, all inheriting the same core functionalities and due to the subject of this article, the focus will be to give a general overview. For details about the individual analyses we refer the reader to the given references.

The applications couple specific specialized codes to be used in the calculations with the SUnCISTT. Any new application can be easily implemented by inheritance from the SUnCISTT core functionalities if the underlying code provides ASCII formatted output files. A graphic sketch of the SUnCISTT applications, that have been implemented by GRS so far, is shown in Figure 4.

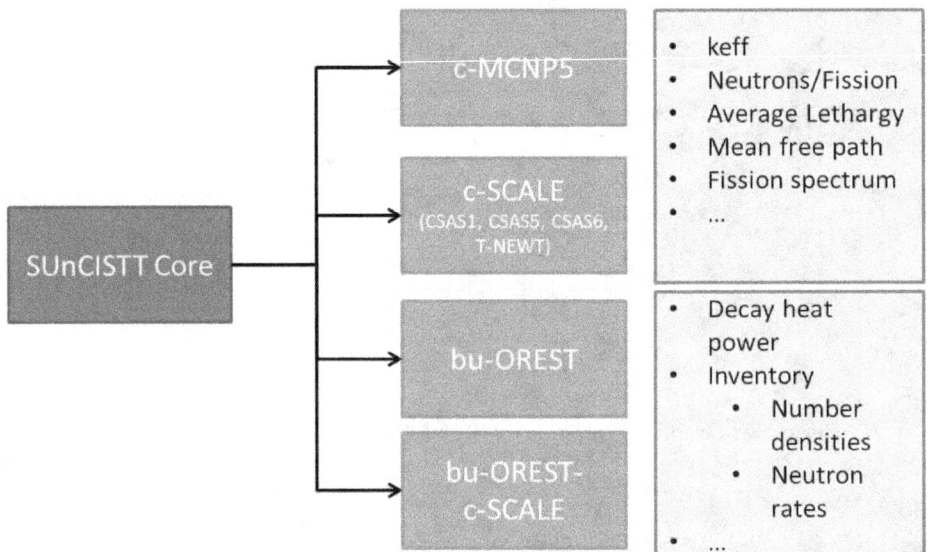

Figure 4: SUnCISTT applications for nuclear criticality safety. Depending on the problem and the physical quantity of interest, the application can be chosen.

The naming scheme for the SUnCISTT applications combines the analysis type to be performed (c = criticality, bu = burnup) with the corresponding code that has been coupled.

3.1. Criticality Assessments with SCALE – the c-scale Application

In the field of criticality safety analyses the main interest is in the determination of the neutron multiplication factor k_{eff}. GRS has implemented several SUnCISTT applications serving this purpose. They are based on the SCALE sequences CSAS1, CSAS5, CSAS6, and T-NEWT, and the MCNP5 application.

We show results for the CSAS5 sequence of the Benchmark Phase IV [8], proposed by the OECD/NEA "Expert Group on Uncertainty and Criticality Safety Assessment" (UACSA), a subgroup of the "Working Party for Nuclear Criticality Safety" [9]. The aim of the benchmark is to determine the impact of correlations between different benchmark experiments on the estimation of the computational bias on k_{eff}. These correlations play a role in the validation of codes for criticality safety assessments, especially in cases where the experimental basis that can be used for the validation is poor and limited to only a few contributing experimental facilities.

To verify a code for a calculated application case, the user can test the code against qualified experimental results. A collection of qualified results for critical experiments is e.g. documented in the "International Handbook of Evaluated Criticality Safety Benchmark Experiments" (ICSBEP) [10] Besides information on the experimental setup it contains additional information about the major sources of uncertainties of each experiment. This can be used to perform Monte Carlo sampling based uncertainty analyses in each experiment of the validation pool. Their results can be used to determine correlation between those benchmark experiments. Since some experimental series share major parts of the experimental setup the manufacturing tolerances of these parts cannot be treated as statistically independent. An add-on of SUnCISTT provides the capability to steer and execute the uncertainty analyses of several benchmark experiments simultaneously. The add-on utilizes the SUnCISTT applications for individual uncertainty analyses, described in the previous chapters. The user can choose if common sources of uncertainties are to be treated identical or individual in the contributing experiments. The results of the uncertainty analyses can be combined and quantities like the Pearson's correlation coefficients can be determined. The benchmark proposed by the UACSA aims to determine the correlation between 21 different criticality experiments, described in the ICSBEP Handbook at LEU-COMP-THERM (LCT) 7, cases 1 to 4 and LCT 39 cases 1 to 17. The experiments share the experimental setup (e.g. fuel rods) and were performed in the same apparatus. An overview of the uncertain parameters common to all experiments are given in Table 1

The experiments consisted of low enriched Uranium in fuel rods with varying pitches and varying formations. The water level in each setup was triggered to ensure a system k_{eff} =1. The water height result is different for each experimental setup and thus the corresponding model parameter has to be treated individually.

Since this articles purpose is to demonstrate the capabilities of SUnCISTT, we will not go into to many details. For an elaborated presentation of the analysis and an interpretation of its results see [11,12,13].

The SUnCISTT add-on is able to steer the uncertainty analyses for the experiments specified in a configuration file at a time. In this case, the file contained 21 entries. For each experiment, the input files necessary for the individual uncertainty analysis has to be provided. The individual uncertainty analyses are then processed corresponding to the description in chapter 2. The add-on provides additional modes, compared to the single-experiment application: a check for missing result files and the possibility to analyze the results for correlation between the individual models. A graphic sketch can be found in Figure 5. The modes marked with an asterisk are those belonging to the single-experiment application that is called from the add-on.

The criticality calculations for the 21 experiments of the UACSA proposed benchmark were performed with the SUnCISTT application c-scale-csas5. It includes the CSAS5 sequence of SCALE version 6.1.2, using the neutron transport code KENO V.a and CENTRM for the resonance self shielding with the 238 group ENDF/B-VII cross section library. Each experiment was sampled 625 times leading to more than 13000 individual calculations. The individual CSAS5 configuration consisted of 100k neutrons per generations and a convergence criterion of 5×10^{-5}. The first 1000 generations were skipped to ensure source convergence.

Parameter	Distribution Model	Model Parameter	
		a or μ	b or σ
Fuel rod inner diameter	U(a,b)	0.81 cm	0.83 cm
Fuel rod thickness	U(a,b)	0.055 cm	0.065 cm
Fuel pellet diameter	N(μ,σ^2)	0.78919 cm	0.00176 cm
Mean linear density of fissile coloumn	N(μ,σ^2)	5.0778 g/cm	0.0282 g/cm
Height of fissile column	N(μ,σ^2)	89.703 cm	0.306 cm
^{234}U content	N(μ,σ^2)	0.0307 At.-%	0.0005 At.-%
^{235}U content	N(μ,σ^2)	4.79525 At.-%	0.002 At.-%
^{236}U content	N(μ,σ^2)	0.1373 At.-%	0.0005 At.-%

Table 1: Parameters, their distribution models, and model parameters common to all experiments. The distribution U(a,b) represents a uniform distribution between a and b, N(μ,σ^2) represents a normal distribution with expectation value μ and standard deviation σ. Additional to these parameters the water heights were varied independently for each experiment.

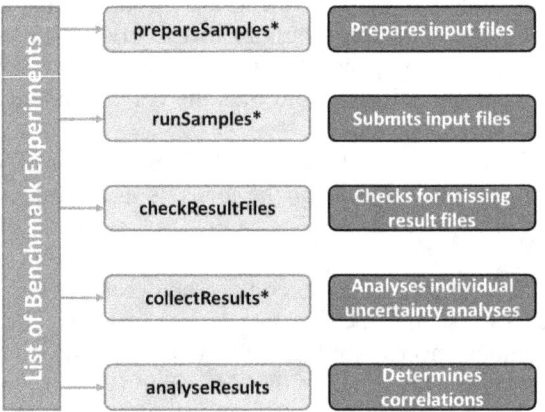

Figure 5: Flowchart of the SUnCISTT add-on for the determination of correlations between calculated k_{eff} values of the benchmark experiments. The asterisks indicate the use of the SUnCISTT core modes.

The mean values and standard deviations calculated for each experiment are shown in Figure 6. Each entry corresponds to the result of one uncertainty analysis as it would have been obtained if the analysis would have been performed with just the c-scale-csas5 application instead of the add-on. The plot was generated automatically in the add-on's *analyseResults* mode from the ROOT TTree object that is prepared during the *collectResults* mode. The ROOT file containing the ROOT TTree object includes also overview plots of other potential result parameters, like the mean free path of neutrons. With this automated visualization it is possible to obtain a quick overview about the overall results, right after the calculations are finished.

Following the flowchart in Figure 5 for our example of the 21 experiments, SUnCISTT determines the correlations between the calculated k_{eff} introduced by sharing the same experimental setup. The result is shown in Figure 7 as a color coded matrix. It shows the high correlations between the calculated k_{eff} values with coefficients close to 1 for almost all experiments. The exceptions are experiments 2 and 4 from the LCT 7 series. These experiments have a significantly larger pitch which influences the neutron spectra. For a detailed discussion of the results see [11,12].

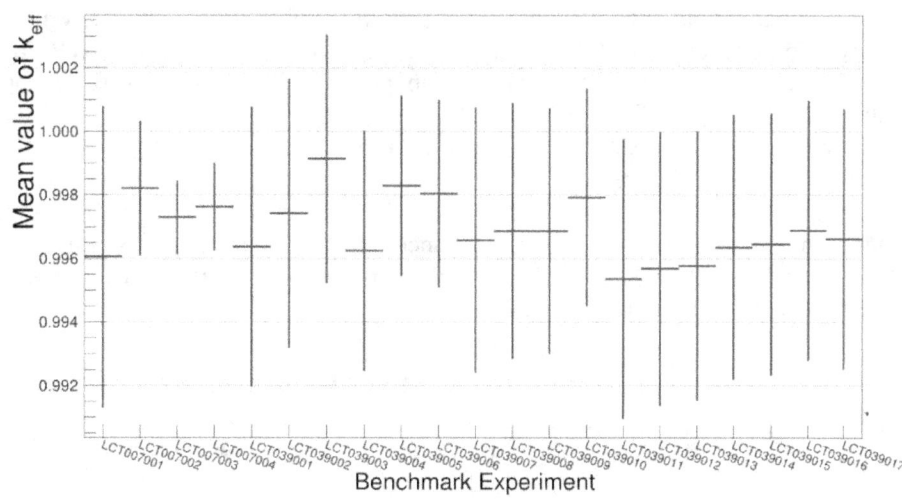

Figure 6: Mean values and standard deviations of k_{eff} of the 625 samples for each of the 21 experiments. All mean values are below k_{eff} of 1.0 and in agreement calculations stated in [10].

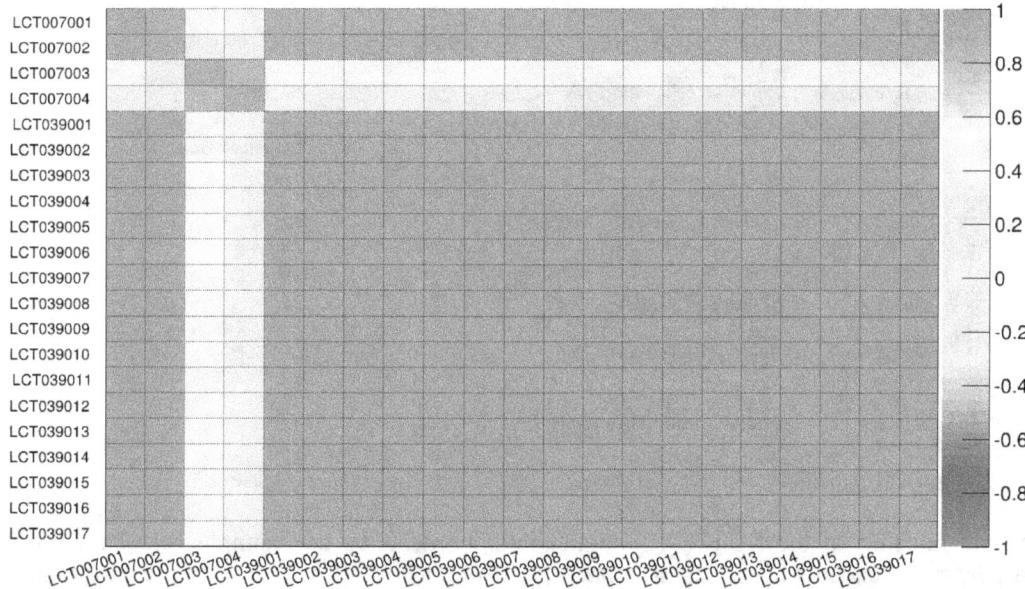

Figure 7: Color coded correlation matrix of the calculated k_{eff} values for the experiments LEU-COMP-THERM (LCT) 7 cases 1 to 4 and LCT 39 cases 1 to 17. All cases are highly correlated, except case number 3 and 4 from the LCT 7 series. These two cases have a significantly larger pitch between the fuel rods, which influences the thermal neutron spectra.

3.2. Burnup Credit Application: bu-orest-c-scale

In the following we show some details of a more complex calculation with SUnCISTT. The goal is to calculate typical physical quantities of interest for a generic transport cask loaded with spent nuclear fuel. The criticality safety calculations include basically two steps: The calculation of the inventory of the spent fuel elements and the criticality calculations of the spent fuel in the cask. In both of these calculations the variation of manufacturing parameters has to be considered in a consistent way. This leads to a complex bookkeeping SUnCISTT needs to handle, in particular the error propagation through the whole calculation chain of burnup and criticality calculation.

When calculating criticality, taking into account the burnup of fuel rods in 3D, one has also to consider axial non-homogeneities in the burnup. Typically, top and bottom of PWR fuel rods are radiated less than the middle leading to axial burnup profiles. The bu-orest-c-scale application is capable of taking these burnup profiles into account.

A schematic overview of the bu-orest-c-scale application, that was developed to perform this type of complex analyses, is shown in **Figure 8**.

The SUnCISTT application bu-orest-c-scale calculates for each defined axial zone the burnup with OREST. In addition to the generated samples of uncertain parameters, the user has to provide a databank of burnup profiles. The resulting Monte Carlo samples for the nuclides of each axial zone are then transferred into the model for criticality calculations.

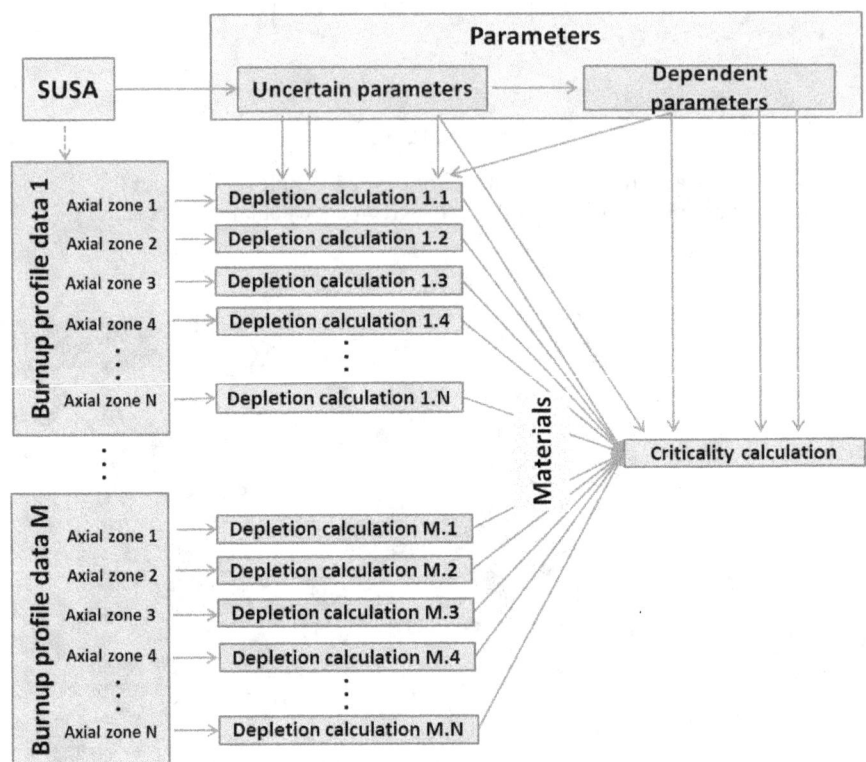

Figure 8: Schematic overview of the SUnCISTT application *bu-orest-c-scale*.

To illustrate the function of the SUnCISTT application, we describe an example calculation of a generic transport cask, loaded with typical irradiated PWR fuel assemblies. This example represents a typical task in the field of criticality safety assessments: the initial enrichment of the fuel assemblies, their average burnup, and their geometry, and the cask itself are known. Criticality safety of the loaded cask has to be ensured for example in interim storage scenarios or for the disposal of cask in a final repository. The Monte Carlo sampling method is an adequate choice for this kind of assessment.

The model of the generic cask used in this example is the GBC-32 transport cask for spent nuclear fuel of Pressurized Water Reactors (PWR) as described e.g in [15]. The design includes a basket with 32 shafts for the fuel elements. Between the shafts neutron absorbers (B_4C/Al) are placed to ensure subcriticality. The whole basket is surrounded by a cylindrical steel body. A horizontal cut through the SCALE model is shown in Figure 9. The fuel elements are modelled in SCALE as 17x17 Westinghouse "Optimized Fuel Assemblies", shown in Figure 10. In the model, each fuel rod is divided into 18 equal axial zones, indicated in the right picture of Figure 10.

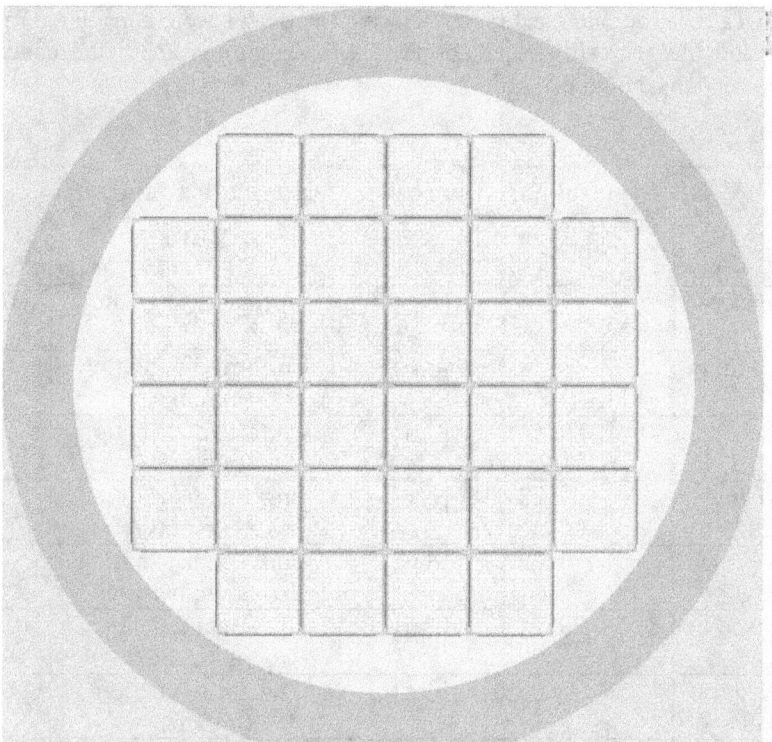

Figure 9: Horizontal cut through the SCALE model of the loaded transport cask GBC-32. Each quadrat represents a complete fuel assembly. For details of the model see e.g. [15]

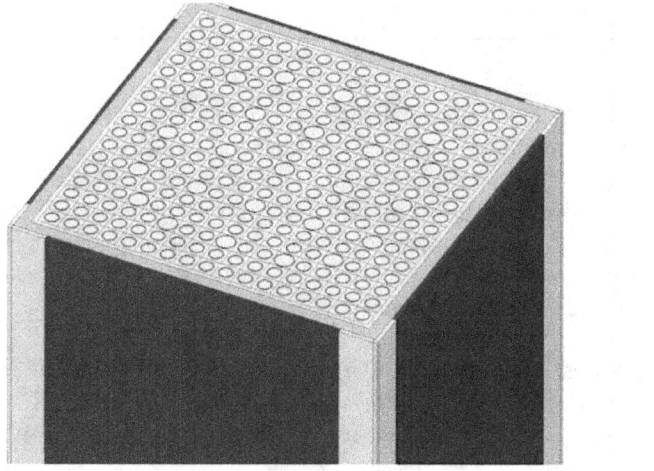 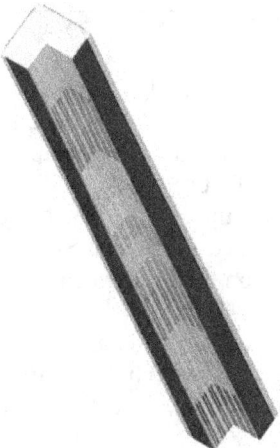

Figure 10: Simulated 17x17 fuel assemblies. On the left a detailed view of a horizontal cut through an assembly. The green spots represent fuel rods, while the bigger, yellow circles represent the guide tubes. For details of the model see e.g. [15]. On the right hand side a ¾ vertical cut through the element shows the 18 different color coded axial zones.

The fuel elements are assumed to be uniform and typical PWR UO_2 fuel elements with an initial enrichment of 4.4% ^{235}U.
The cask is further assumed to be loaded with elements of two different burnups. The outer elements have a burnup of 60 GW/d, while the inner 16 elements should have a burnup of 27 GW/d.

For the SUnCSITT calculation, identical manufacturing tolerances were considered in both calculation steps, the burnup and the criticality calculations. The assumptions about the characteristics of the varied input parameters are shown in Table 2.

Components	Expectation value	Dimension	Distribution function (P1,P2)	P1	P2
Fuel Pellet Diameter	0.7844	[cm]	Normal distribution (μ,σ)	0.7844	0.00176
Inner Diameter Cladding	0.8	[cm]	Uniform distribution (min,max)	0.79	0.8
Outer Diameter Cladding	0.9144	[cm]	Uniform distribution (min,max)	0.8544	0.9744
Inner Dimension Fuel Element Shaft	22	[cm]	Uniform distribution (min,max)	21.95	22.05
Wall Thickness Fuel Element Shaft	0.75	[cm]	Uniform distribution (min,max)	0.7	0.8
Width Neutron Absorber	0.20574	[cm]	2,4-Beta function (min,max)	0.18074	0.23074
Wall Thickness Neutron Absorber Shaft	19.05	[cm]	2,4-Beta function (min,max)	19.04	19.06
Fuel Density	10.198	[g/cm^3]	Normal Distribution (μ,σ)	10.198	0.0435
Boron Concentration	524.568	[ppm]	Normal Distribution (μ,σ)	524.568	26.0776
Enrichment	4.4	[wt-% ^{235}U]	Uniform distribution (min,max)	4.35	4.45

Table 2: Input parameters and their characteristics varied in in the SUnCISTT analyses.

At first, SUnCISTT calculated for each of the 18 axial zones of the two different fuel elements 100 samples with varying boron concentration in the moderator, enrichment, and fuel density. To consider the axial varying burnup, SUnCISTT uses profiles provided from the NEA program ZZ-PWR-AXBUPRO-SNL [CAC 97]. The databank consists of axail burnup distributions of commercial PWRs for varying setups, reactor types, enrichments and more. A preselection of the 3169 profiles has been performed outside of SUnCISTT to ensure the use of only the best fit profiles. The preselection was based on the following parameters: geometry, enrichment, and burnup.

With use of the axial burnup profiles, the bu-orest application in SUnCISTT calculated a total of 3600 inventories. Each of the inventories was then transferred to its corresponding position in the SCALE model for the criticality calculations. A graphical sketch of the bu-orest-c-scale application is shown in Figure 8.

Thus, 100 samples of the loaded cask were created with SUnCISTT, respecting manufacturing tolerances in the complete analysis chain. Although 100 samples might not be sufficient for a profound analysis, they can be used to present the complexity of this kind of analysis and to demonstrate the successful application of the bu-orest-c-scale implementation.

The 100 input files, describing the variations of the loaded cask, were then analysed with the SCALE 6.1.2 version and its 3 dimensional Eigenwert Monte Carlo transport code KENO-V.a using the ENDF/B-VII continuous energy cross section library. 50,000 generations of Neutrons with 100,000 neutrons per generation were calculated, skipping the first 500 generations and using a convergence criterium of 0.0001.

The results for the neutron multiplication factor k_{eff} are shown in Figure 11.

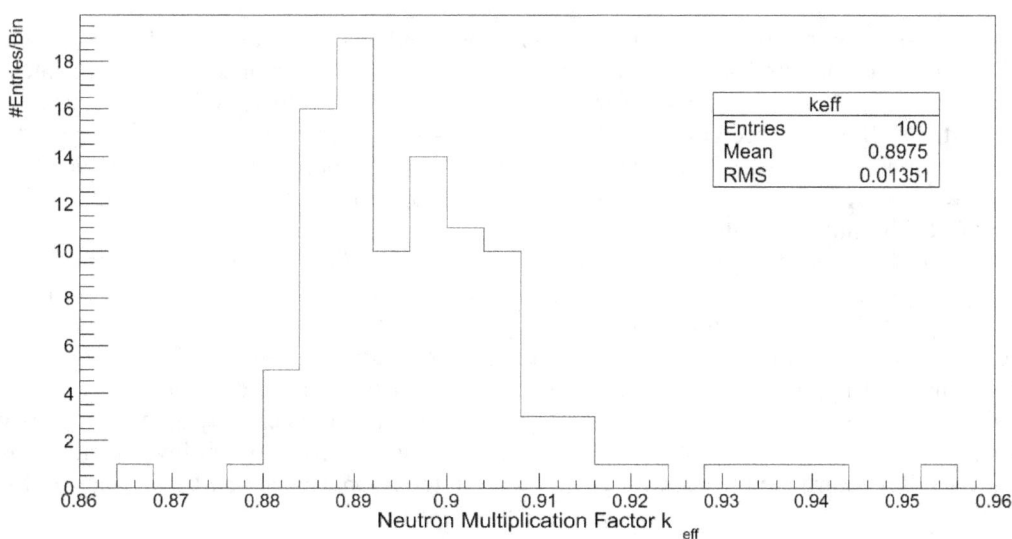

Figure 11: Frequency of the calculated neutron multiplication factor k_{eff} of the 100 sampled transport casks. The plot is also an example of one of the numerous automatically generated ROOT analysis files.

The results for k_{eff} were then transferred back to SUSA for the sensitivity analyses. The sensitivity analysis shows the impact of the uncertainty of the input parameters on the uncertainty of the resulting k_{eff} values. A correct and rigorous propagation of the uncertainties through the calculation chain as done by SUnCISTT is mandatory for an error free sensitivity analysis.
The result of our example indicate that reducing the uncertainty in the production process for the cladding of the fuel rods would have the biggest impact on decreasing the uncertainty of k_{eff}.

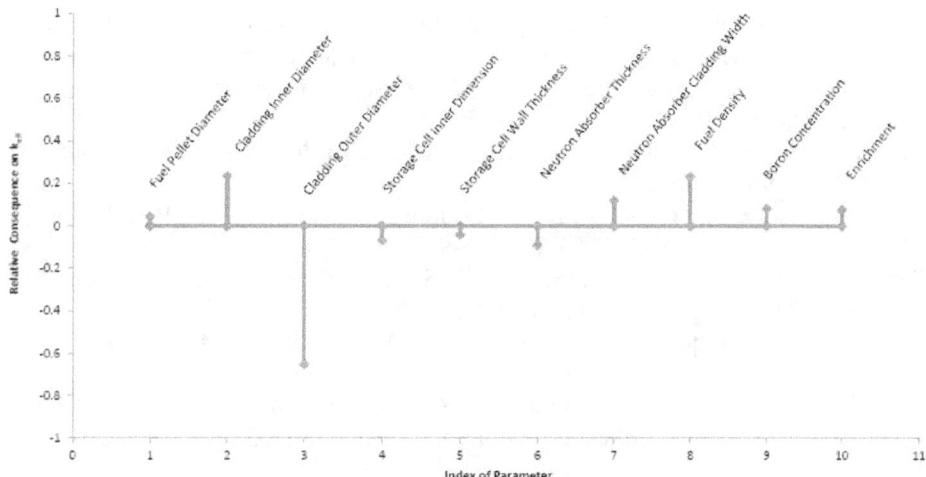

Figure 12: Sensitivity coefficients of the GBC-32 cask analysis. Plotted is the index of the model parameter on the x-axis versus the relative consequence on k_{eff} on the y-axis. The largest impact stems from the cladding diameters.

Probabilistic Safety Assessment and Management PSAM 12, June 2014, Honolulu, Hawaii

4. SUMMARY

Uncertainty analyses based on Monte Carlo sampling methods are state of the art, even in computational challenging applications. With SUnCISTT, GRS has developed an abstract interface tool that is capable to connect statistic software with recognized best-estimate codes. In this article the adoption with the well-established GRS tool SUSA was presented, that offers all statistical ingredients needed for uncertainty and sensitivity analyses. Successful couplings in so called SUnCISTT applications were shown in examples with best-estimate criticality codes from ORNL's SCALE package and GRS' burnup code OREST.

The object oriented approach taken in the development of SUnCISTT with the choice of Python as programming language is the basis for several of its features: the user can interact with the tool during runtime, new applications can be implemented easily by inheritance and it is thus easy to adapt to codes of other fields then the nuclear fuel cycle. With add-ons, taking advantage of the features of the underlying SUnCISTT applications, new analysis opportunities can easily be developed.

We have shown the powerful capabilities of SUnCISTT in individual uncertainty assessments as well as for complex and extensive analyses by presenting selected results from current investigations. With these analyses GRS contributed successfully in several international benchmarks on various topics [8,14,15].

Future developments will concentrate on extending the capabilities of SUnCISTT to a wider range of fields of research, the improvement of the users' experience, and the continuous use of the SUnCISTT applications in challenging analyses.

Acknowledgements

The authors would like to thank V. Hannstein, R. Kilger, and M. Wagner for their contributions during early stages of the SUnCISTT development. The work was funded by the Federal Ministry for the Environment, Nature Conservation, Building, and Nuclear Safety of Germany.

References

[1] M. Kloos, "SUSA Version 3.6, User's Guide and Tutorial", Oct. 2008 (2008).
[2] "SCALE: A Comprehensive Modeling and Simulation Suite for Nuclear Safety Analysis and Design", ORNL/TM-2005/39, Version 6.1.2, Oak Ridge National Laboratory, Oak Ridge, Tennessee, June 2011. Available from Radiation Safety Information Computational Center at Oak Ridge National Laboratory as CCC-785 (2011).
[3] X-5 MonteCarlo Team, "MCNP — A General Monte Carlo N-Particle Transport Code, Version 5", LA-UR-03-1987, April 2003.
[4] U. Hesse and W. Denk, "Orest – A Direct Coupling of Hammer and ORIGEN for the simulation of LWR Fuel" (in german), GRS-63, ISBN 3-923875-12-6, Updated Version of 2008.
[5] V. Teschendorff, H. Austregesilo, G. Lerchl, "Status and plans for development and assessment of the code ATHLET", Proceedings of OECD/CSNI workshop on transient thermal-hydraulic and neutronic codes requirements, Annapolis, USA, Nov. 5-8, NEA/CSNI/R(97)4, pp 112.. See e.g. http://www.grs.de/sites/default/files/fue/ATHLET-Overview.pdf
[6] H.J. Allelein et. al., „Weiterentwicklung der Rechenprogramme COCOSYS und ASTEC, GRS-A-3266, Gesellschaft für Anlagen- und Reaktorsicherheit (GRS)mbH, April 2005. See e.g. http://www.grs.de/sites/default/files/fue/COCOSYS%20Kurzbeschreibung_Shortdescription.pdf
[7] R. Brun, F. Rademakers, "ROOT - An Object Oriented Data Analysis Framework", Proceedings AIHENP'96 Workshop, Lausanne, Sep. 1996, Nucl. Inst. & Meth. in Phys. Res. A 389 (1997) 81-86. See also http://root.cern.ch/.
[8] T. Ivanova, A. Hoefer, "Proposal for a Phase IV UACSA Benchmark, Impact of correlations between different criticality safety benchmark experiments on the estimation of the computational bias of keff", Meeting of the UACSA Expert Group, Paris, France, September 2012 (2012).

[9] "UACSA Expert Group of the Working Party on Nuclear Criticality Safety of the OECD/NEA", http://www.nea.fr/science/wpncs/UACSA/index.html

[10] OECD/NEA Nuclear Science Committee, International Handbook of Evaluated Criticality Safety Benchmark Experiments, NEA/NSC/DOC(95)03 (September 2012).

[11] M. Bock and M. Stuke, "Determination of Correlations Among Benchmark Experiments by Monte Carlo Sampling Techniques", Proceedings of ANS NCSD 2013 Meeting "Criticality Safety in the Modern Era: Raising the Bar." Wilmington, NC, September 2013.

[12] M. Bock and M. Behler, "Impact of Correlated Benchmark Experiments on the Computational Bias in Criticality Safety Assessment", Proceedings of ANS NCSD 2013 Meeting "Criticality Safety in the Modern Era: Raising the Bar." Wilmington, NC, September 2013.

[13] M. Behler, M. Bock, M. Stuke, M. Wagner, „Stochastic Methods for Quantification of Sensitivities and Uncertainties in Criticality Analysis" (in german), GRS-319, ISBN 978-3-939355-98-4, 2014.

[14] J.C.Neuber, "Proposal for an UACSA Benchmark Study on the Reactivity Impacts of Manufacturing Tolerances of Parameters characterizing a fuel assembly configuration", OECD/NEA No. 5435, ISBN 978-92-64-99049-4, 2008.

[15] B.T. Rearden, W.J. Marshall, C.M. Perfetti, J. Miss, and Y. Richet, "Quantifying the Effect of Undersampling Biases in Monte Carlo Reaction Rate Tallies", Benchmark Proposal EGAMCT Phase I, OECD/NEA, Juli 2013.

BWR-club PSA Benchmarking – Bottom LOCA during Outage, Reactor Level Measurement and Dominating Initiating Events

Anders Karlsson[*a], Maria Frisk[b], and Göran Hultqvist[c]
[a] Forsmarks Kraftgrupp AB, Östhammar, Sweden
[b] Risk Pilot AB, Stockholm, Sweden
[c] Havsbrus Consulting, Öregrund, Sweden

Abstract: Benchmarking is an important activity in order to eliminate unjustified differences between PSA models and enable harmonisation. It could also be used in order to understand plant differences. As part of the BWR-club PSA activities benchmarking of bottom LOCA during outage, reactor level measurement and dominating initiating events have been performed. Modelling of bottom LOCA during outage varies between the BWR-club members and work performed within the BWR-club aims at compiling and understanding these differences. When it comes to reactor level measurement modelling varies from a more detailed modelling to more of a "black box" approach. Information has also been collected from the BWR-club members regarding dominating initiating events in their PSA studies. The initiating event frequencies, scope of the PSA studies and risk importance of different initiating events vary between the BWR-club members and work has been performed compiling and understanding these differences. BWR-club reports have been issued for bottom LOCA during outage and reactor level measurement, while the benchmarking of dominating initiating events is yet to be finalised.

Keywords: PSA, Bottom LOCA during Outage, Reactor Level Measurement, Initiating Events

1. INTRODUCTION

This report deals with the benchmarking activities performed within the European network BWR-club. Twelve power plants (representing 17 BWR units) in Europe have filled in a questionnaire regarding bottom LOCA during outage in the PSAs of their boiling water reactors (BWRs). Ten power plants in Europe with 14 BWRs have filled out an extensive form considering reactor level measurements in PSA models. Work is on-going regarding dominating initiating events. The activities were presented at an earlier stage at the PSA 2013 conference in Columbia, South Carolina, USA in September 2013 [1].

The aim of this paper is to identify and highlight similarities and differences and to create a discussion that initiates efforts to harmonise and further develop PSA modelling in these three areas.

Anders Karlsson is working at Forsmark Nuclear Power Plant in Sweden, where Göran Hultqvist also worked before becoming a consultant. Maria Frisk is working at the consultant company Risk Pilot AB based in Stockholm, Sweden. This group has led this benchmarking project with the active participation of representatives from other BWR-club members.

1.1. BWR-club

BWR-club is a network consisting of European boiling water reactor (BWR) owners. The aim of the network is information and experience exchange between its members. This is performed through a yearly European conference, yearly European technical workshops (including one for PSA), benchmarking activities and the creation of BWR-club reports. BWR-club was initiated in 2010 by associated members leaving the BWROG associated program. BWR-club has developed contacts with Japanese utilities (JBOG) and maintains contacts with US utilities through BWROG.

---

[*] ask@forsmark.vattenfall.se

1.2. The Importance of Benchmarking in PSA

There are several reasons for performing PSA benchmarking. One reason is to understand differences in plant design and modelling. Another even more important reason is to identify how and why similar plant designs differ in modelling and to enable harmonisation. For events with no or little data and which at the same time may have large effect on PSA results, harmonisation is vital in order to gain PSA acceptance.

1.3. BWR-club PSA Benchmarking

This benchmarking project has been performed as a BWR-club PSA Workshop activity. The activities started in 2011 and have been on-going periodically since then. BWR-club reports on bottom LOCA during outage and reactor level measurement have been issued in the beginning of 2014 [2,3]. A BWR-club report on dominating initiating events is yet to be finalised and follow-up questions are to be sent out to the BWR-club members.

Bottom LOCA during outage

Modelling of bottom LOCA during outage varies significantly between the European BWRs. While it dominates the results for cold shutdown in some PSAs, and thereby also is major part of the total risk, it is screened out in other PSAs. Internationally some PSAs do not even consider the cold shutdown state at all. Bottom LOCAs during outage, in particular large ones, are typical examples of events with low frequency but with very serious consequences and weak barriers against core damage.

Is it justified to have such differences? What are the motives behind the differences: different plant designs, different PSA traditions, different requirements from regulators, etcetera?

Reactor level measurement

Reactor level measurement controls many systems of importance for plant safety and differences in modelling could therefore have a major impact on the overall PSA results. Different modelling of the reactor level measurement function could therefore lead to significantly different conclusions regarding the plant safety, even though the function itself often is more or less identical between different plants. This demonstrates the importance of benchmarking in the field of reactor level measurement.

Dominating initiating events

Benchmarking has also been initiated for dominating initiating events. The methodology for estimating initiating event frequencies is a much debated issue is Sweden, both when it comes to difference between the PSAs of different plants (Forsmark NPP, Oskarshamn NPP and Ringhals NPP) and when it comes to differences between probabilistic and deterministic analyses at the same plant (e.g. Forsmark NPP). This BWR-club benchmarking activity is partly connected a research and development project regarding initiating events performed within the Nordic PSA Group (NPSAG) called *Common Methodology for Analysis of Initiating Events* [4,5].

2. BOTTOM LOCA DURING OUTAGE

Bottom LOCA during outage is most probably caused be human error and the frequencies for those scenarios are estimated by using human reliability analysis (HRA). It could also occur as a mechanical failure or break in pipes or tubes which are the barrier for the reactor water. Such breaks may result in smaller break flows compared to the possible maximum break flows from human induced failures, which may lead to large open penetrations into the reactor pressure vessel (RPV).

There are a large number of HRA methods and the HRA results may also vary within an HRA method, even if the analysed manual actions are more of less similar [6,7].

The questionnaire sent out to the BWR-club members regarding modelling of bottom LOCA during outage was focusing on:

1. Main recirculation pumps
   a. External
   b. Internal
2. Control rod drives
3. Core instrumentation probes

Below follows a summary of the answers given to the questionnaire. The questionnaire was subsequently iterated in order to obtain as comparable answers as possible.

2.1. HRA Assessment of Maintenance Work

As failure data is scarce or non-existing for human induced failure causing bottom LOCA during outage, the failure data will have to be developed by methods used for assessing human errors.

Work with overhaul or replacement of:

- Valves in recirculation loop for external pump reactors
- Recirculation pumps in internal pump reactors
- Drive control rod drives
- Core instrumentation probes

is performed according to maintenance procedures. These procedures include critical steps that have to be performed correctly not to cause leakages of water out from the RPV. All critical steps include possible ways to stop minor leakages by going back in the procedures or by performing complementary actions. The assessment of risk for getting large openings in the RPV from such maintenance work include the risk for making mistakes and also the risk of not performing the countermeasures listed in the procedures. This means that if one mistake is leading to a small leakage and the countermeasures are successfully performed there is no bottom LOCA event. The initiating event frequencies for these bottom LOCA scenarios include the probability to get to a condition causing large leakages that cannot be stopped by following the maintenance procedures. The event trees do not include any recover actions to stop the flow as there are no such actions described in the manuals. The event trees include the responses from other plant functions and systems.

For units with internal main recirculation pumps there are technical solutions for avoiding leakages during dismounting of the pump e.g. sealing plugs (from above) and tight flanges (from below).

2.2. Main Recirculation Pumps

The function of main recirculation pumps is to cool the reactor core by recirculating the water inside the RPV and thereby maintaining an acceptable margin against dry out. Within certain limits the main recirculation pumps also controls the power of the reactor.

During outage maintenance work is performed on the main recirculation pumps, either with fuel still placed in the RPV or with the fuel removed from the RPV. Incorrect dismantling of main recirculation pumps may lead to very large bottom LOCAs. The main recirculation pumps may either be external, which means that they are connected to the RPV by piping, or internal, which means that they are attached directly to the RPV.

External pump reactors

The answers for units with external pumps are relatively similar regarding assumed flow rates (1895 – 2300 kg/s), but the frequencies vary significantly ($1.20 \cdot 10^{-7} - 1.87 \cdot 10^{-4}$ /year). It could also be noted that in the United States main recirculation pump LOCA during outage is normally not included in the PSAs.

One plant only assesses the risk for mechanical failures with probabilities in the range of 10^{-7} per year. The other plants are assessing the risk for human errors resulting in event frequencies between 10^{-4} down to 10^{-7} per year. The bases for these differences are not described in the responses. It indicates differences in HRA methodology. The break flows are large and the following scenarios will have small margins for avoiding core damage.

The plants have also submitted information on what precautions shall be establish during the period when actual overhaul or replacement are performed in order to avoid leakage from main recirculation pumps or to mitigate the consequences of leakage from main recirculation pumps. For units with external main recirculation pumps unloaded core and isolation valve locked close are common precautions made.

Internal pump reactors

The answers for units with internal pumps differ significantly both for assumed flow rates (circa 300 – 1580 kg/s) and for frequencies ($4.40 \cdot 10^{-8} - 4.52 \cdot 10^{-6}$ /year). It could again be noted that in the United States main recirculation pump LOCA during outage is normally not included in the PSAs.

Some plants have screened out large leakages from the PRV during outage as an event based on the strong administrative barriers. Others have not screened it out and the probabilities are in the level of 10^{-6} to 10^{-7} per year.

The resulting flow rates in the case of human error failure are high and the following scenarios will have small margins for avoiding core damage.

For some of the plants that have screened out large bottom LOCA during outage based on pump dismantling, have instead included bottom LOCA during outage related to mechanical failure (break). It is not clear if others have included these minor breaks in their assessments.

The plants have also submitted information on what precautions shall be establish during the period when overhaul or replacement are performed in order to avoid leakage from main recirculation pumps or to mitigate the consequences of leakage from main recirculation pumps.

For internal main recirculation pump reactors demanding containment airlock(s) to be closed are the precaution normally being done. Some procedures recommend closing the containment airlock(s) only after the failure has occurred. In the latter cases the time available for manual actions in order not to lose too much reactor water outside of the containment is small.

2.3. Control Rod Drives

Control rod drives are typically used for inserting control rods hydraulically during a scram or mechanically with the help of electrical motors (or high pressured water) controlling the power of the reactor. Incorrect dismantling of control rod drives may lead to medium or large bottom LOCAs. Compared to the main recirculation pumps the design of the control rod drives is more similar between the units in the survey, which makes comparisons easier. The answers varied extensively regarding assumed flow rates (7.1 – 210 kg/s) and the frequencies also varied significantly ($1.80 \cdot 10^{-6} - 1.00 \cdot 10^{-2}$ /year). In the United States control rod drive LOCA during outage is normally not included in the PSAs.

Two of the plants have assessed control rod drive LOCAs caused by the mechanical failures that can cause smaller leakages from the vessel in the range of 10 kg/s or lower. These plants have screened out or not assessed the risk for human errors during control rod drive maintenance.

All other plants have assessed human errors during control rod drive maintenance. The frequency varies from 10^{-4} to 10^{-6}. The flow rates are high and seem to be based on similar assessments of the end state of the failure. Different values are depending on plant design. The flow rates are high enough to transfer the plant into a scenario with high demands on safety systems and functions.

The plants have also submitted information on what precautions have been made in order to avoid leakage from control rod drives or to mitigate the consequences of leakage from control rod drives. Closing of containment airlock(s) in case of an event is a precaution normally being done. Other solutions involve technical specifications requiring at least two emergency core cooling trains available and the use of sealing plugs.

2.4. Core Instrumentation Probes

Core instrumentation probes are used for monitoring the reactor core. Incorrect dismantling of core instrumentation probes may lead to significant but much smaller bottom LOCAs than what is assumed for main recirculation pumps and also normally smaller than what is assumed for control rod drives. Like for the control rod drives the designs are relatively similar regarding core instrumentation probes, which make comparisons easier. In the United States core instrumentation probe LOCA during outage is normally not included in the PSAs.

The answers vary somewhat regarding assumed flow rates (5 – 20 kg/s) and frequencies ($3.20 \cdot 10^{-4}$ – $3.40 \cdot 10^{-3}$ /year).

One of the plants has only assessed LOCAs caused by the mechanical failures. The failure rate is very low and this plant has screened out or not assessed the risk for human errors during core instrumentation probe maintenance.

Several plants have screened out the failure scenarios based on the long time needed for getting into a critical core cooling condition depending on the flow rates and the available water above the core during outages.

The other plants have assessed human errors during core instrumentation probe maintenance. The frequency varies from 10^{-3} to 10^{-4} with most in the $3 \cdot 10^{-4}$ to $4 \cdot 10^{-4}$ range. The flow rates are in the range of 10 – 20 kg/s and seem to be based on similar assessments of the end state of the failure, but with different values depending on plant design. The flow rates are such that safety system and functions should be able to handle the scenarios if these functions are working properly. With reduced function of safety system these scenarios will cause core damage.

The risk for core damages will be dependent on the frequencies for getting into the failure state during core instrumentation probe maintenance or if these failure modes are screened out.

The plants have also submitted information on what precautions have been made in order to avoid leakage from core instrumentation probes or to mitigate the consequences of leakage from core instrumentation probes. Closing of containment airlock(s) in case of an event is a precaution normally being made. Other solutions involve technical specifications requiring at least two emergency core cooling trains available.

2.5. Discussion

The lessons learnt from the responses are that the input data for bottom LOCA events during outage are developed based on quite different ways of assessing the risk for different failure modes. Some plants even screen out human induced failures for these category of event. It is unclear whether this is done based on detailed failure mode assessments or on an "engineering judgement" base. The plants that screen out human induced failures instead assess failures based on mechanical failures.

Plants that perform assessment of human induced failures seem to get data with large uncertainties. It is however not acceptable to end up with such large variation in the results. There is thus a need to develop a common methodology among BWR owners on estimating the frequencies for leakage from the RPV during outages.

Another issue to be addressed as a result of this benchmarking is what kinds of water flows that should be considered. Should it be the worst case scenario, the most probable or both? To obtain a deeper understanding it is also important to reflect on PSA and HRA differences regarding the estimation of frequencies. The HRA methodology is also of importance. Other forms of bottom LOCAs during outage are another area to be addressed. This could for example be LOCAs caused dropping heavy equipment or LOCAs affecting spent fuel pools.

For more detailed information see [2].

3. REACTOR LEVEL MEASUREMENT

Reactor level measurement in BWR-reactors is performed by measuring the weight of the water by assessing the pressure difference between the reference chamber and a position that is under water. See Figure 1.

During normal operation with steady state condition this measurement can with high accuracy (in the range of centimetres) measure the water level in the RPV. By assuring that the instrument pipes, sensing lines, connecting the pressure transmitter with the reactor vessel and the reference chamber is filled with degased water and that the difference pressure transmitters are calibrated the operators can rely on the output from these level measurements and the logics steered by the measurements. As the difference pressure is depending on the temperature of the water and on the temperature outside the instrument pipes most plants are equipped with measurement of these temperatures. The measured differential pressure is compensated or corrected by the outputs from these temperature measurements. If this compensation does not work or do not exist the output will not present the actual level correct when these parameter are changed. This happens when the pressure is dropped in the reactor vessel or a pipe-break occurs in the containment.

When the temperature in containment increases (or decrease) this will also affect the weight of the water in the sensing lines and the differential pressure changes. With very high temperature in containment it will also be possible to boil the water in the sensing lines.

Some plants have installed functions to cool the sensing line during scenarios when there is risk for high temperature in containment. Some plants have installed their reference chamber in such a way that degasing of the steam entering into the reference chamber is not transferred back to the reactor vessel. This can lead to an accumulation of non-condensable gases in the reference chamber and also into the water in the instrument lines. With gas in the water the weight of the water changes and effects the measurements.

During a rapid depressurisation the gas in the water will also result in flashing out water from the instrument lines. For plant with reference chamber lacking natural degasing it is of importance handle degasing by other means.

As parts of the reactor level measurement system is connected to the RPV and installed in the containment changes of parameters are transferred to all trains of the systems. If some of these changes introduce failures into one train it is very likely that this will also occur in the other trains and even in parallel measuring systems measuring on other levels.

The output from a level transmitter and its amplifier is assessed by limit switches to initiate safety systems if water level passes certain levels. The operators get information by the instrument in the control room. If the instrument fails it will not affect the limit switch function.

A reactor level measurement system needs power to its transmitters, amplifiers and compensators. The transmitters and amplifiers are at least calibrated once per year after the outage as well as the limit switches. Failure of the reactor level measurement system could result in too early or too late actuation of safety system. This will also happen in case of loss of power to the different sub-components. The worst case will be cases when a transmitter or amplifier is blocked on a certain level and this happens in several trains simultaneously.

To reduce the risk for getting the level signal to freeze in a specific level in many trains at the same time some plants have installed automatic system that continuously (within some seconds) compare the outputs from measurements in different trains. If the outputs deviate in one train from the other the operators are informed and corrective actions can be initiated. This function reduces the risk for common cause failure (CCF) in the reactor level measurement system.

3.1. Analysis

In general the European power plants model the reactor level measurement with high degree of detail. In the United States on the other hand reactor level measurement is modelled as a "black box" with only a few parameters as input. Europe and the United States seem to have different opinions about how reactor level measurement should be modelled in PSA studies.

Beneath the PSA modelling in Europe is assessed by looking into the different parts of the modelling and its data.

Failure modes

When it comes to failure modes the PSAs of the BWR-club members have many similarities, the following failure modes were modelled for a majority of the plants:

- Loss of function of dP (differential pressure) sensor
- No movement in limit switch
- Spurious movements in limit switch
- Power supply for components (modelled as fault tree or basic event)
- Incorrect calibration of dP sensor (HRA)
- Loss of function input relay in safety logics

Figure 1 illustrates a typical reactor level measurement function of a BWR.

Figure 1: Reactor Level Measurement Function

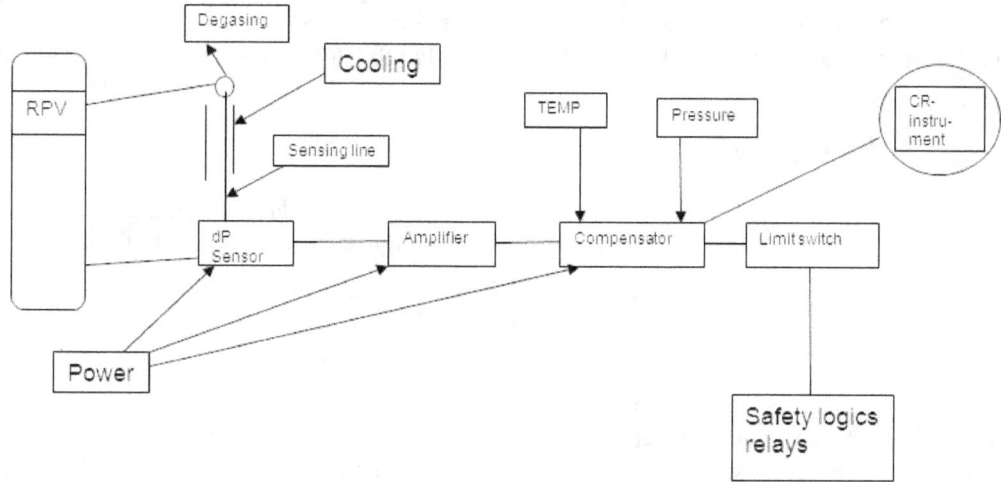

The reactor level measurement function is relatively well covered in the analysed PSAs. There are however a few failure modes that are generally not modelled.

- Loss of function sensing line[†]
- Loss of function amplifier
- Loss of function compensator

There are clearly more similarities than differences regarding reactor level measurement failure modes. The level of detail in modelling of power supply for components varies however.

Some differences can of course be explained by differences in design. But in case of very similar power plants the PSA-studies should be comparable and have similar parameters as input.

It is assumed that most plants calibrate their dP sensors during outage one time per year. The responses however indicate that at least one plant does this every three months.

At one plant the reactor level measurement system is viewed as a monitored system. This effect the failure rate of:

- Loss of function dP sensor
- Spurious movements in limit switch

This view is based on the fact that the system is used in the daily operation of the plant and failures will be detected by the operators without doing any tests.

Repair time for any failure on those components is set to eight hours in the modelling of this plant. This also affects the view of modelling CCF, where this plant does not include CCF on these components.

At another plant the reactor level measurement system is complemented with an automatic function that continuously compare the output of from all level transmitters in all trains. Alarm is given if any transmitter has an output that has a minor deviation from the others. This has been used in the modelling to indicate a test frequency of one per week, even though the comparisons are made every second. This affects the modelling of the dP sensors. These are important differences in the modelling assumptions.

[†] If modelled it is the cooling of the sensing line that is considered.

The responses indicate that the modelling assume that when there is a failure the output will change to a maximum or minimum output. Such a failure mode will in most plant result in actuation of safety logics. No plant have modelled that there should be a failure in the dP sensor that freeze the output to a specific value and that any changes in the water level are not recorded and no safety logics are activated.

Common cause failure (CCF) groups

The answers to the questionnaire show that CCF groups are modelled relatively similar among the BWR-club members, even though there are also differences, mainly regarding the level of ambition to include CCF in the modelling of reactor level measurement.

The following CCF groups are the most common:

- No movement of limit switch – modelled at seven plants
- No signal from dP sensor (loss of function) – modelled at six plants
- Incorrect calibration of dP sensor (HRA) – modelled at three plants

As indicated above one plant does not model any CCF regarding reactor level measurement, as the system is monitored during normal operation.

Plant design regarding number of trains and the diversification of trains varies between the plants. Often the reactor level measurement is divided into coarse and narrow range.

Should all trains (coarse range as well as narrow range) be seen as one large CCF group or should coarse range and narrow range be separate CCF groups? The responses indicate that all members have separate CCF groups. Work is however on-going at one plant to assess a large CCF covering all measuring ranges.

Data and modelling

When it comes to data and modelling there are differences. Parameter values vary since different parameter sources are used. In the Nordic countries the T-book [8] is used, while other failure data sources are used in other countries. The test intervals also vary between the different plants.

The different data sources give a difference in failure data in the level of one decade, see Table 1. For these failure modes with almost similar components the failure rates should be within one third of a decade to be able to develop PSAs with low uncertainties. This implies that a work needs to be done regarding harmonisation of failure data.

Table 1: Reactor Level Measurement Failure Rates

Failure mode	Highest failure rate	Lowest failure rate
Loss of function sensing line	$2.4 \cdot 10^{-6}$	$6.00 \cdot 10^{-7}$
Loss of function dP sensor	$1.00 \cdot 10^{-6}$	$1.21 \cdot 10^{-7}$
No movement in limit switch	$1.94 \cdot 10^{-7}$	$4.91 \cdot 10^{-8}$
Spurious movements of limit switch	$3.80 \cdot 10^{-7}$	$4.6 \cdot 10^{-7}$
Loss of function input relay in safety logics	$1.76 \cdot 10^{-8}$	$8.40 \cdot 10^{-8}$

The test intervals are relevant for the following data is presented in Table 2.

Table 2: Reactor Level Measurement Test Intervals

Test	Typical test interval	Variation
Function dP sensor	1 year	1 month for one plant
Movement in limit switch	3 months	Between once a 1 month to 1 year
Function input relay in safety logics	3 months	No variation between the plants

Modelling of power supply for components of the reactor level measurement function varies. It is mainly a difference on level of detail. Power supply for level measurement components is often modelled in detail with the use of failure trees, considering several failure modes and considering CCF. The power supply could however also be modelled as a "black box" using only a few basic events. The power used for components in the reactor level measurement function is all DC power systems with battery back-up. Events with failures of the DC power systems will affect different parts of the measuring system. With a failure tree modelling it will be possible to catch the connections between loss of power and loss of measurement. The importance of this cannot be evaluated as no data have been delivered. When power fails to power level measurement functions it will give alarm or safety actuations therefore the safety risk will be low.

Total failure rates of a train

In table 3 data of unavailability of a reactor level measurement system is presented.

Table 3: Reactor Level Measurement – Total Failure Rates of a Train

Plant	Description	Result
1	One train	$4.38 \cdot 10^{-2}$
2	RPS, one train	$1.4 \cdot 10^{-5} - 8.003 \cdot 10^{-4}$
3	RPS, one train	$5.091 \cdot 10^{-4}$
4	RPS, one train	$7.6 \cdot 10^{-4}$

Failure of the reactor level measurement function and failure to actuate safety logics is depending on the ways the reactor level measurement is built into different functions. The data above indicate different outputs from existing studies.

A spread of unavailability in the range of 10^{-2} to 10^{-5}, with a lot of data at 10^{-5}, indicates a too large uncertainty of data for such an important function in a BWR plant.

Perhaps more importantly the modelling of power supply for components of the reactor level measurement function varies. It is mainly a difference regarding level of detail. Power supply for level measurement components is often modelled in detail with the use of fault trees, considering several failure modes and considering CCF. The power supply could however also be modelled as a "black box" using only a few basic events.

3.2. Discussion

Failure modes and the use of CCF are relatively similar between the plants, while parameter data vary to a larger extent, which at least to some extent is inevitable since different parameter sources are used. Harmonisation in the field of failure data is outside the scope of this paper. Such work has however been performed [9] but focused on harmonisation of methodologies not harmonisation of data.

The most important finding in this investigating of reactor level measurement are the differences in modelling of power supply for components of the reactor level measurement function varies. Should it be modelled in detail (as most plants do) or is "black box" modelling also acceptable?

For more detailed information see [3].

4. DOMINATING INITIATING EVENTS

Benchmarking regarding dominating initiating events is on-going. A questionnaire was sent out and representatives from the BWR-club members submitted information from their PSAs. This information has then been compiled. Some follow-up questions will however be sent out during spring 2014.

This benchmarking deals with power operation as well as shutdown and the plants have been asked to give information about their top 20 events with the highest initiating event frequency (all events as well as external events excluded) as well as the initiating events that are part of the 20 most dominating core damage frequency cut sets in their PSAs (all events as well as external events excluded).

One early conclusion is that the scope and the impact of external events vary significantly between the plants. Seismic events might for example dominate in some PSAs, even in areas with low seismicity, while it is screened out in other PSAs. The importance of common cause initiators (CCIs) also vary significantly. It is clear that there are different plant designs, but also different "PSA traditions" and different requirements from regulators, all leading up to different risk profiles in the PSAs.

The work will continue during 2014 and lead to a BWR-club report.

5. CONCLUSIONS

Different methodologies may lead to unjustified differences affecting safety prioritisations at the plants, even if the plant designs are similar. One obvious conclusion of this benchmarking activity is thus that methodologies, regarding for example failure modes, should be harmonised when plant designs are similar.

The lessons learnt from the responses are that the input data for bottom LOCA events during outage are developed based on quite different ways of assessing the risk for different failure modes. Some plants even screen out human induced failures for this category of event. It is unclear whether this is done based on detailed failure mode assessments or on an "engineering judgement" base. The plants that screen out human induced failures instead assess failures based on mechanical failures.

Plants that perform assessment of human induced failures seem to get data with large uncertainties. It is however not acceptable to end up with such large variation in the results. There is thus a need to develop a common methodology among BWR owners on estimating the frequencies for leakage from the RPV during outages.

Another interesting outcome of the benchmarking activities is what kinds of water flows that should be considered for bottom LOCAs during outage. Should it be the worst case scenario, the most probable or both? To obtain a deeper understanding it is also important to reflect on PSA and HRA differences regarding the estimation of frequencies. The HRA methodology is also of importance.

For reactor level measurement failure modes and the use of CCF are relatively similar between the plants, while parameter data vary to a larger extent, which at least to some extent is inevitable since different parameter sources are used.

An important finding is the differences in modelling of power supply for components of the reactor level measurement function. It is mainly a difference regarding level of detail. Power supply for level measurement components is often modelled in detail with the use of fault trees, considering several failure modes and considering CCF. The power supply could however also be modelled as a "black box" using only a few basic events. Should it be modelled in detail or is "black box" modelling also acceptable?

It is in general of interest to investigate on which bases events are screened out in some PSAs, which they dominate the results in other PSAs. Seismic events and other external events as well as bottom LOCA during outage are examples of that. It illustrates the problem of analysing events which have low frequencies and high degrees of uncertainty, which at the same time have very serious consequences and weak barriers against core damage.

It is clear that different plant designs, but also different "PSA traditions" and different requirements from regulators, lead up to different risk profiles in the PSAs. There is in other words a large need for harmonisation. To obtain harmonisation is however a long process and work must be performed in close cooperation with more resources than what is possible in this benchmarking activity.

References

[1] A. Karlsson, G. Hultqvist and M. Frisk. "*BWR-club PSA Benchmarking - Bottom LOCA during Outage, Reactor Level Measurement and Dominating Initiating Events*", PSA 2013, (2013).

[2] M. Frisk, G. Hultqvist and A. Karlsson. "*BWR-club PSA Benchmarking Regarding Failure Data for Bottom LOCA Scenarios during Outage*", BWR-club, (2014).

[3] M. Frisk, G. Hultqvist and A. Karlsson. "*BWR-club PSA Benchmarking Regarding Reactor Level Measurement – Modelling and Data*", BWR-club, (2014).

[4] G. Johanson and M. Håkansson. "*Kick Off: Framtagning av harmoniserade metoder för skattning av inledande händelser*", proceeding of project meeting (2013).

[5] G. Johanson and M. Håkansson. "*Avstämning 1: Framtagning av harmoniserade metoder för skattning av inledande händelser*", proceeding of project meeting (2013).

[6] A. Bye et al. " *International HRA Empirical Study – Phase 2 Report*", NUREG/IA-0216, Volume 2 (2011).

[7] G. Johanson et al. "*Evaluation of Existing Applications and Guidance on Methods for HRA - EXAM-HRA Phase 2 Summary Report*", ES-konsult report 2011001:015 1.0 (2012).

[8] TUD Office. "*T-book – Reliability Data of Components in Nordic Nuclear Power Plants*", 7th edition. TUD Office, 2010, Stockholm.

[9] V. Hedtjärn Swaling. "*Function Oriented Classification of Components – Pilot Study – Comparison of the German and Nordic Classification Systems*", Scandpower report 210266-R-001 (2011).

Effects of an Advanced Reactor's Design, Use of Automation, and Mission on Human Operators

Jeffrey C. Joe[a*] and Johanna H. Oxstrand[a]
[a] Idaho National Laboratory, Idaho Falls, USA

Abstract: The roles, functions, and tasks of the human operator in existing light water nuclear power plants (NPPs) are based on sound nuclear and human factors engineering (HFE) principles, are well defined by the plant's conduct of operations, and have been validated by years of operating experience. However, advanced NPPs whose engineering designs differ from existing light-water reactors (LWRs) will impose changes on the roles, functions, and tasks of the human operators. The plans to increase the use of automation, reduce staffing levels, and add to the mission of these advanced NPPs will also affect the operator's roles, functions, and tasks. We assert that these factors, which do not appear to have received a lot of attention by the design engineers of advanced NPPs relative to the attention given to conceptual design of these reactors, can have significant risk implications for the operators and overall plant safety if not mitigated appropriately. This paper presents a high-level analysis of a specific advanced NPP and how its engineered design, its plan to use greater levels of automation, and its expanded mission have risk significant implications on operator performance and overall plant safety.

Keywords: Human Factor Engineering, Advanced NPPs, Automation, Human Reliability Analysis

1. INTRODUCTION

Since the construction of the first commercial nuclear power plants (NPPs) in the United States (U.S.), there has been considerable variability in the amount of interest and momentum behind the development of new NPPs. Economic and political factors (e.g., the fluctuating cost of producing energy from fossil fuels, and the political stalemate over Yucca Mountain), as well as high profile events (e.g., Deepwater Horizon and Fukushima Daiichi) and the debate over climate change appear to cause interest in nuclear power to wax and wane over time. Despite these externalities, a number of entities have worked ardently to develop new and advanced NPPs, including small modular liquid-metal cooled reactors, and have prepared their designs for licensing review by the U.S. Nuclear Regulatory Commission (NRC).

General Electric-Hitachi is one such entity developing an advanced sodium cooled small modular reactor (SMR), called the Power Reactor Inherently Safe Module reactor (PRISM) [1]. Other entities have also been pursuing their own advanced NPP designs [2, 3, 4], but as of the writing of this paper, PRISM is, to our knowledge, the only sodium cooled SMR design that has both human factors engineering (HFE) and risk related information (e.g., probabilistic risk assessment) publicly available in an unredacted form, thereby allowing us to perform this analysis. Specifically, General Electric (GE), prior to partnering with Hitachi, submitted a preliminary safety information document [5] to the NRC as part of its license application. The NRC reviewed GE's submission [6], and notably included the following comment, "On the basis of the review performed, the staff, with the ACRS [Advisory Committee on Reactor Safeguards] in agreement, concludes that no obvious impediments to licensing the PRISM design have been identified." (pg. xxiv) However, [6] also specified that the NRC reviewed only a conceptual design of PRISM, and that their review, "did not, nor was it intended to, result in an approval of the design". (pg. C-1). Given the conceptual state of design for this reactor, and other advanced reactors, we believe there are a number of HFE issues (as well as other design related questions), which have not been considered in sufficient detail from a risk analysis perspective.

---

[*] Corresponding Author: Jeffrey.Joe@inl.gov

That is, advanced NPPs whose engineering designs differ considerably from existing light-water reactors (LWRs) will affect the conduct of operations and operator performance in ways that may not be adequately addressed without an explicit HFE analysis. Changes to the engineered design of the reactor, however, are only one of a number of factors that can affect human performance in advanced NPPs. Some advanced NPPs are designed to operate with more automation in their digital instrumentation and control (I&C) system, and a reduced operating staff in one central control room operating multiple units. More automation will make the operator's role more supervisory in nature, but controlling multiple modules or units may introduce additional workload challenges. Similarly, some advanced reactors are touted as being capable of producing other commodities in addition to electricity in an economically competitive manner. These changes to the mission of the advanced NPP will also likely affect the operator's performance. In short, the role the operator(s) play (e.g., their function and the tasks they perform) are slated to change, given changes to the engineered design, planned use of automation, and change in plant mission for advanced NPPs. In our view, it is unlikely that suitable solutions to mitigate the risk significance of these issues will be found easily, such as by adopting best practices from operational experience of existing LWRs. Rather, a new HFE technical basis for the human performance requirements for advanced NPP operators will need to be established.

Other researchers, in particular O'Hara, Higgins, and Pena [7], studying human performance issues related to the design and operation of SMRs have made this point previously in their study of multiple classes of advanced NPPs. In this paper, however, we are explicating some risk implications for a specific advanced reactor design. We will show how specific changes in reactor design, such as the core design and type of coolant used, and other aspects (e.g., greater use of automation and change in plant mission) change the safety and risk impacts on the operator. By doing this, we will explicate what some of the human performance issues operators will face given a specific design.

This paper presents a high-level analysis of the PRISM SMR, and how its engineered design, its anticipated use of greater levels of automation, and its expanded mission have significant risk implications on the human operator's roles, functions, and tasks. From this analysis we further conclude that formal HFE approaches, such as Functional Requirements Analysis (FRA) and cognitive work analysis (CWA), as well as human reliability analysis (HRA), are helpful in considering how to mitigate the risk impacts of these changes on human performance, particularly because these approaches are helpful in understanding the complex interactions of these factors.

2. DIFFERENCES BETWEEN PWRs AND PRISM

2.1. Design

PRISM is an example of an advanced NPP in that its engineered design differs from LWRs in a number of important ways, including but not limited to, the type of coolant and fuel it uses, and its core design. PRISM is a sodium cooled, metallic fuel, pool-type, fast breeder reactor. By way of comparison, pressurized light water reactors (PWRs) in the U.S. are water-cooled, oxide fuel, loop-type, thermal reactors. Perhaps the most important difference is the fact that the design of PRISM's reactor core and the use of metallic fuel are designed not to require operators or automation to intervene in order to shut down safely given certain initiating events (i.e., passive safety features). Upon reactor trip, a PWR operator must be actively involved in performing certain actions (e.g., performing decay heat removal actions) that a PRISM operator would not. Clearly, these differences have significant implications on the operator and their role (e.g., what functions and tasks operators perform). There are also a number of important differences, and associated advantages and disadvantages, in using sodium versus water as the primary coolant. Some of these differences are highlighted in Table 1, which was adapted from Bays, Piet, Soelberg, Lineberry, and Dixon [8].

Table 1: Differences in Coolant Characteristics Between PWRs and PRISM

Coolant Characteristic	PWRs	PRISM
Stability *(Single-phase can be easier to control)*	Two-phase fluid *(Coolant can be in liquid or gas state)*	Single-phase fluid *(Coolant will only be in liquid state during normal operations)*
Pressure *(Lower pressure has safety benefits)*	15 MPa	0.1 MPa
Chemical Inertness	Moderately	Not inert *(Reacts with air and water)*

Sodium has the benefits of remaining in a single phase during normal operations, and does not need to be under high pressure to serve its purpose of heat transfer/cooling. Additionally, sodium manages the core's fissile inventory more effectively than water in that there is very little excess reactivity available for power excursions to occur. Sodium also has a high heat capacity that makes high-power-density cores feasible and facilitates a very slow thermal response [8, 9]. However, since the coolant sodium is chemically reactive, one key safety feature that PRISM has that PWRs do not need is double-walled steam generator piping where the secondary sodium-potassium exchanges heat with water. Overall, the coolant characteristics of sodium means that the operator does not have to be concerned with a number of issues that they would with water as the coolant (e.g., steam voiding), but would need to be concerned about sodium's chemical reactivity.

While there are differences in reactor core design, type of fuel used, and coolant characteristics, both PWRs and PRISM are NPPs that at a fundamental level operate on the same underlying thermohydraulic processes. As such, there are a number of similarities with respect to major systems, structures, and components (SSCs), such as reactor vessel, containment structure, reactivity control, reactor cooling system, primary heat transport system, steam generators, and balance of plant systems. All of these SSCs are central to both (1) the generation of electricity or other commodities, and (2) the protection of people, workers and the environment. However, there are a few obvious differences in SSCs between PWRs and PRISM. For example, PWRs have a pressurizer and PRISM does not, due to the fact that the sodium coolant does not need to be under high pressure. PWRs have a chemical volume control system that introduces boron into the reactor coolant system. The chemical volume control system is also the main source of water for the reactor coolant system. Conversely, PRISM has a coolant purification chemistry system that removes impurities from the primary sodium coolant. Additionally, according to Hylko [10], PWRs have an emergency core cooling system that PRISM does not need given that its coolant is in a large pool and is not under pressure. For example, PWRs have emergency diesel generators to help remove decay heat from the reactor core when on-site or off-site power is lost, while PRISM has a standby power supply system used to provide power to help with the orderly and controlled shutdown of systems in order to avoid equipment damage, and not for removal decay heat under abnormal conditions [1, 5]. PRISM has both a primary and secondary sodium system (i.e., intermediate heat transport system) as a means to isolate the radioactive primary sodium from the tertiary water/steam based balance of plant, whereas PWRs have only a primary reactor coolant system, and secondary water-based balance of plant. Overall, though we recognize the specific effects will depend on the details of the final as-built design of PRISM, all of these general differences in SSCs can change the operator's roles, functions, and tasks, and these changes further have potential risk important impacts on the overall plant.

Figure 1 provides high-level conceptualizations of PWRs and PRISM displaying some key similarities and differences between the designs of these two reactor types as a function of (1) the generation of electricity or other commodities, and (2) the mitigation of hazards to people, workers and the environment. Again, though the specific risk impacts on human performance will depend on the as-built design of PRISM, we assert that these are some examples of fundamental design changes that will likely have some risk relevant effects on operator performance, and as a result, overall plant safety.

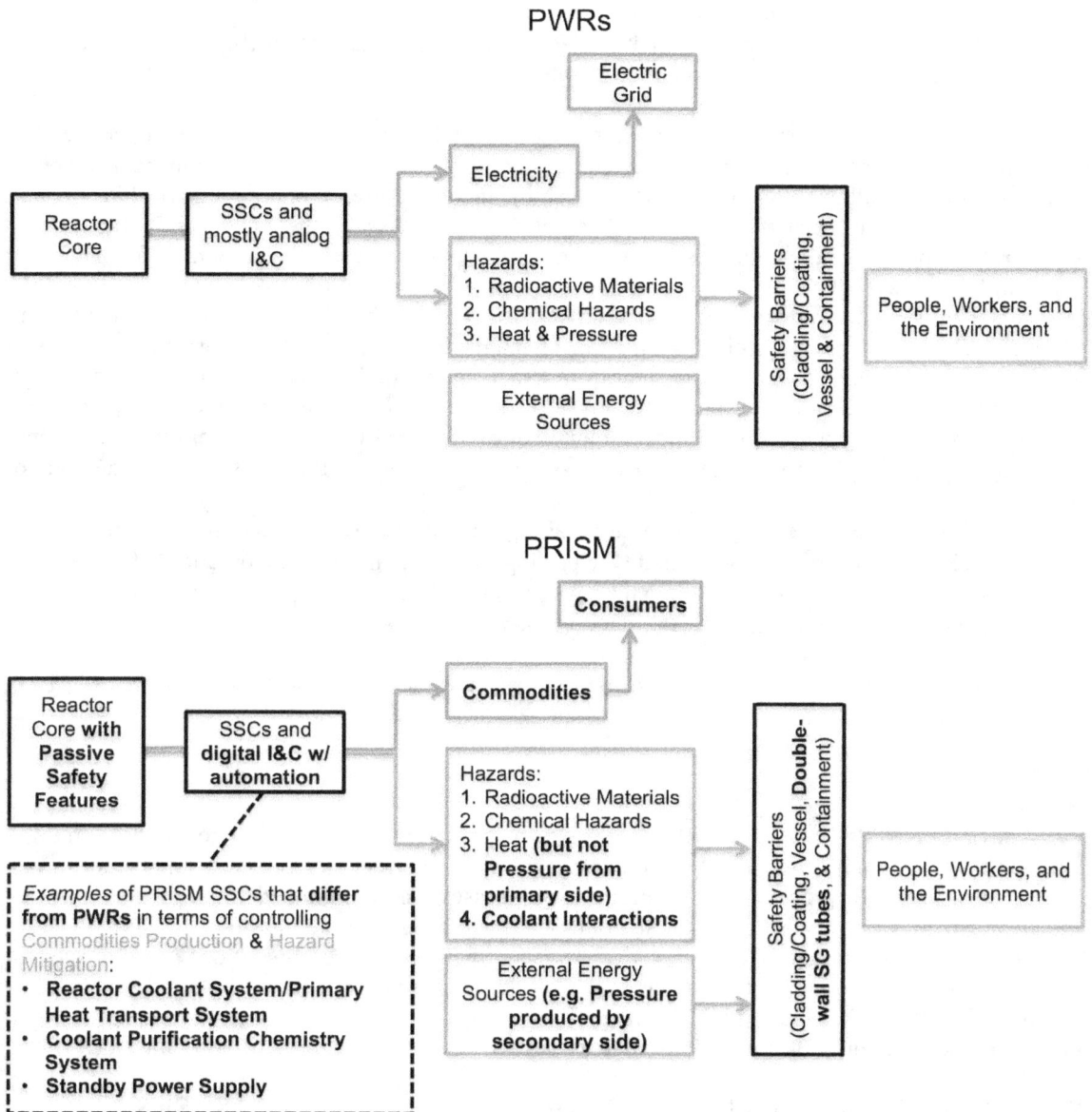

Figure 1. High-level Conceptualizations Showing Differences Between PWRs and PRISM.

2.2. Use of Automation

Many have noted, including the International Atomic Energy Agency [11], that the digital I&C system in advanced NPPs will use greater levels of automation to operate these plants as one of many different means to reduce human operator staffing levels. This is driven in large part by the need for advance NPPs to be economically competitive with existing LWRs and other sources of electrical power generation. In the specific case of PRISM using greater levels of automation, [6] states, "Normal reactor operations are conducted using the plant control system (PCS). The PCS contains a high level of automation for plant control, protection of plant investment, and data handling transmission." (pg. 7-1). Furthermore, according to PRISM documentation the primary role of the operator is to:

- Monitor and verify performance of safety systems, though operators will have the capability to initiate reactor shutdown by manual scram or manual activation of the ultimate shutdown system
- Maintain communication with appropriate onsite and offsite personnel

- Initiate recovery actions following an event
- Serve as an important source of knowledge concerning plant status, design, and behavior, especially during the management of off-normal conditions.

The role of the PRISM operator as described is a significant departure from the role the operator plays in existing LWRs. That is, the U.S. nuclear energy industry has historically had humans as a central component of controlling LWR plant operations, so the increased use of automation will likely have a number of significant impacts on the roles, functions, and tasks of the operator. Furthermore, a considerable amount of human factors research has found that increasing the amount of automation in complex human sociotechnical systems (e.g., aviation) and operations (e.g., military) can have detrimental impacts on human performance (for a review, see Sheridan [12] and Lee [13]). Some relatively recent high profile accidents involving human factors issues with automation include the 2009 crash of Air France 447 on its journey from Brazil to France [14], and the 2009 Metrorail (i.e., commuter train) accident in Washington, DC [15]. The common theme underlying both of these accidents is the finding that the design of the automation hindered the human's ability to collaborate with the automation effectively. Examples of some deficiencies in the design of the automation include:
- Not adequately supporting the operator's ability to maintain awareness of the system's state
- Increasing the operator's workload when trying to recover from the automation's failure
- Contributing to a loss of the operator's abilities to skilfully perform manual control tasks that they had to perform when the automation failed.

These impacts on advance NPP operators, explained in more detail in Section 4 below, need to be evaluated for their risk significance.

2.3. Mission

Many advanced NPPs, including PRISM, have been promoted as being able to produce multiple commodities safely and economically, such as electricity, oxide fuels for PWRs from weapons grade nuclear materials [10], process heat for industrial processes, and desalinized water (Ingersoll, [16]). The change from a single mission (e.g., electricity generation) to multiple missions will be a departure from the conduct of operations at LWRs. Coupled with the need to keep operations and management costs economical, this change in mission will have implications for what is expected of operators, including specific HFE issues such as operator training and workload. The risk impact of these changes to the operator's roles, functions, and tasks on overall plant risk need to be evaluated.

3. CHANGES TO THE OPERATOR'S ROLES, FUNCTIONS, AND TASKS

Section 2 showed that the advanced NPP PRISM is designed differently from PWRs, that it is intended to be operated primarily by automation, and is designed to accomplish multiple missions. With these differences from existing LWRs, there are a number of specific changes to the operator's roles, functions, and tasks that are readily apparent, and are summarized here. In reviewing PRISM documentation [1, 5, 6, 10] some examples of additional tasks for PRISM operators, relative to PWR operators, include:

- Supervising the safe production of both electricity and other commodities, such as oxide fuel for PWRs [10]
- Monitoring for sodium coolant interactions with air and water.

According to [6], if a sodium coolant interaction occurs, it will require the operator to monitor for pressure incursions on the intermediate heat transport system from the tertiary side, ensure actuation of isolation valves, and verify the integrity of double-wall piping.

Additionally, according to [1, 5], some examples of tasks PRISM operators will not have to perform that PWR operators do include:

- Monitoring for leaks in the reactor coolant system
- Performing reactor shutdown actions
- Performing decay heat removal actions
- Performing post-accident containment cooling actions
- Managing boron concentration levels in primary reactor coolant system (e.g., primary heat transport system) during normal operations.

Thus, according to the published PRISM documentation, the operator will have virtually no role, functions, or tasks associated with control of normal operations. Their role will be primarily to monitor and verify (i.e., supervise) the performance of safety systems [5]. However, the PRISM operator will have some additional tasks upon failure of the PCS, depending on the severity of the failure of this automated control system. But none of these additional tasks deal directly with maintaining core cooling, or controlling what are considered to be the primary safety functions that exist in LWRs. Post-accident, operators will be responsible for monitoring variables associated with reactivity control, core cooling, and reactor vessel integrity. Clearly, PRISM's engineered design, extensive use of automation, and change in the plant's mission are significant departures from the conduct of operations in PWRs, change the performance parameters of operators, and will likely have risk related implications for HFE issues, including but not limited to, staffing levels and operator workload [6].

4. RISK SIGNIFICANT IMPACTS

Given the differences and complexity of the design of PRISM, the planned use of higher levels of automation, and the proposed changes in the plant's mission, there are a myriad of ways in which these factors can have risk significant impacts on the operator and the plant. However, since we presented only a high level analysis of these factors, we do not provide a comprehensive or exhaustive analysis of the risk significant ways in which these factors affect operator performance. Instead, this paper highlights what we considered to be the most obvious impacts (both positive and negative) to risk.

With respect to the design differences between PRISM and PWRs, the most significant change for human operators in our view is the passive safety design of the reactor core and use of metallic fuel. There are numerous accident scenarios for PWRs that require the operator to actively intervene to mitigate the consequences of those accidents. However, for a number of important accident scenarios, the PRISM operators do not have to perform any actions related to reactor shutdown, decay heat removal, or post-accident containment cooling. As [17] states, "In the event of a worst-case-scenario accident, the metallic core expands as the temperature rises, and its density decreases slowing the fission reaction. The reactor simply shuts itself down. PRISM's very conductive metal fuel and metal coolant then readily dissipates excess heat without damaging any of its components." In short, this design uses the laws of physics to achieve passive safety, and this drastically changes the roles, functions, and tasks of the human operator and their risk significance.

The fact that PRISM is a pool type cooled design (vs. loop cooled) means that the PRISM operator will not have roles, functions, or tasks associated with monitoring for leaks in the reactor coolant system (e.g., primary heat transport system). Many loss-of-coolant accidents scenarios that can occur in loop type PWRs are no longer feasible given the pool type design. The fact that PRISM uses sodium as a coolant means that the operator and/or automation will not have roles, functions, or tasks associated with monitoring the pressure of the reactor coolant system, or be concerned about accidents initiated by the primary system being under high pressure. The fact that PRISM has a standby power supply system that is not needed to remove decay heat is another example of an SSC that has risk implications for the operator. In this case, the operator's roles, functions, and tasks associated with a loss of onsite power event will likely be related to the orderly shutdown of powered systems and the protection of plant assets, and not the removal of decay heat. The operator will also have roles, functions, and tasks associated with monitoring for sodium interactions with air and water. If not automated, the operator may also have some responsibilities related to the coolant purification chemistry system that somewhat overlap with the PWR operator's responsibilities related to the chemical volume control system.

We hope that it is also obvious that PRISM's highly automated digital I&C system prompts a number of additional questions related to the human factors aspect of human-automation collaboration. As previously mentioned, other researchers have asked many of these questions before [7], and there are numerous examples in other high-risk and complex industries outside of the nuclear domain where poorly designed automation has had a risk significant impact on operations [12, 13, 14, 15]. Nevertheless, highly automated advanced NPPs such as PRISM will need to be designed to address the potential risk significance of issues including, but not limited to:

- Operators having difficulty understanding what the automated control system is doing
- Operators losing important skills that they will need when required to intervene (i.e., loss of proficiency or 'deskilling')
- The extent to which a reduced number of operators, potentially monitoring and controlling more than one reactor, will be able to ensure safe operations
- The extent to which there may be additional workload placed on operators when they must control both the NPP thermohydraulic processes and interact with the automated control system

And finally, given that PRISM is designed to produce more than one commodity may mean that there are some risk significant impacts on operator performance and overall plant safety, such as an increase in workload leading to difficulties in controlling plant processes.

In summary, advanced NPP designs differ from existing LWRs in the U.S. with respect to core design (i.e., passive versus active safety features), the type of coolant used (i.e., gas or liquid metal versus water) and/or type of fuel it uses (i.e., metal or triso/prismatic/pebble versus oxide fuel). It is also anticipated that many advanced NPPs will be highly automated, and have multiple missions. While we have identified many of the risk impacts of these changes on human performance in this paper, the specific risk impacts of all changes in advanced NPPs on the roles, functions and tasks of operators needs to be identified, preferably early in the design life cycle of the advanced NPP. Doing so will help the designers include the human performance and operational requirements in their design, including documentation such as technical specifications and procedures.

Furthermore, HFE approaches such as FRA and CWA, as well as HRA, can and should be used to help determine the potential risk impacts. FRA, function analysis, and function allocation help identify, define, and distribute as appropriate the functions the NPP must have to accomplish its goals of safe and reliable production of commodities. FRA not only helps the advanced NPP designer figure out the operational and human performance requirements of their NPP, FRA is also a part of the NRC's HFE program review model [18]. CWA [19] is a complimentary approach to FRA in that it is a framework that models a complex sociotechnical system based on the various constraints the system's design imposes on how functions are defined and work is accomplished. CWA is very appropriate for the problem defined here in that PWRs and advanced NPPs are complex sociotechnical systems that fundamentally accomplish the same work or mission (e.g., generate commodities, protect people and the environment), but both their systems and functions, and the constraints those systems and function place on human operators, vary significantly. CWA is central to identifying how the constraints of the as designed advanced NPP and the cognitive abilities of the human interact such that the risk-significant impacts of the system design on the human operator can be mitigated. Finally, using HRA to further analyze the importance of human action and inaction, including diagnosis and problem solving activities, after risk significant initiating events and/or hardware failures (e.g., secondary sodium water interaction, failure of the PCS, and toxic gas release from an additional tertiary chemical process using process heat from the reactor) is another important aspect to understanding how changes to the design and mission of advanced NPPs can affect human operator performance, and how successful operator performance and operator errors can contribute to overall plant risk.

Acknowledgements

INL is a multi-program laboratory operated by Battelle Energy Alliance LLC, for the United States Department of Energy under Contract DE-AC07-05ID14517. This work of authorship was prepared as an account of work sponsored by an agency of the United States Government. Neither the United States Government, nor any agency thereof, nor any of their employees makes any warranty, express or implied, or assumes any legal liability or responsibility for the accuracy, completeness, or usefulness of any information, apparatus, product, or process disclosed, or represents that its use would not infringe privately-owned rights. The United States Government retains, and the publisher, by accepting the article for publication, acknowledges that the United States Government retains a nonexclusive, paid-up, irrevocable, world-wide license to publish or reproduce the published form of this manuscript, or allow others to do so, for United States Government purposes. The views and opinions of authors expressed herein do not necessarily state or reflect those of the United States government or any agency thereof. The INL issued document number for this paper is: INL/CON-14-31341. The lead author also thanks Allison C. Joe for her help in preparing this article.

References

[1] B. Triplett, E. Loewen, and B. Dooies, *"PRISM: A Competitive Small Modular Sodium-Cooled Reactor,"* Nuclear technology, 178(2), 186-200, (2012).

[2] Y. Tsuboi, K. Arie, N. Ueda, T. Grenci, and A. Yacout, *"Design of the 4S Reactor,"* Nuclear Technology, 178(2), 201-217, (2012).

[3] A. Shenoy, J. Saurwein, M. Labar, H. Choi, and J. Cosmopoulos, *"Steam Cycle Modular Helium Reactor,"* Nuclear Technology, 178(2), 170-185, (2012).

[4] J. Halfinger and M. Haggerty, *"The B&W mPower™ Scalable, Practical Nuclear Reactor Design,"* Nuclear Technology, 178(2), 164-169, (2012).

[5] General Electric, *"PRISM™: Preliminary Safety Information Document,"* (GEFR-00793, UC-87Ta), 1986, San Jose, CA.

[6] J. Donoghue, J. Donohew, G. Golub, R. Kenneally, P. Moore, S. Sands, E. Throm, and B. Wetzel, *"Preapplication Safety Evaluation Report for the Power Reactor Innovative Small Module (PRISM) Liquid-Metal Reactor: Final Report,"* (NUREG-1368). U.S. Nuclear Regulatory Commission, 1994, Washington, DC.

[7] J. O'Hara, J. Higgins, and M. Pena, *"Human-Performance Issues Related to the Design and Operation of Small Modular Reactors,"* (NUREG/CR-7126), U.S. Nuclear Regulatory Commission, 2012, Washington, DC.

[8] S. Bays, S. Piet, N. Soelberg, M. Lineberry, and B. Dixon, *"Technology Insights and Perspectives for Nuclear Fuel Cycle Concepts,"* (INL/EXT-10-19977), Idaho National Laboratory, 2010, Idaho Falls, ID.

[9] J. Tester, E. Drake, M. Golay, M. Driscoll, and W. Peters, *"Sustainable Energy: Choosing Amongst Options,"* MIT Press, 2005, Cambridge, MA.

[10] J. Hylko, *"PRISM: A Promising Near-Term Reactor Option,"* Power, 155(8), 68-74, (2011).

[11] International Atomic Energy Agency, *"Staffing Requirements for Future Small and Medium Reactors Based on Operating Experience and Projections,"* (IAEA-TECDOC-1193), 2001, Vienna, Austria.

[12] T. Sheridan, *"Humans and Automation: System Design and Research Issues,"* John Wiley & Sons, Inc., 2002, New York, NY.

[13] J. Lee, *"Human Factors and Ergonomics in Automation Design,"* In G. Salvendy (Ed.), *Handbook of Human Factors and Ergonomics* (3rd Ed.). John Wiley & Sons, Inc., 2006, New York, NY.

[14] Bureau d'Enquêtes et d'Analyses, *"Final Report on the Accident on 1st June 2009 to the Airbus A330-203 Registered F-GZCP Operated by Air France Flight AF 447 Rio de Janeiro – Paris,"* 2012, Paris, France.

[15] S. Vedantam, *"Metrorail Crash May Exemplify Automation Paradox,"* The Washington Post, Section A, page 9, 29 June 2009.

[16] D. Ingersoll, *"Deliberately Small Reactors and the Second Nuclear Era,"* Progress in Nuclear Energy, 51(4), 589-603, (2009).

[17] http://gehitachiprism.com/what-is-prism/how-prism-works/. Retrieved on 24 March 2014.

[18] J. O'Hara, J. Higgins, S. Fleger, and P. Pieringer, *"Human Factors Engineering Program Review Model,"* (NUREG-0711, Rev. 3), U.S. Nuclear Regulatory Commission, 2012, Washington, DC.

[19] K. Vicente, *"Cognitive Work Analysis: Towards Safe, Productive, and Healthy Computer-Based Work,"* Lawrence Earlbaum & Associates, 1999, Mahwah, NJ.

For the completeness of the PRA Implementation standard

Yoshiyuki Narumiya[a]*, Akira Yamaguchi[b], Takayuki Ota[a], Haruhiro Nomura[a]

[a]The Kansai Electric Power Co., Inc, Osaka, Japan
[b]Osaka University, Suita, Osaka, Japan

Abstract: Risk Technical Committee in the Standards Committee of Atomic Energy Society of Japan has formulated the standards related to PRA procedure, data, and its utilization. It provides the standards for the internal events in every operation condition usable for the risk assessment up to environmental effects (Level 3). As for the external events PRA also, the standards are expanded to earthquake, internal flooding, and tsunami. Also fire PRA or complex events PRA, which are especially induced by earthquake, are being examined for standardization. While accumulating the formulated content of the standards, usage experience and noticed matters of the PRA standards are to be feedback to the process of standard formulation.

Keywords: PRA, PSA, Standard, Japan.

1. INTRODUCTION

The paper presents a current situation of the Standards for related to the Probabilistic Risk Assessment (hereafter, PRA) procedure and its utilization undertaken by the Risk Technical Committee (hereafter, RTC) in the Standards Committee of Atomic Energy Society of Japan (hereafter, AESJ). The RTC has been positively promoting the formulation of various kinds of PRA Standards to provide "technical basis" which plays an important role in ensuring the quality of PRA. In addition to this, the RTC occasionally holds the workshops for the Standards to enhance understandings of PRA Standards and widely inform PRA method and its concept, while developing The task group on risk assessment study to share their discussion among researchers and technical experts about risk.

2. BACKGROUND AND PURPOSE OF DEVELOPING PRA STANDARDS

2.1. Background of the academic society standards development

In European countries and the US, nongovernmental standards are developed in various technical fields by members publically fairly selected from neutral organizations including academy, associations, and international conventions, so as not to benefit only a specific organization.

In Japanese nuclear energy field, for the purpose of swiftly reflecting new findings and operation experiences on the regulations, Nuclear Safety Commission (hereafter, NSC) has examined the formulation of the guideline performance and public utilization of the standards through the discussion on the systematization of the guideline. Nuclear and Industrial Safety Agency (hereafter, NISA) has also worked on the formulation of the technical standards performance and public utilization.

In the AESJ, Japan Society of Mechanical Engineers, and Japan Electric Association respectively developed the organizations aimed to formulate nongovernmental standards and began to define them. This approach was practiced with the principle of fairness, impartiality, and transparency. In September 1999, the AESJ developed the Standards Committee to formulate "standards" describing the consented matters about techniques used in a wide range of activities of design, construction, operation, and decommissioning measures for the nuclear power plants. Four Technical Committees (see Fig.1) including the RTC are organized under the Standards Committee according to the specific fields.

2.2. Organization of the Risk Technical Committee

The development of PRA Standards was originally conducted by Power Reactor Technical Committee, but after revision of the organization in 2008, the RTC was founded to specifically work on the PRA Standards.

The organization of the Standards Committee of the AESJ is shown in Figure1. Under the Standards Committee, the RTC is organized. Currently nine subcommittees are organized for the RTC. Each subcommittee is responsible for development, maintenance, revision and education of the individual standard. Also involved is the Steering Taskforce for planning the future development of standards.

Here, the title of the standard in the RTC is explained. The RTC has decided to use PRA, not PSA (Probabilistic Safety Assessment), since the development of Tsunami PRA Standards in 2011. Both PRA and PSA have the same meaning. However, both titles are used parallel for the moment as a transitional period, because existing standard titles will be changed when they are revised. Thus, in this paper, PRA is used for all standard titles except the existing standards.

The PRA was a distinctive method to show comprehensive safety by means of quantitative indexes with uncertainty. Each PRA standard has some specific steps of methodology, e.g. seismic fragility analysis in Seismic PRA. These were causes of developing separated subcommittees. Thus, the effects of separated subcommittee were to incorporate more expertise and to speed-up the process, and to perfect the discussion. On the other hand, because the initial events used in the PRA procedure include not only internal hazards, such as facility malfunction and human error, but also external hazards, such as earthquake, tsunami, and fire, the technical experts related to these events were also requested to precipitate to the subcommittee. Such kind of a variety of special knowledge is necessary for the discussions in the meeting, architecture and fuel cycle facility specialists were also called. As a result, the RTC include almost 30 members.

Currently, there are nine subcommittees, as each subcommittee formulates one standard respectively. However the scope of the Level 1 PRA subcommittee was expanded to formulate the level1 PRA standard, the PRA parameter estimation standard, and the shutdown PRA standard. The purposes of this irregular role sharing are two advantages. One of them is sharing opinions or information about three PRA standards among one subcommittee. Other one is revising three standards sequentially.

As for the seismic PRA procedure, it is not effective to assemble all experts into one subcommittee for discussion, because, for example, seismic PRA procedure is comprised of three processes of seismic hazard evaluation, building/components fragility evaluation, and accident sequence evaluation, and because the seismic hazard evaluation requires the experts from earthquake and geotechnical engineering, etc. for the discussion and the fragility evaluation does architectural and mechanical engineering experts. To cope with this problem, three working groups are organized to share the formulation activities. Their smoother cooperation is enhanced through participation of some members in multiple subcommittees.

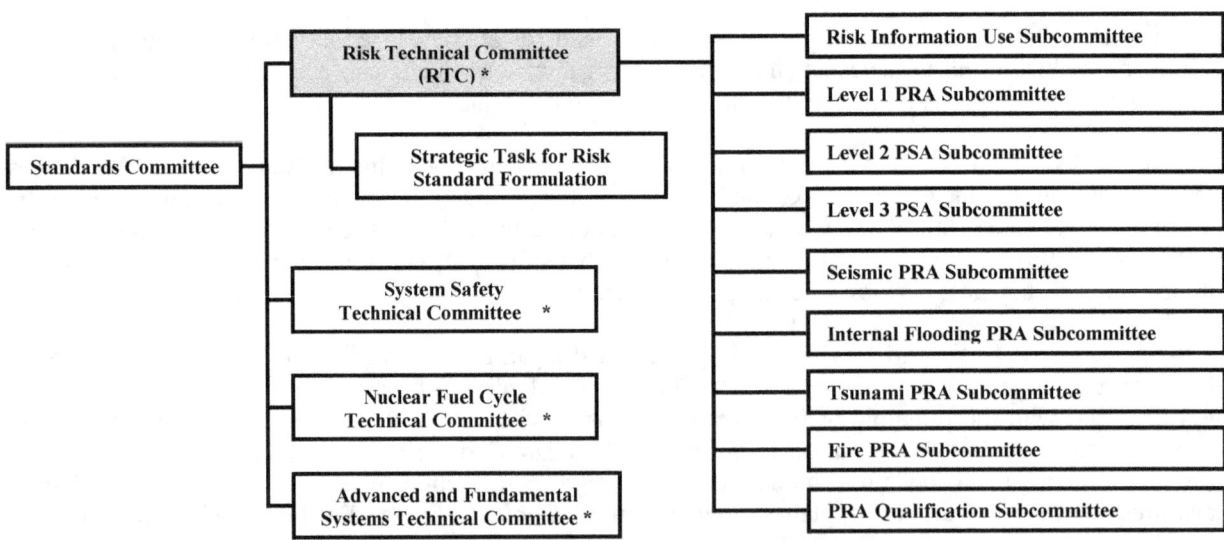

Figure1. Standards Committee organization chart for PSA standards development
*: These committees were "Four Technical Committee"

2.3. Purpose and Position of PRA Standards

The first Standard undertaken by the Power Reactor Technical Committee (of the day) was the shutdown PSA procedure (the title of the day). At that time, the implementation of the shutdown risk assessment was being introduced in the Periodic Safety Review (PSR), so that the standard was formulated to ensure its quality. Later, along with the NSC's examination on safety goals[1] and then performance objectives[2], the formulation of Level 1 PSA Standard (evaluation on up to the core damage, hereafter referred to as "L1 PSA Standard") and Level 2 PSA Standard (evaluation on up to emission of radioactive substances, here after "L2 PSA Standard") was undertaken, and then Level 3 PSA Standard (evaluation on up to environmental effects assessment, hereafter "L3 PSA Standard") followed to those. In this way, the standards have been developed preferentially from the required fields. When the standards were extended to L3 PSA Standard, the PRA methods for the internal events ranged from core damage to emission of radioactive substances causing environmental influence in all operation status (during power generating operation, shutdown) were developed.

On the other hand, NISA and Japan Nuclear Energy Safety Organization (hereafter, JNES) published the quality guideline [3] which defines technical validity of PRA to prepare for the utilization of risk information. This guideline describes the requirements regarding the quality provided in PRA, so that the AESJ's PRA Standards become the specification code which materializes the performance codes described in this guideline. There are several ways to utilize PRA. If the utilities desire to assess the safety or to know the weak points of NPPs, the objective validity of the assessment can be ensured by conducting PRA according to the AESJ PRA Standard procedure. Regulating authority can also examine the conformity to PRA standards and demonstrate the achievement of technical evaluation/endorsement in PRA reviews.

In the revision of NSC's "Regulatory Guide for Reviewing Seismic Design of Nuclear Power Reactor Facilities (revised on September 2006)", necessity of quantitative risk assessment on earthquake was discussed, and as the result of this, Seismic PSA subcommittee was developed to extend PRA Standards to the external events.

The purpose consistent to these PRA Standards is "to demonstrate the matters examined and consented under the principles of fairness, impartiality, and transparency for the purpose of utilizing PRA procedures and risk information obtained from them". By using the PRA Standards developed according to this purpose, the following advantages can be obtained.

(1) The latest PRA technique having the suitable quality will be available.

(2) The validity of conducted PRA can be demonstrated objectively by demonstrating its conformity to the academic society's standard.

(3) When reviewing the PRA validity, the part conformed to the Standard can be simplified in the review.

(4) If newly developed or revised methods or data are available, standardization can be achieved by submitting to the subcommittees.

3. KINDS OF PRA STANDARDS AND OVERVIEW

3.1. Kinds of PRA Standards

Currently, ten standards have been developed or revised and three standards are being developed through the Risk Technical Committee and Power Reactor Technical Committee (of the day).

Since the accident at Fukushima Daiichi Nuclear Power plant caused by the 2011 off the Pacific coast of Tohoku Earthquake, there has been growing demands for assessing the effects of external hazards, such as earthquake and tsunami, and taking counter measures to address those external hazards. The newly developed Japanese regulatory requirements claim design considerations associated with external hazards. The primary objective of the risk assessment for external hazards is to develop countermeasures against such hazards rather than grasping the risk profile. Therefore, applying detailed risk assessment methods, such as probabilistic risk assessment (PRA), to all the external hazards is not always required. Risk assessment methods can vary in types including qualitative evaluation, hazard analysis (analyzing hazard frequencies or their influence), margin assessment, and deterministic core damage frequency (CDF) evaluation.

The Risk Technical Committee comprehensively identified the external hazards which had potential risks, and has developed "the implementation standard for the identification of assessment methods for risks associated with external hazards." This implementation standard will help to understand plant safety against all the objective external hazards and develop appropriate countermeasures against individual hazards.

The list of the AESJ standards regarding the PRA methods and the risk informed approach is given in Table 1.

Table 1. AESJ Standards regarding risk assessment methods and the risk informed approach

Standard	The Date of Issue
The Standard for Procedures of Probabilistic Risk Assessment of Nuclear Power Plants during Power Operation (Level 1 PRA):2013	December 2013
The Standard for Procedures of Probabilistic Safety Assessment of Nuclear Power Plants during Power Operation (Level 2 PSA):2008	March 2009
The Standard for Procedures of Probabilistic Safety Assessment of Nuclear Power Plants (Level 3 PSA):2008	March 2009
Standard for Procedures of Probabilistic Safety Assessment of Nuclear Power during Shutdown State (Level 1 PSA):2010 (revision 1)	November 2011
Implementation Standard Concerning the Estimation of Parameters for Probabilistic Safety Assessment of Nuclear Power Plant : 2010	June 2010
A Standard for Procedures of Seismic Probabilistic Safety Assessment for nuclear power plants:2007	September 2007 (Under revision)
The Standard of Implementation on Use of Risk Information in Changing the Safety Related Activities in Nuclear Power Plants:2010	October 2010
Implementation Standard Concerning the Tsunami Probabilistic Risk Assessment of Nuclear Power Plants: 2011	February 2012
Implementation Standard Concerning the Internal Flooding Probabilistic Risk Assessment of Nuclear Power Plants:2012	November 2012
Terms and Definitions used Commonly in the Probabilistic Risk Assessment Standards for Nuclear Power Plants:2011	January 2012 (Under revision)
Implementation Standard Concerning the Internal Fire Probabilistic Risk Assessment of Nuclear Power Plants:201*	Under development
Implementation Standard Concerning the Risk Analysis Methodology Selection for the External Hazard:201*	Under development
A Standard for Ensuring the Quality of Probabilistic Risk Assessment for Nuclear Power Plants:2013	March 2014

3.2. Contents of PRA Standards

AESJ Standards are comprised of the "Main body" and "Annex (normative)". In addition to this, "Annex (informative)", which describes evaluation examples as a reference, and "Description", which describes discussion background and commentary, can be added in some cases.

The sections of "Scope", "Normative references", "Terms, definitions, and abbreviated terms", and "Documentation" are described in the main body of all PRA Standards as a common section of PRA Standards (AESJ Standards use the term of "Section" as "Clause"). The section of "Scope" specifies the scope of PRA defined in the PRA Standard. The section of "Normative references" enumerates the title of other standards cited in the main body and the annex (normative). The section of "Terms, definitions, and abbreviated terms" describes the terms and their definitions used in the Standards in addition to the list of abbreviations such as LOCA. In PRA Standards, there are terms commonly used in plural PRAs, like "event tree". Such terms are defined in each Standard, but their definitions may differ due to kinds and an developed period of the standards. Accordingly, PRA common glossary was formed in 2011 so as to avoid confusion and inconvenience to the users and organizations (hereafter, users) of PRA Standards. Since then, the terms uniquely used in a specific PRA are defined in each PRA Standard, which will possibly enhance better understandings to PRA terms.

As the beginning of the Section specifying the content of PRA procedure, the PRA process is defined. Though there is a basic order in each PRA process, feedback and particularization of the already evaluated results may be necessary for the repeated adjustments. This kind of practical process is specified in this section to cope with various types of usages. Next, the section of "Surveys of the configuration and characteristic of the plant" is developed. It is necessary in PRA to comprehend not only the specifications of components in a nuclear power plant but also the system configuration, properties, and functions. The practice of plant walk down is also stipulated in addition to the internal survey and collection of flow diagrams, layouts, and operation manuals. Actual visit of the site will supplement insufficiency of drawings and procedure manuals, because relative positioning or difficulty in operation of the site facility cannot be recognized only from the drawings. Especially in Tsunami PRA and Internal Flooding PRA, checking and comparing the inlet port of water flow at a drawing and an actual site, for example, may help to find out the serious scenarios.

The last section of "Documentation" stipulates for not only the content to be described in PRA reports but storage of documents to facilitate easy understanding in PRA usages and reviews.

Moreover, three items of exploitation of experts' opinions, peer review, and quality assurance activity are described in the annex (normative) as the provisions to ensure PRA validity. Though these items are defined as the provisions conducted to demonstrate PRA validity, they are currently attached to all PRA Standards as a form of annex (normative) because practical experiences are still insufficient.

3.3. Characteristics of PRA Standards

In general, the standards prescribe specifications. PRA Standards also stipulate for the concrete method but not always do for the procedures of a specific method. PRA is used for various purposes. Preciseness of PRA may differ, for example, in the cases that their purposes are to examine which system/component is important for the risk control or to comprehend the general risk in whole plant. In the former case, modeling of the component or the part of interest is necessary, and it is desirable to use parameters including failure rate of each component. To cope with such various purposes, PRA Standard provides several measures so that users can choose, devise, and extend.

3.3.1. Provisions for offering several methods

Depending on the assessment accuracy or the purposes of PRA, it is preferable, in some cases, to provide several methods to which the users can make a decision. In other words, the evaluated results can be too conservative by using only one method, or serious scenarios cannot be analyzed enough because too much time has been spent on accident scenario analysis having small influence.

For example, fragility analysis in the Seismic PSA Standard defines to use either one of the following three methods, "method using actual fragility and actual response", "method using actual fragility and response factors", and "method using fragility factors and response factors", or their combinations. Based on this, the fragility of the non risk-dominant facility can be determined by the method using factors, so that more resources can advantageously be allocated to the device having high risk importance.

In the section "6. Development of success criteria" of the L1 PRA Standard, while the use of thermal-hydraulic analysis and/or structural analysis is basically designated to the success criteria analysis (analysis to determine the conditions such as number of mitigation equipment to achieve safety function), conservative data can be used depending on the purpose and the available data can be used for the analysis in the application document for the permission of reactor installment license.

In the Tsunami PRA standard, as a method to determine actual fragility or response of the fragility analysis, four methods of those based on experiments, experience, analysis, and technical judgment are provided.

3.3.2. Provisions for users to examine/determine

Though it is important for the standards to be easily usable and understandable, the standards like an analysis manual may exclude user's judgment from PRA process. Besides, the existent findings from experiments or analysis are not always available, though the major feature of PRA is quantification of uncertainty. Wide and deep consideration is always required for users, as technical judgment is appropriately necessary for many points of the PRA process. To cope with this matter, PRA Standards provide several methods so that users can choose the best method with considering the purpose of PRA. Even in the case of showing an explicit method, they require the users to list the points to be considered before examining and making decisions.

For example, the section of "5. Selection of initiating events and evaluation of occurrence frequency" of the L1 PRA standard does not define the list of initiating events to be considered. It just requires to identify the initiating events without missing anything. Also in the hazard analysis and fragility analysis in the Seismic PSA Standard, it specifies the method of forming data from investigation results and experiment/analysis results, not by limiting the available input data.

In the section of "6.6.3 Method using fragility factors and response factors" of the Seismic PSA Standard, the calculation method to obtain a median of each coefficient and logarithmic standard deviation is specified, but it is necessary to adjust the coefficient with imagining the phenomenon, as is in the compensation of response in nonlinear field. Therefore, a number of examples are shown for users in the description for their consideration.

The Section 6 of "Identification of an accident scenario" of the Tsunami PRA Standard defines to widely extract the possible influence caused by tsunami, with estimating the process of tsunami attack and its spreading by referring the past damage.

3.3.3. Provisions applicable by users

The result obtained by the Standard should attain a certain degree of quality. However, in the case of using a method different from the one defined in the standard, it is necessary to separately demonstrate the validity. Because "PRA deals with all accident sequences theoretically considerable", it is necessary for the users to investigate/analyze the actual facility conditions and widely comprehend the system/equipment design information and maintenance information.

For example, in the section of "5. Classification of Plant Operating State (POS)" of the Shutdown PSA Standard (rev.2), the method of grouping with using pre-POS or that of subdividing by the condition of facility configuration is usable for the POS classification. This allows the users to conduct more detailed PRA.

In fragility analysis in the Tsunami PRA Standard, for the influence of the case that flood measures are taken, step function is applicable to the actual fragility in submerged and soaked modes. In this case, it is required that the actual fragility is not just set by a water level but considering the flooding routes and the height and shape of opening ports. The users can correct the fragility curves according to the property/condition of component/system depending on the condition of flooding measures.

3.3.4. Showing a number of various kinds of examples

As mentioned in (3.3.1) to (3.3.3.), PRA Standards can be expressed as "a guidance of thinking process" for users. However, in order to ensure a certain degree of PRA quality, the standards cannot be just user-oriented. Then, a number of various kinds of analysis examples and parameter examples are described in the annex (informative).

For example, more than 400 pages of evaluation examples including hazard analysis and fragility analysis are attached in the Seismic PSA Standard for the convenience of user's reference.

In the Tsunami PRA Standard, evaluation examples will be published as a separate volume. The evaluation examples are more useful if they have taken the latest tsunami damage into consideration. Though setting examples of the parameters for use are described in the annex (informative) of Tsunami PRA Standard, to provide the latest evaluation examples reflecting the findings of the Great East Japan Earthquake is considered to be important. By separating the volume, it is expected that the evaluation examples are appropriately updated and the tsunami PRA results can be obtained based on the latest findings.

4. IMPROVEMENT PROGRAM OF PRA IMPLEMENTATION STANDARD

4.1. Selection of the schemes for Standard development

Currently, ten standards have been developed or being revised and three standards are being newly developed by RTC. These standards can cover not only the internal event PRA at-power, but also main external event PRAs. However, they have not yet covered all possible events, e.g. a fire caused by earthquake level 2 shutdown PRA, it is not necessary to develop all standard newly, because it may be able to use the existing standard by expanding or adding for assessment. Moreover, it is also thought that the risk can be guessed with the combination of other PRA results. Based on the above, RTC considered about the events or combination of events, which combined operational state, and the levels 1-3, to make a decision of the priority for development. The forty-eight schemes given in Table.2 are target of formulation and there are four internal event Standards have been developed, and also four external event Standards have been developed or under development, so forty schemes are target of consideration.

Table.2 Consideration of Standard development

		Internal Events	External Events						
			Internal Flooding	Internal Fire	Tsunami	Earthquake	Internal Flooding caused by Earthquake	Internal Fire caused by Earthquake	Tsunami with Earthquake
At-power	L1	*	*	*	*	*			
	L2	*							
	L3	*							
Shutdown	L1	*							
	L2								
	L3								

*: Have been developed or under revision
: Target of consideration

4.2. Prioritizing development of Standards

4.2.1 Criteria for Priority

The following thing was taken into consideration in the determination of the priority of PRA standard development. One thing is influence degree exerted on a plant. Priority is given to event, which has complicated accident scenario, and cannot judge the important sequence for finding out a risk reduction measure if not based on PRA. The other thing is workload required to develop PRA standard. Priority is given from the standard which has many workloads and requires many time for development.

4.2.2 Classification by availability

PRA is a technique which can evaluate the combination of a lot of scenarios, and can show risk quantitatively. Therefore, it is effective to adapt PRA for events which have various accident scenarios, and cannot be grasped peculiar risk of plant, without evaluation of whole accident scenario analysis.
The schemes are classified into three groups by availability for PRA evaluation. The availability of PRA means reduction effect of risk and urgency.

The scheme which have complicated risk profile, and have difficulty to be assessed without PRA classified into group "A".
The scheme which may be able to be assessed by using other PRA assessment, classified into group "B".
Furthermore, the scheme, which is expected more availability, classified into group "A^+".

Moreover, in present, the scheme is classified into group "A", but it will be group "B" after development of other new standard is written as B (A). A classification result is given in table.3.

Table.3 Arrangement from validity

		Internal Events	External Events						
			Internal Flooding	Internal Fire	Tsunami	Earthquake	Internal Flooding caused by Earthquake	Internal Fire caused by Earthquake	Tsunami with Earthquake
At-power	L1	*	*	*	*	*	A^+	A^+	A^+
	L2	*	A	A	A^+	A^+	B(A)	B(A)	B(A)
	L3	*	A	A	A^+	A^+	B(A)	B(A)	B(A)
shutdown	L1	*	A	A	A^+	A^+	A	A	A
	L2	B	B	B	B	B	B	B	B
	L3		B	B	B	B	B	B	B

*: Have been developed or under revision

4.2.3 Classification by Workload

In addition to the classification by availability, schemes are classified into following four groups by viewpoint of the amount of workload.

   a: What needs to decide a method newly. (Workload: large)
   b: What can be evaluated by the combination and extension of the existing practice standard. However, the thing which needs to examine the concrete method newly (workload: medium)
   c: What can be evaluated by adding the points of concern by basing on the existing practice standard. (Workload: small)
   d: What can be evaluated by the existing practice standard.

Moreover, in present, the scheme is classified into group "b", but it will be group "c" after development of other new standard is written as c (b). A classification result is given in table.4.
The priority of work is "a" to "c".

Table.4 Arrangement from a workload

		Internal Events	External Events						
			Internal Flooding	Internal Fire	Tsunami	Earthquake	Internal Flooding caused by Earthquake	Internal Fire caused by Earthquake	Internal Flooding with Tsunami
At-power	L1	*	*	*	*	*	b	b	b
	L2	*	c	c	c	c	c	c	c
	L3	*	d	d	b	b	d	d	d
shutdown	L1	*	b	b	a	a	c(b)	c(b)	c(b)
	L2	c	c	c	c	c	c	c	c
	L3		d	d	d	d	d	d	d

*: Have been developed or under revision

4.2.4 The Judgment of the Priority

Based on the classifications which are mentioned in section (4.2.2) and (4.2.3), the schemes are arranged as follows.

(1) The scheme which validity is especially high
    (1.1) : A^+a   Shutdown   L1   Earthquake, Tsunami
    (1.2) : A^+b   At-power   L2   Tsunami with Earthquake, Internal flooding with Earthquake, Internal fire with Earthquake

		A⁺b	At-power	L3	Earthquake, Tsunami
(1.3) :		A⁺c	At-power	L2	Earthquake, Tsunami

(2) The scheme which validity is high
- (2.1) :　Ab　　Shutdown　L1　Internal flooding, Internal Fire
- (2.2) :　Ac　　At-power　L2　Internal flooding, Internal Fire
- (2.3) :　Ac　　Shutdown　L1　Tsunami with Earthquake, Internal flooding with Earthquake, Internal fire with Earthquake

RTC basically undertake development of the standard earlier which has high validity, and much amount of workload. Moreover, the regular revision term in a subcommittee is also taken into consideration, and RTC advances so that two or more practice standard decision work can be performed simultaneously.

4.3. Plan for Future

4.3.1 Multiunit suffering a hazard

Moreover, this project gives some action plans for important problems, such as the multi-unit core damages caused by earthquake and tsunami, risk of multi-hazard, treatment of the uncertainty for decision making, risk literacy, and development of non-PRA methods for risk evaluation.
Since some sites are geographically close in Japan, it is necessary to assume also about a multisite simultaneous accident.

4.3.2 Treatment of Uncertainty for risk informed decision making

Occurrence of frequency of external hazards such as seismic which may affect core damage and containment failure, has large uncertainty. Then, a PRA result will also have large uncertainty.
Therefore, treatment of uncertainty is needed to be considered.

4.3.3 Non-PRA methods for risk evaluation

In The standard for Procedures of external hazard Probabilistic Risk Assessment of Nuclear Power Plant, procedure of the risk assessment technique for an external hazard is shown. It is necessary to offer the implementation standard for the external hazard assessment by include Non-PRA methodology.
Therefore, RTC search also about the Non-PRA methods, such as qualitative evaluation, hazard analysis (analyzing hazard frequencies or their influence), margin assessment, and deterministic core damage frequency evaluation.

5. CONCLUSION

The RTC has the mission to develop PRA standards including estimation of parameters using in PRA, and utilization of risk information. It has been providing the standards for the internal event PRA in every operation condition usable for the risk assessment up to environmental effects. As for the external events also, the standards are expanded to earthquake, internal flooding, and tsunami, and also fire and complex events are being examined for standardization. The method regarding system analysis, such as initiating events, human error rate, and common-caused failure, is also to be deepened further. The results of external events PRA can naturally include wide range of uncertainty. The treatment of uncertainty is to be examined because it is critical element for risk informed decision making. Further, PRA Standards which can ensure the quality high enough to be applied to the actual facility and appreciated in various kinds of decision makings are to be achieved.

Then, accumulating the content of the standards, usage experience and noticed matters of the PRA standards will be reflected to the standard formulation. RTC considered the standard preparation plan and decided the priority of standard development. By taking such a variety of measure, we expect that the standards will have completeness.

References

[1] Nuclear Safety Commission, Interim Report on the Investigation and Review on Safety Goals, NSC, December 2003
[2] Nuclear Safety Commission, Performance Objectives for Light Water Nuclear Power Reactor Facilities, NSC, March 2006
[3] Nuclear Industry Safety Agency, Guideline for Quality of PSA for Nuclear Power Plants (trial version), NISA, JNES, April 2006

Nuclear Safety Design Principles & the Concept of Independence: Insights from Nuclear Weapon Safety for Other High-Consequence Applications

Jeffrey D. Brewer[*]
Sandia National Laboratories, Albuquerque, NM, USA

Abstract: Insights developed within the U.S. nuclear weapon system safety community may benefit system safety design, assessment, and management activities in other high consequence domains. The approach of *assured nuclear weapon safety* has been developed that uses the Nuclear Safety Design Principles (NSDPs) of *incompatibility*, *isolation*, and *inoperability* to design safety features, organized into subsystems such that each subsystem contributes to safe system responses in *independent* and *predictable* ways given a wide range of environmental contexts. The central aim of the approach is to provide a robust technical basis for asserting that a system can meet quantitative safety requirements in the widest context of possible adverse or accident environments, while using the most concise arrangement of safety design features and the fewest number of specific adverse or accident environment assumptions. Rigor in understanding and applying the concept of independence is crucial for the success of the approach. This paper provides a basic description of the *assured nuclear weapon safety* approach, in a manner that illustrates potential application to other domains. There is also a strong emphasis on describing the process for developing a defensible technical basis for the independence assertions between integrated safety subsystems.

Keywords: System Safety Design, Safety Assessment, Independence, Nuclear Weapon Safety.

1. INTRODUCTION

Insights developed within the U.S. nuclear weapon system safety community may benefit system safety design, assessment, and management activities in other high consequence domains. The approach of *assured nuclear weapon safety* has been developed that uses the Nuclear Safety Design Principles (NSDPs) of *incompatibility*, *isolation*, and *inoperability* to design safety features, organized into subsystems such that each subsystem contributes to safe system responses in *independent* and *predictable* ways given a wide range of environmental contexts. The *assured nuclear weapon safety* approach strives toward use of a concise arrangement of safety design features and a limited number of specific adverse or accident environment assumptions. Simplicity of safety features, passive safe responses, and a systematic allocation of basic features among engineered features and human actions[1] are emphasized in the implementation, and an innovative *inside out* process for hazard identification, which is described in this paper, is also applied throughout iterative system design phases to support the process of NSDP integration. In essence, this approach claims to be an efficient method for engineering bounded system safety-related responses.

Appropriate independence assertions are essential given that multiple safety subsystems, each providing safe responses for all relevant environments in independent and predictable ways must be integrated into the system to form a basis for meeting stringent qualitative and quantitative safety requirements. Overreliance on the concept of independence for asserting levels of safety without providing a sufficient technical basis is tempting, and must be avoided. In addition, it is recognized that humans do a poor job both of conceiving the many ways things may fail and estimating

---

[*] Sandia National Laboratories is a multi-program laboratory managed and operated by Sandia Corporation, a wholly owned subsidiary of Lockheed Martin Corporation, for the U.S. Department of Energy's National Nuclear Security Administration under contract DE-AC04-94AL85000. This paper is designated as SAND2014-1832C at Sandia National Laboratories. The author may be contacted at: jdbrewe@sandia.gov.

[1] The primary human actions to consider are those designed to provide an unambiguous indication of intent to achieve a nuclear detonation. Other human actions include those related to ensuring safety during weapon production, assembly, testing, transportation, maintenance, and disassembly.

probabilities for conceivable failures [1-3]. Particularly in the domain of low probabilities and high stakes it is helpful to move beyond the common 'model' and 'parameter' uncertainty approach and consider opportunities for flaws across the spectrum of theories, models, and calculation techniques [3]. For high-consequence systems it can be argued that the approach taken ought to be more *possibilistic* than *probabilistic* [2]. Therefore, the foundations for any (hopefully few) probabilistic statements regarding safety must be clear and defensible.

Particular emphasis in this paper is placed on implementing the NSDPs using independence assumptions founded upon the distinctions of functional, temporal, and physical differences. Although the concepts of function, time, and physical properties do not provide mutually exclusive categories for describing dependence, they are proposed as helpful concepts when striving to increase the independence of safe responses to ensure the control of hazards in high-consequence applications. The motivation for using all three of these conceptual distinctions is increased in situations where data are sparse or non-existent for demonstrating that a focus on only one or two is sufficient. The paucity of data is typical, thus far (fortunately), for extremely high-consequence events where observing data may involve observing massive disasters.

For example, no inadvertent nuclear weapon detonations have been observed. This is due in part to the care with which such systems are handled, but the weapons must still meet stringent requirements if an accident were to occur—and there are many ways in which credible accidents may occur. As another example, no pandemic illness resulting from an accidental release of a biological organism from a research facility has been observed. No doubt this has been aided by the lack of man-made or natural disasters in the vicinity of such facilities, but the facilities must still be able to provide a safe response to these externally imposed conditions if such an event were to occur.

This paper provides a basic description of the *assured nuclear weapon safety* approach, incorporating the NSDPs, and describes the process for developing a defensible technical basis for the independence assertions between integrated safety subsystems. The contribution of functional, temporal, and physical differences to achieving independence are described. Clear and distinct definitions for *common-cause failure* and *common-mode failure* (with examples) are provided. Attributes of *predictability* that strengthen the technical basis for safe system responses are given. It is hoped that insights presented here, developed within the U.S. nuclear weapon system safety community over the course of more than 50 years, may benefit system safety design, assessment, and management activities in other high consequence domains.

2. NUCLEAR WEAPONS

To elucidate the concepts in this paper, it is beneficial to: (1) describe some specific notion of a system, (2) identify the hazards to safety associated with the system, and (3) develop a grouping strategy for sources of energy that can contribute to the release of the hazards. In this discussion the system of interest is a nuclear weapon. A nuclear weapon is a device in which the explosion results from the energy released by reactions involving fission or fusion (of atomic nuclei) [4]. It is not unusual for the total energy release of a nuclear weapon, given that it is detonated in the intended mode, to be expressed in the range of hundreds of kilotons or even in the vicinity of a megaton of trinitrotoluene (TNT) equivalent, i.e., the energy equal to exploding two billion pounds of TNT. The greatest hazard[2] to safety that a nuclear weapon can pose is an inadvertent nuclear detonation[3] (IND), especially an IND that approaches the designed yield for the weapon—this may well be the archetype of a high consequence safety hazard.

---

[2] In addition to a nuclear detonation, hazards to safety include high explosive detonation/deflagration as this will disperse special nuclear material. There are certainly other hazards to consider with the materials inside nuclear weapons (e.g., fire, toxic chemicals, and various radionuclide concerns); however, nuclear detonation and special nuclear material dispersal are the generally considered the greatest hazards.
[3] Nuclear Detonation – An energy release through a nuclear process, during a period of time on the order of 1 microsecond, in an amount equivalent to the energy released by detonating 4 or more pounds of TNT [5].

The simplified conceptual nuclear weapon design of interest here is that of a sealed pit, implosion type nuclear weapon. In this design, a mass of fissile material (the pit) is surrounded by high explosives (HE) and a detonation system is also included that initiates the HE. When the HE detonation system is operated, the explosives compress the fissile material until a nuclear detonation results. Figure 1, A notionally represents the weapon configuration before HE detonation and Figure 1, B notionally represents the configuration immediately after HE detonation, just prior to the nuclear explosion.

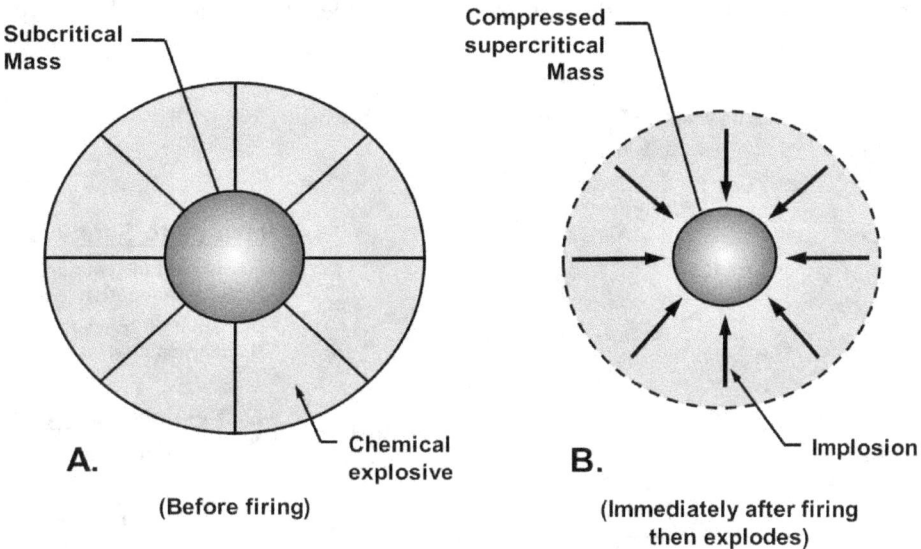

Figure 1: Illustration of sealed pit, implosion assembled weapon.

In the context of a nuclear weapon, a helpful grouping strategy for sources of energy that can contribute to the release of the hazards to safety (i.e., IND or dispersal of special nuclear material) is that of front, side, and back door energy scenarios. These scenario groups are associated with distinct energetic materials and/or distinct collections of energy storage/production devices that are essential for producing a nuclear detonation or dispersal of nuclear material when the weapon is fully assembled or partially assembled either due to adverse or accident environment exposure or due to assembly or dismantlement actions. It is important to note that this grouping strategy for the nuclear weapon example is focused on distinguishing the *energy sources* associated with release of the hazard(s). It is not focused on distinguishing all of the *specific pathways* through which energy may travel. The specific energy pathways are considered, as needed, during the safety design/analysis process relative to the level of assembly of the weapon and a decomposition of types of adverse or accident environments. The main benefit of this approach can be described in simple terms: "if the *energy sources* respond safely in an adverse or accident environment, then consideration of the possible *energy pathways* is not required."

Here are descriptions of the front, side, and back door energy scenarios for the nuclear weapon example: *Front Door Energy Scenarios* involve operating the HE detonation system using the intended energy storage/production devices designed to operate the HE detonation system or another internal energy storage/production device that is compatible with operating the HE detonation system. For example, the intended energy storage/production devices for operating the HE detonation system are certainly compatible with operating it and achieving a nuclear detonation. However, another energy device may also be able to provide an energy input to the HE detonation system along an unintended energy pathway.

Side Door Energy Scenarios involve operating the HE detonation system in any way that does not involve the intended energy storage/production devices. For example, a lightning strike to a damaged weapon may provide sufficient energy to operate the HE detonation system. *Back Door Energy Scenarios* involve direct initiation of the high explosive material required to achieve a nuclear detonation; the HE detonation system is either irrelevant or simply plays a secondary role in achieving a nuclear detonation. For example, a shock environment or thermal environment that causes initiation of the explosives would be a back door energy scenario. Mixed-door scenarios are also possible such as a case in which a lightning strike event is able to provide the energy needed to energize intended energy storage/production devices and operate the HE detonation system. As soon as HE is assembled with the fissile material then back door energy scenarios are possible. When the HE detonation system is included then both back door and side door energy scenarios are possible. Finally, once the intended energy storage/production devices are installed then back, side, and front door energy scenarios are possible (see Figure 2).

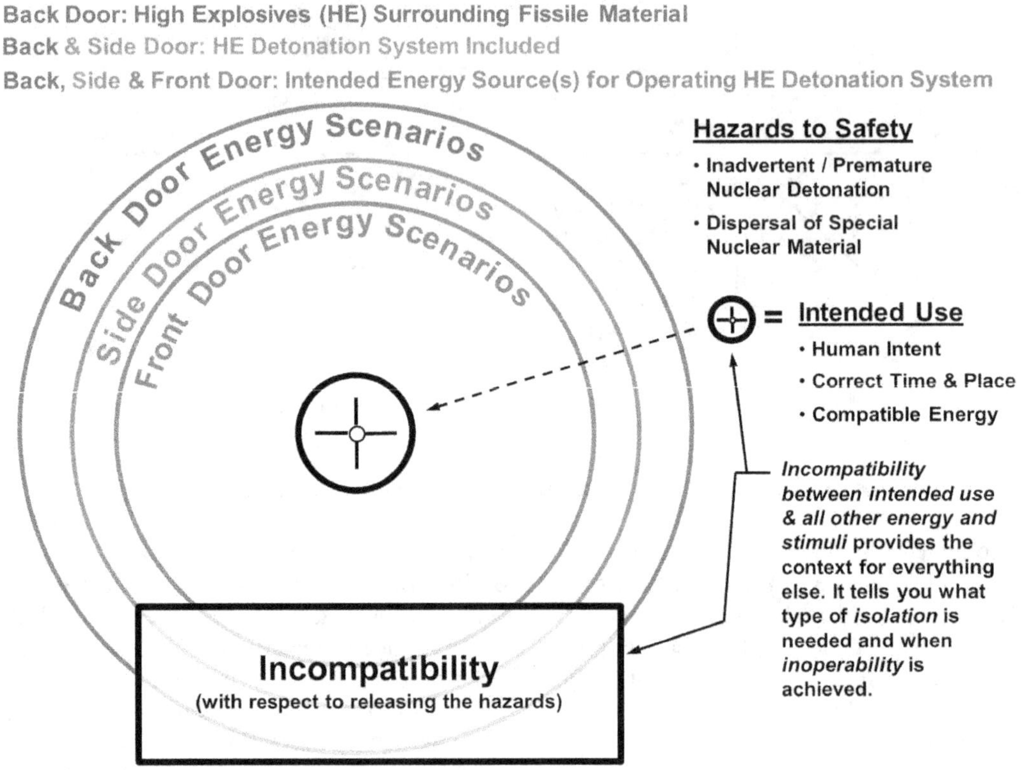

Figure 2: Illustration of energy scenarios for nuclear weapon example.

The front, side, and back door energy scenario taxonomy for the nuclear weapon example was inspired by the *inside out* (IO) approach to identifying hazards to safety. The inside out approach involves: (1) identifying the simplest configuration of system elements that present a safety hazard of concern, (2) evaluate all hazards to safety and the possible energy types, energy sources, energy pathways, and information that may facilitate unsafe consequences, (3) add the next design feature to the simplest configuration, (4) evaluate hazards, energy, and information as in step 2, and (5) repeat the process until the full system design configuration has been analyzed. The inside out (IO) approach, when repeated at various times in the system design process, is an excellent means of identifying hazards and incorporating an arguably complete set of features to control the hazards throughout the lifecycle of the system. The IO approach proves beneficial for deciding how to manage hazards during the production process of a weapon system, as well as assessing levels of safety when the weapon system is fully assembled or partially assembled either due to adverse or accident environment exposure or due to assembly or dismantlement actions. The same approach can be used when managing the radionuclide hazards of a nuclear power reactor or biological hazards of a research facility.

Nuclear weapons are indeed 'special' due to their potentially devastating consequences in terms of loss of life, injury, and damage to property and the environment. Therefore, highly stringent safety requirements, including *stringent numerical inadvertent nuclear detonation safety requirements* are appropriate. U.S. Department of Defense (DoD) Directive 3150.02 [6] and U.S. Department of Energy (DOE) Order 452.1D [5] both reflect this perspective. The following nuclear detonation safety design requirements have been established by the U.S. DOE for nuclear weapons delivered to the U.S. DoD (DOE O 452.1D section 4.f.(3)(a)) [5]:

1. <u>Normal Environment</u>[4]. Prior to receipt of the enabling input signals and the arming signal, the probability of a premature nuclear detonation must not exceed one in a billion (1E-09) per nuclear weapon lifetime.
2. <u>Abnormal Environment</u>[5]. Prior to receipt of the enabling input signals, the probability of a premature nuclear detonation must not exceed one in a million (1E-06) per credible nuclear weapon accident or exposure to abnormal environments.
3. <u>One-Point Safety</u>. The probability of achieving a nuclear yield greater than 4 pounds of TNT equivalent in the event of a one-point initiation of the weapon's high explosive must not exceed one in a million (1E-06).

It is noted that the normal environment requirement is for the lifetime of a weapon, and the abnormal environment requirement is given an accident or exposure, and that both requirements are for a single weapon.

The only way to confidently assert that such stringent numerical nuclear detonation safety requirements have been met is to use a decomposition approach in which multiple, independent elements work together in the system to provide the desired level of safety. DoD 3150.2-M [4], U. S. Navy and U. S. Air Force nuclear safety requirements documents (e.g., AFMAN 91-118 [7]), DOE O 452.D [5], and requirements used at the U. S. nuclear weapon laboratories all reflect this perspective.

3. ASSURED NUCLEAR WEAPON SAFETY APPROACH

Sandia National Laboratories (SNL) has established a nuclear safety design philosophy, called assured nuclear weapon (NW) safety, that is used to ensure that the quantitative safety requirements for U. S. nuclear weapons are met or exceeded. This section provides the author's 'summary perspective' of several key aspects of that philosophy. The foundation for the safety design philosophy consists of the three irreducible nuclear safety design principles (NSDPs) of incompatibility, isolation, and inoperability. *Incompatibility* is the use of energy or information that will not be duplicated inadvertently. *Isolation* is the predictable separation of weapon elements from compatible energy. *Inoperability* is the predictable inability of weapon elements to function. Incompatibility is the dominant NSDP as it provides the context for both isolation and inoperability features. That is, unless you know what energy and stimuli are compatible with operation of all or part of the nuclear weapon, you will not know what to isolate, nor will you know the point at which inoperability is achieved. The definition of predictability is 'certain[6] to happen based on knowledge and experience.' To be "predictable," knowledge and experience should be based on attributes that are identifiable, analyzable, testable, controllable, and verifiable[7].

---

[4] Normal Environment – In DOE operations, the environment in which nuclear explosive operations and associated activities are expected to be performed. In DoD operations, the expected logistical and operational environments, as defined in a weapon's stockpile-to-target sequence and military characteristics, that the weapon is required to survive without degradation in operational reliability [5].

[5] Abnormal Environment – In DOE operations, an environment that is not expected to occur during nuclear explosive operations and associated activities. In DoD operations, as defined in a weapon's stockpile-to-target sequence and military characteristics, those environments in which the weapon is not expected to retain full operational reliability [5].

[6] In practical applications 'certain to happen' is often implicitly understood to mean that a safe response is known to occur with a high degree of confidence.

[7] *Identifiable* – capable of being distinguished and named, specifically, the safety-critical characteristics and parameters of the safety critical components and other elements. *Analyzable* – amenable to available analysis techniques. *Testable* – can be demonstrated by measurement from experimentation and test activities.

3.1. Four Step Process of Implementing the Nuclear Safety Design Principles

The four step process of implementing the NSDPs is as follows[8]:

1. Develop a nuclear weapon that is **incompatible** with all forms and levels of energy except the correct sequence of intended, authorized, and unambiguous energy and stimuli.

    Begin with examination of incompatibility with respect to back door energy scenarios from the point at which the HE is assembled with the fissile material. Then examine incompatibility with respect to back and side door energy scenarios from the point that the HE detonation system is included. Then examine back, side, and front door energy scenarios from the point that energy storage/production devices capable of operating the HE detonation system are added.

2. For any part of the nuclear weapon that is compatible with unintended energy or stimuli, provide **isolation** from that energy or stimulus that could lead to an accidental explosion of any kind, and/or provide a reversible **inoperability** feature to eliminate or minimize exposure to safety hazards. The inoperability feature must be incompatible with all forms and levels of unintended energy and stimuli.

3. For any isolation feature that also blocks intended HE detonation system energy, provide a reversible **isolation** feature (a.k.a., a *stronglink*) to allow only intended energies to propagate to the HE detonation system. The stronglink must be **incompatible** with all forms and levels of energy and stimuli except the correct sequence of intended, authorized, and unambiguous energy and stimuli.

4. For any of these stronglinks, isolation features, reversible inoperability features, or incompatibilities that are subject to failure, provide an irreversible inoperability feature (a.k.a., a *weaklink*) that passively renders the nuclear weapon **inoperable** and incapable of producing an accidental explosion of any kind before such failure.

    For example, prior to melting through an energy isolation barrier, the material within a weaklink changes its physical state due to the rise in the temperature and renders the system irreversibly inoperable.

For a specific system, a high-level, concise expression of what will be isolated, inoperable and/or incompatible to provide assured safety is generated and captured in a *safety theme*. A *safety theme implementation* is a detailed explanation of how independent safety subsystems and associated safety design features are integrated into a system to provide assured safety [2]. The safety design features should provide their safety function in an *inherently safe*[9] or *passively safe*[10] manner. *Active safety*

Controllable – can be produced in a repeatable fashion. *Verifiable* – critical parameters can be shown to have met their requirements.

[8] This version of the process was adapted from an elegant articulation of it made by a Principal Member of the Technical Staff at SNL, Robert G. Hillaire, Ph.D., PE — provided via personal communication to the author.

[9] An *inherent safety feature* achieves its safety function by the elimination of a specific hazard by means of the choice of material and design concept. Therefore, no changes of any kind, such as internally or externally caused changes of physical configuration can possibly lead to an unsafe condition. An inherent safety feature represents conclusive or deterministic safety, not probabilistic safety [adapted from ref. 8].

[10] A *passive safety design feature* provides a safety function that does not depend on external mechanical and/or electrical power, signals, or forces to operate. That is, passive safety design features provide their safety function without having to sense/detect an undesirable environment and then actuate to a safe state. Passive safety design features rely on natural laws and properties of materials. It is important to note that passive safety design features, while not relying on human action or other power, can fail due to mechanical or structural failure or willful human interference. Thus, passive safety is not synonymous with inherent safety or absolute reliability [adapted from ref. 8]. One example of a passive safety design feature would be driver restraints in a

design features[11] should only be used when inherent or passively safe features cannot be used. In other words, the safety theme implementation should be designed to rely mainly upon *first principles of physics and chemistry*. These principles are the fundamental characteristics inherent in physics and chemistry that provide a predictable response when the subsystem in question is subjected to specified stimuli [2]. Nuclear weapons tend to be amenable to the incorporation of inherent or passive safety features given that they are "single-use" devices which are designed to spend most of their existence in a passive or dormant state, i.e., they are unpowered and stored in secured areas[12]. A largely passive system such as a nuclear weapon may be contrasted with an active system such as a commercial passenger aircraft that exercises its repeatable engineered functions (e.g., takeoff, safe flight, landings) many times. An active system will necessarily incorporate more active safety design features during its operational life.

Figure 3 provides a graphical illustration of how the NSDPs may be applied in a safety theme for controlling the hazards associated with the notional nuclear weapon. Some features may provide isolation or inoperability only for particular energy scenarios. Others may be built into the weapon in a safe state and must be actively transitioned to an unsafe state to allow intended use. Other features such as weaklinks (WL) may be built into the weapon in an unsafe/operable state, but they will irreversibly transition, in a passive manner, to a safe/inoperable state if certain adverse or accident environments occur. Isolation features such as stronglinks (SL) may be able to transition between safe and unsafe states upon receipt of unambiguous information resulting from human actions.

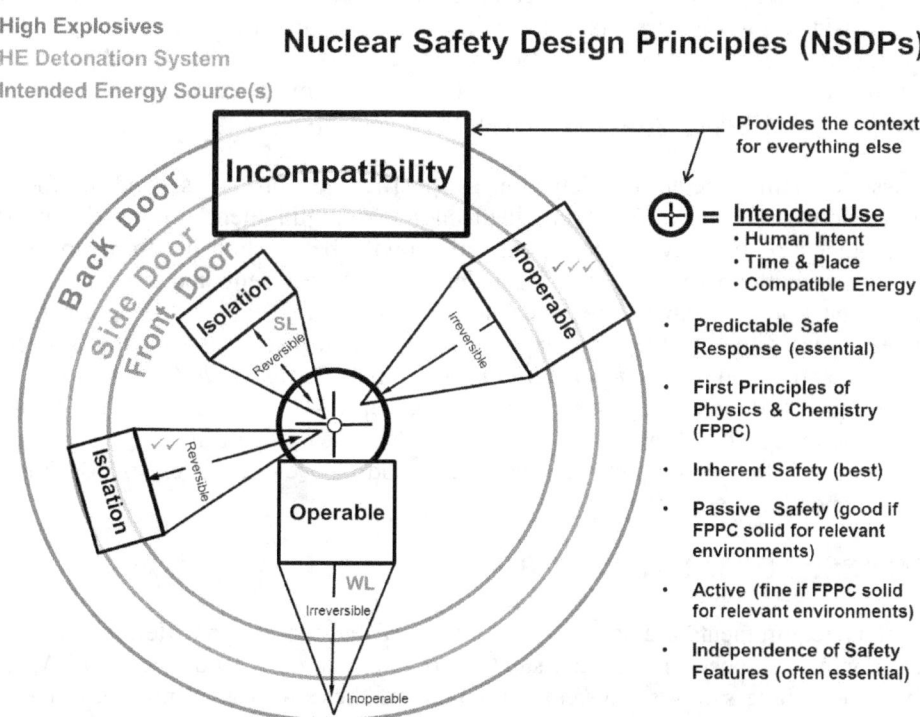

Figure 3: Illustration of NSDP application to energy scenarios for weapon example.

---

race car provide their safety function without having to change state in an accident. Another example would be an energy storage/production device in a weapon that becomes irreversibly inoperable due to a physical material state change when the weapon is subjected to a thermal environment such as in an aircraft fuel fire.

[11] An *active safety design feature* is anything that does not provide its safety function in an inherent or passive manner. That is, an *active safety design feature* must successfully sense/detect an undesirable environment **and** then actuate to a safe state (e.g., an airbag deployed in a motor vehicle crash); **or** it must continue to **act** in a safe manner (e.g., the functioning engines on an aircraft maintaining safe flight).

[12] Ensuring the safety of nuclear weapons includes designing systems which strive to meet stringent IND requirements given AEs while simultaneously striving to prevent weapons from exposure to AEs [6].

3.2. Levels of System Safety & Assessment of the 'Level' for a Specific Weapon Design

In the practice of implementing the SNL safety design philosophy, there are three levels[13] of safety that can be identified: (1) the Safety Design Ideal, (2) SNL Assured Safety, and (3) SNL Asserted (Implemented) Safety. The three levels are described below:

1. **Safety Design Ideal** — Isolation of compatible energy from detonation-critical components of an operable weapon until after the weapon becomes irreversibly inoperable is assured by first principles of physics and chemistry for all levels of relevant environments.

2. **SNL Assured Safety** — Isolation of compatible energy from detonation-critical components of an operable weapon until after the weapon becomes irreversibly inoperable to the levels required in a sufficiently predictable manner. Adequate requirements and technical bases exist to support compliance with system level requirements. This is the highest possible level of system safety that can be implemented given currently available knowledge, experience, and technology. This approach to engineering system safety changes over time with increases in knowledge, experience, and available technology.

3. **SNL Asserted (Implemented) Safety** — this is the same as assured safety in that system level requirements are asserted to be met, but with specific exceptions noted as necessary. This is the level of system safety that is actually achieved in light of "as safe as reasonably practicable" (ASARP) considerations that are balanced against security, reliability, operational capabilities, and overall resource constraints (e.g., time, money, skilled personnel). This level of safety may be expected to change over time due to factors such as aging, maintenance, operational conditions, and increases in knowledge, experience, and available technology.

The SNL Asserted (Implemented) Safety for a specific weapon is asserted in the combined engineering judgment of SNL. This 'combined engineering judgment' means the informed and authoritative judgment of executive management, based on the best technical information and opinions of relevant subject matter experts, including independent assessment and others who may have differing professional opinions about a technical solution or its adequacy. Two major parties involved in this judgment process include the *weapon project* (WP) team that is responsible for designing and producing the weapon, and the *independent nuclear safety assessment* (INSA) team that is an independently funded and managed group of experts that reports assessment results to SNL executive management. The independence and autonomy of the INSA team from the WP team is an essential element for ensuring that a complete and balanced body of evidence exists for informing the authoritative judgment of executive management.

4. THE CONCEPT OF INDEPENDENCE

Since IND safety requirements are so stringent (i.e., < 1E-09 per weapon lifetime and < 1E-06 per abnormal environment exposure), multiple safety subsystems are incorporated into NW systems to avoid reliance on a single safety subsystem. Accounts of those who were present at the time these stringent criteria were developed (circa 1968) have stated that the reason for selecting these numbers was that they reflected the following perspectives: (1) the intention to never have an inadvertent nuclear detonation (IND); (2) the intention to ensure that the probability of an IND is vanishingly small; and (3) the belief that *independent* components and subsystems could be engineered such that combinations of them could attain those numerical system safety levels [2]. The U. S. NW safety community has determined that two safety subsystems are the optimal number relative to the quantitative abnormal environment (AE) safety requirements and three are optimal[14] relative to the quantitative normal environment (NEs) safety requirements [9].

---

[13] This section provides the author's 'summary perspective' of the levels of safety that may be identified within the implementation of the SNL nuclear safety design philosophy.

[14] Optimal here indicates a balancing of safety, security, reliability, and operational considerations.

If two safety subsystems each provide a safe response to all AEs, independently, such that the failure to provide a safe response is < 1E-03 per abnormal environment exposure for either subsystem, then the probability that the overall system would respond safely to all AEs could be asserted to be bounded by the product of the subsystem responses. The reductionist approach of designing multiple safety subsystems, each of which are believed to *independently* contribute to achieving the numerical system safety requirements, while mathematically expedient, requires great confidence in the degree of independence achieved. Note also that when decomposing inadvertent nuclear detonation safety subsystem requirements down to the components that comprise a subsystem, the approach has typically been to provide a technical basis that demonstrates an assurance that each component will provide their contribution to the safety function of the subsystem without fail to a level of < 1E-04 to <1E-05 per weapon lifetime or per abnormal environment exposure as appropriate. It is recognized that it is very difficult to provide assurance that the < 1E-03 numerical requirements for a safety subsystem have been met, let alone the < 1E-04 or 1E-05 assignments given to elements or components within a safety subsystem. In fact, it is not possible to amass quantitative data that supports such assertions with a high degree of statistical confidence across all relevant environmental conditions. In other words, it is not possible to conceive of all possible AEs, nor can all environments which can be conceived be tested exhaustively in a repeated fashion to generate overwhelming statistical certainty of weapon response[15] [2]. Thus, since it cannot be rigorously 'proved' that requirements are met, a technical basis is demanded to support assertions that requirements are met.

4.1. Definition & Discussion of Independence and Related Concepts

In recent years, the author of this paper has worked to improve understandings of the concept of independence and how it applies to the *assured nuclear weapon safety* approach. This section summarizes some of that work and may be of use in designing safety features and developing a technical basis to justify safety-related independence assumptions associated with those features. This is the definition of independence that has emerged:

Definition of Independence: The occurrence of a state of one or more things does not provide any information regarding the likelihood of occurrence of another state of one or more things, or sequences thereof.

Discussion: The concept of independence provides meaning only when describing the attributes of the relationship between two or more states,[16] or sequences of states, of one or more things.[17] This independence concept is a fundamental pillar in the domains of probability theory and statistics [2, 10-16]. In probability theory or statistics settings, these "states" of "things" are typically described as events, or outcomes of a process or experiment [14]. Therefore, the definition of independent events would be—the occurrence of one event does not provide any information regarding the likelihood of occurrence of another event. Mathematically, this may be written in the form[18]: event (A) and event (B) are independent events if $P(A|B) = P(A)$ and $P(B|A) = P(B)$, thus $P(A \cap B) = P(A)P(B)$. Recall that in general, $P(A \cap B) = P(A) \cdot P(B|A) = P(B) \cdot P(A|B)$, given $P(A) \neq 0$, $P(B) \neq 0$. To reiterate, independent events are defined when the occurrence of one event does not provide any information regarding the likelihood of occurrence of another event, *because* $P(A|B) = P(A)$ and $P(B|A) = P(B)$. It is important to realize that independence is defined by the absence of information.

---

[15] It should be noted that with respect to the aforementioned 'one-point safety' requirements the probability distribution functions describing weapon response behavior are known; therefore, probabilities can be calculated. However, the relevant normal and abnormal environmental contexts extend beyond situations in which the one-point safety probabilities represent bounding probabilities for weapon response.

[16] "State" is defined here as a mode or condition of being [17]. State may be interpreted here to be a single state or a sequence of states.

[17] The definition of "thing" ranges from ideas and concepts to physical objects or processes [17].

[18] Notation: P(A) is the probability of event A occurring. P(A|B) is the conditional probability of "A given that B has occurred." $P(A \cap B)$ means the probability of "A intersected with B," i.e., all points common to both subset "A" and subset "B," or simply the probability of both event A and event B occurring.

Note that the formal definitions of independence are founded upon verification using a sufficient amount of observed data for all relevant contexts. They do not provide much assistance for verification in situations where data are sparse or non-existent [2]. "***Declaring*** events independent for reasons other than those prescribed in (the formal definition) is a necessarily subjective endeavour. In practice, all we can do is look at each situation on an individual basis and try to make a ***reasonable judgment*** (emphasis added) as to whether the occurrence of one event is likely to influence the outcome of another" [15, p. 74].

When discussing independence the terms of "common-cause failure" and "common-mode failure" are often used in a synonymous fashion. However, in the development of a technical basis for independence assertions it is beneficial to give these terms the distinct meanings provided below.

Definition of Common-Cause Failure: Failures involving multiple design attributes of a system that fail as a result of the same causal event or the same type of causal event. The mode of failure across the design attributes may be different (see also the definition for common-mode failures below).

Discussion: Consider a situation in which a flooding event results in the submersion of two components within a subsystem. One component fails to maintain an isolation barrier due to a chemical reaction involving component materials and water. Another component fails to operate due to an electrical short. Both components failed due to a common cause yet the modes of failure were different. Examples of common cause failures include mechanical, electrical, thermal, chemical, biological, moisture, or radiological insults, or failures of common energy connections, materials, power sources, barriers, maintenance, calibration, testing, design, manufacturing, or operational practices. The distinction in terminology between "same causal event" and "same type of causal event" given in the common-cause failure definition is important, since the "same type of causal event" may occur at vastly different times to multiple design attributes within the same system. An environmental example of this situation would be chemical corrosion that slowly leads to penetration of the isolation barrier of one subsystem. Months or years later, the same chemical corrosion process (having breached one barrier) now penetrates the isolation barrier of another subsystem that is nested within the first subsystem. In this case, both the causal events leading to failure were common and the modes of failure were common for the two barriers. A human-related example would involve the performance of the same type of incorrect maintenance or test activity, due to an incorrectly written procedure, on two redundant components within the same system during maintenance or testing events occurring months or years apart. In this case, as in the previous example, the redundant components fail the same way (common-mode) due to the same type of maintenance or testing event (common-cause).

Definition of Common-Mode Failure: Failures involving multiple design attributes of a system that fail in a similar manner as a result of the same causal event or the same type of causal event.

Discussion: Consider a situation in which a combined thermal and crush environment causes two energy isolation barriers associated with two subsystems to rupture in a similar manner. Both components failed in the same mode (rupture) due to the same cause (thermal and crush environment). Common-mode failures can occur across different types of system design attributes or across identical, redundant components such as redundant switches relying on the same conductive electrical energy pathway. Note that the concept of "failure" here refers to a failure to perform an intended function. Failure does not always involve damage to or destruction of an object. For example, two different types of electrical switches may fail in different ways due to a single shock environment. One may violently disassemble (shatter) and another may simply actuate to a closed state. The actuated switch was not damaged, but it may now be in the incorrect state relative to the system-level function it is intended to perform.

To summarize, the distinction made here between common-cause and common-mode makes it easy to conceptually separate the modes of failure manifested across multiple parts of a system given exposure to the same cause. In other words, parts of a system may break or fail to perform their intended function in different ways given exposure to the same adverse, accidental, or otherwise abnormal environment. In many cases this distinction is not necessary, such as when a power plant maintenance worker uses the same faulty procedure to maintain or test a series of identical valves and leaves them all in the open state when they are normally closed valves. The cause of the failure and the mode of the failure are the same for the series of identical valves. In some cases the distinction between common-cause and common-mode is beneficial, as in the case where the same electrical energy insult (a lightning strike event) subjects the system to high-voltage source with a fast rising current. One part of the system fails due to a minor electrical short, another part of the system deflagrates, another part of the system melts, and another part of the system containing a solenoid actuates to an improper state due to the electromagnetic flux of the event. All of these parts of the system experienced different modes of failure due to a common-cause environment.

Now that some key definitions have been provided, let's return to a central question: How can one develop a rigorous, defensible technical basis justifying independence assertions in situations where data are sparse or non-existent? Independence assumptions can be founded upon the distinctions of *functional*, *temporal*, and *physical* differences. Although the concepts of function, time, and physical properties do not provide mutually exclusive categories for describing dependence, they are proposed as helpful concepts when striving to increase independence between states of one or more things.

Functional independence[19] between two or more states of one or more things is increased when the states are achieved by functions that use different types of energy, logical relationships, materials, mechanisms, and/or methods of operation. Therefore, across a wide range of unintended energy or stimuli best described as functionally variant[20], the independence of the state changes observed between two or more states of one or more things subjected to the same energy or stimuli may be anticipated to increase as the functional differences increase between the possible states of the one or more things. Stated differently, to unintentionally remove energy isolation provided by functionally independent safety features, an accident environment would be required to inadvertently accomplish the same overarching objective (e.g., unintentionally remove electrical isolation barriers) via different means. An example of functional independence in the engineered hardware domain would be to have both hydraulic mechanical actuators and electric motor actuators to control aircraft flight control surfaces. They use different types of energy and mechanisms to accomplish the same overarching goal so a failure in one system will not affect the other and the potential for common-mode and common-cause failures is minimized.

Temporal independence between two or more states of one or more things is increased by manipulating the time when states may be achieved. Therefore, across a wide range of unintended energy or stimuli best described as time variant[21], the independence of the state changes observed between two or more states of one or more things subjected to the same energy or stimuli may be anticipated to increase as the time-related differences[22] increase between the possible states of the one or more things. Typically, but not always, this involves maximizing the time separation of states. Stated differently, to unintentionally remove energy isolation provided by temporally separated safety features, an accident environment would be required to inadvertently exhibit stability for longer periods of time to use design features of the functional system as a significant contributor to compromising a safety subsystem. Temporal independence may involve *minimizing* the time of exposure of energy or information to the system, or *maximizing* the time separation of packets of

---

[19] Functional independence as used here is basically synonymous with the term *diversity* as may be used when describing engineered safety systems [18]. It is acknowledged that true diversity is difficult to achieve due to common-cause/common-mode failures; this often results in redundancy but not diversity in practice.

[20] For example, thermal energy, high or low voltage electrical energy, direct current or alternating current electrical energy, mechanical shock or vibration, spurious digital messages or data words on a communication bus, corrosive chemicals.

[21] For example, large fluctuations in the rate at which energy or stimuli is imparted to the one or more things.

[22] For example, time-related differences between states could be achieved by limiting times of mechanical operation, using timer-based functions, and manipulating speeds at which digital information can be communicated.

energy, enabling stimuli, or packets of information provided to the system, or some combination of each approach depending upon the environmental context.

Physical independence between two or more states of one or more things is increased when the states must be achieved on different sides of one or more barriers or there are significant intervening structures.[23] Therefore, across a wide range of unintended energy or stimuli best described as physically variant[24], the independence of the state change responses observed between two or more states of one or more things subjected to the same energy or stimuli may be anticipated to increase as the physical differences increase between the possible states of the one or more things. Physical independence is very similar to functional independence and can also be distinguished using the above examples. However, physical independence emphasizes maximizing separation imposed by various types of barriers or structures.

It is important to note that the functional, temporal, and physical separation strategies for increasing independence are different from the concept of redundancy as commonly used when describing engineered safety systems. The concept of redundancy typically involves providing multiple safety-system components or subsystems of the same type in series (e.g., isolation valves) or parallel (e.g., emergency coolant supply valves) as a hedge against failure modes which manifest primarily due to random, independent failure mechanisms associated with the redundant items [18]. Redundancy may be used, for example, when the time between failures of critical design attributes in the redundant components is exponentially distributed and there are no known common-cause or common-mode failure mechanisms (e.g., those associated with aging or wear) that may degrade the redundant components simultaneously.

Ideally, a technical basis for assuring independence between events would be established that is not greatly affected by all three factors of function, time, *and* physical properties, but achieving such a technical basis is difficult for a system that is designed to execute specific goal-directed behaviors such as a nuclear weapon system[25]. Examples of design features that can create unsafe dependencies between safety subsystems in accident environments include: conductor assignments in cables, choices of materials, printed wiring board layouts, computer programming algorithms, types of power supplies, switch designs, collocated isolation barriers, types of enabling stimuli. The designed-in tendencies toward particular responses resulting from such design features are not random. They are neither equally likely nor independent *a priori*, and it is not clear how to generate a credible technical basis for asserting that such independent responses would result across a wide range of adverse, abnormal, or otherwise unintended environments.

---

[23] Here a "barrier" means the presence or absence of any type of matter or energy within a defined space. For example, an air gap maintained by a predictable barrier would itself be a physical barrier to some types of phenomena, a steel plate barrier would be a barrier to some types of phenomena, and separation of a certain distance across the vacuum of space would be a barrier to some types of phenomena. Note however, that barriers used to implement a nuclear weapon safety theme must be predictable across the range of relevant environments. For example, wire separation between the input and output of a safety device maintained merely by a thin layer of non-conductive insulation does not provide predictable separation since there are failure mechanisms which may readily defeat this type of separation (e.g., certain types of vibration-induced damage that may occur in normal environments).

[24] For example, energy or stimuli that varies with respect to the spatial location at which it is applied to the one or more things.

[25] If the only goal was to provide a uniform distribution of well-defined and easily achieved outcomes the design activity becomes much simpler, but still not trivial. For example, it is possible to design a machine that randomly selects numerical "events" within a certain range—these are called lottery machines. Designing a properly functioning lottery machine requires skill, but it is far more difficult to design a lottery machine that is guaranteed to provide you with one particular sequence of ball draws for all "operationally desired" environmental conditions in a highly reliable fashion, and also provides completely random ball draws for all other environmental conditions.

4.2. Practical Tools to Aid in Generating and/or Identifying Independence Features

The process of testing independence assertions should include development of specific propositions, structured in both positive and negative frames of reference, which are tested using theoretical, analytical, and experimental models to provide sufficient knowledge and experience to defend the claim of predictable response in normal and accident environments. The practical tools below provide a method for constructing independence-related propositions to guide construction of a defensible technical basis for independence assertions.

When applying the definition of independence to a specific analysis, it is helpful to complete the following statement employing a *positive frame of reference*:

_____ is independent of _____ with respect to _____.

For example, the <u>energy isolation features of safety subsystem 1</u> are independent of the <u>energy isolation features of safety subsystem 2</u> with respect to <u>exposure to a high voltage accidental electrical environment</u>.

For example, the <u>steel isolation barriers for subsystem 1</u> provide a safe response independently of the <u>aluminium isolation barriers for subsystem 2</u> with respect to exposure to the chemical _____.

Altering the terminology to create a *negative frame of reference* and requiring open-ended responses for justification is also helpful:

<u>Inadvertent operation of safety interlock 1</u> does not lead to <u>inadvertent operation of interlock 2</u> due to the following independence attributes:
- Enablement energy is different (hydraulic versus electric)
- Enablement stimuli is different (two different uncorrelated patterns of 12 binary pulses)
- ...

Unintentionally circumventing the _____ does not increase the likelihood of unintentionally removing energy isolation provided by the _____ due to the following attributes of independence:
- ...

In accordance with the functional, temporal, and physical decomposition of the independence concept, technical justifications should be required in each of those three areas:

The attributes of independence between the _____ and the _____ with respect to _____ which are best described as (**functional, temporal, physical**) include the following:
- ...

For example, the attributes of independence between the <u>subsystem 1</u> and <u>subsystem 2</u> with respect to <u>energy isolation</u> which are best described as **temporal** include the following:
- Viscous damped interlock in subsystem 1 requires 10 seconds to actuate, the spring pin interlock on subsystem 2 re-latches if not operated within 300 ms.
- Subsystem 1 interlock must operate before subsystem 2 interlock

For example, the attributes of independence between <u>interlock 1</u> and <u>interlock 2</u> with respect to <u>isolating movement of material</u> which are best described as **physical** include the following:
- Interlock 1 is located 25 meters from interlock 2
- The power supply cable for interlock 1 travels through the east-west cable tray, the power supply for interlock 2 travels through the north-south cable tray.

In addition, one question that can aid in probing for dependencies between safety features is:

"If the electrical or information enabling energy or stimuli were changed, would one or both of the safety features need to be modified to accommodate the change to maintain a safe response in a predictable manner?"

If both safety features must be changed then it may represent a lack of independence between the features.

This same type of question can be adapted to investigate independence-related impacts of any change involving the key elements of the safety theme that implement any of the NSDPs.

Another essential question that probes for independence between an intended weapon environment experienced in normal use and any other environment experienced in an accident is:

"If the mechanical enabling energy or stimuli were changed, what design features need to be modified to accommodate the change to maintain a safe response in a predictable manner?"

If any safety features must be changed, then it may indicate pre-storage of some portion of enabling stimuli information that should only originate from the intended human actions, or only from the physical stimuli experienced by the weapon during normal anticipated operations; thus it is *information or energy that should never be pre-stored* in the system.

In summary, independence assertions can be founded upon the distinctions of *functional*, *temporal*, and *physical* differences. Thus the recommended approach is to implement independence between safety subsystems using functional diversity and the separation of energy and information both physically and temporally. In other words, achieve independence by preventing common-cause and common-mode failures when implementing the relevant NSDPs (incompatibility, isolation, inoperability) by scrutinizing the safety subsystems in terms of functional, temporal, and physical differences. This insight is particularly valuable in situations where a technical basis supporting independence assertions must be created when data are lacking to sufficiently demonstrate compliance with the formal definition of independence.

5. CONCLUSION

This paper provided a basic description of the *assured nuclear weapon safety* approach that uses the Nuclear Safety Design Principles (NSDPs) of *incompatibility*, *isolation*, and *inoperability* to design safety features, organized into subsystems such that each subsystem contributes to safe system responses in *independent* and *predictable* ways given a wide range of environmental contexts. Simplicity of safety features, passive safe responses, and a systematic allocation of basic features among engineered features and human actions are emphasized in the implementation, and an innovative *inside out* process for hazard identification, which was described in this paper, is also applied throughout iterative system design phases to support the process of NSDP integration. The central aim of the *assured nuclear weapon safety* approach is to achieve a robust technical basis for asserting that a system can meet stringent quantitative safety requirements in the widest context of adverse or accident environments, while using the most concise arrangement of safety design features and the fewest number of specific adverse or accident environment assumptions. In essence, this approach claims to be an efficient approach for engineering bounded system safety-related responses.

Particular emphasis in this paper was placed on implementation of the NSDPs using independence assumptions founded upon the distinctions of *functional*, *temporal*, and *physical* differences, which are proposed as helpful concepts when striving to increase the independence of safe responses across multiple safety design features within a system. Clear and distinct definitions for *common-cause failure* and *common-mode failure* (with examples) were provided, as were the attributes of *predictability* that strengthen the technical basis for safe system responses. It is hoped that insights presented here, developed within the U. S. nuclear weapon system safety community over the course of more than 50 years, may benefit system safety design, assessment, and management activities in other high consequence domains.

Acknowledgements

In summarizing the 'assured nuclear safety concept' the author has condensed hard-won insights earned by many over a span of more than 50 years. Numerous individuals have helped the author understand the development of U. S. nuclear weapon safety, the most important have been two of the 'founding fathers'—Stanley D. Spray and the late J. Arlin Cooper. With respect to recent refinements in the concept of independence and application of the nuclear safety design principles, thanks are due to the author's cadre of colleagues/sounding boards at SNL and across the nuclear security enterprise. The author expresses particular thanks to the seven reviewers of an earlier version of this paper.

References

[1]   C. Perrow, "*Normal Accidents: Living with High-Risk Technologies,*" Princeton University Press, 1999, Princeton, NJ.
[2]   J. D. Brewer. "*The concept of independence in weapon safety: foundations and practical implementation guidance (SAND2009-4216C)*", Proceedings of the 27th International System Safety Conference, System Safety Society, 2009, Huntsville, AL.
[3]   T. Ord, R. Hillerbrand, and A. Sandberg. "*Probing the improbable: methodological challenges for risks with low probabilities and high stakes*", Journal of Risk Research, 13, pp. 191-205, (2010).
[4]   DoD, "*Department of Defense Manual 3150.2-M: DoD Nuclear Weapon System Safety Program Manual,*" Assistant to the Secretary of Defense for Nuclear and Chemical and Biological Defense Programs, United States Department of Defense, 1996, Washington, DC.
[5]   DOE, "*Department of Energy Order 452.1D: Nuclear Explosive and Weapon Surety Program,*" United States Department of Energy, 2009, Washington, DC.
[6]   DoD, "*Department of Defense Directive 3150.02: DoD Nuclear Weapons Surety Program,*" United States Department of Defense, 2013, Washington, DC.
[7]   USAF, "*Air Force Manual 91-118: Safety Design and Evaluation Criteria for Nuclear Weapon Systems,*" United States Air Force Safety Center (AFSC/SEWN), 2011, Kirtland AFB, NM.
[8]   IAEA, "*IAEA-TECDOC-626: Safety related terms for advanced nuclear plants,*" International Atomic Energy Agency, 1991, Vienna, Austria.
[9]   P. D'Antonio. "*Surety Principles Development and Integration for Nuclear Weapons (SAND98-1557)*", High Consequence Operations Safety Symposium II, Sandia National Laboratories, 1998, Albuquerque, NM.
[10]   A. J. Duncan, "*Quality Control and Industrial Statistics,*" Richard D. Irwin Inc., 1974, Homewood, IL.
[11]   I. Miller and J. E. Freund, "*Probability and Statistics for Engineers,*" Prentice-Hall, 1977, Englewood Cliffs, NJ.
[12]   W. W. Hines and D. C. Montgomery, "*Probability and Statistics in Engineering and Management Science,*" John Wiley & Sons, 1990, New York, NY.
[13]   E. Kreyszig, "*Advanced Engineering Mathematics, 7th ed.,*" John Wiley & Sons, 1993, New York, NY.
[14]   W. J. Conover, "*Practical Nonparametric Statistics,*" John Wiley & Sons, 1999, New York, NY.
[15]   R. J. Larsen and M. L. Marx, "*An Introduction to Mathematical Statistics and Its Applications,*" Prentice-Hall, 2001, Upper Saddle River, NJ.
[16]   D. C. Montgomery, "*Design and Analysis of Experiments,*" John Wiley & Sons, 2001, New York, NY.
[17]   Merriam-Webster, "*Definitions of 'state' and 'thing',*" Merriam-Webster Online Dictionary <http://www.merriam-webster.com/dictionary/state>., Retrieved April 30, 2010.
[18]   R. A. Knief, "*Nuclear Engineering: Theory and Technology of Commercial Nuclear Power, 2nd ed.,*" American Nuclear Society Inc., 2008, La Grange Park, IL.

Advancing Human Reliability Analysis Methods for External Events with a Focus on Seismic

Jeffrey A. Julius, Jan Grobbelaar, & Kaydee Kohlhepp
Scientech, a Curtiss-Wright Flow Control Company, Tukwila, WA, U.S.A.

Abstract: The reliability of operator actions following an external initiating event is a topic that has increased importance following the 2011 seismic-induced tsunami at the Fukushima Daiichi site in Japan. This event has prompted licensees in the U.S.A., and internationally, to reexamine their plant's risk profile and the plant's ability to prevent and/or mitigate damage following external initiating events (external hazards). In support of the industry initiatives to evaluate and prepare for external initiating events, the Electric Power Research Institute and Scientech have developed a preliminary approach to analyze the reliability of operator actions following external initiating events, with a specific focus on seismic events. The preliminary approach has been published in EPRI 1025294, A Preliminary Approach to Human Reliability Analysis for External Events with a Focus on Seismic, in December 2012. Since the development of the 2012 report, the approach and methods suggested in the report have been applied in the development and in the review of seismic PRAs that are currently in development. This paper summarizes the development of the current external events human reliability analysis (HRA) methods and guidance, and summarizes recent insights from applying this approach to seismic PRAs.

Keywords: HRA, Seismic HRA, External Events, External Hazards

1. INTRODUCTION

The purpose of EPRI report 1025294 [1] is to provide methods and guidance for the human reliability analysis of external events PRAs based on the current state-of-the-art in both PRA and in HRA modeling. Prior to the development of this report, substantial research has been performed to develop and improve Human Reliability Analysis (HRA) methods in support of Probabilistic Risk Assessment (PRA). The development of existing HRA methods, however, was limited primarily to internal events PRA, specifically to initiating events that did not involve spatial impact. These methods often contain underlying assumptions that may or may not be applicable to area/spatial impacts, especially those affecting the plant site such as the regional impact following a seismic event, external flood or hurricane (external initiating event or external hazard). Recent HRA advances that culminated in the publication of Fire HRA methods and guidance in NUREG-1921 [2] were considered in the development of EPRI report 1025294.

Additionally, the state-of-the-art in seismic and external events PRA models and issues was surveyed in order to understand existing external events HRA guidance [3, 4]. The results of this review were not surprising. As is common in HRA, there was a wide variation in existing methods for external events HRA. Variation existed between methods used for different hazard types as well as plant-to-plant variation for evaluation of a given hazard type. In addition to reviewing current external events PRA models and methods, a review of historical operating experience was conducted. The relevant insights from the review of operational experience were incorporated into the development of the various steps of the 2012 external events HRA process. The operating experience review task primarily built upon previous, published work conducted by EPRI (Post-Earthquake Investigation Program from 1985 to 2012), as well as a review of utility presentations and LERs. Additionally, interviews were conducted with personnel from nuclear plants impacted by recent seismic events. The review of historical data focused on real-world seismic events at nuclear power plants and other industrial facilities, and it was performed in order to identify potential failure modes and performance shaping factors (PSFs) that should be considered when developing external events HRA

EPRI report 1025294 provides a framework for external events HRA, a general screening approach, and a detailed quantification approach which can be applied consistently across a variety of external events were developed. The report was written to provide methods and guidance for all external events, but included specific guidance for HRA in a seismic PRA, including operator actions to recovery from relay chatter. The detailed quantification approach provided in EPRI report 1025294 is an adaptation of the "EPRI HRA Methodology", also known as the "EPRI HRA Approach" for internal events [5, 6, 7, and 8]. The specific objectives of EPRI report 1025294 are listed below.

- Provide a consistent framework for analysts to perform HRA for all external hazards.

- Provide hazard-specific guidance for consideration of relevant performance shaping factors (PSFs) based on operational experience and existing research.

- Provide a general screening approach and detailed quantification method that can be applied consistently to a variety of external events.

- Provide seismic-specific guidance that reflects, to the extent possible, current research and relevant operational experience.

Paper organization. Section 2 of this paper describes the HRA process as it supports external hazards PRA. Sections 3 through 8 summarize the treatment of external events HRA in EPRI report 1025294. Additional information on the external events HRA approach of EPRI report 1025294 has been described in earlier conference papers [9, 10, and 11]. Section 9 summarizes insights, including areas of potential future research, and conclusions.

2. HRA PROCESS

As with recent HRA guidance, such as NUREG 1921 [2], the external events HRA process is often appears as a linear process with the following elements.

1. Identification and Definition

2. Qualitative Analysis

3. Quantification

4. Model Integration:

    a. Cut set Review and HEP Reasonableness Check

    b. Recovery

    c. Dependency

    d. Uncertainty

Although this process is often depicted as sequential steps, in the practical application to developing an HRA these steps are iterative. EPRI 1025294 [1] presents the guidance in the order which an HRA analyst is likely to use the various elements of the guidance, accounting for the iteration between screening and detailed assessments. Figure 1 provides a mapping between the external events HRA process and the sections of EPRI 1025294. This figure shows the iterative relationship between the PRA process, the HRA process and allows both tasks to proceed in parallel.

EPRI report 1025294 provides two approaches for quantification, first a screening quantification and then a detailed quantification. The screening approach is intended to require fewer resources and be more conservative than the detailed quantification.

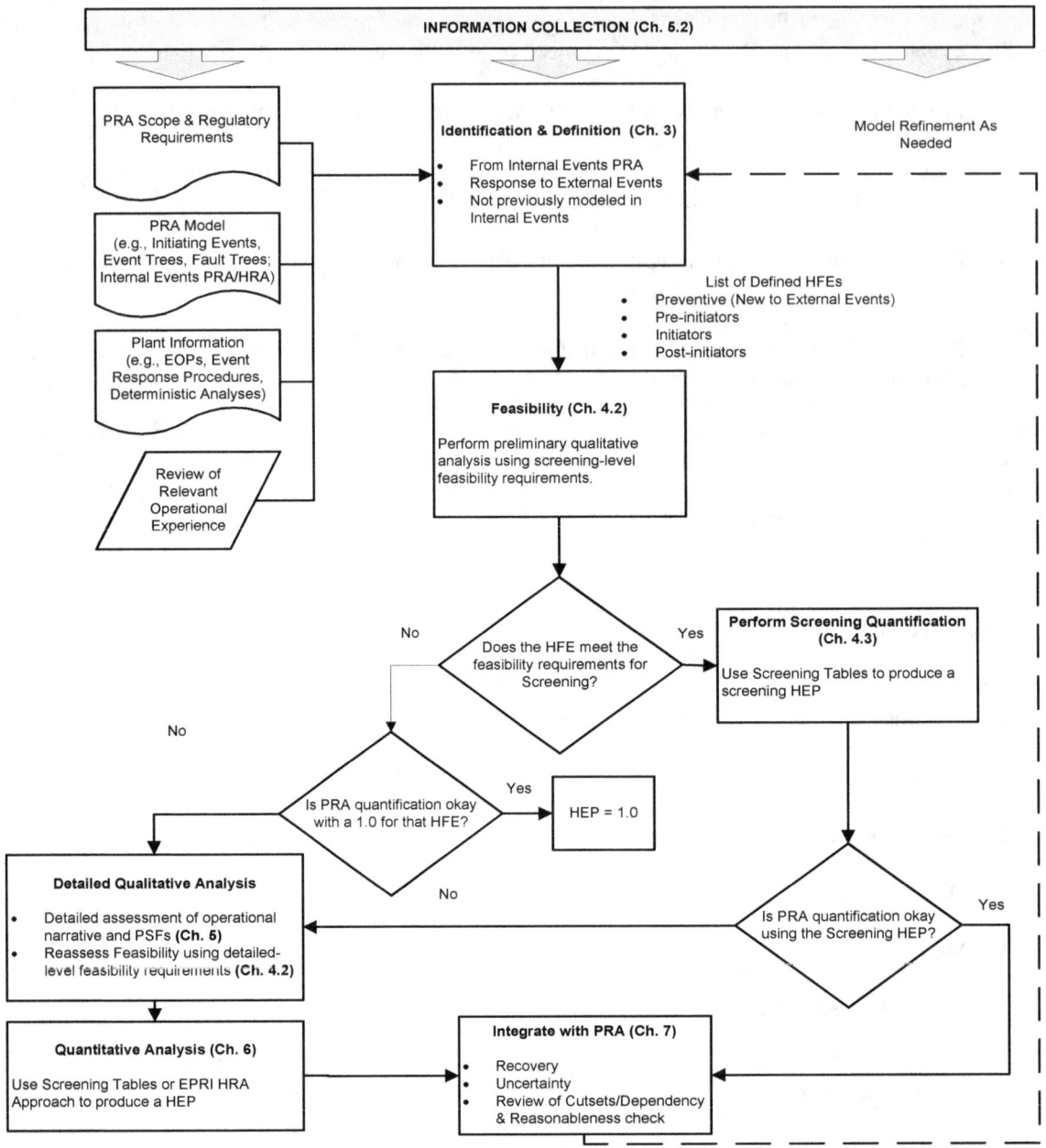

Figure 1.
Mapping of HRA Steps To Sections Within EPRI 1025294 [1]

3. IDENTIFICATION AND CLASSIFICATION

The identification process in an external events PRA follows the same approach as in an internal events PRA – to understand the plant response (including the procedures to be implemented) and to understand how the plant response is captured in the PRA model. Within the PRA, operator actions typically come from one of two sources: 1) HFEs already existing in the PRA (usually from the internal events PRA); or 2) procedure review in conjunction with modeled accident sequence review to identify new operator actions.

In general the following groups of procedures are reviewed for applicability to external events.

- Preventive procedures – procedures for preparing for high winds, hurricane or other external events where the onset of the event is known beforehand, typically when the hazard is imminent.

- Response procedures – Those procedures used in response to an initiating event. Response procedures include: Emergency Operating Procedures (EOP), Abnormal Operating Procedures (AOPs), Alarm Response Procedures (ARP), Severe Accident Management Guidelines (SAMGs), fire procedures, and seismic or other external event procedures (including FLEX procedures).

- Normal operating procedures (NOP, also known as operating procedures). Those procedures used in day to day plant operation. These procedures include normal shutdown and start up procedures, system alignment procedures, and test and maintenance procedures. These plant specific procedures are well trained on and the wording is standardized across the complete procedure set.

For external events HRA, there are three types of post-initiating event operator actions.

- Internal events operator actions
- Preventive operator actions
- External event response operator actions

The internal events operator actions associated with these HFEs are actions required in response to a plant initiating event and/or reactor trip, typically directed by the EOPs, ARPs, AOPs, and/or NOPs. Because internal events operator actions have been identified, their HFEs defined, and their HEPs quantified as part of the internal events HRA, it is not necessary to repeat the internal events HRA identification process. All that is required for the external events PRA identification process is to determine which of these HFEs could occur in external events scenarios.

Preventive actions would be plant and external event specific, and the identification of these actions would be performed by a review of procedures and discussions with plant operations. These actions would typically be included in the external events PRA on as-needed bases. Preventive operator actions are an area of ongoing study, and while they are not explicitly within the scope of EPRI 1025294 [1], they are subject to the same feasibility criteria described in that report. Example of preventive actions could include:

- Closing doors or placing flood barriers, such as sand bags or drain plugs, prior to flood damage

- Transporting additional diesel fuel on site prior to an expected prolonged loss of offsite power such as a hurricane.

- Staging portable equipment (e.g., preparing to implement FLEX options)

External events response actions are new post-initiating event operator actions used to mitigate the effects of an external event. This category of HFEs is typically not included in the EOP/AOP network of procedures. These operator actions are identified by review of the external event response procedures in conjunction with the modeled PRA functions and sequences.

Response actions are sometimes called *recovery actions*, and may appear in event trees or the fault tree portions of the PRA. Response actions consist of the following types of actions.

- Terminating the impact of the external initiating event – actions taken to identify and protect components that are operating in an undesired state or are threatened after the external event has occurred. These are somewhat analogous to preventive actions described above, but often have a shorter time window. For example, if a power-operated relief valve spuriously opens due the initiating event, the failure may be able to be recovered by de-energizing the valve.

- Mitigation of external initiating event consequences using the affected SSC – actions taken to recover failed SSCs by providing an alternate success path. For example, actions taken in response to a seismically-induced LOSP and SBO due to relay chatter preventing load sequencers from loading EDGs and equipment loads onto vital AC buses. The HFE models operators resetting circuits/relays from the control room or in the switchgear rooms, restarting the EDGs and loading equipment manually if the load sequencer remains unavailable. Note – human reliability analysis does not address repair of failed components.

- Mitigation of external initiating event consequences using alternate components – actions taken to recover failed SSCs by providing an alternate success path. For example, restoration of power to an electrical bus by aligning an alternate component such as a standby rectifier/inverter or a source (such as a skid-mounted diesel generator used for FLEX). Note – human reliability analysis does not address repair of failed components.

Regardless of how the operator action is identified, the corresponding HFE must be defined for use in the external events PRA. The human failures are defined to represent the impact of the human failures at the function, system, train, or component level as appropriate.

For new actions, the definition should start with the collection of information from PRA and engineering analyses. For actions carried over from the internal events, the existing definition should be reviewed and modified as-needed to account for the new context of the external event.

4. FEASIBILITY

The HRA for most spatial analyses is typically performed in conjunction with the PRA development. Because tasks of the PRA are typically developed concurrently, not all of the information required to perform a detailed HFE quantification will be known initially and the PRA will need screening HEP values initially to develop and quantify the risk model.

For a screening analysis, the HFE definition and feasibility assessment are conducted simultaneously as part of the initial qualitative analysis. If a more detailed analysis is needed, then the initial qualitative analysis should be further developed. Prior to performing the qualitative analysis, if the operator action did not pass the screening-level feasibility assessment, the feasibility should be reassessed after gathering more details.

Regardless of when the feasibility assessment is conducted or the level of detail of the current PRA, the feasibly assessment needs to consider the following, at a minimum.

- Timing
- Manpower
- Cues

- Procedures and training
- Accessible Location & Environmental Factors
- Tools and equipment operability.

If the operator action is feasible, the analyst can proceed to perform either a screening or a detailed quantification. If the analyst finds the screening to be too conservative or limiting, the analyst is encouraged to apply the detailed HRA method. EPRI 1025294 [1] provides additional detail on considerations for each of these feasibility criteria.

5. SCREENING ANALYSIS & QUANTIFICATION

The screening process is optional, but it provides a set of HEPs for the initial PRA model quantification and helps identify the important sequences. The ranking can be used to determine which sequences might be further analyzed to reduce the calculated risk by detailed modeling.

The screening method provided was initially developed specifically for application in developing a seismic risk assessment. However, with the current state-of-knowledge, it is reasonable to use the described screening approach in EPRI 1025294 [1] as a screening method for other external events, with the caveat that future research on other external events may require this approach to be modified to incorporate relevant operating experience.

5.1 Step 1- Identify Damage State of the Plant Following the External Initiating Event

Both the screening HRA method and the detailed HRA method start by asking the analysts to identify the damage state of the plant following the external initiating event. The damage state is intended to account for the overall context resulting from the external event beyond the specific failures dictated by the cut set, including impact to local infrastructure and non-safety related systems, level of heightened stress, general increase in level of coordination and workload, and quality of working environment. These damage states, described in Table 1, were selected based on the definitions provided in EPRI NP-6695 and its update EPRI 1025288 [4], but have been adjusted here to correlate more closely with the impact of the context on operator performance. Because the design basis for the range of external events can vary substantially from plant to plant, the bins selected here reflect the effect of the external event on the plant rather than providing absolute values (e.g., PGA values). Bin definition is generic for seismic because the seismic hazard is not straightforward, and there is not a direct correlation between hazard level and damage state. Recommendations for seismic HRA provided in Table 1 may not fit the damage state definition appropriately for every plant, and is provided only as a starting place when no other information is available; it is expected that the HRA analyst will have to interface with the PRA analyst to correlate the damage states provided here with the hazard bins (e.g., ground motion intervals) used in the PRA. Note: The SSE is a convenient, but generally very conservative value; higher values could be justified. HCLPF recommendations here may also be overly conservative.

Table 1
Damage State Definitions For Screening

Bin #	External Event Damage State Description	Recommended Link to Seismic Hazard
1	No damage to the plant safety-related SSCs or non-safety SSCs required for operation. Limited damage to non-safety, non-seismic designed SSCs like residences and office buildings.	Below the SSE.
2	No expected damage to the plant safety-related SSCs or to rugged industrial type non-safety SSCs required for operation. Damage may be expected to non-safety SSCs not important to plant operations and to the switchyard (e.g., LOOP expected). Falling of suspended ceiling panels.	At or above the SSE, up to HCLPF of most fragile safety-related SSC. (e.g., 2011 North Anna event)
3	Widespread damage to non-safety related SSCs and/or some damage expected to safety related SSCs. Significant number of vibration trips and alarms requiring resetting.	Above the HCLPF of most fragile safety-related SSC to HCLPF of critical instrumentation or HCLPF level of 25th percentile component, whichever is lower. (e.g., 2007 Kashiwazaki-Kariwa, 2011 Onagawa events)
4	Substantial damage to safety related and non-safety SSCs. The threshold of this damage state is such that it produces a cliff-edge effect in the likelihood of operator response.	Wide-spread damage to critical instrumentation. (2011 Fukushima Daiichi and Daini events)

5.2 Step 2 - Plant Damage Assessment

Another consideration that appears in the screening trees is based on the plant damage assessment. Following an external initiating event often times the entire site is affected and the effects (such as flood water obstructing access, high radiation areas, and/or damaged equipment) can impact human performance far after the event is over. Thus the external initiating event can have impacts on both cognition and execution that last a considerable amount of time. There is expected to be an overall reduction in workload and complexity once the site has been assessed and the extent of the damage understood.

The damage assessment has been defined in the external events HRA as a break-point for both cognition where not only is the damage to the plant known (after the break-point), but also the workload and distractions associated with determining the impact of the event are reduced. The cognitive load on the operators is reduced because they have a clearer picture of the damage inflicted on the plant due to the initiating event, including an understanding of what equipment is damaged but may not have failed yet. By this point in the scenario it is also expected that the crew have had the opportunity to implement basic "working solutions" to compensate for issues caused by the external event (e.g., determining usable pathways, establishing alternate means of communication, etc.). For seismic events, a detailed plant damage assessment consists of a post-event walkdown and is usually performed within 4 to 8 hours of the event. This initial walkdown should not be confused with the more involved, formal damage assessment required prior to restart to identify issues which might degrade long-term reliability of components, typically called "post-shutdown inspections and tests".

5.3 Step 3 – Assessment of Time Margin

A review of the operating experience suggests that the PSFs associated with seismic events manifest themselves most often as a delay, rather than direct failure, of the operator action. Therefore, the level of credit assigned at the screening level is dependent upon the amount of time margin – or tolerance for unexpected delays – available.

5.4 Step 4a - Quantify Screening HEP For HFEs From Internal Events PRA

For operator actions carried over from the internal events PRA into the external events PRA, the internal events qualitative and quantitative analysis can be used as the starting point for the external events PRA quantification. A simple decision tree has been developed to show how to determine a multiplier to apply to the internal events HEP. The event tree considers the following headings in the development of the HEP. The end state of the decision tree branches are either a screening HEP or a multiplier for the internal events HEP. Multipliers range from 2 to 50, and screening HEPs range from the internal events HEP value to 1.

- Immediate, memorized action (or not)
- Action location
- Damage state
- Time margin consideration
- Cue before or after plant damage assessment

5.5 Step 4b - Quantify Screening HEP For New External Events HFEs

For new operator response actions that were not carried over from the internal events PRA into the external events PRA, multipliers are not applicable, but the same factors are used to determine a screening HEP. Screening HEPs were developed by selecting a base human error probability (BHEP) then applying the same multipliers used in the internal events HFEs described above; the screening values were given no more credit than 1.0E-2.

6. DETAILED ANALYSIS & QUANTIFICATION

By the time the analyst has reached the stage requiring a detailed analysis of the HFE, the HFE has been defined and the basic feasibility has been assessed. The HFE definition and feasibility comprise the foundation of the qualitative analysis. The feasibility criteria for screening is more stringent than that required for detailed analysis (e.g., the screening requirement for an action to be considered feasibility is that the *primary* cue must be available, whereas the detailed analysis stipulates that either the *primary* or *secondary* cues must be available). Therefore, additional data gathering and analysis may need to be performed to satisfy the feasibility criteria for a detailed analysis if the feasibility criteria for the screening analysis were not met.

Qualitative analysis is an essential part of an HRA. The objectives of qualitative analysis are to: understand the modeled PRA context for the HFE, understand the actual "as-built, as-operated" response of the operators and plant, and translate this information into factors, data, and elements used in the quantification of human error probabilities.

Recent experimental studies have shown that the quality of the quantitative analysis is strongly impacted by the quality of the qualitative analysis, even for fairly prescriptive methods such as the EPRI HRA Methodology. ERPI 1025294 [1] provides detailed guidance on performing a thorough qualitative analysis, using insights from seismic operating experience.

The EPRI HRA Methodology (also known as the EPRI HRA Approach) is based on EPRI's SHARP and SHARP1 [5] HRA framework. After the qualitative analysis has been performed, a detailed quantification is performed using methods recommended by EPRI within the HRA approach. Specifically one or more of the following methods:

- Cognitive Methods. The Human Cognitive Reliability/Operator Reliability Experiment (HCR/ORE) and/or Cause-Based Decision Tree Method (CBDTM) [6,7] for cognition.
- Execution. Technique for Human Error Rate Prediction (THERP) [8] for execution.

One advantage of using existing methods for external events HRA is that, at a minimum, the same fundamental aspects and factors affecting human performance apply to Level 1 internal events PRA as well as external events PRA —therefore, applying these methods to external events scenarios should yield a good first-order approximation of operator failure and would further be consistent with the modeling for non-external events scenarios at many nuclear power plants. Although the methods used for external events HRA modeling are the same as those used for Level 1 internal initiating events, EPRI 1025294 provides guidance on how to make the relevant selections within the EPRI HRA Methodology to appropriately account for the impacts of the external events defined in the qualitative analysis.

7. FLOOR HEP FOR HIGH DAMAGE STATES

For extremely high damage states, the uncertainties dominate, so EPRI report 1025294 [1] recommends that a floor HEP to reflect the uncertainty associated with the plant damage. The floor HEP is treated as a lower bound. If the external events HRA calculates HEPs below the lower bound then they will not be used and the floor HEP will be used instead. Based on accounts of historical events, operational experience data has shown that it is possible for operators to become confused or distracted by multiple, conflicting indications such as spurious instruments or alarms or many failures caused by a highly damaging event. In theory, operators should be focused only on the safe shutdown paths, associated equipment, and instruments and alarms as directed by the applicable procedures. However, in a complicated scenario such as following a spatial event like a seismic event, maintaining this focus might be difficult. In addition, good reasons might exist for the operators to have a wider scope of attention (e.g., secondary-side systems or equipment that is commonly important during normal operations and systems or equipment of recent concern as a result of current plant configurations and preexisting conditions).

Sensitivity studies could be conducted to identify whether the applied lower bound limit has little (or no) effect, a significant effect, or perhaps a moderate effect. Effects might be represented and evaluated simply as different values of HEPs to represent the HEPs associated with different conditions for the same HFE.

8. MODEL INTEGRATION

Once the HEPs have been quantified at the appropriate level, the operator actions and associated HEPs must be appropriately integrated into the PRA model. Model integration consists of various tasks, depending on the PRA model, including: cut set review, HEP reasonableness check, recovery, dependency and uncertainty. EPRI 1025294 [1] provides guidance on these elements of model integration in the context of external events.

9. INSIGHTS AND CONCLUSIONS

EPRI report 1025294 [1] provides methods and guidance for the human reliability analysis of external events PRAs. EPRI report 1025294 was developed using insights from recent HRA advances from Fire HRA [2], as well as considering the state-of-the-art in seismic and external events PRA models and issues [3, 4]. EPRI report 1025294 provides a framework for external events HRA, a general screening approach, and a detailed quantification approach which can be applied consistently across a variety of external events were developed. The report was written to provide methods and guidance for all external events, but included specific guidance for HRA in a seismic PRA. The detailed quantification approach provided in EPRI report 1025294 is an adaptation of the "EPRI HRA Methodology", also known as the "EPRI HRA Approach" for internal events [5, 6, 7, and 8].

In 2013, initial testing was conducted on the proposed guidance in EPRI report 1025294. The testing was limited to pilot plants that were in the process of developing seismic PRA and also to testing the concepts on plants that were conducting seismic PRA peer review. The objective of the testing was to obtain insights that would be used to refine the external events HRA methods and guidance. As part of the testing, a gap analysis was conducted on the ability of current human reliability analysis (HRA) methods to support current requirements of external flooding risk assessments [11]. The gap analysis started with a review of the requirements from the ASME/ANS PRA Standard [12] and the requirements of the Interim Staff Guidance (ISG) for the Flooding Integrated Assessment [13].

The insights from the gap analysis between the current external flood risk assessment requirements and the current state-of-practice, and the insights from the current testing, are summarized below.

- The external events HRA process (including external flood and seismic) is identical to the Fire HRA process, which includes a screening step that the internal events PRA (IEPRA) does not typically require for the IEPRA post-initiator HRA.

- External flooding requires a new category of HRA events for external flooding – those operator actions taken as preventive measures. This category includes preparatory measures such as building isolation. This new category of actions should have the same engineering treatment and modelling considerations as post-initiating event actions.

- Feasibility of operator actions applies to all types of operator actions in all hazard groups, including external flooding. A strong qualitative analysis represents the best means for supporting a detailed HRA evaluation in order to demonstrate compliance with PRA Capability Category II [3], specifically supporting requirement HR-G3 for the incorporation of plant-specific and scenario-specific factors.

- Quantitative methods for external flooding actions will likely have difficulty demonstrating compliance with PRA Capability Category II due to limitations in existing methods such that it can be difficult to tell if the resultant human error probability is conservative (Capability Category I) or best-estimate (Capability Category II) for supporting requirement HR-G1.

- Quantification of human error probabilities has the following issues related to each category of operator action: 1) Preventive actions – in general, qualitative analysis and quantitative methods need to be developed and refined, 2) Post-initiating event actions – in general, qualitative and quantitative methods essentially follow the guidance of NUREG-1921 (Chapter 4), which is being updated in the preliminary EPRI External Events HRA Guidance document [2].

- Uncertainty and dependency considerations are the same, although new sources of uncertainty are introduced.

A pilot of the EPRI 1025294 approach is underway as part of a seismic PRA development. At the same time that this preliminary guidance is being tested and the pilot being conducted, EPRI is updating its methodology for seismic PRA. This seismic HRA method will be updated and finalized based on the lessons learned from this pilot and based on changes in the general seismic PRA guidance. The general objectives of the pilot study are listed below.

1. better understand the seismic HRA issues and how they interact with the SPRA
2. test the screening method guidance for usability and reasonableness
3. test the detailed method guidance for usability and reasonableness
4. identify any gaps in the method

EPRI 1025294 [1] is considered to be a preliminary draft as it is expected that this guidance will be updated in the future. The update is expected to include a review of operational experience and additional guidance specific to other external events (e.g., External Flooding and High Winds).

10. REFERENCES

[1] *A Preliminary Approach to Human Reliability Analysis for External Events with a Focus on Seismic.* EPRI, Palo Alto, CA: 2012. EPRI 1025294.
[2] *EPRI/NRC-RES Fire Human Reliability Analysis Guidelines.* U.S. NRC, Washington DC: 2012. NUREG-1921/EPRI 1023001.
[3] *Pre-Earthquake Planning and Immediate Nuclear Power Plant Operator Post-Earthquake Actions.* U.S. Nuclear Regulatory Commission Regulatory Guide 1.166, 1997
[4] *Guidelines for Nuclear Plant Response to an Earthquake.* EPRI, Palo Alto, CA: 1989. NP-6695. This reference includes the Technical Update, published October 2012 under the same title (EPRI Report 1025288).
[5] *Operator Reliability Experiments Using Nuclear Power Plant Simulators.* EPRI, Palo Alto, CA: 1990. NP-6937, as supplemented by EPRI TR 100259 [6].
[6] *An Approach to the Analysis of Operator Actions in Probabilistic Risk Assessment.* EPRI, Palo Alto, CA: 1992. EPRI TR-100259.
[7] *Handbook of Human Reliability Analysis with Emphasis on Nuclear Power Plant Applications (THERP)*, A. D. Swain and H. E. Guttman, U.S. NRC, Washington DC: 1983. NUREG/CR-1278.
[8] *Systematic Human Action Reliability Procedure (SHARP) Enhancement Project: SHARP1 Methodology Report.* EPRI, Palo Alto, CA: 1992. EPRI TR-101711.
[9] *A Review of Seismic Operating Experience with Implications for Human Reliability*, Mary Presley et al, presented at PSA 2013, American Nuclear Society sponsored Probabilistic Safety Assessment Conference, Columbia, SC, September, 2013.
[10] *A Preliminary Approach to Human Reliability Analysis for External Events with a Focus on Seismic HRA*, Mary Presley et al, presented at PSA 2013, American Nuclear Society sponsored Probabilistic Safety Assessment Conference, Columbia, SC, September, 2013.
[11] *Gap Analysis Insights Between External Events HRA Requirements and Current HRA Methods*, Mary Presley et al, presented at PSA 2013, American Nuclear Society sponsored Probabilistic Safety Assessment Conference, Columbia, SC, September, 2013.
[12] *Standard for Level 1/Large Early Release Frequency Probabilistic Risk Assessment for Nuclear Power Plant Applications*, American Society of Mechanical Engineers (ASME), 2009, ASME/ANS RA-Sa-2009.
[13] *Guidance for Performing the Integrated Assessment for Flooding*, United States Nuclear Regulatory Commission (USNRC), 2012c, JLD-ISG-2012-05, Rev. 0 draft, September 2012.

Expected maintenance costs model for time-delayed technical systems in various reliability structures

Anna Jodejko-Pietruczuk, Sylwia Werbińska-Wojciechowska
Wroclaw University of Technology, Wroclaw, Poland

Abstract: In the article, there is presented the mathematical model of expected maintenance costs of two- and multi-unit system in a single cycle of operation (between (i-1)th and ith time moments of inspection action performance), provided that at the beginning of the inspection cycle system elements are in the same age and show no signs of forthcoming failure. The mathematical modelling of maintenance decisions for such a system is provided with the use of delay-time analysis. Moreover, there was examined the compatibility of developed analytical model with a simulation model. The directions for further research work are defined.

Keywords: delay-time modelling, multi-unit systems, analytical model, simulation.

1. INTRODUCTION

As machine become more complex, the possibility that failures or deterioration of machine systems can result in economic losses increases. Much importance has been, therefore, attached in maintenance to reduce such losses.

One of the main functions of maintenance is to control the condition of facilities. A technique called delay time analysis has been developed for modeling the consequences of an inspection policy for any systems [8]. The central concept here is a delay time h, of a fault, which is the time elapsed from when a fail could first be noticed until the time, when its repair can be delayed no longer because of consequences (see e.g. [6, 7]) This concept, which provide useful means of modelling the effect of periodic inspections on the failure rate of repairable technical systems, was developed by Christer et al., see e.g. [8, 11, 12, 14, 31].

Known in the literature technical systems maintenance models which base on delay-time concept implementation, can be divided into two main groups [34]:
- inspection models for single-unit or complex systems;
- inspection models for multi-unit systems.

The maintenance modelling issues for single-unit and complex systems have been extensively analyzed in the literature see e.g. [20]. In the case of a repairable component, it is possible to model the reliability, operating cost and availability functions when pdf of delay time $f_h(h)$ and pdf of initial point u $g(u)$ are known. In the case of multi-component or complex system, the arrival pattern of defects within the system is modelled by an instantaneous arrival rate parameter $\lambda(u)$ at time u. If $\lambda(u)$ is constant, the model is a Homogeneous Poisson Process type (HPP), otherwise it is of a Non-Homogeneous Poisson Process type (NHPP) [27].

The inspection models for multi-component systems are widely known in the literature. The basic multi-component system delay-time model is given in e.g. [9, 13, 14, 35]. The basic assumptions include perfect inspection case, independent system components, the number of defects arising over T following a HPP with a constant rate λ per unit time, known pdf $f_h(h)$, regular inspection actions performance which requires a constant time d. Following these assumptions, probability of defect arising as a failure $P_b(T)$ and downtime per unit time $E_d(T)$ may be estimated. The basic extensions for analyzed delay time model regard to the non-perfect inspection case (see e.g. [4, 32, 33]), or multiple nested inspections case (see e.g. [36, 31]).

Moreover, in the literature there can be also found multi-component inspection models, which assume that the defect arrival process is non-homogeneous (see e.g. [7, 15, 16, 35]). Aspects of testing for trend and fitting a NHPP process to data are discussed e.g. in [3]. Moreover, the imperfect inspections case is analyzed in e.g. [4, 32].

Another important aspect which cannot be neglected is the problem of delay time model parameters estimation. This research issues are investigated e.g. in [7, 33, 34, 38].

The presented above maintenance models regard to PM policy consideration. The implementation of condition monitoring policy in delay-time models for complex systems is given e.g. in [5, 18, 37].

Many works have been carried out on the DT modelling to production plants (e.g. [1, 26]). Other works include the application to gearbox failure process (see e.g. [28]), modelling PM for a vehicle fleet (see e.g. [12, 17, 19, 30]), PM process for a coal face machinery [5], wind turbine maintenance optimization (see e.g. [2]), modelling maintenance of fishing vessel equipment (see e.g. [29]), or medical equipment (see e.g. [10]).

To sum up, most of the known in literature maintenance models regard to complex or multi-unit systems performance optimization issues. However, when assuming that the working elements are in non-series reliability structure, the analytical delay-time models are almost not investigated. Following this, authors focus on the development of analytical maintenance model of technical systems performing in various reliability structures. Thus, in the next Section, there is presented the investigated maintenance model for two-element and n-element system. Following this, in the next step, authors examine the compatibility of developed analytical model with simulation results, obtained from the model presented e.g. in [23]. Based on this research, it was possible to analyze the possibility of the presented model use for time between inspections period optimization issues. The work ends up with summary and directions for further research.

In conclusion, this article is to be the continuation of consideration about mathematical modelling of technical objects with time delay maintenance, presented e.g. in [20]. Moreover, it is also a continuation of the considerations about future research, connected with delay time modelling for complex and multi-unit systems in given reliability structure, developed e.g. in [21, 23, 24, 25, 39, 40].

2. DELAY-TIME MODEL FOR MULTI-UNIT SYSTEMS

In this Section, first there is considered a basic delay-time model for two-unit systems. It is assumed, that system elements working independently under the same conditions. Moreover, components are prone to become defective independently of each other when the system is in operating. In the second step of analytical model formulation, there is made an assumption that system is comprised of n elements. Moreover, there is investigated system elements performance in three main reliability structures – series, parallel and k-out-of-n.

The performed maintenance policy bases on Block Inspection policy (*BI*) which assumes, that inspections take place at regular time intervals of T, and each requires a constant time. The inspections are assumed to be perfect. Thus, any component's defect, which occurred in the system till the moment of inspection, will be identified. All elements with identified defects will be replaced within the inspection period.

In the analyzed system may be performed one of the two maintenance operations: failure repair or inspection together with replacement of elements with defects. Following this, it is assumed that when a system failure occurs, there is only performed replacement of failed components without additional inspection action performance. However, in the case of planned inspection action performance, the replacement will be performed only for those elements with visible symptoms of their future damage.

The performance of the investigated system is also defined by the additional assumptions:

- the system is a three state system where, over its service life, it can be either operating, operating acceptably or down for necessary repair or planned maintenance,
- failures of the system are identified immediately, and repairs or replacements are made as soon as possible,
- inspection action performance begins the new maintenance cycle for the analyzed system,
- maintenance actions restores system to as good as new condition,
- the system can remain functioning in an acceptable manner until breakdown (despite having defects),
- system incurs maintenance costs of: new elements, when they are replaced c_r, inspection costs c_i, and some additional, consequence costs, when system fails c_f;

- time of defects occurrence defined by u is a random variable described by its probability distribution functions $g(u)$ and $G(u)$,
- the length of the delay time before element's failure is random and its probability distributions are given by $f_h(h)$ and $F_h(h)$.

Moreover, there should be underlined here, that the developed mathematical model gives the possibility for estimation of expected maintenance costs for system, which elements are as good as new at the beginning of the maintenance cycle (e.g. first maintenance cycle performance). The maintenance cycle is here understood as the time between the two consecutive inspection actions performance.

2.1. Reliability models of system with delay time working in various reliability structures

When system performs in a series reliability structure, there is a possibility to define the *C.d.f.* of time to failure, $F(x)$, as the convolution of u and h such that $u + h \leq x$:

$$F(x) = 1 - \left[1 - \int_{u=0}^{x} g_1(u) F_{h1}(x-u) du\right] \cdot \left[1 - \int_{u=0}^{x} g_2(u) F_{h2}(x-u) du\right] \quad (1)$$

And the system reliability function: $R(x) = 1 - F(x)$.

The formulae (1) for multi-unit systems may be defined as:

$$F(x) = 1 - \prod_{j=1}^{n} \left[1 - \int_{u=0}^{x} g_j(u) F_{hj}(x-u) du\right] \quad (2)$$

When the system elements work in a parallel reliability structure, the *C.d.f.* of time to failure, $F(x)$, given by the formulae (1) may be estimated as:

$$F(x) = \int_{u=0}^{x} g_1(u) F_{h1}(x-u) du \cdot \int_{u=0}^{x} g_2(u) F_{h2}(x-u) du \quad (3)$$

and respectively for multi-unit systems as:

$$F(x) = \prod_{j=1}^{n} \int_{u=0}^{x} g_j(u) F_{hj}(x-u) du \quad (4)$$

Taking into account n-element system, when its elements work in the most general, k-out-of-n reliability structure, the system reliability function may be estimated as:

$$R(x) = \sum_{l=1}^{m} R^l(x) \quad (5)$$

where:
$R^l(x)$ – probability of system correct operation related to lth combination of up-stated elements providing the system being up state
m – number of possible combinations of up-stated elements which provide the up-state of the system (number of system up states)

And the system *C.d.f.* of time to failure: $F(x) = 1 - R(x)$.

The probability of system correct operation for lth combination of system elements being in up-state in order to system up-state providing, may be estimated as:

$$R^l(x) = \prod_{j=1}^{n} [R_j(x)]^{e_j} [1 - R_j(x)]^{(1-e_j)} \quad (6)$$

where:
e_j – indicator defined as follows:

$$e_j = \begin{cases} 1, & \text{if } j\text{th element in } l\text{th combination is in up state} \\ 0, & \text{if } j\text{th element in } l\text{th combination is failed} \end{cases} \quad (7)$$

$R_j(x)$ – element's reliability function, given by the formulae:

$$R_j(x) = 1 - F_j(x) = 1 - \int_{u=0}^{x} g_j(u) F_{hj}(x-u) du \quad (8)$$

As a result, formulae (6) can be expressed as:

$$R^l(x) = \prod_{j=1}^{n} \left[1 - \int_{u=0}^{x} g_j(u) F_{hj}(x-u) du\right]^{e_j} \left[\int_{u=0}^{x} g_j(u) F_{hj}(x-u) du\right]^{(1-e_j)} \quad (9)$$

2.2. Maintenance model for system with delay time

The expected costs of two-element system maintenance in one inspection cycle are defined as:

$$C(T_i) = C(t_{i-1}, t_i) = \frac{c_f F(t_i) + c_r(1-F(t_i))(G_1(t_i)+G_2(t_i)) + c_i(1-F(t_i))}{(1-G_1(t_{i-1}))(1-G_2(t_{i-1}))} \quad (10)$$

The maintenance cost expressed in the equation (10) presents the sum of possible cost: of a system failure, replacement cost of working elements with observable defects and inspection costs per a single inspection period. The formulae may be used under the condition that all the system elements were as good as new at the beginning of the cycle. The model may be developed to the form usable for multi-unit systems:

$$C(T_i) = C(t_{i-1}, t_i) = \frac{c_f F(t_i) + c_r(1-F(t_i))\left(\sum_{j=1}^{n} G_j(t_i)\right) + c_i(1-F(t_i))}{\prod_{j=1}^{n}(1-G_j(t_{i-1}))} \quad (11)$$

3. CONVERGENCE OF THE MODEL WITH LITERATURE FINDINGS

The majority of literature findings deal with results of simulation maintenance models of multi – unit system with delay time. In order to present the convergence of the foregoing model and simulation ones, presented in the literature, authors focus on the general case for system in *k-out-of-n* reliability structure. The analysis results for other reliability structures were given in [22].

The cost results for chosen case of *k-out-of-n* system are depicted in the figures 1-4. The costs are presented in the function of the length of inspection period (*T*). The *Monte Carlo* simulation was conducted for the first inspection cycle to realize assumption regarding all new elements at its beginning. The table 1 presents system parameters assumed in the simulation model.

Table 1: Simulated system parameters

Notation	Description	Value
k out of n	Reliability structure of a system	*2 out of 3*
c_e	The unit cost of new element	5
c_i	The unit cost of inspection	1
c_c	The unit cost of consequences pertaining a system failure	100
$G(u)$	Probability function of *u* period (the time that elapses from the moment when a defect arises in a new element)	$G(u) = 1 - e^{-(u/75)^{3.5}}$
$F_h(h)$	Probability function of *h* period (the time between first signals of defect and element's failure)	$F_h(h) = 1 - e^{-(h/35)^{3.5}}$

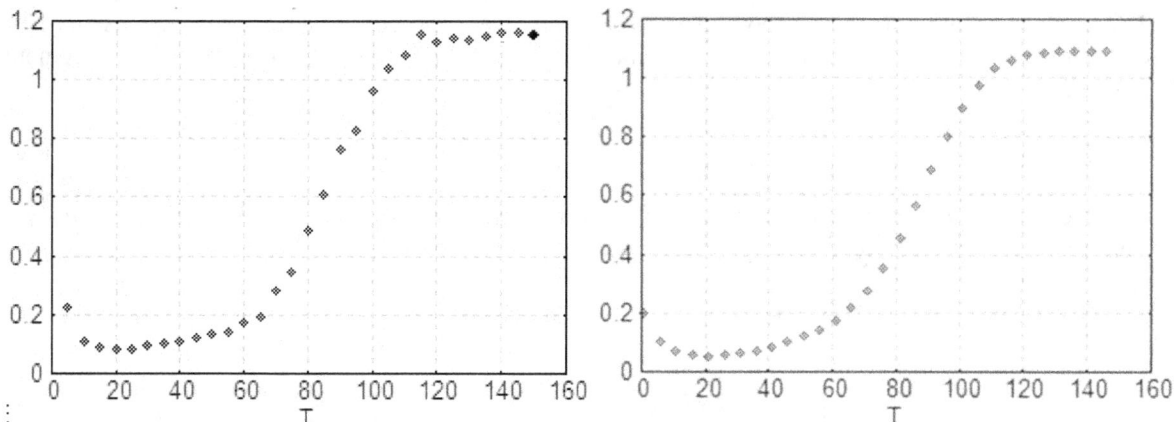

Fig. 1 The simulated expected maintenance costs of *2-out-of-3* system in a single inspection cycle

Fig. 2 The expected maintenance costs of *2-out-of-3* system in a single inspection cycle obtained according to Eq. (11)

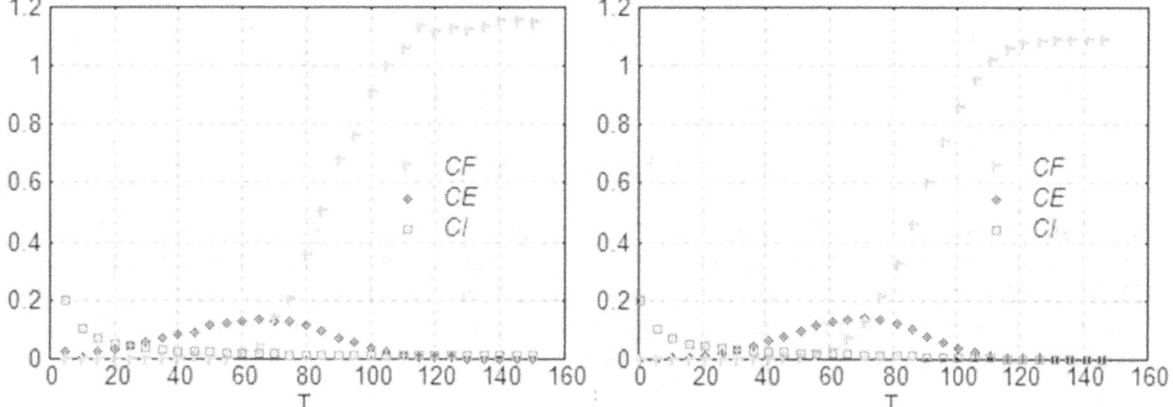

Fig. 3 The simulated costs of a system failure (*CF*), elements' replacement (*CR*) cost and inspection costs (*CI*) in *2-out-of-3* system per a single inspection cycle

Fig. 4 The costs of a system failure (*CF*), elements' replacement (*CR*) cost and inspection costs (*CI*) in *2-out-of-3* system per a single inspection cycle obtained according to Eq. (11)

The figures 1- 4 confirm the strong convergence of simulation and analytical results of the presented maintenance model. The greatest divergence is observable when an inspection period becomes longer than the mean time to failure of system components ($T > 110$) and is a result of almost zero reliability level of the system. Both the models, simulation and analytical one yield the same results, what may be the foundation to confirm their correctness.

4. SENSITIVITY ANALYSIS

The general maintenance model of *k-out-of-n* systems with delay time, presented in the paper, gives the possibility to use it when parameters of the BIP policy are to be optimized. Even though the minimum of the total maintenance cost for the studied system is placed about $T \approx 20$, the optimum length between inspections should not be set directly on this base. One has to remember of the strong assumption of system components "as good as new" at the beginning of the cycle. That makes model directly usable only for maintenance optimization of systems with a series reliability structure. In the case of *k-out-of-n* systems, when $k \neq n$, a system should be inspected on the base of partial maintenance costs. It seems to be reasonable to inspect system components when:
- the probability of a system failure is low – in order to avoid failure costs,
- the probability of elements replacement is high – to avoid costs of unnecessary inspections.

Thus, the optimum (or at list good) period between two consecutive system inspections may be determined on the base of minimization of the difference of the failure and elements' replacement costs [22]:

$$\qquad(12)$$

The plot of the K value in the function of period between inspections (T) for various $k\text{-}out\text{-}of\text{-}n$ systems is presented in the figure 5.

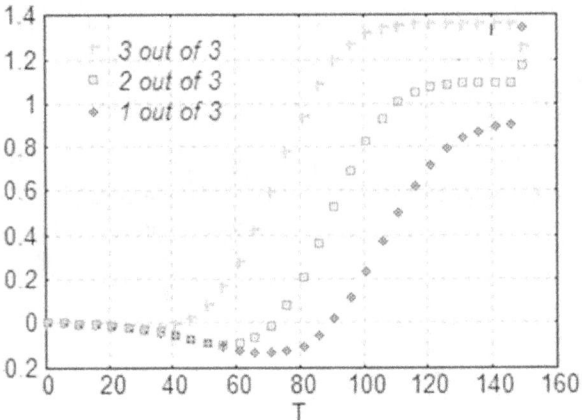

Fig. 5. Function K according to Eq. (12)

The results presented in the figure 5 shows different optimum periods between inspections for $k\text{-}out\text{-}of\text{-}n$ system, dependently on the k value. When system is liable to every component failures (3-out-of-3), it should be inspected much more often than systems being more resistant for elements' unreliability ($k < 3$). The presented effect exemplifies the usability of the analytical maintenance model for the general case of system with $k\text{-}out\text{-}of\text{-}n$ reliability structure.

4. CONCLUSIONS

In the area of technical systems modeling maintenance modelling using the concept of DT main task is to optimize the length of time between successive inspection actions performance. In the presented article, this problem is solved by developing an analytical model of the expected maintenance cost of technical objects operating in the three main structures reliability ($k\text{-}out\text{-}of\text{-}n$, series and parallel ones). The authors are continuing their research on the mathematical modeling with the use of DT approach for technical systems presented e.g. in [20, 22]. On the other hand, the work continues considerations associated with modeling the maintenance of multi-unit technical systems, presented in [21, 23, 24, 25, 39, 40].
In the next step, authors plan to develop DT model with assumptions being fitted to the real-systems performance (e.g. imperfect maintenance case). The main research efforts are to define some rules how to choose a PM policy from an engineering point of view.

References

[1] A. Akbarov, A. H. Christer and W. Wang. *Problem identification in maintenance modelling: a case study*. International Journal of Production Research, 46(4), pp.1031-1046, (2008).

[2] J. A. Andrawus, J. Watson and M. Kishk. *Wind turbine maintenance optimization: principles of quantitative maintenance optimization*. International Journal of Wind Engineering, 31, 2, pp. 101-110, (2007).

[3] H. E. Ascher and K. A. H. Kobbacy. *Modelling preventive maintenance for deteriorating repairable systems*. IMA Journal of Mathematics Applied in Business & Industry, 6, pp. 85-99, (1995).

[4] J. Cai and L. Zhu. *A delay-time model with imperfect inspection for aircraft structure subject to a finite time horizon*. Proc. of IEEE International Conference on Grey Systems and Intelligent Services 2011. 15-8 Sept. 2011, Nanjing, China.

[5] J. B. Chilcott and A. H. Christer. *Modelling of condition based maintenance at the coal face*. International Journal of Production Economics, 22, pp. 1-11, (1991).

[6] A. H. Christer. *A Review of Delay Time Analysis for Modelling Plant Maintenance*. in: Stochastic Models in Reliability and Maintenance, Osaki S. (ed.), Springer 2002.

[7] A. H. Christer. *Developments in delay time analysis for modelling plant maintenance*. Journal of the Operational Research Society, 50, pp. 1120-1137. (1999).

[8] A. H. Christer. *Modelling inspection policies for building maintenance*. Journal of the Operational Research Society 33, pp. 723-732, (1982).

[9] A. H. Christer and D. F. Redmond. *Revising models of maintenance and inspection*. International Journal of Production Economics, 24, pp. 227-234, (1992).

[10] A. H. Christer and P. A. Scarf. *A robust replacement model with applications to medical equipment*. Journal of the Operational Research Society, 45(3), pp. 261-275, (1994).

[11] A. H. Christer and W. M. Waller. *A Descriptive model of capital plant replacement*. Journal of the Operational Research Society, 38, 6, pp. 473-477, (1987).

[12] A. H. Christer and W. M. Waller. An *operational research approach to planned maintenance: modelling P.M. for a vehicle fleet*. Journal of the Operational Research Society, 35(11), pp. 967-984, (1984).

[13] A. H. Christer and W. M. Waller. *Reducing production downtime using delay-time analysis*. Journal of the Operational Research Society, 35(6), pp. 499-512, (1984).

[14] A. H. Christer and W. M. Waller. Delay *Time Models of Industrial Inspection Maintenance Problems*. Journal of the Operational Research Society, 35(5), pp. 401-406, (1984).

[15] A. H. Christer and W. Wang. *A delay-time-based maintenance model of a multi-component system*. IMA Journal of Mathematics Applied in Business & Industry, 6, pp. 205-222, (1995).

[16] A. H. Christer, W. Wang and K. Choi. *The delay-time modeling of preventive maintenance of plant given limited PM data and selective repair at PM*. IMA Journal of Mathematics Applied in Medicine and Biology, 15, pp. 355-379, (1998).

[17] M. I. Desa and A. H. Christer. *Modelling in the absence of data: a case study of fleet maintenance in a developing country*. Journal of the Operational Research Society, 52, pp. 247-260, (2001).

[18] S. G. Jalali Naini, M. B. Aryanezhad, A. Jabbarzadeh and H. Babei. *Condition based maintenance for two-component systems with reliability and cost considerations*. International Journal of Industrial Engineering & Production Research, 20(3), pp. 107-116, (2009).

[19] A. K. S. Jardine and M. I. Hassounah. An *optimal vehicle-fleet inspection schedule*. Journal of the Operational Research Society, 41(9), pp. 791-799, (1990).

[20] A. Jodejko-Pietruczuk, T. Nowakowski and S. Werbińska-Wojciechowska. *Time between inspections optimization for technical object with time delay*. Journal of Polish Safety and Reliability Association, Summer Safety and Reliability Seminars vol. 4, nr 1, pp. 35-41, (2013).

[21] A. Jodejko-Pietruczuk, T. Nowakowski and S. Werbińska-Wojciechowska. *Block inspection policy model with imperfect inspections for multi-unit systems*. Reliability: Theory & Applications vol. 8, nr 3, pp. 75-86, (2013).

[22] A. Jodejko-Pietruczuk and S. Werbińska-Wojciechowska. *Model of expected maintenance costs for multi-unit systems with time delay* (in Polish). Proc. of XLII Winter School of Reliability, Szczyrk, 12-18 January 2014. Warsaw: Warsaw University of Technology, pp. 1-16, (2014).

[23] A. Jodejko-Pietruczuk and S. Werbińska-Wojciechowska. Block *Inspection policy for non-series technical objects*. Proc. of the European Safety and Reliability Conference, ESREL 2013, Amsterdam, The Netherlands, 29 September-2 October 2013. Leiden: CRC Press/Balkema, pp. 889-898, (2014).

[24] A. Jodejko-Pietruczuk and S. Werbińska-Wojciechowska. *Economical effectiveness of Delay Time approach using in Time-Based maintenance modelling.* Proc. of PSAM 11 & ESREL 2012 Conference, 25-29 June 2012, Helsinki, Finland.

[25] A. Jodejko-Pietruczuk and S. Werbińska-Wojciechowska. *Analysis of Block-Inspection Policy parameters from economical and availability point of view.* Proc. of PSAM 11 & ESREL 2012 Conference, 25-29 June 2012, Helsinki, Finland.

[26] B. Jones, I. Jenkinson and J. Wang. *Methodology of using delay-time analysis for a manufacturing industry.* Reliability Engineering and System Safety, 94, 111-124, (2009).

[27] C. Lee. *Applications of delay time theory to maintenance practice of complex plant.* PhD work. T.I.M.E. Research Institute. University of Salford, Salford, UK 1999.

[28] F. Leung and M. Kit-leung. *Using delay-time analysis to study the maintenance problem of gearboxes.* International Journal of Operations & Production Management, 16(12), pp. 98-105, (1996).

[29] A. Pillay, J. Wang, A. D. Wall and T. Ruxton. *A maintenance study of fishing vessel equipment using delay-time analysis.* Journal of Quality in Maintenance Engineering, 7(2): 118-127, (2001).

[30] P. A. Scarf and O. Bouamra. On the *application of a capital replacement model for a mixed fleet.* IMA Journal of Mathematics Applied in Business & Industry, 6, pp. 39-52, (1995).

[31] W. Wang. *An overview of the recent advances in delay-time-based maintenance modeling.* Reliability Engineering and System Safety, 106, pp. 165-178, (2012).

[32] W. Wang. *Modeling planned maintenance with non-homogeneous defect arrivals and variable probability of defect identification.* Eksploatacja i Niezawodnosc - Maintenance and Reliability, 2, pp. 73-78, (2010).

[33] W. Wang. *Delay time modelling for optimized inspection intervals of production plant.* In: Handbook of Maintenance Management and Engineering. Ben-Daya, M., Duffuaa, S. O., Raouf, A., Knezevic, J., Ait-Kadi, D. (eds.). Springer, London 2009.

[34] W. Wang. *Delay time modelling.* In: Complex system maintenance handbook. Kobbacy, A. H., Prabhakar Murthy, D. N. (eds.). Springer 2008.

[35] W. Wang and A. H. Christer. *Solution algorithms for a nonhomogeneous multi-component inspection model.* Computers & Operations Research, vol. 30, pp. 19-34, (2003).

[36] W. Wang. *A model of multiple nested inspections at different intervals.* Computers & Operations Research, 27, pp. 539-558, (2000).

[37] W. Wang. *Modelling condition monitoring inspection using the delay-time concept.* PhD thesis. University of Salford, U. K, 1992.

[38] L. V. Wen-yua and W. Wang. *Modelling preventive maintenance of production plant given estimated PM data and actual failure times.* Proc. of International Conference on Management Science and Engineering 2006, 5-7 October 2006, Lille.

[39] S. Werbińska-Wojciechowska. *Time resource problem in logistics systems dependability modelling.* Eksploatacja i Niezawodnosc - Maintenance and Reliability, vol. 15, nr 4, pp. 427-433, (2013).

[40] S. Werbińska-Wojciechowska. *Problems of logistics systems modelling with the use of DTA approach.* Logistics and Transport, 2, pp. 63-74, (2012).

Modeling the Reliability and the Performance of a Wind Farm Using Cyclic Non-Homogenous Markov Chains

Theodoros V. Tzioutzias[a], Agapios N. Platis[a]*, Vasilis P. Koutras[a]
[a]University of the Aegean Department of Financial and Management Engineering, Chios, Greece

Abstract: Reliability issues concerning wind power installations are of prior interest due to the increasing production of electricity through wind energy around the world. Attaining the highest performance of a wind farm lies on two factors: wind intensity and mechanical failures. Wind intensity is related to the geographical characteristics of the farm, while mechanical failures are related to the wind turbines reliability. The latter issue is the one studied in this paper. In order to achieve higher performance levels when assuming the wind capacity as constant for a certain installation and a certain period, we can increase reliability by reducing mechanical failures and increasing repair rates. The objective of this paper is to model reliability's impact on the overall performance of the wind farm. To this direction, a Continuous Time Markov Chain (CTMC) and a Cyclic Non-Homogenous Markov Chain (CNHMC) are used to model the system. CNHMCs are adopted in order to capture the periodicity of the wind intensity. The results of the above models are compared in order to identify which one fits better the characteristics of the system.

Keywords: Wind Power, Reliability, Cyclic Non-Homogenous Markov Chains (CNHMC), Asymptotic Probability Distribution.

1. INTRODUCTION

The production of electricity using renewable energy resources is an urgent matter. The continuous CO_2 emissions from industry aggravate the global warming phenomenon. That is the main reason of EU's promise that by 2020 the 20% approximately of the total demand of electricity will come from renewable energy resources, with wind energy being the leader with a percentage between 14-18%.

The technology that uses the power of wind in order to produce electricity is very growing. Wind generator companies come up with new ideas that can achieve better performance year after year. But the performance of a wind farm is a compromise between the wind that blows in the area where the generators are located and the reliability of that system. Modeling such a system is of prior importance in evaluating properly the reliability indices. A simple and easy to understand method for modelling is the deterministic criteria, but for the evaluation of the reliability, probabilistic methods seem to be a more accurate and reliable method. Due to the nature of the wind and its changes over time, modelling becomes more interesting and difficult.

In this paper, we evaluate the reliability of a wind park considering not only weather data from a certain area, but also the mechanical failures. In order to model our problem we use classical probabilistic methods, with which we will be able derive our results for different intensities of the wind taking also into account the mechanical failures. Our model is based on the categorization of wind intensity in four categories and on the transitions between them using Markov models [1]. Beyond the classical CTMC approach, we expand our study by using CNHMC in order to capture the periodicity revealed due to seasonal wind changes.

In our study, we consider wind data provided by the Hellenic National Meteorological Service (HNMR), from the meteorological station which is located in the island of Chios in the area of airport. The data refer to the last five years (from January 2008 to July 2013). We had a measurement about the speed of the wind once every three hours, so we had 8 measurements per day. Because of the volatile nature of wind and the different intensities over the time we categorize the wind speed into four categories, including also a no-wind state. Every transition from a state to another is possible,

Contact Author: Agapios N. Platis platis@aegean.gr

even more from a state with a strong wind to no-wind state. Combining wind intensity categories with the state of a system regarding machine failures, we develop a model with eight states in total; the first four refer to the states without any machine failure, and the rest to the states where a failure occurs

The application of probabilistic methods for modeling of wind power was addressed by different authors in the past. Gouveia and Matos in [1], inspired from the previous researches, develop a new methodology by which the wind speed was categorized in four categories, no-wind state (0% production), light wind (30% production), strong wind (70% production), and very strong wind (100% production). In their model, the transitions need not to be only in adjacent wind levels, but it is possible to have a transition between the states excessive wind to no wind state.

This paper is organized as follows: The methodologies used are described in Section 2; we model the same problem using Homogenous Markov Chains and Cyclic Non-homogenous Markov Chains. In section 3 we apply the above methodologies using real data that HNMR provide us, in section 4 we present an example and finally in section 5 we come up with our concluding remarks.

2. HOMOGENEOUS AND NON-HOMOGENEOUS MARKOV MODELS

A continuous-time Markov chain is a probability model which takes values in countable set. The time spent in each state takes only real and non-negative values according to an exponential distribution. The evolution of the system in time, according to the Markov property, does not depend on the historical behavior and the previous system states, but only on the current system state [1-2, 5].

Non-homogenous Markov chains are more complex, and the reason is why every transition rate between the states is a function of time and more accurate of a global clock [2, 3, 5].

In order to have a better understanding of the two cases lets assumes a system with two possible states, operation and failure. The rate of transition from the operational state to the failure state is called failure rate (l), while the transition from failure to operation repair rate (m). Using a CTMC to model the system indicates constant rates l, m, whereas for the NHMC the transition rates $l(t)$ and $m(t)$ are time dependent. In addition, NHMC modeling is preferable when the systems to model have hazard rates that evolve with time; it is generally the case of systems that hazard rates depend on environmental parameters such as wind farms where the performance and the failures depend mainly on the wind intensity, and weather phenomena like thunders.

2.1. Transition probability and transition function

Let us assume that $S = \{0,1,2,3,\ldots s\}$ is the state space of the system, and $X = \{X_t; t \in \text{IN}\}$ is a stochastic process on a probability space X_t which is a Markov chain if, for all $k \in E$ and all $t > 0$, the following equation holds true [5]:

$$\Pr\{X_{t+1} = k \mid X_t = i_t, \ldots, X_0 = i_0\} = \Pr\{X_{t+1} = k \mid X_t = i_t\}$$

As a transition probability of a Markov chain, we define the probability that a transition from state i to state j occurs in the time interval $[t-1, t]$. The probability $p_{ij}(t) = \Pr\{X_t = j \mid X_{t-1} = i\}$ for all $i, j \in S, t \in \text{IN}$ is called transition function of the chain X and it is a one-step transition probability. We can also define multiple steps transition probabilities by $p_{s,t}(i, j) = \Pr\{X_t = j \mid X_s = i\}$ and the transition probability function can be then derived [2]:

$$p_t(i, j) = p_{t,t+1}(i, j) \quad \text{and} \quad p_{tt}(i, j) = \mathbf{I}$$

The transition function follows the property known as *Chapman-Kolmogorov* equation, for all $i, j \in S$

$$p_{t,s}(i,j) = \sum_{l=0}^{k} p_{t,l}(i) \cdot p_{l,s}(j) \quad \text{for all} \quad t < l < s$$

A transition function $p_t(i,j)$ with $t \in \mathbb{N}$ and $i, j \in S$ is called cyclic or periodic with period $h, (h > 1)$ if h is the smallest integer verifying the equation $p_{th+r} = \Pr(i,j)$ with $t, r, \in \mathbb{N}$, and $i, j \in E$ [2, 4].

2.2. State Probability Vector

Let us now consider the initial distribution probability $\boldsymbol{\alpha} = (\alpha_i) \, i \in E$ with, $\alpha_i = \Pr(X_0 = i)$ and transition probability matrix $\mathbf{p}_t$. The Markov chain X is completely defined by the initial distribution $\boldsymbol{\alpha}$ and the transition probability matrix p_t with $t \in \mathbb{N}$, and the transition probability from state i to j at time t is given by

$$p_j(t) = \Pr(X_t = j) = \sum_{i \in E} \Pr(X_t = j \mid X_0 = i) \Pr(X_0 = i)$$

$$= \sum_{i \in E} \alpha_i p_{i,j}(t) = (ap_{0,t})(j) = \left[a \left(\prod_{k=0}^{t-1} p_k \right) \right] (j)$$

$p_{ij}(t)$ is called state probability at time t. The vector $\mathbf{P}(t) = (P_1(t), ..., P_s(t))$ is called state probability vector at time t. We can also write $\mathbf{P}(t) = \boldsymbol{\alpha} \mathbf{p}_{0,t}$.

2.3. Asymptotic Behavior

Let's assume that S_t is an embedded homogenous chain from X_t. As a result we have that $S_t = X_{td}$ for every $t > 0$. If $\mathbf{p}_t$ is the transition matrix of the NHMC X_t, then $\mathbf{p}_{0,d}$ is the constant transition matrix of the HMC S_t. The asymptotic behavior of the transition matrixes is closely related with the known as ergodic properties of the Markov chain. The term ergodicity is related to the convergence of probability distributions p_0 in time, $\mathbf{p}^n \to \mathbf{\Pi}$ as $n \to \infty$ $n \to \infty$, with $\mathbf{\Pi}$ ergodic and $\mathbf{P}$ the transition matrix probability, and assumes aperiodicity as a necessary condition. The above result can be used for the embedded HMC Y_t, in order to obtain the asymptotic results concerning the NHMC X_t.

3. MODELING THE SYSTEM OF WIND GENERATORS

3.1. Model description

The general model adopted and studied is a system consisting of *n* wind generators that are able to operate in different wind levels. Because of wind nature, we assume that its intensity can vary between 4 categories: the first category is the no-wind state and the last one is the excessive wind state. In our approach, the possible transitions are the transition from a wind category to another, without being necessary for the new category to be adjacent to the previous, and the transition from a state without failure to a state with a failure.

Fig. 1 Markov chain model for four categories of wind

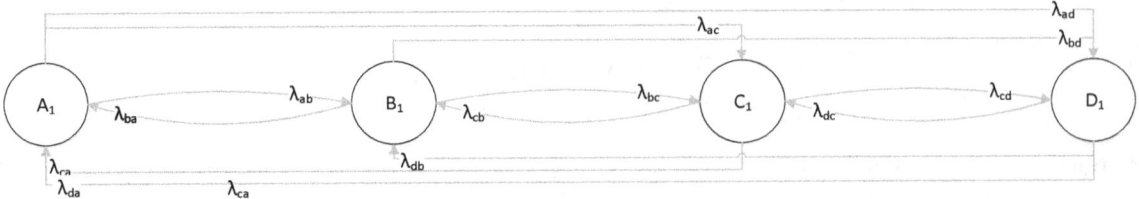

Two different probability models are described below, a CTMC and a CNHMC, and we will use both models for our study. Firstly, we use a CNHMC to model a wind turbine and then a wind farm, where the hazard rates evolve periodically. Moreover, we will model the same case using a CTMC, and after deriving the results we will compare them with the results of the CNHMC model.

As previusly above, the performance of a wind park depends on the intensity of the wind and the reliability of the generators. In order for the wind farm to achieve the best performance, the blowing wind in the area of interest should be able to give full production without any risk. Moreover the system is designed to operate without a mechanical failure that can cause loss of performance. The high repair rate and the very low failure rate are the keys to achieve the better performance. In addition, the failures of a system are highly dependent on weather conditions, while on the other hand the repairs depend on both the weather conditions and the working hours of the units. It is clear enough that seasonality plays also an important role for the repair and failure rates, and through our study we find out that it is not needed to keep a constant repair rate through all the seasons of the year, but if we reduce the repair rate in some periods of time the overall performance of the system will be the same as keeping the repair rate constant. For example, we may need fewer repairs in winter than in summer time without having lower production.

In our approach we are interested only the variation of the transition rate among the wind intensity states and we assume that not only the failure but also the repair rates are constant. This is because, initially we are interesting in studying the effects of the time dependent transition rates on system's asymptotic probability distribution. In the future we intend to extend our study by taking into account time dependent failure and repair rates.

3.2. Modelling of the wind transitions using Markov chains

In our case, we will consider the wind variation through a season (spring, summer, autumn or winter). We rely on the data we got from the HNMS from January 2008 to June 2013, for every day, every three hours, from the meteorological station which is located in the area of Chios Island airport. Through appropriate data processing, we categorized the data according to the season and the year. As a result we have four seasons for every year from 2008 to 2012 and for the year 2013 we have only the spring. The next step was to categorize the wind according to the intensity into four categories. State A refers to the scenario where there is no wind or very weak wind unable to give any electricity production. States B and C are intermediate states and in state D we have winds able to provide full electricity production. In Fig.1 the Markov model for all the possible transitions between the four states of wind is represented. We consider that all the transitions are possible, even from state A without wind, to state D in which excessive winds, and the opposite [1].

3.3. Modeling a wind park using Markov chains

In order to completely model the operational scenario of a wind farm, apart from the wind states that are presented in Fig. 1, we should consider the number of the wind generators n in addition with the repair rates μ and failure rate λ.

In Fig. 2 the final Markov model for a wind farm consisting of n turbines is shown, considering four different categories of wind intensity. A transitions occurs in three possible cases. Firstly when there is

Fig. 2 Markov model of a wind farm consisting of n turbines

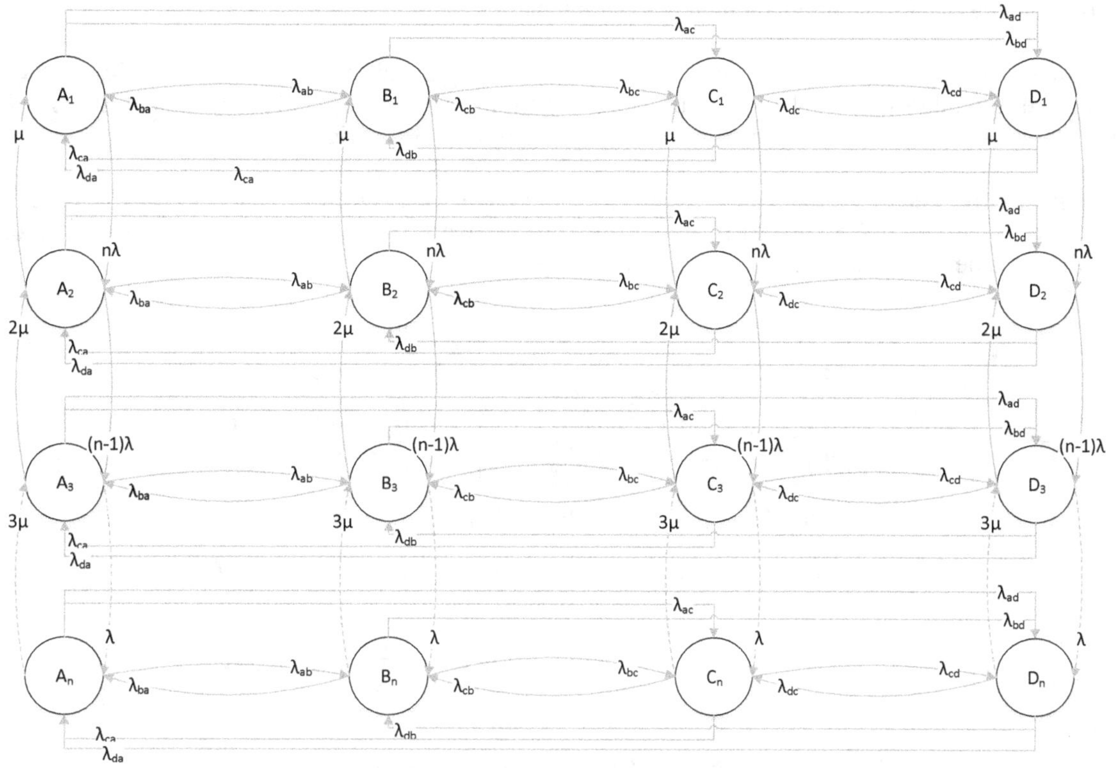

Fig. 3 Simplified Markov model considering aggregation

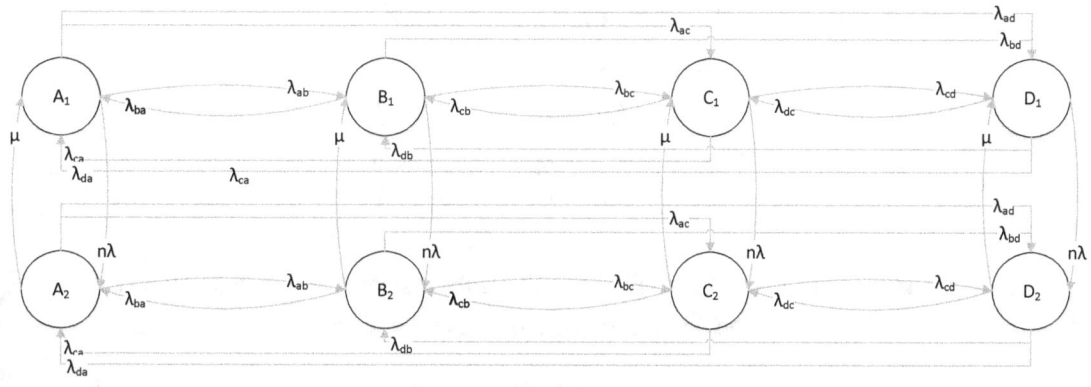

a wind variation, secondly, when we have a mechanical failure of a unit and thirdly when we have a repair of a unit. As it is mentioned in [1], if we want to receive more accurate results we should consider more than four wind states.

As it is stated in [1], the aggregation of wind farms is very common, in order to have a simpler model without great number of states, and the main reason for that is the high loads of wind power that is being produced. A simplified diagram consisting of the eight states that are needed in order to combine the state of a unit with the state of other turbines' state is presented in Fig.3

In order to have a better understanding of the system and the time dependency, we study its behavior for every season from spring 2008 to summer 2013, and we set as a period a full year, or the four season of a year, spring, summer, autumn and winter.

The main aim of our work is to calculate the steady state probability of a wind park using CNHMC since we are interested in the long run behavior of the system. Initially the system starts at an operational state, and hence it can convert all the wind power to electricity.

A transition between two states is possible when we have variation in the intensity of the wind, for example from state B (low wind) to state D (extensive wind), in case of a failure, and in case of a repair. Considering the above, the generator matrix is:

$$\mathbf{Q} = \begin{bmatrix} y_{1,k} & \lambda_{ab,k} & \lambda_{ac,k} & \lambda_{ad,k} & n\lambda & 0 & 0 & 0 \\ \lambda_{ba,k} & y_{2,k} & \lambda_{bc,k} & \lambda_{bd,k} & 0 & n\lambda & 0 & 0 \\ \lambda_{ca,k} & \lambda_{cb,k} & y_{3,k} & \lambda_{cd,k} & 0 & 0 & n\lambda & 0 \\ \lambda_{da,k} & \lambda_{db,k} & \lambda_{dc,k} & y_{4,k} & 0 & 0 & 0 & n\lambda \\ \mu & 0 & 0 & 0 & y_{5,k} & \lambda_{ab,k} & \lambda_{ac,k} & \lambda_{ad,k} \\ 0 & \mu & 0 & 0 & \lambda_{ba,k} & y_{6,k} & \lambda_{bc,k} & \lambda_{bd,k} \\ 0 & 0 & \mu & 0 & \lambda_{ca,k} & \lambda_{cb,k} & y_{7,k} & \lambda_{cd,k} \\ 0 & 0 & 0 & \mu & \lambda_{da,k} & \lambda_{db,k} & \lambda_{dc,k} & y_{8,k} \end{bmatrix}$$

where

$$y_{1,k} = -(\lambda_{ab} + \lambda_{ac} + \lambda_{ad} + n\lambda)$$
$$y_{2,k} = -(\lambda_{ba} + \lambda_{bc} + \lambda_{bd} + n\lambda)$$
$$y_{3,k} = -(\lambda_{ca} + \lambda_{cb} + \lambda_{cd} + n\lambda)$$
$$y_{4,k} = -(\lambda_{da} + \lambda_{db} + \lambda_{dc} + n\lambda)$$
$$y_{5,k} = -(\mu + \lambda_{ab} + \lambda_{ac} + \lambda_{ad})$$
$$y_{6,k} = -(\mu + \lambda_{ba} + \lambda_{bc} + \lambda_{bd})$$
$$y_7,k = -(\mu + \lambda_{ca} + \lambda_{cb} + \lambda_{cd})$$
$$y_{8,k} = -(\mu + \lambda_{da} + \lambda_{db} + \lambda_{dc})$$

$k = 1, 2, 3, 4$ stands for season k, λ_{ab},k is the transition from state A to the state B in season k, λ is the failure rate, n the number of the wind turbines and μ is the repair rate [1,3-4].

One of the advantages of modelling using CNHMC is that we can determine the steady state probabilities π of the k-th season from the equation ([3]):

$$\pi_k = \pi_{k-1} \mathbf{p}_{k-1} \qquad (1)$$

where $\mathbf{p}_{k-1}$ is the probability matrix for the $(k-1)$-th season which is derived from the equation

$$\mathbf{p}_k = \mathbf{Q}_k h + \mathbf{I} \qquad (2)$$

with h a constant number near 1 and $\mathbf{I}$ the identity matrix.

In order to derive the steady state probability of the first period we should solve the linear system

$$\pi_1 = \pi_0 \mathbf{p}_0 \text{ with } \pi_0 = \boldsymbol{\pi} \qquad (3)$$

with $\boldsymbol{\pi}$ the steady state probability vector and p_0 is the probability matrix for the first season.

The stationary probability distribution can be computed by solving the linear system

$$\pi_0 = \pi_0 \mathbf{p}_0 \mathbf{p}_1 \mathbf{p}_2 \mathbf{p}_3$$

with the constrains that $\sum_{1 \le i \le s} \pi_0(i) = 1$ which results to $\pi_0 (\mathbf{p}_0 \mathbf{p}_1 \mathbf{p}_2 \mathbf{p}_3 - \mathbf{I}) = \mathbf{0}$ with the additional condition $\boldsymbol{\pi}_0 I_s = 1$, with I_s is being a $(s+1)th$-dimension vector with ones.

Finally we define the following matrix

$$\mathbf{A} = \mathbf{p}_0 \mathbf{p}_1 \mathbf{p}_2 \mathbf{p}_3 - \mathbf{I}, \qquad (4)$$

and the following system of linear equation

$$\mathbf{u}\mathbf{A}' = \mathbf{b}$$

where $\mathbf{A}'$ is the matrix $\mathbf{A}$ where the last column has been replaced by ones and $\mathbf{b}$ is an 8-dimention zero vector with one in the last column,

$$\mathbf{b} = \begin{bmatrix} 0 & 0 & 0 & 0 & 0 & 0 & 0 & 1 \end{bmatrix}$$

and the stationary probability $\mathbf{u}^{CNHMC}$ can be calculated solving

$$\mathbf{u}\mathbf{A}' = \mathbf{b} \qquad (5)$$

4. NUMERICAL ILLUSTRATION

The first season for which the HNMS provide us data is the spring of 2008. For this season the generation matrix is the following

$$\mathbf{Q}_{spg2008} = \begin{bmatrix} 0.001444 & 0.000926 & 0.000463 & 0.000000 & 0.000055 & 0.000000 & 0.000000 & 0.000000 \\ 0.001389 & -0.015333 & 0.012963 & 0.000926 & 0.000000 & 0.000055 & 0.000000 & 0.000000 \\ 0.000000 & 0.012500 & -0.059777 & 0.047222 & 0.000000 & 0.000000 & 0.000055 & 0.000000 \\ 0.000000 & 0.001852 & 0.046296 & 0.048203 & 0.000000 & 0.000000 & 0.000000 & 0.000055 \\ 2.000000 & 0.000000 & 0.000000 & 0.000000 & -2.001389 & 0.000926 & 0.000463 & 0.000000 \\ 0.000000 & 2.000000 & 0.000000 & 0.000000 & 0.001389 & -2.015278 & 0.012963 & 0.000926 \\ 0.000000 & 0.000000 & 2.000000 & 0.000000 & 0.000000 & 0.012500 & -2.059722 & 0.047222 \\ 0.000000 & 0.000000 & 00.000000 & 2.000000 & 0.000000 & 0.001852 & 0.046296 & -2.048148 \end{bmatrix}$$

Fig. 4 Asymptotic probability distribution comparison

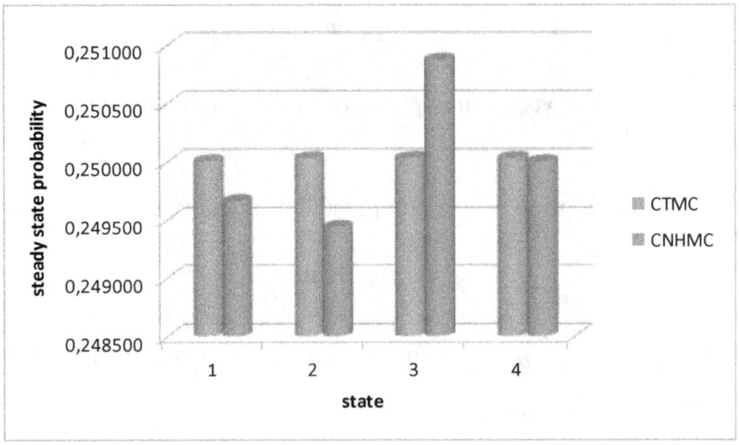

and the transition probability according to (2) is

$$\mathbf{P}_{spg2008} = \begin{bmatrix} 0.999299 & 0.000450 & 0.000225 & 0.000000 & 0.000027 & 0.000000 & 0.000000 & 0.000000 \\ 0.000674 & 0.992556 & 0.006294 & 0.000450 & 0.000000 & 0.000027 & 0.000000 & 0.000000 \\ 0.000000 & 0.012500 & -0.059777 & 0.047222 & 0.000000 & 0.000000 & 0.000055 & 0.000000 \\ 0.000000 & 0.001852 & 0.046296 & 0.048203 & 0.000000 & 0.000000 & 0.000000 & 0.000055 \\ 2.000000 & 0.000000 & 0.000000 & 0.000000 & -2.001389 & 0.000926 & 0.000463 & 0.000000 \\ 0.000000 & 2.000000 & 0.000000 & 0.000000 & 0.001389 & -2.015278 & 0.012963 & 0.000926 \\ 0.000000 & 0.000000 & 2.000000 & 0.000000 & 0.000000 & 0.012500 & -2.059722 & 0.047222 \\ 0.000000 & 0.000000 & 00.000000 & 2.000000 & 0.000000 & 0.001852 & 0.046296 & -2.048148 \end{bmatrix}$$

By the time we have the transition probability of the four season of the year, $\mathbf{P}_{spg2008}$, $\mathbf{P}_{sum2008}$, $\mathbf{P}_{aut2008}$, $\mathbf{P}_{win2008}$ we can calculate matrix $\mathbf{A}$ from (3):

$$\mathbf{A} = \begin{bmatrix} 0.005874 & 0.005144 & 0.000686 & 0.000017 & 0.000027 & 0.000000 & 0.000000 & 0.000000 \\ 0.004862 & -0.044692 & 0.038090 & 0.001713 & 0.000000 & 0.000026 & 0.000001 & 0.000000 \\ 0.000938 & 0.035896 & -0.099709 & 0.062848 & 0.000000 & 0.000001 & 0.000025 & 0.000002 \\ 0.000046 & 0.003407 & 0.061340 & -0.064821 & 0.000000 & 0.000000 & 0.000002 & 0.000026 \\ 0.994126 & 0.005143 & 0.000686 & 0.000016 & 0.999972 & 0.000000 & 0.000000 & 0.000000 \\ 0.004862 & 0.955308 & 0.038089 & 0.001713 & 0.000000 & -0.999974 & 0.000001 & 0.000000 \\ 0.000938 & 0.035895 & 0.900291 & 0.062848 & 0.000000 & 0.000001 & 0.999975 & 0.000000 \\ 0.000046 & 0.003406 & 0.061340 & 0.935179 & 0.000000 & 0.000000 & 0.000002 & -0.999974 \end{bmatrix}$$

Now, we can calculate the stationary probability for the year 2008 using (5)

$$\mathbf{u}_{2008}^{CNHMC} = \begin{bmatrix} 0.249662 & 0.249438 & 0.250873 & 0.249999 & 0.000007 & 0.000007 & 0.000007 & 0.000007 \end{bmatrix}$$

We use the same methodology also for the years from 2009 to 2012, and we obtain the stationary probability for each year. Our next step is to calculate the average stationary probability for that five years, which is the following

$$\mathbf{u}^{CNHMC} = \begin{bmatrix} 0.249890 & 0.249908 & 0.250188 & 0.249987 & 0.000007 & 0.000007 & 0.000007 & 0.000007 \end{bmatrix}$$

In contrast, when using a CTMC instead of CNHMC, considering the same data set, firstly we do not assume any periodicity for the transition rates. Instead, the transition rates are calculated as the number of the transitions between the four states of wind through the period of time from January 2008 to May 2013. Hence the stationary probability distribution for the wind park system can be derive by solving the system of linear equations:

$$\mathbf{u}^{CTMC}\mathbf{Q} = \mathbf{0}, \sum_{i=0}^{i=n} u^{CTMC}(i) = 1$$

where $\mathbf{u}^{CTMC}$ is the steady state probability vector for the continuous time the Markov chain and the (**8x8**) generation matrix is ([4]):

$$\mathbf{Q} = \begin{bmatrix} y_1 & \lambda_{ab} & \lambda_{ac} & \lambda_{ad} & n\lambda & 0 & 0 & 0 \\ \lambda_{ba} & y_2 & \lambda_{bc} & \lambda_{bd} & 0 & n\lambda & 0 & 0 \\ \lambda_{ca} & \lambda_{cb} & y_3 & \lambda_{cd} & 0 & 0 & n\lambda & 0 \\ \lambda_{da} & \lambda_{db} & \lambda_{dc} & y_4 & 0 & 0 & 0 & n\lambda \\ \mu & 0 & 0 & 0 & y_5 & \lambda_{ab} & \lambda_{ac} & \lambda_{ad} \\ 0 & \mu & 0 & 0 & \lambda_{ba} & y_6 & \lambda_{bc} & \lambda_{bd} \\ 0 & 0 & \mu & 0 & \lambda_{ca} & \lambda_{cb} & y_7 & \lambda_{cd} \\ 0 & 0 & 0 & \mu & \lambda_{da} & \lambda_{db} & \lambda_{dc} & y_8 \end{bmatrix}$$

The steady state probability distribution can be then computed:

$$\mathbf{u}^{CTMC} = \begin{bmatrix} 0.250003 & 0.250003 & 0.250003 & 0.250003 & 0.000007 & 0.000007 & 0.000007 & 0.000007 \end{bmatrix}$$

Apart from the limiting probabilities of the last for states, it is clearly shown in Fig. 4, that there is a difference between the state probabilities of the first four states in the case of the CNHM model. This is due to the fact that CNHMC despite the complexity can give us more accurate results because the model is not generalized but it is analyzed in periods of time and finally we have a total result for the time of interest, while in CTMC the model is not analyzed and we have only a general point of view for the overall time of interest.

5. CONCLUSION

In this paper, we calculated the steady state probabilities of a wind park using two different types of probabilistic methods, CTMC and CNHMC. Our main purpose was to compare these two approaches and find out which of the two is the more appropriate and accurate to model our problem.

We received weather data from the HNMS for a particular place in the island of Chios, we created four categories of wind according to the intensity, A was the no wind state, B was the intensity of wind able to give us 30% electricity production, C was the state which was able to give 70% production and finally, D was the state in which we have full production of electricity. We studied the reliability of the system using Markov Chains, pointing out and analyzing the use of CNHMC, which is proven to be a more accurate approach for our case in comparison with the classical CTMC model. The accuracy of the results when considering multiple wind states lies on the fact that the CNHMC approach can capture the dependencies of the transition rates on time. Despite the fact that modeling with CNHMCs is more complicated it is worthwhile to apply the corresponding analysis in order to achieve the limiting distribution for all the seasons for each year and additionally the overall limiting distribution for the period of interests.

At last, assessments of reliability can also be studied using much more simple probabilistic methods, but the use of CNHMC is more accurate method able to model systems like the system we study or even more complicated.

Although we studied only the variation of the transition rates among the wind intensity states by assuming constant failure and repair rates, we intend to extend our study by taking into account time dependent failure and repair rates. In authors' opinion, by this we will manage to provide a more detailed model for the operational scenario of a wind park.

References

[1]. E. M. Gouveia, M. A. Matos. *"Evaluating operational risk in a power system with a large amount of wind power"*, Electric Power Systems Research, 79, pp. 734-739, (2009)

[2]. A.N. Platis, N.E. Limnios, M.L. Du. *"Asymptotic Availability of Systems Modelled by Cyclic Non-Homogenous Markov Chains"*, Annual Reliability and MAINTAINABILITY Symposium (1997)

[3]. V.P. Koutras, A.N. Platis, G.A. Gravvanis. "Optimal server resource reservation policies for priority classes of users under cyclic non-homogenous markov modelling", European Journal of Operational Research, 198, pp. 545-556, (2009).

[4]. V.P. Koutras, A.N. Platis, G.A. Gravvanis. "On the optimization of free resources using non-homogenous Markov chain software rejuvenation model", Reliability Engineering and system safety, 198, pp. 545-556, (2009).

[5]. A. Papoulis, S. U. Pillai, "Probability, Random variables and Stochastic Processes" McGraw-Hill, 2002

Performance and Reliability of Bridge Girders Upgraded with Post-tensioned Near-surface-mounted Composite Strips

Yail J. Kim[a*], Jae-Yoon Kang[b], and Jong-Sup Park[b]

[a]University of Colorado Denver, Denver, CO, USA
[b]Korea Institute of Construction Technology, Ilsan, Korea

Abstract: This paper deals with a research program concerning the performance and reliability of bridge girders strengthened with post-tensioned near-surface mounted (NSM) carbon fiber reinforced polymer (CFRP) composite strips. The advantages of CFRP application include non-corrosive characteristics, prompt execution on site, reduced maintenance expenses, favorable strength-to-weight ratio, and good chemical or fatigue resistance. NSM CFRP technologies are emerging in the infrastructure rehabilitation community because of several benefits such as enhanced bond performance and durability. For the present study, computational and analytical approaches are employed to examine the behavior of CFRP-strengthened girders, including 51 finite element models. Preliminary reliability analysis is conducted for evaluating the level of safety associated with the strengthened girders.

Keywords: bridge, composite, performance, rehabilitation

1. INTRODUCTION

Constructed bridge structures require significant attention due to the degradation of constituent members. Typical attributes causing structural deterioration include increased service load and environmental distress. Carbon fiber reinforced polymer (CFRP) composites have demonstrated promising performance in terms of upgrading the behavior of existing structural members [1]. Two types of CFRP application may be used for practice:

- Externally bonded (EB) CFRP: CFRP sheets/laminates are bonded to the substrate of the structural element using an adhesive
- Near-surface mounted (NSM) CFRP: CFRP strips/rods are inserted into precut slits along the member and bonded permanently

The EB method has been used since the late 1990s, while the NSM method has relatively short history [2]. The former is usually susceptible to premature CFRP-debonding, whereas the latter shows enhanced bond resistance. Additional benefits of using NSM CFRP are prompt installation with reduced labor, enhanced durability, and satisfactory aesthetics [3-5]. Prestress may be applied to augment the efficacy of NSM CFRP. The reason is that only a certain range of CFRP strength can be used when the strengthened structure fails, provided the capacity of CFRP materials is substantially high [6]. Several prestressing methods (i.e., post-tensioning hereafter) have been proposed previously such as external jacking device [7,8], brackets [9], and embedded anchors [10]. Research projects have been reported in the area of post-tensioned NSM CFRP for upgrading concrete members. Nordin and Taljsten [11] reported test results concerning the behavior of reinforced concrete beams retrofitted with post-tensioned NSM CFRP that was tensioned using external jacking device. The strengthened beams revealed improved cracking and yield capacities in comparison to an unstrengthened control. Badawi and Soudki [12] predicted the flexure of reinforced concrete beams upgraded with post-tensioned NSM CFRP. Sectional analysis was employed based on force equilibrium, material nonlinearity, and strain compatibility. Although the capacities of the strengthened beams increased

---

[*] jimmy.kim@ucdenver.edu

(i.e., yield and ultimate loads), ductility of these beams decreased with an increasing level of post-tensioning. Choi et al. [8] tested reinforced concrete T-beams retrofitted with post-tensioned NSM CFRP. The beams had partial CFRP-bonding induced by use of a thin plastic duct (i.e., $0.4L$ to $0.75L$ in which L is the beam span). Various levels of post-tensioning were applied from 40% to 60% of the CFRP capacity. Enhanced deformability was noticed because of the partial bonding scheme and this effect became more obvious when a post-tension force increased. El-Hacha and Gaafar [9] compared the behavior of concrete beams strengthened with NSM CFRP with and without post-tensioning. The beam with post-tensioning demonstrated better serviceability such as lower deflection when compared with the beam without post-tensioning. It was also noted that the loss of post-tension force was not significant. Wahab et al. [13] reported the fatigue response of reinforced concrete beams post-tensioned with NSM CFRP. The fatigue range implemented was up to 70% of the control capacity. CFRP-slip was observed while the beam was submitted to fatigue cycles. Failure of the fatigue beams was resulted from CFRP-debonding and steel rupture. The surface condition of the CFRP affected the performance of the test beams such that sand-coated CFRP showed longer fatigue life than spirally wound CFRP did.

As discussed above, post-tensioned NSM CFRP is a promising technique that can improve the behavior of existing concrete members. It is, however, important to note that this emerging strengthening method is still in an early stage and further development is necessary. This paper deals with a numerical approach predicting the behavior of prestressed concrete bridge girders strengthened using post-tensioned NSM CFRP. A three-dimensional finite element model was developed to examine the flexure of the strengthened girders. An analytical approach (i.e., strain compatibility) was employed to estimate the capacity of the girders. A preliminary reliability study was reported.

2. RESEARCH SIGNIFICANCE

Effort is required to better understand the behavior of existing concrete members when upgraded with post-tensioned NSM CFRP. The current state of research remains in laboratory-scale investigations. Limited endeavors have been made in terms of design guidelines that can lead to the full-scale site application of this technology. For example, the effect of CFRP-bonded length and post-tensioning levels that will influence the performance of strengthened members is not known. The reliability of the girders strengthened with post-tensioned NSM CFRP is also of interest from a practice point of view. The research program addresses these technical challenges associated with post-tensioned NSM CFRP technologies.

3. PROTOTYPE BRIDGE GIRDERS AND STRENGTHENING SCHEMES

This section explains the background of the numerical study conducted to examine the behavior of bridge girders upgraded with post-tensioned NSM CFRP strips. Below is a summary of material properties, girder details, and strengthening schemes.

3.1. Materials

The specified 28-day compressive strength of concrete was 40 MPa and its elastic modulus was 30 GPa with a Poisson's ratio of 0.2. The ultimate strength of 7-wire prestressing steel strands was 1860 MPa with a modulus of 190 GPa and a Poisson's ratio of 0.3. The unidirectional CFRP used included the following material properties [14]: ultimate tensile strength = 2500 MPa, modulus = 165 GPa, rupture strain = 1.48%, and Poisson's ratio = 0.25. The level of post-tensioning force was up to 981 kN to upgrade the performance of constructed bridges (details follow) according to a preliminary study: a cross sectional area of 700 mm^2 was required for the NSM CFRP to satisfy the provision of the ACI.440.4R-04 document [15].

3.2. Details of Prototype Girders

Prototype girders were taken from the standard girders used in Korea. Three types of girders were used in this research program (L = 25, 30, and 35 m), as shown in Fig. 1. Three to five bundled strands were draped along the girder span and their cross-sectional areas were 3560 mm^2, 4750 mm^2, and 5940 mm^2 with prestressing levels of 64%, 65%, and 68% as per the Korean Standard [16] for the girders with L = 25, 30, and 35 m, respectively.

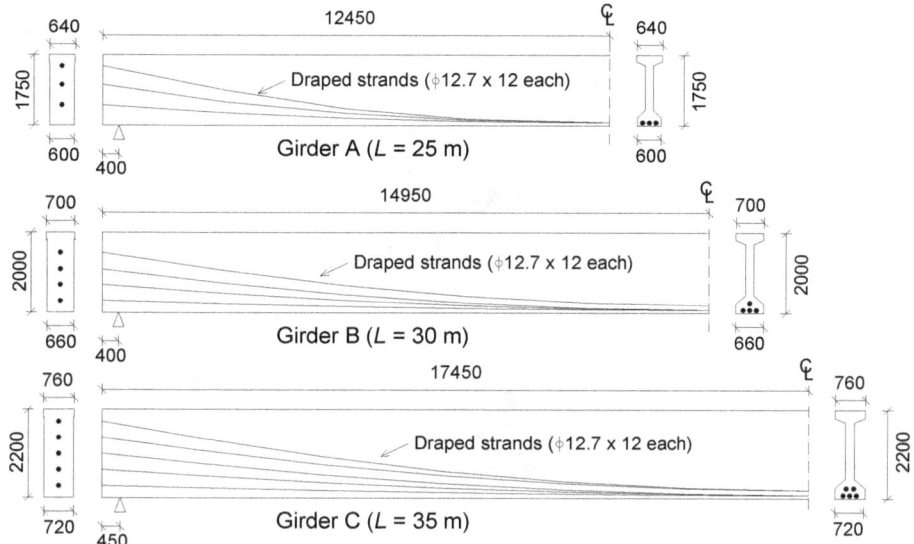

Figure 1: Prototype bridge girders

3.3. Strengthening Plan

The following was intended to strengthen the prototype girders shown in Fig. 1: a girder carrying a design load of 318 kN (designated DB-18 in Korean Standard) is upgraded to the girder resisting a load of 424 kN (DB-24). The details of DB-24 truck load are provided in Fig. 2.

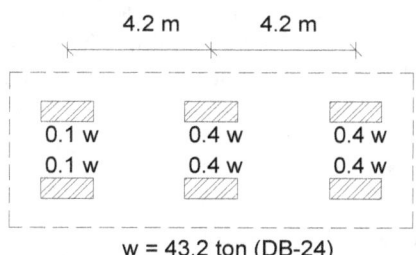

Figure 2: Design live load DB-24

Figure 3 shows the proposed strengthening plan based on the patented technology. The sequences of implementing this method are as follows:

(a) *Cutting a groove*: a narrow groove is cut for inserting NSM CFRP along the girder span. Additional space is required for mounting anchorage at both ends of the groove.
(b) *Installing anchorage*: anchor bearing blocks are positioned inside the precut anchorage space and fixed with anchor bolts.
(c) *Installing jacking apparatus*: NSM CFRP is located along the groove and connected with jacking apparatus for post-tensioning.

(d) *Jacking operation*: a hydraulic jack is placed within the jacking apparatus and a pressure is applied to the system until a desired post-tensioning force is achieved.

(e) *Transferring post-tension force*: once the post-tensioning force is achieved, the jacking apparatus is removed and the force is transferred to the bearing-anchor blocks.

(f) *Grouting*: all the precut regions are grouted with a cementitious material to improve the aesthetics of the strengthened member.

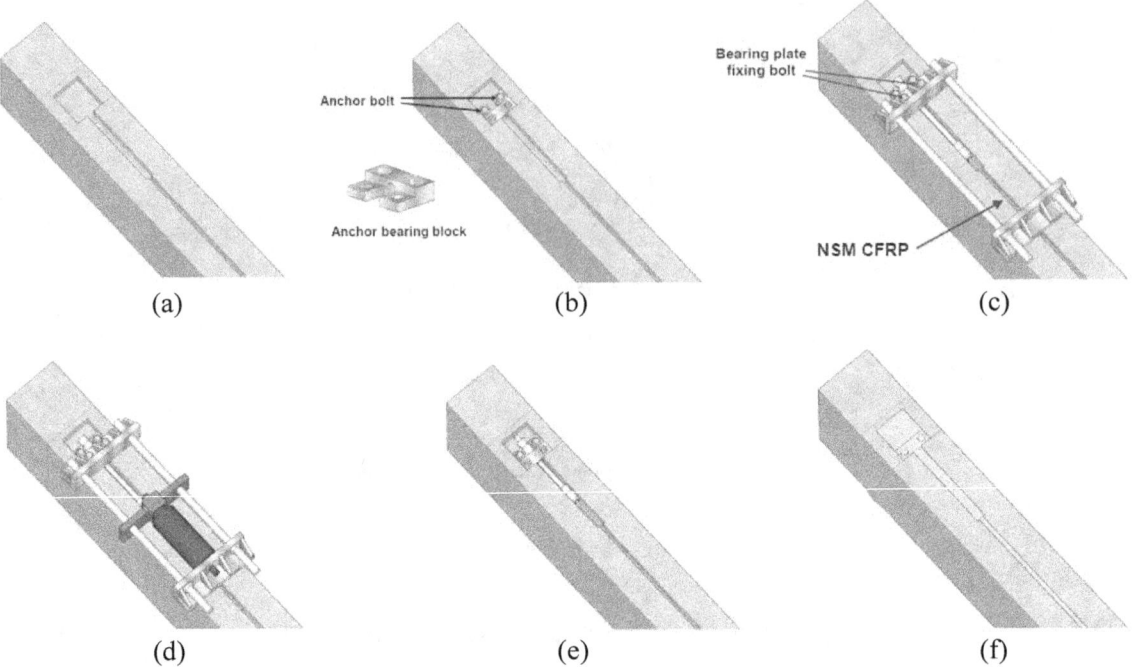

Figure 3: Proposed strengthening plan (patent numbers 10-0653632, 10-1005347, and 10-1083626): (a) cutting a groove; (b) installing anchorage; (c) installing jacking apparatus; (d) jacking operation; (e) transferring post-tension force; (f) grouting

To examine the effect of post-tensioning, various forces were planned from $0\%f_{fu}$ to $60\%f_{fu}$, where f_{fu} is the ultimate capacity of the CFRP, which were equivalent to post-tensioning forces from 0 kN to 981 kN. An expression was proposed to assist examining the effectiveness of the NSM CFRP for upgrading an existing bridge girder: $\alpha = L_p/L$ where α is the strengthening coefficient and L_p is the bond length of the NSM CFRP.

4. RESEARCH APPROACH

The research approach taken was two-fold: i) three-dimensional finite element analysis for global response investigations and ii) a preliminary reliability examination.

4.1. Finite Element Modeling

The commercial finite element package ANSYS was used. Figure 4 shows a constructed model with elements. Four-node elastic shell elements were used to represent the girder concrete. This element has six degrees of freedom at each node (i.e., three rotational and three translational degrees of freedom). Three-dimensional link elements were employed for modeling the prestressing strands and the CFRP, which would show unidirectional behavior in tension and compression. To facilitate computational modeling, the multiple prestressing strands shown in Fig. 1 were simplified to one representative tendon with the same center of gravity (Fig. 4). A function called INISTATE was used to apply a prestressing force to the steel and the CFRP: the link element was subjected to an initial strain equivalent to the desired prestressing force levels explained earlier.

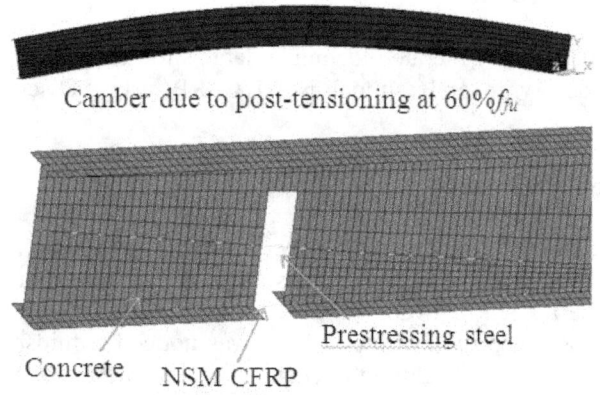

Figure 4: Finite element model developed (cutaway view)

Displacement compatibility was assumed among all constituent elements and the material properties discussed in Sec. 3.1 were input. The preprocessed bridge girder models were then constrained to simulate a simply supported condition (one end was hinged and the other end was rollered as shown in Fig. 1). The live load model (Fig. 2) was positioned to induce the maximum bending moment along each girder. The number of simulated models was 51, including three loading scenarios (i.e., camber loading only without live load, truck load for unstrengthened girders, and truck load for strengthened girders).

4.2. Validation of Modeling Approach

The proposed modeling approach was validated with theoretical and experimental data. For the case of unstrengthened girder models, camber and net deflections were evaluated against structural analysis solutions as shown in Fig. 5(a). The validation approach for strengthened girders was accomplished based on the experimental program conducted by Taljsten and Nordin [17]. The test program included a reinforced concrete T-beam strengthened with three NSM CFRP rods with a total cross-sectional area of 300 mm^2. The CFRP rods had an ultimate strength of 2600 MPa and a modulus of 150 GPa. The 5.6 m simply supported beam was monotonically loaded and its camber and deflection were obtained. Provided the developed model was valid before concrete-cracking takes place, the experimental data were compared with the model prediction up to a decompression load [Fig. 5(b)]. Overall, the modeling approach demonstrated reliable results.

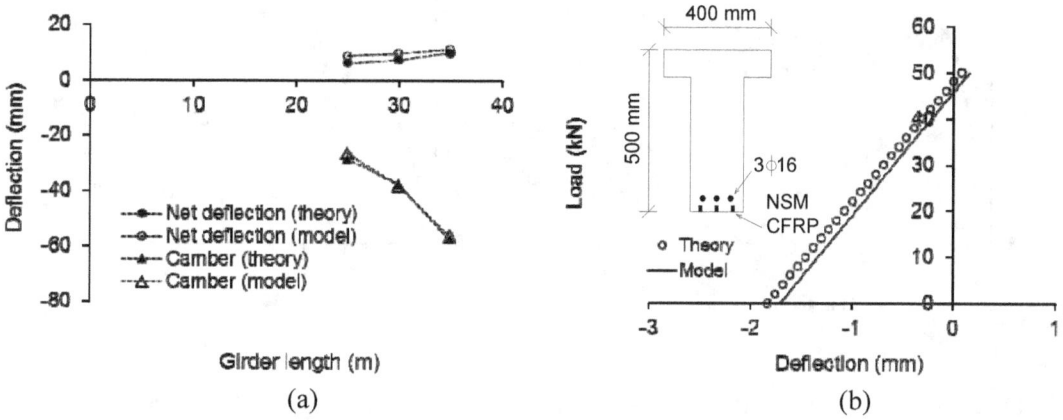

Figure 5: Validation of modeling approach: (a) unstrengthened case; (b) strengthened case

5. TECHNICAL DISCUSSION

The solved finite element models generated technical data as to the flexural behavior of the prototype prestressed concrete girders with and without NSM CFRP strips. Discussion is provided in this section.

5.1. Response of Reinforcement

The strain profiles of the steel reinforcement of selected girders are shown in Fig. 6. The development of steel strain was rapid near the end of the girders and stabilized as approaching midspan (i.e., the development length of the $L = 25$ m girder was 2,063 mm [Fig. 6(a)]). The presence of NSM CFRP reduced steel strain; for instance, a reduction of 2.2% was noticed at midspan of the $L = 25$ m girder [Fig. 6(b)]. It should be noted that no dramatic change in steel strain was predicted because the load applied was a service load (i.e., DB-24). The effect of post-tensioning levels was depicted in Fig. 6(c). Insignificant differences in strain were noticed within the development-length zone, whereas the level of post-tensioning exhibited an apparent effect beyond the zone. This observation corroborates a stress-sharing mechanism between the prestressing steel and the NSM CFRP. The strain variation of the prestressing steel was influenced by the bonded length of the NSM CFRP, as shown in Fig. 6(d). Strain concentrations were found near the termination point of the CFRP, which may induce premature failure of the installed strengthening system in service. To avoid this technical concern, the bond length of the NSM CFRP is recommended to be sufficiently long (i.e., between 60% and 90% of the span length, $\alpha = 0.6$ to 0.9).

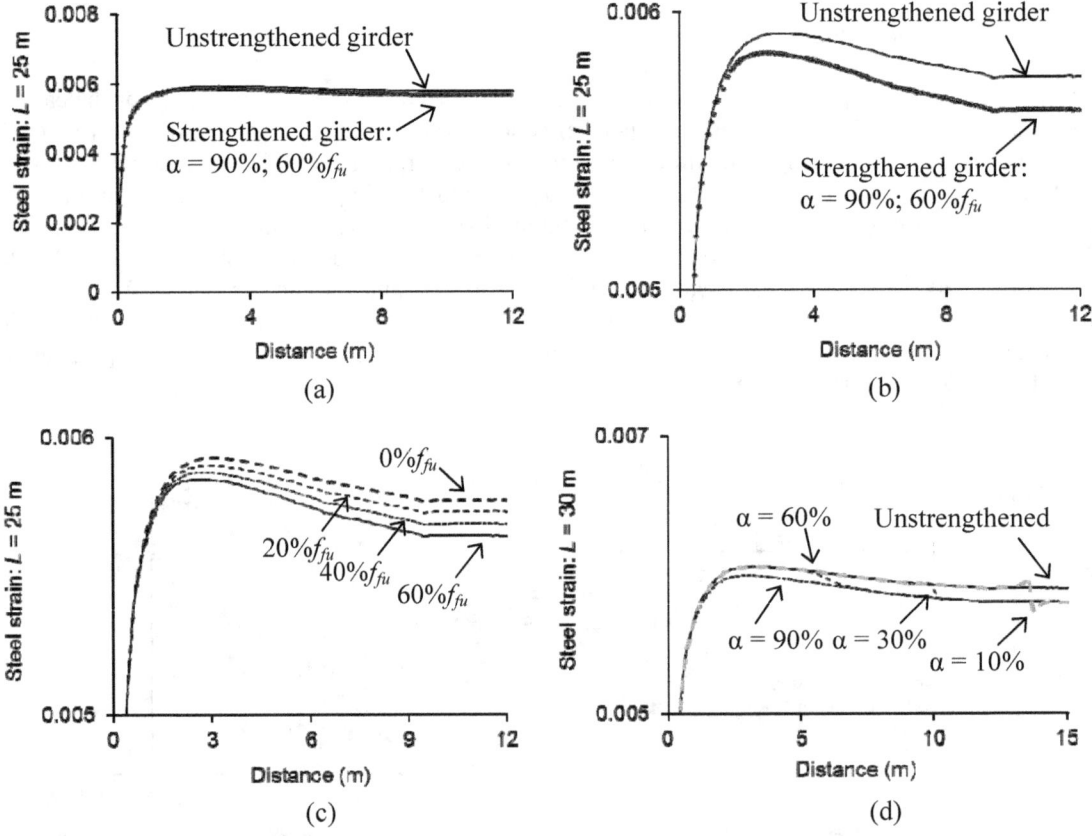

Figure 6: Response of steel reinforcement in strengthened girders under DB-24 load: (a) effect of NSM CFRP strengthening ($L = 25$ m); (b) effect of NSM CFRP strengthening ($L = 25$ m) close-up view; (c) effect of post-tensioning level with $\alpha = 90\%$; (d) effect of CFRP bond length (strengthening coefficient α) with $60\% f_{fu}$

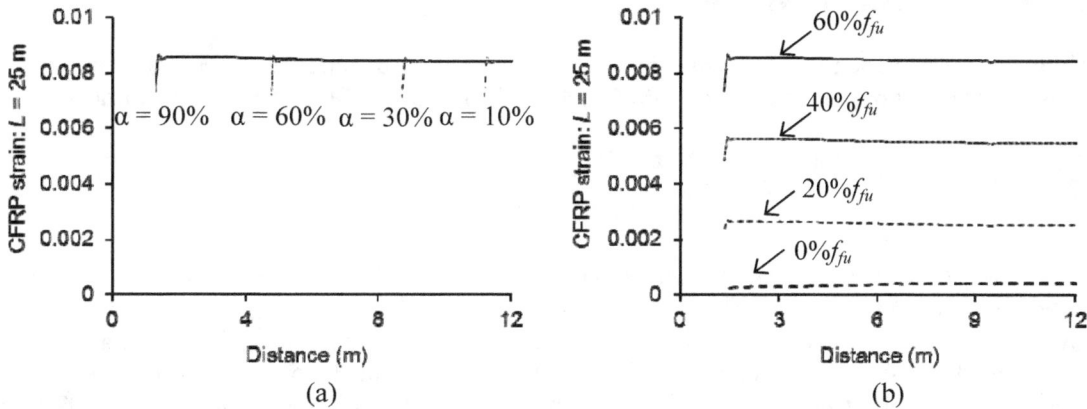

Figure 7: Response of NSM CFRP: (a) effect of CFRP bond length (strengthening coefficient α) with 60%f_{fu}; (b) effect of post-tensioning level

The profile of CFRP strain for the $L = 25$ m girder is exhibited in Fig. 7(a), depending upon the strengthening coefficient α (i.e., bonded length of the NSM CFRP). As in the case of the steel strain (Fig. 6), the CFRP strain significantly developed from its termination point and became stable. Regional strain softening was noticed immediately after the strain-development zone. Such a local strain softening effect tended to be reduced with a decrease in post-tensioning force [Fig. 7(b)].

5.2. Deflection of Girders

Figure 8(a) reveals the deflection profile of the girders with various strengthening coefficients. The camber of the prestressed concrete girders was predicted (i.e., negative deflection). With an increasing CFRP-bonded length (or an increase in the strengthening coefficient), the upward deflection proportionally increased up to $\alpha = 60\%$ beyond which such an increasing tendency in camber was reduced (in other words, the camber of the girders with $\alpha = 60\%$ and 90% was similar). Figure 8(b) shows the effect of a post-tensioning level on enhancing girder deflections. The normalized deflection was defined by a ratio between the upward deflection of a strengthened girder and that of an unstrengthened counterpart. The difference between the unstrengthened girder and the strengthened girder without post-tensioning (0%f_{fu}) was not observed, which means that the serviceability of the strengthened girders without post-tensioning was not improved. Substantial enhancement in deflection was, on the other hand, noticed when the post-tensioning level increased, as reported in Fig. 8(b).

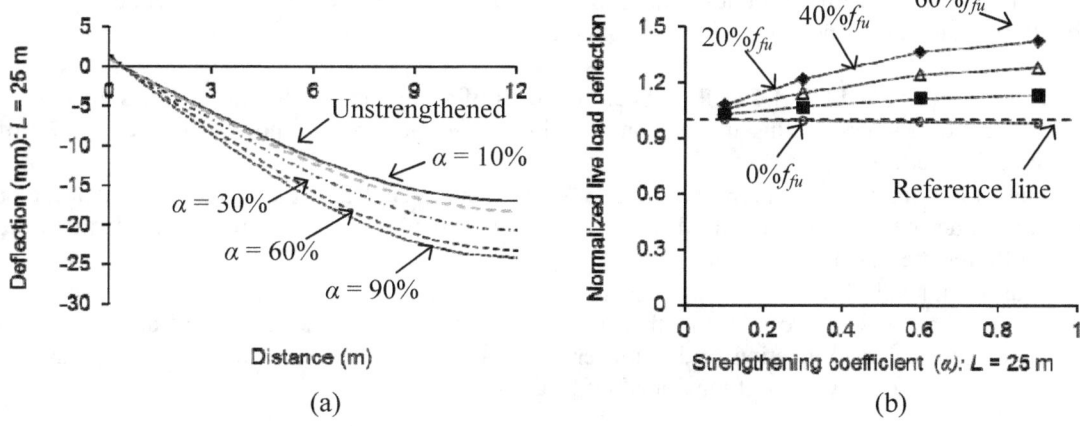

Figure 8: Response of girders: (a) deflection profile under live load with a post-tensioning level of 60%f_{fu}; (b) normalized live load deflection with post-tensioning level

6. RELIABILITY EXAMINATION

A preliminary investigation into structural reliability was conducted using the concept of safety index (β) to assess the effect of post-tensioned NSM CFRP on the response of the strengthened girders:

$$\beta = \frac{M_n - M_E}{\sqrt{\sigma_n^2 + \sigma_E^2}} \tag{1}$$

where M_n and M_E are the nominal flexural resistance and the load effect induced by DB-24 (wheel load), respectively; and σ_n and σ_E are their standard deviations. The strain compatibility method was employed to determine the nominal flexural resistance of each girder (M_n), while structural analysis theory provided the unfactored load effects (M_E). The standard deviations were estimated as per previous research: coefficients of variation for prestressed concrete girders and live load are 0.075 and 0.18, respectively [18]. As shown in Fig. 9, the safety indices of all the strengthened girders were positioned at around $\beta = 12$. Such a high safety index value is attributed to the fact that the girders were loaded in service. Another thing to note is that the post-tensioned NSM CFRP resulted in a uniform level of reliability, irrespective of post-tensioning levels.

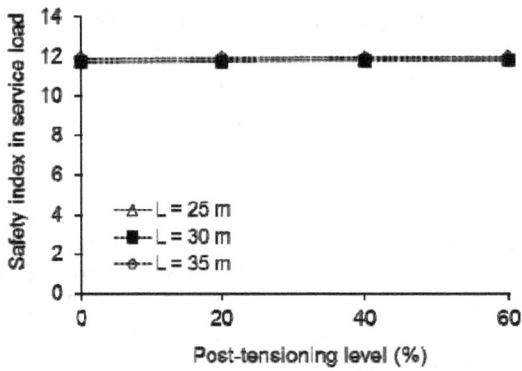

Figure 9: Safety index with post-tensioning level

7. CONCLUSIONS

This paper has discussed the flexure of prestressed concrete girders strengthened with post-tensioned NSM CFRP strips under a service design load. A three-dimensional finite element approach was proposed and validated with theoretical and experimental responses. A preliminary reliability study was conducted. The following conclusions are drawn:

- The post-tensioned NSM CFRP reduced the level of stress in the internal prestressing strands because of a stress-sharing mechanism. Strain concentrations were noticed in the vicinity of the CFRP termination.
- The CFRP-bonded length represented by a strengthening coefficient α was an important factor to consider when implementing this strengthening method on site. It is recommended that the coefficient be in between 0.6 and 0.9.
- A uniform level of reliability was observed for all the girders strengthened with post-tensioned NSM CFRP. Further research in this area is recommended to develop load and resistance factors dedicated to existing bridge girders, with or without the presence of damage, and to those strengthened with post-tensioned NSM CFRP.

Acknowledgements

The authors gratefully acknowledge financial support from the Strategic Research Project (Development of a Bridge Strengthening Method Using Prestressed FRP Composites) of the Korea

Institute of Construction Technology that owns the patented strengthening method presented in this study.

References

[1] Bakis, C.E., Bank, L.C., Brown, V.L., Cosenza, E., Davalos, J.F., Lesko, J.J., Machida, A., Rizkalla, S.H., and Triantafillou, T.C. "*Fiber-reinforced polymer composites for construction- state-of-the-art review*", Journal of Composites for Construction, 6(2), 73-87, (2002).

[2] De Lorenzis, L. and Teng, J.G. "*Near-surface mounted FRP reinforcement: an emerging technique for strengthening structures*", Composites Part B, 38, 119-143, (2007).

[3] Hassan, T. and Rizkalla, S. "*Investigation of bond in concrete structures strengthened with near surface mounted carbon fiber reinforced polymer strips*", Journal of Composites for Construction, 7(3), 248-257, (2003)

[4] Barros, J.A.O. and Fortes, A.S. "*Flexural strengthening of concrete beams with CFRP laminates bonded into slits*", Cement and Concrete Composites, 27, 471-480, (2005).

[5] Seracino, R., Jones, N.M., Ali, S.M., Page, M.W., and Oehlers, D.J. "*Bond strength of near-surface mounted FRP strip-to-concrete joints*", Journal of Composites for Construction, 11(4), 401-409, (2007).

[6] ACI. "*Report on fiber-reinforced polymer (FRP) reinforcement for concrete structures (ACI-440R-07)*", American Concrete Institute, Farmington Hills, MI, USA (2007).

[7] Wu, Z., Iwashita, K., and Sun, X. "*Structural performance of RC beams strengthened with prestressed near-surface-mounted CFRP tendons*", American Concrete Institute Special Publication (ACI SP-245), 143-164, (2007).

[8] Choi, H.T., West, J.S., and Soudki, K.H. "*Effect of partial unbonding on prestressed near-surface-mounted CFRP-strengthened concrete T-beams*", Journal of Composites for Construction, 15(1), 93-102, (2011).

[9] El-Hacha, R. and Gaafar, M. "*Flexural strengthening of reinforced concrete beams using prestressed near-surface-mounted CFRP bars*", PCI Journal, 56(4), 134-151, (2011).

[10] De Lorenzis, L., Micelli, F., and La Tagola, A. "*Passive and active near-surface mounted FRP rods for flexural strengthening of RC beams*", 3rd International Conference on Composites in Infrastructure, San Francisco, CA, USA (CD-ROM), (2002).

[11] Nordin, H. and Taljsten, B. "*Concrete beams strengthened with prestressed near surface mounted CFRP*", Journal of Composites for Construction, 10(1), 60-68, (2006).

[12] Badawi, M. and Soudki, K. "*Flexural strengthening of RC beams with prestressed NSM CFRP rods- experimental and analytical investigation*", Construction and Building Materials, 23, 3292-3300, (2009).

[13] Wahab, N., Soudki, K.A., and Topper, T. "*Experimental investigation of bond fatigue behavior of concrete beams strengthened with NSM prestressed CFRP rods*", Journal of Composites for Construction, 16(6), 684-692, (2012).

[14] Jung, W.-T., Park, Y.-H., Park, J.-S., Kang, J.-Y., and You, Y.-J. "*Experimental investigation on flexural behavior of RC beams strengthened by NSM CFRP reinforcements*", American Concrete Institute Special Publication (ACI-SP-230), 795-806, (2005).

[15]   ACI. *"Prestressing concrete structures with FRP tendons (ACI440.4R-04)"*, American Concrete Institute, Farmington Hills, MI, USA (2004).

[16]   KHC. *"Standard drawing of highway bridges, Korea Highway Corporation"*, Seongnam-si, Gyeonggi-do, Korea, (2001).

[17]   Taljsten, B. and Nordin, H. *"Concrete beams strengthened with external prestressing using external tendons and near-surface-mounted reinforcement (NSMR)"*, American Concrete Institute Special Publication (ACI SP-245), 143-164, (2007).

[18]   Nowak, A. S. *"Calibration of LRFD bridge design code"*, NCHRP 12-33, Transportation Research Board, Washington, D.C., USA.

A Quantitative Method for Assessing the Resilience of Infrastructure Systems

Cen Nan[ab], Giovanni Sansavini[bc], Wolfgang Kröger[c] and Hans Rudolf Heinimann[ac]

[a] Land Using Group, ETH Zürich, Switzerland
[b] Reliability and Risk Engineering, ETH Zürich, Switzerland
[c] ETH Risk Center, ETH Zürich, Switzerland

Abstract: Resilience is a dynamic multi-faceted term and complements other terms commonly used in risk analysis, e.g. reliability, availability, vulnerability, etc. The importance of fully understanding system resilience and identifying ways to enhance it, especially for infrastructure systems our daily life depends on, has been recognized not only by researchers, but also by the public. During recent years, several methods and frameworks have been proposed and developed to explore applicable ways to assess and analyse system resilience. However, they are tailored to specific disruptive hazards/events mainly for other than technological systems, or fail to properly include all the phases, e.g., mitigation, adaptation and recovery. In this paper, after defining the term, a generic quantitative method for the assessment of the system resilience is proposed, which consists of two components: a hybrid modelling approach and an integrated metric for resilience quantification. The feasibility and applicability of the proposed method is tested using an electric power supply system as the exemplary system.

Keywords: Resilience, Critical Infrastructure, Agent-based Modeling, Reliability

1. INTRODUCTION

Engineered infrastructure systems have always been "complicated", but in recent years, they have witnessed growing integration and interconnectedness, which have turned them into complex systems [1, 2]. To better understand the performance of these systems, especially their behaviors during and after the occurrence of disturbances (e.g., natural hazards or technical failures), a great effort has been devoted by researchers with emphasis on different phases and aspects, e.g. availability assessment during the initial loss phase, evaluation of restoration efforts during recovery phase, etc. However, these assessments are challenged by the diversity of the physical flow in the infrastructure systems, by the lack of comparable indexes for quantifying system performances, and by the multiplicity of loss scenarios. A unifying method to analyze and strengthen system performance as responses to disturbances is still missing. To this aim, resilience analysis [3-6] is a proactive approach to enhance the ability of the infrastructure systems to prevent/avoid damage before disturbance events, mitigate losses during the events and improve recover capability after the events.

2. RESEARCH STREAMS AND PROPOSED METHOD

The term resilience is still evolving and has been developing in various fields. The first definition is given by an ecologist, C. S. Holling, who described resilience as "*a measure of the persistence of systems and of their ability to absorb change and disturbance and still maintain the same relationships between populations or state variables*" [7]. Since then, others have put forward domain-specific resilience definitions [8, 9]. The concept of resilience is also introduced to engineered technical systems. From an engineering perspective, resilience can be defined as "*the ability of the system to withstand a major disruption within acceptable degradation parameters and to recover within an acceptable time and costs*" [10]. In recent years, assessing and engineering resilience of infrastructure systems has emerged as a fundamental concern for researchers [11, 12]. Up to now, resilience still lacks a comprehensive description, calling for further developments to frame its definition. A broader definition for this term is then proposed in this paper, which describes resilience as "the ability of a system or a so-called 'System of Systems' (SoS) to resist effects of a disruptive internal or external event/force, either shocking or creeping, and the ability to reduce both magnitude and duration of deviation of the system performance level between original (or target) state and new steady state due to internal and external efficient efforts".

The proposed definition can be further interpreted as the ability of the system or SoS to withstand a change or a disruptive event by reducing the initial negative impacts (absorptive capability), by adapting itself to them (adaptive capability) and by recovering from them (restorative capability). These capabilities can be regarded as three essential resilience features: enhancing any of them will enhance system resilience. They focus on the system response during and after the occurrence of disruptive events. It is important to further understand and find ways to quantify them that contribute to characterization of the system performance [13]. **Absorptive capability** refers to an endogenous ability of the system to reduce the negative impacts caused by disruptive events and minimize consequences. In order to quantify this capability, *robustness* can be used, defined as strength of the system to resist initial impacts [14]. An example of enhancing it is to improve system redundancy, which provides an alternative way for system to operate. **Adaptive capability** refers to an endogenous ability of the system to adapt to disruptive events through its self-organization capabilities in order to minimize consequences. It is the dynamic ability of the system to adjust itself throughout the recovery period. Emergency systems can be used to enhance it. **Restorative capability** refers to an exogenous ability of the system to be repaired by external actions throughout the recovery period. For example, installing real-time monitoring systems (e.g., SCADA (Supervisory Control and Data Acquisition system) for most infrastructure systems) enhances the system restorative capability because it allows the automatic detection or even prediction of disruptive events and, therefore, shortening the total disruption period. Both adaptive and restorative capabilities describe the system's ability during the recovery phase, and it is not straightforward to distinguish their effects on system performance, which can be improved by enhancing the restorative capability during recovery phase after the deployment of repair actions and by enhancing adaptive capability before and after repair actions. Therefore, the simultaneous quantification of both capabilities is given same indicators, i.e. *rapidity* (*RAPI*) and *performance loss* (PL).

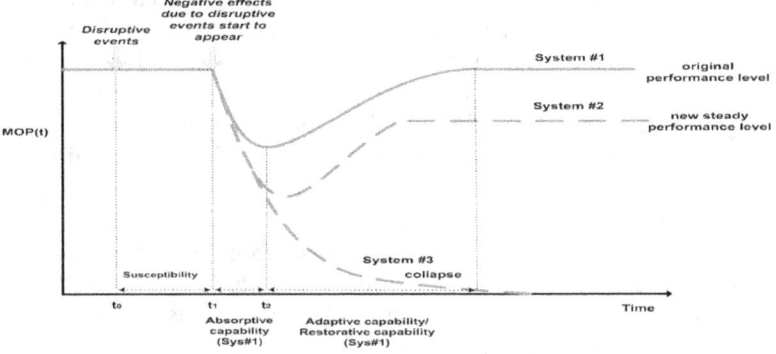

Figure 1 Illustration of essential resilience capabilities

Figure 1 provides a general illustration of these essential resilience capabilities. The y-axis represents the measurement of performance (*MOP*). Examples of MOP include availability of critical facilities, the number of customers served, connectivity of a network, the level of economic activities, etc. The selection of the appropriate MOP depends on the specific service provided by the infrastructure system under analysis. For generality, in the following we assume that the value of MOP is normalized between 0 and 1 where 0 is total loss of operation and 1 is the target MOP value in the steady phase. It is assumed that the disruptive event occurs at t_0, and that the *MOP* values starts dropping at t_1. It should be noted that in many cases t_0 might not be equal to t_1 and the $t_1 - t_0$ delay depends on the selection of the *MOP* and on the disruptive event. For instance, it could take several hours for customers to lose electricity services due to maintenance mistakes, while it might only takes seconds for same customers to lose services due to natural hazards such as earthquake, hurricane, etc. System susceptibility can be used to characterize the system performance during the time between t_0 and t_1. The focus of this paper is related to resilience quantification after the appearance of the negative effects, and therefore, susceptibility will not be considered. The system capabilities that have an effect on system resilience can be exemplified with respect to system performance variations following a disruptive event. As seen in Figure 1, system#1 performance returns to its original steady level after recovering from the lowest level at t_2. System#2 performance reaches a new steady level, which is lower than its original steady level. It should be noted that the new steady performance level could also be higher than its original steady level. System#3 performance drops significantly and finally collapses to zero. System#1 and system#2 outperform system#3 with respect to the three essential resilience capabilities. Therefore, system#3 can be considered the least resilient system. On the other hand, system#1 seems more robust against the disruptive event than system#2, i.e. the lowest performance level is higher for system#1 than for system#2. However, it takes more time for system#1 to reach the new steady level. Therefore, system#2 is more adaptive and restorative than system#1. The qualitative assessment and comparison among resilience capabilities call for the development of an approach that can be used to quantify them and integrate them into one system resilience index.

During last decade, researchers have proposed different methods and frameworks to quantify/assess system resilience. In 2003, the first conceptual framework is proposed by Bruneau et al. in [14]. The purpose of this framework is to measure the seismic resilience of a community to an earthquake by estimating the expected degradation regarding the quality of community infrastructure. In this pioneering research work, the concept of *Resilience Loss* (RL), later also refereed as so called "resilience triangle", is introduced, which has been widely used afterwards as a fundamental guidance for system resilience assessment. Based on this framework, more research works have been carried out from different aspects using various approaches. In 2004, Chang and Shinozuka propose a probabilistic approach for measuring seismic resilience after earthquake events [15]. In 2007, Rose at al., develop static and dynamic metrics based on the resilience loss concept to measure economic resilience. In recent years, the importance of improving the resilience of interdependent infrastructure systems or at least minimizing negative impacts caused by unexpected disruptive events has been recognized and accepted by the public. Therefore, a variety of research works have been developed targeting interdependent systems. In 2008, McDaniels et al. develop a knowledge-based approach using decision flow diagrams to improve the understanding of the resilience of infrastructure systems [16]. Similar approach is also proposed by Argonne National Lab [5, 17]. This type of knowledge-based approach is straightforward and easy to understand. However, it is a pure data-driven approach and the quality of the collected information could have significant effects on the accuracy of final results. To overcome these limitations, more comprehensive analytical approaches have been developed with the help of the advanced modeling techniques. In [18], System Dynamic (SD) is applied to assess the degree of socio-ecological system resilience. This approach combined with Complex Network Theory (CNT) is later applied by Filippini and Silva as part of the framework of qualitative resilience analysis of infrastructure systems [19].

Resilience is a dynamic multi-faceted term and its assessment should cover all the phases, e.g., disruption and recovery phase, and include all the essential resilience capabilities using an integrated metric. Most of existing methods for resilience quantification lack the ability to cover all the phases, and to include all resilience capabilities within all integrated metric and even overlap with other concepts such as robustness, vulnerability, fragility etc [20, 21]. Some quantitative methods for resilience measurement are not consistent with the concept of resilience [22]. Furthermore, these methods rely on the modeling approaches that partially capture the complex behavior of infrastructure systems. All this makes clear that there is a pressing need to develop a quantitative method for the assessment of the resilience of different infrastructure systems subjected to various hazards, which is built on the aforementioned challenges and should be able to provide an over-arching and cross disciplinary vision of integrated risk management. Therefore, a generic quantitative method targeting the holistic analysis of today's infrastructure systems is proposed and presented in this paper. The method consists of two components: 1) a hybrid modelling approach to achieve a close representation of the system and analyze its dynamics during and after the occurrence of the disruptive events. 2) an integrated metric for resilience quantification which can incorporate all essential resilience capabilities and characterizing resilience as the system ability. The focus of this paper is mainly related to the second component of the method.

3. METHOD PART 1-A HYBRID MODELLING APPROACH

Developing a comprehensive modeling framework with the capabilities of achieving a closer representation of infrastructure systems and gaining insights into interactions within and among them is vital for improving our understanding of these systems. Currently, a broad scale of modeling approaches have been developed, e.g., Input-output Inoperability Modeling (IIM), Complex Network Theory (CNT), Agent-based Modeling (ABM), System Dynamic (SD), Petri-Net (PN), etc (see [23] and [24] for details about these approaches). The lack of coherent modeling approaches hinders the possibilities of analyzing dynamic system behaviors in a sufficient way. Therefore, it is essential to develop a comprehensive modeling approach with the capabilities of achieving in-depth insights of infrastructure system behaviors. In practice, there is still no "silver bullet approach". Instead, it has proven necessary to integrate different types of modeling approaches into one simulation tool in order to fully utilize the advantages of each approach and optimize the efficiency of the overall simulation. In [25], the PN approach is integrated with the ABM approach to model interdependent infrastructure systems. In [26], authors combine the CNT with SD approach in order to develop a resilience indicator. However, all of these hybrid approaches are not capable to solve one of key challenges for developing such type of simulation tool: the required ability to create multiple-domain models, and effectively exchange data among them [27]. One solution for handling these technical difficulties and meeting these challenges is to distribute different simulation components by adopting the concept of modular design [28]. Through this approach, the overall simulation tool is divided into different simulation modules, which are domain-specific or sector-specific simulation components. The modules are then combined in a distributed simulation platform [24, 29].

4. METHOD PART 2-AN INTEGRATED RESILIENCE METRIC

Resilience is a complex concept that can not to be adequately addressed considering one single system capability[16]. One solution is to develop corresponding measures assessing various essential resilience capabilities in different phases, and then integrate them into a unique resilience metric.

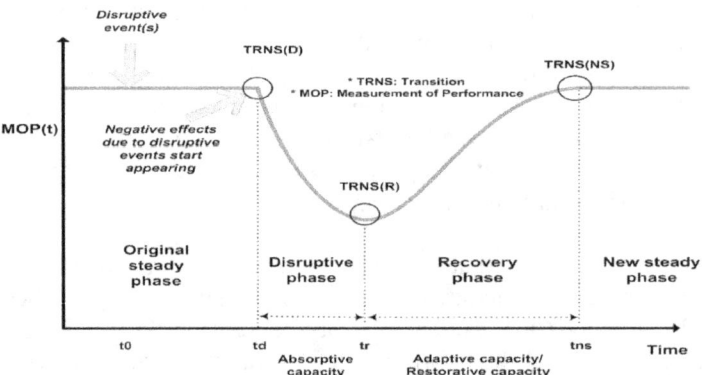

Figure 2 System resilience transitions and phases

The performance of the system#1 shown in Figure 1 is further illustrated in Figure 2, and it can be characterized by four phases and three transitions. The first phase is the original steady phase ($t<t_d$), in which the system performance assumes its target value. The second phase is the disruptive phase ($t_d \leq t < t_r$), in which the system performance starts dropping until reaching the lowest level at time t_r. During this phase, the system absorptive capability can be assessed by developing appropriate measures. As discussed in Section 1, *Robustness* (R) is one measure to assess this capability, which represents the minimum *MOP* value. This measure is able to identify the maximum impact caused by disruptive events; however, it is not sufficient to reflect the ability of the system to absorb the impact. Two additional complementary measures are further developed: *Rapidity* ($RAPI_{DP}$) and *Performance Loss* (PL_{DP}) during disruptive phase. In [30], the term rapidity is referred as "the capability to meet priority and achieve goals in a timely manner in order to contain losses and avoid future disruption". This term can be quantified mathematically as the slope of performance level. To improve the accuracy of the estimation of the measure rapidity, the method of ramp detection can be adopted. In general, a ramp is a change with a large enough amplitude and over a relatively short period [31]. According to [32], a ramp is assumed to occur if the difference between the measured value at the initial and final points of a time interval Δt is greater than a predefined ramping threshold value. The system rapidity can then be calculated as the average of slope of each ramp. Compared to the calculation of average the rapidity, this method is more comprehensive in term of capturing the system performance during different phases. The performance loss, using the system illustrated in Figure 2 as an example, can be interpreted can be quantified as the area of the region bounded by the graph of the measurement of performance before and after occurrence of negative effects caused by disruptive events, which can also be referred as the system impact area. A new measure, i.e. the time averaged performance loss (*TAPL*), is introduced. Compared to the measure PL, it encompasses the time of appearance of negative effects due to disruptive events up to full system recovery and provides a time-independent indication of both adaptive and restorative capabilities as responses to the disruptive events. A system that experiences less performance loss has larger resilience. The third phase is the recovery phase ($t_r \leq t < t_{ns}$), in which the system performance starts increasing until the new steady level. During this phase, the system adaptive and restorative capability can be assessed by developing appropriate measures: *Rapidity* ($RAPI_{RP}$) and *Performance Loss* (PL_{RP}). As shown in Figure 2, the newly attained steady level may equal to the previous steady level. But it may also reach a lower level. In order to take these situations into consideration, a simple quantitative measure *Recovery Ability* (RA) is also developed. Different system phases and related system capabilities are summarized in Table 1.

Table 1 Summary of different resilience phases

Phases	Time Scope	Transition Point	Capacities (features)	Measurements
Original steady phase	$t < t_d$		Susceptibility	Susceptibility
Disruptive phase	$t_d \leq t < t_r$	TRNS(D)	Absorptive capacity	Robustness (R)
				Rapidity in disruptive phase ($RAPI_{DP}$)
				Performance Loss in disruptive phase (PL_{DP})
Recovery phase	$t_r \leq t < t_{ns}$	TRNS(R)	Adaptive capacity	Rapidity in recovery phase ($RAPI_{RP}$)
			Restorative capacity	Performance Loss in recovery phase (PL_{RP})
New steady phase	$t \geq t_{ns}$	TRNS(NS)	Recovery ability	Recovery ability (RA)

Although the measurements introduced above are useful in assessing system behavior during and after disruptive events, an integrated metric with the ability of combining these capabilities is needed in order to assess the system resilience with an overall perspective and to allow comparisons among different systems and system configurations. Therefore, a general resilience metric is further developed. This metric differs from existing ones in that it is time-dependent and able to incorporate all three essential capabilities. Furthermore, it is not system-specific. The resilience metric builds on the quantification of the system capabilities and is calculated as:

$$GR = R \times \left(\frac{RAPI_{RP}}{RAPI_{DP}}\right) \times (TAPL)^{-1} \times RA$$

$$= R \times \left(\frac{\left|\sum_{i=1}^{K_{RP}} \frac{MOP(t_i)-MOP(t_i-\Delta t)}{\Delta t}\right|}{\left|\sum_{i=1}^{K_{DP}} \frac{MOP(t_i)-MOP(t_i-\Delta t)}{\Delta t}\right|} \frac{K_{RP}}{K_{DP}}\right) \times \left(\frac{\int_{t_d}^{t_{ns}}[MOP(t_0)-MOP(t)]dt}{t_{ns}-t_d}\right)^{-1} \times \left|\frac{MOP(t_{ns})-MOP(t_r)}{MOP(t_0)-MOP(t_r)}\right| \quad (1)$$

Where TAPL represents time average performance loss; K_{DP} and K_{RP} represent number of detected ramps in disruptive phase and recovery phase; $MOP(t_0)$ represents performance level at original steady phase. The GR provides an integrated way to measure the system resilience by considering all essential capabilities, which is consistent with original definition of the term of resilience. This approach of measuring system resilience is neither model nor domain specific. For instance, historical data can also be used for the resilience analysis. It only requires the time series data that represents system output during whole time period, making the selection of the MOP very important. The GR is a non-negative metric and its value equals to zero if 1) the performance level drops to zero after the disturbance (R=0), 2) after the disturbance events, the system performance immediately drops to its lowest level (RAPI$_{DP}$ → ∞, i.e. no absorptive capability), 3) the system performance maintains at the lowest level, R, which is the new steady state (RAPI$_{DP}$=0, i.e. no adaptive and restorative capability). Furthermore, the GR value is dimensionless and is most useful in a comparative manner. For instance, it can be used to compare the resilience of various systems to the same disruptive event. More resilient system results in higher GR value. It can also be used to compare resilience of same system under different disruptive events. Higher GR value indicates that the system is more resilient to certain disruptive events. Furthermore, the GR value can be used to compare resilience of a system to a specific disturbance under different improvement strategies. A more effective improvement strategy should increase the GR value.

5. CASE STUDY

Electric power systems are among most prominent representatives of engineered infrastructure systems and the need for their reliable and resilient performance during disruptive events becomes essential [33, 34]. In this paper, the high voltage Swiss electric power supply system (EPSS) is selected as an exemplary application to demonstrate the feasibility and applicability of the proposed quantitative method for resilience assessment.

5.1. Modelling Exemplary System

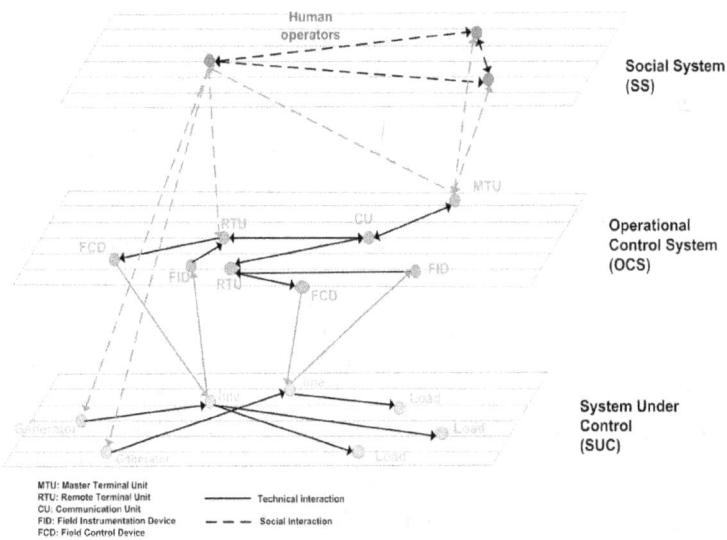

Figure 3 Representing EPSS in three subsystems (layers)

The modelling of CI is a whole research field by itself. Modelling entire infrastructure system as a whole is usually impractical [35-37]. Choosing relevant subsystems and modelling them efficiently for the intended purposes seem more promising. For each subsystem, appropriate modelling approaches/methods can be determined to fully represent its behaviour and functionalities. This type of modelling approach can also be referred as an example of the hybrid modelling approach, introduced in Section 2. Using this approach, the EPSS can be viewed as three interrelated subsystems in corresponding layers, i.e., system under control (SUC), operational control system (OCS), and social system (SS). Both SUC and OCS can be regarded as technical systems, while the SS is regarded as non-technical systems. The SUC represents technical systems that are mainly time-based. In order to fully model this type of systems, both functionality (physical laws) and structure (topology) need to be considered. The OCS represents technical systems that are mainly responsible to control and monitor corresponding SUC, e.g., SCADA system. In general this type of systems is event-driven / service-oriented. The developed model needs to be able to process messages among components efficiently and functionality of the system needs to be represented in detail. The SS represents non-technical systems that are mainly related to social factors. e.g., human performance, etc, that intend to have influences on the overall system performance. The developed model representing this subsystem need to be able to evaluate and quantify the effects of these social factors. The approach of HRA (Human Reliability Analysis) seems adequate. Figure 3 shows a simplified multi-layer representation of the EPSS with three interacting subsystems and associated elements (components): parallel planes represent different subsystems in corresponding layers while nodes represent various elements together with some of interconnections among them. The elements of the various layers depend on each other, depicted by various horizontal (inside a layer) and vertical (between layers) links.

A two-layer ABM-based modelling approach has been developed for purpose of integrating stochastic time-dependent factors into the resilience assessment of the EPSS [38]. Within the two-layer concept, the lower layer represents the separate modelling of the physical components by means of conventional, deterministic techniques such as power flow calculations, whereas the upper layer represents the abstraction of the electric power system with all its components as individual agents. Based on this approach, a time-stepped and agent-based model is developed to simulate the SUC. In total, 585 agents are created to model corresponding components, i.e., transmission lines, generators and loads. Similar to the modelling approach used to model the SUC, a failure-oriented two-layer ABM-based modelling approach is also used to model the SCADA system. In total, 587 agents are created to corresponding components, i.e., FCDs, FIDs, and RTUs [24]. In order to model SS, i.e. human operator performance in this case, the CREAM method combined with the ABM and Fuzzy logic is implemented [24]. It is the first effort to implement a human operator performance model capable assessing Performance Shaping Factors (PSFs) dynamically using the ABM approach. During the simulation, if there is a request for the operator take actions (e.g., handle an alarm), the PSFs will be assessed based on current simulation environment, e.g., time of day, simultaneous goals, etc, and corresponding Human Error Probability (HEP) ([0,1]) will then be calculated. The lower HEP value indicates better performance by human operators. To decide whether or not there is an error by human operators, it is necessary to set a threshold value (HEP_A) representing the maximum acceptable HEP value. If a calculated HEP is more than HEP_A, then it is assumed that a human error occurs (the human action fails to perform). The higher HEP_A indicates less human errors. All the developed models of three subsystems are integrated in a High Level Architecture (HLA)-compliant experimental simulation platform.

5.2. Design of Experiment

Although little damages caused by natural hazards have been observed throughout last century in Switzerland, historical records reveal that hazards such as earthquakes and winter storms were the cause of significant damage in at least 9 events over the past 1000 years [39]. According to [40], the estimated frequency of natural hazards, i.e., winter storms, which have the potential to result in the simultaneous disconnection of 20 transmission lines is about $6E^{-4}$ to $7E^{-4}$ per year. The impact of this disruptive event will result in large negative effects, and should not be underestimated even if the frequency of its occurrence is relatively low (low frequency, high consequence event). Therefore, in this paper, it is assumed that a natural hazard, e.g., winter storm, impacts central region of Switzerland, where power transmission lines are located; as a result of this event, 17 transmission lines are disconnected. This region is selected for the system resilience experiment based on the Lothar winter storm occurred in 1999 [41]. To simulate the performance response of the infrastructure under study (EPSS) subjected to the disruptive event, all three models (SUC, SCADA, SS) are included in the experiment. The number of available transmission lines is selected as the MOP. The simulation starts at the $t = 0$ h. It is assumed that the disruptive event occurs at time 3 h. Before this time, all the modeled systems are in the original steady state and operate under normal conditions. At the $t = 3$ h, the failure generator triggers the disconnection of 17 predefined transmission lines within region affected by the storm. In order to model the quasi-simultaneous disconnection of the affected lines, it is assumed that the interval between disconnections of two lines follows a normal distribution N(35,3) seconds. The sudden disconnection of a transmission line will be first detected by the corresponding RTU (Remote Terminal Unit) component of the SCADA system and an alarm will be sent to the control center (MTU), which is referred as the *abnormal line disconnection alarm*. After receiving the alarm, repair actions will be determined by the operators in the control center in order to restore the disconnected lines. It is assumed that the general response time (Response$_G$) for this type of alarm follows a normal distribution N(80,5) seconds. The sudden disconnections of many transmission lines may overwhelm the operators in control center, possibly resulting in the delay of response and repair actions. In order to simulate this situation, the formula below is used to calculate actual response time (Response$_A$):

$$\text{Response}_A = \text{delay factor} * \text{Response}_G$$

$$\text{Delay factor} = \begin{cases} \text{weighting factor} * (\text{HEP} - \text{HEPA}) + 1 & (\text{HEP} \geq \text{HEPA}) \\ 1 & (\text{HEP} < \text{HEPA}) \end{cases} \quad (2)$$

The actual response time (Response$_A$) determined by the delay factor should close to reality and therefore, need to remain in a rational range. In this experiment, weighting factor is set to 100 based on results from after several trial runs. It is assumed that repair actions for the abnormal disconnected lines are always performed successfully and the repair time is assumed to follow exponential distribution with the mean value equal to MTTR (mean time to repair). The sudden disconnections of transmission lines could also overload other transmission lines, especially their neighboring ones [42], and have the potential for knock-on effects with cascading consequences. If a transmission line is overloaded, an overload alarm will be generated and sent to the operator in the control center (MTU) by the corresponding RTU component. If the operator recognizes this alarm and handles it successfully (HEP<HEP$_A$), the corrective actions will be performed, i.e., power load re-dispatch. However, if no action is taken after a certain time past the overload alarm, it is considered that the operator has failed to react to the overload alarm (HEP≥HEP$_A$), and the protection devices, e.g., circuit breakers, will automatically disconnect the overloaded line to prevent permanent damages to the infrastructure. In general, abnormal line disconnection alarms are triggered when the system is in disruptive phase ($t_d \leq t < t_r$) and the handling of this alarm completes when the system is in recovery phase ($t_r < t < t_{ns}$). Conversely, overload alarms can be triggered either at the disruptive or recovery phase because overloads. The simulation stops when the systems reach the final steady state. Then, the performance measures, including the GR, are evaluated. Furthermore, a reliability metric is also calculated, i.e. ASSAI (Average Substation Service Availability Index), which is the ratio of the total number of hours that service is provided by all available substations during a given time period to the total demanded hours (Eq 2).

$$\text{ASSAI} = \frac{N_S \times (\text{number of hours}) - \sum_{i=1}^{N_S} \text{Res}_i}{N_S \times (\text{number of hours})} \quad (2)$$

where Res$_i$ represents the restoration time for ith substation if service interruption exists and N$_S$ represents the total number of substations.

Two hypothetical strategies are considered in order to compare system's resilience to the same disruptive event, i.e. improvement of the human operator performance (increasing HEP$_A$) and efficiency of line reparation (decreasing MTTR). In total, 18 experiments are set: HEP$_A$ ∈ {0.03,0.3,1} and MTTR(h) ∈ {0.5,1,1.5,2,2.5,3}. The number of simulation runs (N) for each experiment is determined by the coefficient of variation (CV) of the resilience measure of the corresponding target system. In this simulation, $CV_{GR} \leq 0.13$ is the criteria to determine the number of runs for each computer experiment which are needed to estimate system resilience.

5.3. Simulation Results

Figure 4 shows the system performance following the disruptive event under various simulation scenarios, in which the MTTR varies from 0.5 h to 3 h and HEP_A is set to 0.3. In Figure 4, the y-axis denotes the MOP (the number of available transmission lines) and the x-axis denotes the simulation time. At time t = 3 h, the disruptive event is triggered. The negative effects caused by this event start to appear immediately, i.e. the MOP value starts dropping. After about 12 minutes, the MOP value reaches its lowest level, i.e. 92.1%. The MOP value then begins to increase as the result of the repair actions. Similar results are observed in Figure 5, in which HEP_A value varies from 0.03 to 1 and MTTR value is set to 2.5 h. Compared to the results shown in Figure 4, human operator performance has influence not only on the system adaptive and restorability capability during recovery phase, but also on the absorbability capability during disruptive phase, although the effects on the latter are less significant. It can be seen that system robustness is enhanced by improving human operator performance.

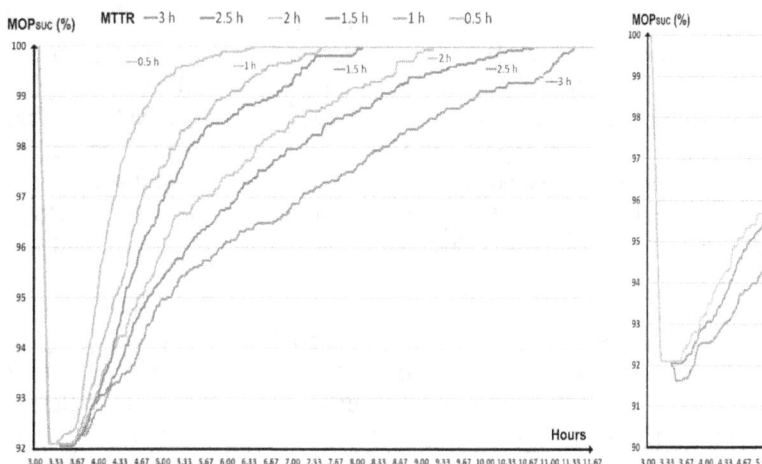

Figure 4 The system performance under different experiments with varying MTTR (HEP_A=0.3, N={18,10,11,10,13,11} for MTTR from 0.5 h to 3 h)

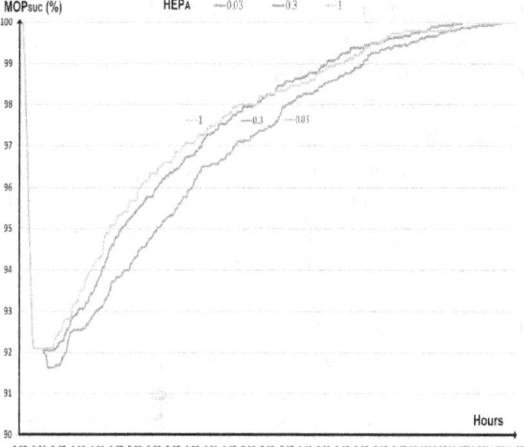

Figure 5 The system performance under different experiments with varying HEP_A (MTTR=2.5 h, N={14,13,10} for HEP_A from 0.03 to 1)

Figure 4 and 5 provide parts of simulation results demonstrating the feasibility of the developed method for the system resilience analysis. The overall simulation results are summarized in Table 2, including the coefficient of variation for each measure. It should be noted that the CV value for R varies very little for each case and is not included in Table 2. The strategy of improving the repair efficiency has little effects on the absorptive capability of the infrastructure. As seen from the Table 2, all the measures related to this capability, i.e., R, PL_{DP}, $RAPI_{DP}$, vary not significantly if comparing results from case studies with same HEP_A value but different MTTR value.

Table 2 Summary of the overall simulation results

MTTR (h)	HEP_A	GR	CV	ASSAI	CV	Disruptive Phase					Recovery Phase			
						R	PL_{DP}	CV	$RAPI_{DP}$ (/h)	CV	PL_{RP}	CV	$RAPI_{RP}$ (/h)	CV
0.5	0.03	16.33	0.13	0.993	0.0015	0.916	0.026	0.053	0.443	0.04	0.09	0.16	0.292	0.068
	0.3	19.15	0.13	0.995	0.0018	0.921	0.0076	0.097	0.446	0.062	0.079	0.13	0.315	0.057
	1	20.82	0.13	0.995	0.0012	0.921	0.0075	0.093	0.451	0.063	0.079	0.14	0.323	0.089
1	0.03	15.90	0.13	0.986	0.0032	0.916	0.027	0.042	0.437	0.051	0.14	0.17	0.273	0.082
	0.3	18.67	0.12	0.990	0.0022	0.921	0.0073	0.07	0.458	0.049	0.12	0.13	0.299	0.06
	1	20.40	0.13	0.991	0.0024	0.921	0.0073	0.065	0.457	0.047	0.12	0.16	0.296	0.056
1.5	0.03	14.98	0.12	0.982	0.0039	0.916	0.026	0.05	0.446	0.037	0.18	0.17	0.278	0.066
	0.3	16.29	0.087	0.984	0.004	0.921	0.0072	0.053	0.467	0.037	0.15	0.19	0.293	0.043
	1	19.26	0.12	0.985	0.0044	0.921	0.0072	0.075	0.453	0.043	0.16	0.18	0.292	0.048
2	0.03	14.04	0.12	0.977	0.0042	0.916	0.026	0.079	0.441	0.042	0.22	0.17	0.260	0.12
	0.3	15.98	0.097	0.984	0.0036	0.921	0.0072	0.051	0.466	0.075	0.18	0.15	0.295	0.075
	1	18.03	0.12	0.985	0.0033	0.921	0.0072	0.028	0.473	0.002	0.17	0.11	0.291	0.055
2.5	0.03	13.76	0.12	0.976	0.0036	0.916	0.026	0.07	0.447	0.036	0.24	0.19	0.272	0.12
	0.3	15.74	0.055	0.980	0.0039	0.921	0.0075	0.077	0.457	0.047	0.20	0.15	0.282	0.022
	1	17.16	0.10	0.981	0.0032	0.921	0.0072	0.061	0.466	0.037	0.23	0.19	0.282	0.024
3	0.03	13.73	0.08	0.973	0.0057	0.916	0.026	0.062	0.448	0.039	0.27	0.18	0.25	0.058
	0.3	14.86	0.089	0.975	0.0038	0.921	0.0072	0.065	0.466	0.039	0.28	0.16	0.266	0.028
	1	16.38	0.088	0.975	0.0058	0.921	0.0075	0.069	0.449	0.046	0.28	0.12	0.267	0.042

MTTR: Mean Time to Repair HEP_A: Maximum acceptable Human Error Probability GR: General Resilience CV: Coefficient of Variation
ASSAI: Average Substation Service Availability Index R: Robustness PL_{DP}: Performance Loss in disruptive phase PL_{RP}: Performance Loss in recovery phase
$RAPI_{DP}$: Rapidity in disruptive phase $RAPI_{RP}$: Rapidity in recovery phase

The R value remains unchanged, while the difference among values of PL_{DP} and $RAPI_{DP}$ is relatively low. For example, if HEP_A is set to 0.03, i.e. the human performance is poor, PL_{DP} value remains at the range of [0.026, 0.027] and $RAPI_{DP}$ value remains at the range of [0.437, 0.439] even if the repair efficiency is improved. Compared to the improvement of the repair efficiency, improving the human operator performance (higher HEP_A value) is able to enhance the absorptive capability more significantly. For example, if HEP_A is set to 0.3, i.e. the human performance is average/acceptable, the PL_{DS} value drops to the range [0.0072, 0.0076]. This indicates that system performance is less impacted during disruptive phase if the human operator performance is improved. A similar trend is also observed for the measure of R, which increases from 91.6% to 92.1% when the HEP_A value is increased from 0.03 to 0.3. However, both R and PL_{DP} do not vary significantly if the HEP_A value is increased from 0.3 to 1, indicating that this strategy becomes inefficient when the operator performance is at an average/acceptable level. Compared to effects on the enhancement of absorptive capability, the strategy of improving the repair efficiency is able to enhance system adaptive and restorative capability more significantly. The PL_{RP} value rises from 0.09 to 0.27 when MTTR value is increased from 0.5 to 3 h and the HEP_A value is set to 0.03. This indicates that the system performance is less impacted during recovery phase if the efficiency of the repair actions is improved. A similar trend can also be observed in other cases (HEP_A = 0.3 and 1). Improving the repair efficiency also has positive effects on the value of $RAPI_{RP}$, which increases from 0.25 to 0.292 (1/h), when MTTR value is increased from 0.5 to 3 h and the HEP_A value is set to 0.03, indicating that more time is needed to recover to a new steady state. Compared to improving repair efficiency, improving the human operator performance has little effects enhancing the adaptive and restorative capability. For example, the PL_{DP} value remains within the range [0.08, 0.09] when HEP_A value is increased from 0.03 to 1 and the MTTR value is set to 0.5 h.

Both strategies have positive effects on the enhancement of resilience capabilities. Improving repair efficiency is a more efficient strategy to enhance the system adaptive and restorative capability during the recovery phase. On the other hand, improving human operator performance is a more efficient strategy to enhance the system absorptive capability during the disruptive state. In order to assess that whether these strategies are able to enhance the overall resilience capability, the integrated GR metric must be used. Figure 6 illustrates the value of GR, i.e. the system resilience to the disruptive event, with respect to the two improvement strategies using the 3D surface diagram. When both improving strategies are performed simultaneously, the system resilience can be enhanced significantly. Figure 6 indicates 51.6% increase of GR value from 13.73 to 20.82 when MTTR value decreases from 3 hours to 0.5 h and HEP_A increases from 0.03 to 1.

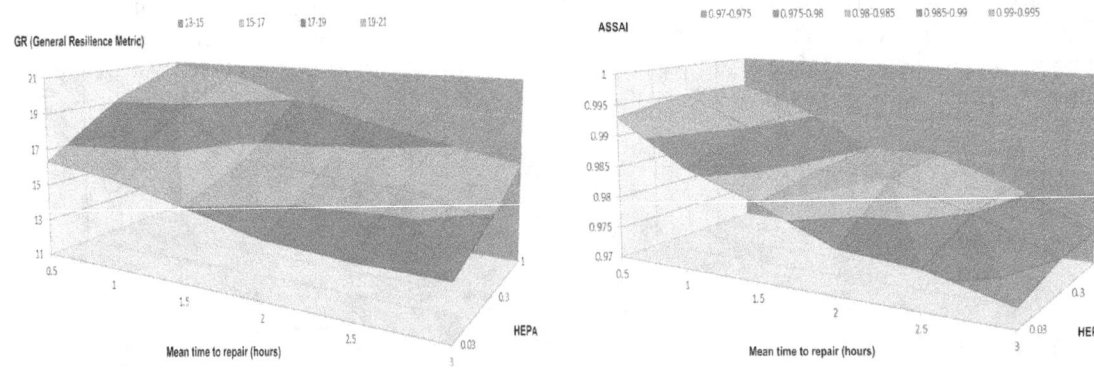

Figure 6 The GR value under different simulation scenarios   Figure 7 The ASSAI value under different simulation scenarios

Figure 7 illustrates the ASSAI value, i.e. system reliability after the disruptive event, with respect to the two improvement strategies using the 3D surface diagram. Reliability and resilience are correlated, although they characterize system performances from different aspects. Reliability is a property of the system; conversely, resilience is an ability of the system. The reliability metric shows a trend which is similar to the trend of resilience, indicating that both improvement strategies have non-negative effects on system reliability and resilience. Nonetheless, according to the Figure 6 and 7, the implementation of both improving strategies improves significantly the system's ability to withstand the negative effects caused by the disruptive event (GR_{SUC} increases 51.6% from 13.73 to 20.82); but on the other hand, it is not capable of improving the system's safety property significantly (ASSAI value only increases 2.3% from 0.973 to 0.995 under same scenario). The comparison between these two metrics exemplifies the advantage of the proposed resilience metric, which is capable of quantifying the behavior of systems in a more comprehensive way by integrating the information related to the system performance in various phases of system loss, adaptive, and recovery.

6. CONCLUSION

Research work on modern infrastructure systems is of great importance for our societies and protection of their assets. In recent years, the infrastructure system evaluation and analysis have broadened to the development of systemic approaches to analyse and understand their behaviours in a holistic way. This paper presents a quantitative method including a novel hybrid modeling approach and the time dependent quantifiable metric for resilience measurement in the context of engineered infrastructure systems. Within this method, three resilience capabilities as well as different measurements and phases related to these capabilities are identified. In order to demonstrate the feasibility and applicability of the proposed quantitative method, it is applied to evaluate behaviors of an electric power supply system (EPSS) after the occurring of a natural hazard, e.g., a winter storm, using two hypothetical improvement strategies (i.e., improvement of the repair efficiency an improvement of human operator performance). The results show that although both strategies are able improve the system resilience significantly, their effects on different resilience capabilities vary. These results also indicate that besides its capability of providing the final outcome quantifying the system behaviors, the resilience metric is also capable of gaining insights into different phases by evaluating corresponding system capabilities, providing a more flexible and comprehensive way to analyze system behaviors compared to the reliability metric (e.g., ASSAI) by gaining insights into system performance in different phases and highlights the limits of reliability indicators in capturing resilience. The ultimate goal is to apply the proposed method in order to develop efficient mitigation and protection strategies by decision makers to maintain and retrofit resilience of infrastructure systems in the long run. Future research work will intend to explore possibilities of expanding the capabilities of this method in other domains, e.g., organizational and social domain. Furthermore, the concept of resilience costs, e.g., service lost cost due to disruptive events and recovery costs, will also be considered in order to improve the applicability and usefulness of the resilience metric. Systems that are resilient to a disruptive event will have lower costs than systems that are less resilient to same event.

References

[1] I. Eusgeld, W. Kröger, G. Sansavini, M. Schläpfer, and E. Zio, "The role of network theory and object-oriented modeling within a framework for the vulnerability analysis of critical infrastructures," *Reliability Engineering and System Safety*, vol. Vol.94, no. 5, pp. p.954-963, 2009.
[2] W. Kröger, "Critical infrastructure at risk: A Need For A New Conceptual Approach and Extended Analytical Tools," *Reliability Engineering and System Safety*, vol. Vol.93, no. 1, pp. p.1781-1787, 2008.
[3] M. Ouyang, and L. Dueñas-Osorio, "Time-dependent resilience assessment and improvement of urban infrastructure systems," *Chaos: An Interdisciplinary Journal of Nonlinear Science*, vol. 22, no. 3, 2012.
[4] A. M. Madni, and S. Jackson, "Towards a Conceptual Framework for Resilience Engineering," *IEEE Systems Journal*, vol. 3, no. 2, pp. 181-191, 2009.
[5] R. Fisher, G. Bassett, W. Buehring, M. Collins, D. Dickinson, L. Eaton, R. Haffenden, N. Hussar, M. Klett, and M. Lawlor, *Constructing a Resilience Index for the Enhanced Critical Infrastructure Protection Program*, 2010.
[6] I. Linkor, F. Creutzig, J. Decker, C. Fox-Lent, and W. Kröger, "Risking Resilience: Changing the Paradigm," *Nature Climate Change*, 2014 (accepted).
[7] C. S. Holling, "Resilience and stability of ecological systems," *Annual review of ecology and systematics*, pp. 1-23, 1973.
[8] W. N. Adger, "Social and ecological resilience: are they related?," *Progress in Human Geography*, vol. 24, no. 3, pp. 347-364, September 1, 2000, 2000.
[9] R. Pant, and K. Barker, "Building Dynamic Resilience Estimation Metrics for Interdependent Infrastructures."
[10] Y. Y. Haimes, "On the Definition of Resilience in Systems," *Risk Analysis*, vol. 29, no. 4, pp. 498-501, 2009.

[11] E. Hollnagel, *Resilience engineering*: Ashgate, 2006.
[12] R. Alliance. http://www.resalliance.org/.
[13] J. Fiksel, "Designing Resilient, Sustainable Systems," *Environmental Science & Technology*, vol. 37, no. 23, pp. 5330-5339, 2003/12/01, 2003.
[14] M. Bruneau, S. E. Chang, R. T. Eguchi, G. C. Lee, T. D. O'Rourke, A. M. Reinhorn, M. Shinozuka, K. Tierney, W. A. Wallace, and D. von Winterfeld, "A Framework to Quantitatively Assess and Enhance the Seismic Resilience of Communities," *Earthquake Spectra*, vol. 19, no. 4, pp. 733-752, 2003.
[15] S. E. Chang, and M. Shinozuka, "Measuring Improvements in the Disaster Resilience of Communities," *Earthquake Spectra*, vol. 20, no. 3, 2004.
[16] T. McDaniels, S. Chang, D. Cole, J. Mikawoz, and H. Longstaff, "Fostering resilience to extreme events within infrastructure systems: Characterizing decision contexts for mitigation and adaptation," *Global Environmental Change*, vol. 18, no. 2, pp. 310-318, 2008.
[17] R. Fisher, G. Bassett, W. Buehring, M. Collins, D. Dickinson, L. Eaton, R. Haffenden, N. Hussar, M. Klett, and M. Lawlor, *Constructing a resilience index for the enhanced critical in Frastructure Protection Program*, Argonne National Laboratory (ANL), 2010.
[18] N. P. Bueno, "Assessing the resilience of small socio-ecological systems based on the dominant polarity of their feedback structure," *System Dynamics Review*, vol. 28, no. 4, pp. 351-360, 2012.
[19] R. Filippini, and A. Silva, "A modeling framework for the resilience analysis of networked systems-of-systems based on functional dependencies," *Reliability Engineering & System Safety*, no. 0, 2013.
[20] R. Francis, and B. Bekera, "A metric and frameworks for resilience analysis of engineered and infrastructure systems," *Reliability Engineering & System Safety*, vol. 121, no. 0, pp. 90-103, 2014.
[21] A. Alessandri, and R. Filippini, "Evaluation of Resilience of Interconnected Systems Based on Stability Analysis," *Critical Information Infrastructures Security*, Lecture Notes in Computer Science B. Hämmerli, N. Kalstad Svendsen and J. Lopez, eds., pp. 180-190: Springer Berlin Heidelberg, 2013.
[22] D. Henry, and J. Emmanuel Ramirez-Marquez, "Generic metrics and quantitative approaches for system resilience as a function of time," *Reliability Engineering & System Safety*, vol. 99, no. 0, pp. 114-122, 2012.
[23] E. Zio, and G. Sansavini, "Modeling Interdependent Network Systems for Identifying Cascade-Safe Operating Margins," *Reliability, IEEE Transactions on*, vol. 60, no. 1, pp. 94-101, 2011.
[24] C. Nan, I. Eusgeld, and W. Kröger, "Analyzing vulnerabilities between SCADA system and SUC due to interdependencies," *Reliability Engineering and System Safety*, vol. Vol.113, no. 0, pp. 76-93, 2013.
[25] R. Klein, E. Rome, C. Beyel, R. Linnemann, and W. Reinhardt, *Information Modelling and Simulation in large interdependent Critical Infrastructures in IRRIIS*, 2007.
[26] A. Alessandri, and R. Filippini, "An Approach to Resilience of Interconnected Systems based on Stability Analysis."
[27] R. Bloomfield, N. Chozos, and P. Nobles, *Infrastructure interdependency analysis: Introductory research review*, Research Report, 2009.
[28] W. Kröger, and E. Zio, *Vulnerable Systems*: Springer, 2011.
[29] C. Nan, and I. Eusgeld, "Adopting HLA standard for interdependency study," *Reliability Engineering and System Safety*, vol. Vol.96, no. 1, pp. p.149-159, 2010.
[30] M. Bruneau, S. E. Chang, R. T. Eguchi, G. C. Lee, T. D. O'Rourke, A. M. Reinhorn, M. Shinozuka, K. Tierney, W. A. Wallace, and D. von Winterfeldt, "A framework to quantitatively assess and enhance the seismic resilience of communities," *Earthquake Spectra*, vol. 19, no. 4, pp. 733-752, 2003.
[31] C. Ferreira, J. Gama, V. Miranda, and A. Botterud, "Probabilistic Ramp Detection and Forecasting for Wind Power Prediction," *Reliability and Risk Evaluation of Wind Integrated Power Systems*, Reliable and Sustainable Electric Power and Energy Systems Management R. Billinton, R. Karki and A. K. Verma, eds., pp. 29-44: Springer India, 2013.
[32] C. Kamath, "Understanding wind ramp events through analysis of historical data." pp. 1-6.
[33] E. Zio, and G. Sansavini, "Vulnerability of Smart Grids With Variable Generation and Consumption: A System of Systems Perspective," *Systems, Man, and Cybernetics: Systems, IEEE Transactions on*, vol. 43, no. 3, pp. 477-487, 2013.
[34] E. I. Bilis, W. Kroger, and C. Nan, "Performance of Electric Power Systems Under Physical Malicious Attacks," *Systems Journal, IEEE*, vol. 7, no. 4, pp. 854-865, 2013.
[35] J. Ring, "Systems of the Third Kind: Distinctions, Implications, and Initiatives," *INCOSE INSIGHT*, vol. 15, no. 2, pp. 9-12, 2012.
[36] R. Bloomfield, N. Chozos, and P. Nobles, *Infrastructure interdependency analysis : Requirements, capabilities and strategy*, 2009.
[37] R. Filippini, and A. Silva, "Resilience analysis of networked systems-of-systems based on structural and dynamic interdependencies," in PSAM 11 & ESREL 2012, Helsinki (Finland), pp 10, 2012, pp. 10.
[38] M. Schläpfer, T. Kessler, and W. Kröger, "Reliability Analysis of Electric Power Systems Using an Object-oriented Hybrid Modeling Approach," in 16th power systems computation conference, Glasgow, 2008.
[39] E. Bilis, M. Raschke, and W. Kroeger, "Seismic response of the swiss transmission grid," in Proceedings of ESREL 2010, 2010, pp. 5-9.
[40] M. Raschke, E. Bilis, and W. Kröger, "Vulnerability of the Swiss electric power transmission grid against natural hazards," *Applications of Statistics and Probability in Civil Engineering*, pp. 1407-1414: CRC Press, 2011.
[41] M. Bründl, and C. Rickli, "The storm Lothar 1999 in Switzerland–an incident analysis," *For. Snow Landsc. Res*, vol. 77, no. 1, pp. 2, 2002.
[42] W. R. Lachs, "Transmission-line overloads: real-time control," *Generation, Transmission and Distribution, IEE Proceedings C*, vol. 134, no. 5, pp. 342-347, 1987.

Use of Reliability Concepts to Support Pas 55 Standard Application to Improve Hydro Power Generator Availability

Gilberto F. M. de Souza[a(*)], Erick M.P. Hidalgo[a], Claudio C. Spanó[b], and Juliano N. Torres[c]

[a] University of São Paulo, São Paulo, Brazil
[b] ReliaSoft Brasil, São Paulo, Brazil
[c] AES Tietê, Bauru, Brazil

Abstract: The electric power generation industry seeks for high equipment availability to fulfill the requirements of regulatory agencies contracts and demands. The availability of electric power generation equipment is affected not only by equipment design characteristics but also by maintenance policies. The recently developed British Standard PAS 55 presents requirements to improve asset integrity aiming at increasing availability and reducing safety risk. The present paper presents a discussion regarding the advantages of application of PAS 55 requirements to increase operational availability of hydro power generator and its relation with traditional reliability techniques used to develop equipment maintenance policy. An assessment is presented to link operational information, standards requirements and the associated reliability analysis techniques are linked to requirements of PAS 55. An example of application is presented for a 30 MW hydro power generator showing the advantages of applying reliability requirements to improve equipment availability.

Keywords: Degradation Analysis, Monitoring System, Reliability and Probabilistic Models.

1. INTRODUCTION

The rapid increase in world energy prices from 2003 to 2008, combined with concerns about the environmental consequences of greenhouse gas emissions, has led to renewed interest in alternatives to fossil fuels—particularly, nuclear power and renewable resources.

According to [8] from 2007 to 2035, world renewable energy use for electricity generation will grow by an average of 3.0 percent per year, as shown in Figure 1, and the renewable share of world electricity generation will increase from 18 percent in 2007 to 23 percent in 2035.

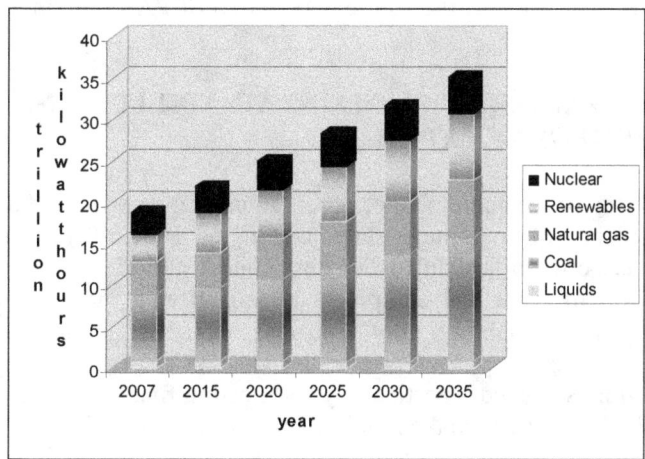

Figure 1. Forecast of World Net Electricity Generation by Fuel, 2007-2030, [8]

The category liquids include petroleum based fuels, such as Diesel oil or crude oil, and the category renewable includes hydroelectric, wind and other renewable electric power generation Those

*gfmsouza@usp.br
Probabilistic Safety Assessment and Management PSAM 12, June 2014, Honolulu, Hawaii

projections are based on a business-as-usual trend estimate, given known technology and technological and demographic trends, [8].

Most of the world's electricity is produced at thermal power plants (TPP), which use traditional fuels, coal, gas and fuel oil, and up to 20% of the world's electricity is produced by hydroelectric power plants (HPP). In countries with well-to hydropower, the figure is much higher: Norway (99%), Brazil (85%), Austria, Canada, Peru, New Zealand - over 50%, [8].

Process companies are adopting a consolidated approach to performance improvement based upon the use of a key performance indicator known as Overall Equipment Effectiveness (OEE), [9]. OEE can be calculated as:

$$OEE = (Availability) \times (Performance) \times (Quality) \qquad (1)$$

Availability refers to the process equipment being available for production when scheduled. Performance is determined by how much waste is created through running at less than optimal speed. Quality focuses on identifying time that was wasted by producing a product that does not meet quality standards.

PAS 55 is the British Standards Institution's (BSI) Publicly Available Specification for the optimized management of physical assets, [6]. The specification states that organizations must establish, document, implement, maintain, and continually improve their asset management system. In this context, asset management system refers collectively to the overall policy, strategies, governance, plans and actions of an organization regarding its asset infrastructure. Reliability and maintenance analysis play an important role in asset management according to that standard.

Traditionally, reliability analysis of equipment, or its components, is performed by the selection of a probability distribution that best models the times between failures recorded for the devices in question. The record of the time between failures is usually performed by maintenance staff, feeding the data into a computerized maintenance management system. This probability distribution models the reliability of the equipment/component.

This paper presents a methodology for reliability analysis used to describe the reliability of hydro generator, including the basic concepts of reliability and maintenance, which served as the basis for the development of the research. Furthermore an application of this methodology for reliability analysis is presented for a generating unit of a specific hydroelectric plant located at São Paulo State, Brazil. The analysis is used to support the future application of PAS 55 concepts to hydro generator management.

2. RELIABILITY, AVAILABILITY, MAINTAINABILITY AND SAFETY (RAMS) ANALYSIS OF HYDRO GENERATOR

Reliability is the probability of the equipment or process functioning without failure, when operated as prescribed for a given interval of time, under stated conditions. Reliability in power plants is affected by operating periods, i.e. between scheduled outages; budget periods; and peak-production periods. Measuring the reliabilities of plant and equipment by quantifying the annual cost of unreliability incurred by the facility puts reliability into a business context. Higher-plant reliability reduces equipment failure costs. Failure decreases production and limits gross profits. Failure is a loss of function when that function is needed – particularly for meeting finance goals. Failure requires a clear definition for organizations striving to make reliability improvements, [3].

General calculations of reliability are based on considerations of the initial failure-mode, which may be termed "infant" mortality (decreasing failure rates then with time) or wear-out mode (i.e. increasing failure rates then with time). Key parameters describing reliability are mean time to failure, mean time between/before repairs, mean life of components, failure rate and the maximum number of failures in a specific time-interval, [4].

High reliability (i.e. corresponding to relatively few failures) and ease of maintainability of the system influence availability, which is related to both frequency and duration of outages as follows:

$$A = \frac{MTBF}{MTBF + MTTR} \qquad (2)$$

where MTBF means 'mean time between failure' and MTTR means 'mean time to repair'.

Thus, the availability goal can be converted into reliability and maintainability requirements in terms of acceptable failure rates and outage hours for each component as explicit design objectives.

PAS 55 is the British Standards Institution's (BSI) Publicly Available Specification (PAS) for the optimized management of physical assets. It provides clear definitions and a 28-point requirements specification for establishing and verifying an integrated, optimized and whole-life management system for all types of physical assets. One of the key aspects of PAS 55 is to connect the company's strategic objectives, including short-term and long-term objectives, with day-to-day asset management activities.

All PAS 55 requirements stress the importance of having a good data collection system, because the core of a good and successful asset management program is based on the analysis of data aiming at supporting reliability and maintainability analysis. The standard sets that the reliability and maintainability analysis of critical equipment support the Life Cycle Cost Analysis of the asset.

Based on reliability analysis, the standard requires the development of root cause analysis aiming at defining the basic cause (usually defined by the failure of a component) of an undesirable behavior of equipment of a processing plant. This analysis should support the application of maintenance plans developed to maximize plant availability, [5].

The analysis of reliability and maintainability of hydro generator proposed in the present paper is based on application of reliability concepts to define critical items of a hydropower electric generation system from the point of view of failure consequences and maintenance planning, having as objective to achieve the availability planned for the system, minimizing downtime for corrective maintenance or even reducing the downtime associated with preventive maintenance, supporting PAS 55 application to manage hydro generator life cycle cost and performance.

To achieve this goal, the method is based on steps. As first step the functional tree of the hydropower generator must be developed, which aims to define the key equipment that compose the system as well as the functional relationship between them. This study analyzes in depth the hydro generator.

From the functional tree a FMEA analysis of all systems that comprise the generating unit can be developed, seeking to define the system main components, their failure modes associated with specific operating conditions and what are the consequences of these failures on the operation of the equipment. Additionally, the analysis seeks to define whether the occurrence of a failure mode presents symptoms, allowing the maintenance action before the occurrence of the machine failure.

This analysis allows the evaluation of the equipment considered critical for the interruption of power generation, either from the point of view of the excessive time to repair the failure or due to the high frequency of occurrence of a fault. This equipment considered critical should have their maintenance prioritized. With the aid of the functional tree and estimated reliability for the most critical equipment the hydro generator reliability may be characterized by building a block diagram.

Finally, to select the most appropriate maintenance policy for specific equipment, a decision process that allows defining the most appropriate maintenance practices for the same should be formulated, keeping in mind the characteristics of its fault modes and maintenance practices employed in mechanical or electrical equipment, which are: corrective, preventive and predictive. This decision is based on RCM technique.

3. EXAMPLE OF ANALYSIS

Aiming to run an example demonstrating the applicability of the methodology proposed in this paper, it is necessary to define a hydroelectric plant that will be taken as the basis of analysis in order to complete characterize their systems, allowing the application of failure modes and effects analysis, which forms the basis for a future implementation of RCM based maintenance policies and asset management techniques according to PAS 55 requirements. The hydro power plant in analysis is located in São Paulo State, Brazil, and has four hydro power generators each one equipped with Francis turbine and presenting 27 MW nominal output

3.1. Functional tree analysis

Initially, in Figure 2, the functional diagram has been proposed for the plant. It is proposed that the plant be subdivided into systems: dam, water intake, auxiliary services of alternating current and direct current, synchronization system, other auxiliary equipment and generating units.

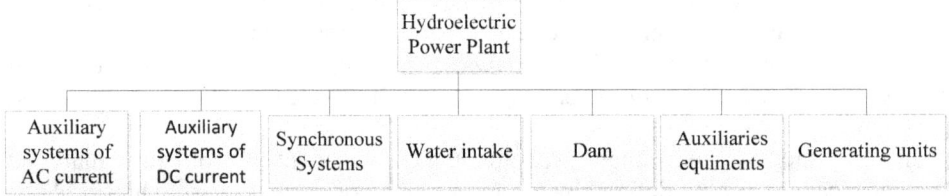

Figure 2. Hydroelectric Power Plant Functional Tree

Figure 3 shows the functional diagram of a generating unit. This piece of equipment is divided into eight subsystems, which are: turbine, generator, thrust and guide bearings, draft tube and suction, power substation, and control/monitoring system.

Throughout the study functional trees of each of subsystems were developed. In Figure 4 is shown as an example the turbine functional tree. According to the diagram in Figure 4, the turbine system is composed of several components having primary function of transforming the kinetic energy of fluid flow into mechanical energy. Therefore the turbine must have a control system, which maintains the frequency of its rotation, regardless of the magnitude of energy transformed, and must have components that act directly on the transformation and transmission of energy to the generator, such as the rotor, axis, the guide bearing and shaft coupling. Additionally, this system has components that support its operation, but do not act directly in the main function, such as the shaft seal, the aeration system, monitoring and protection equipment.

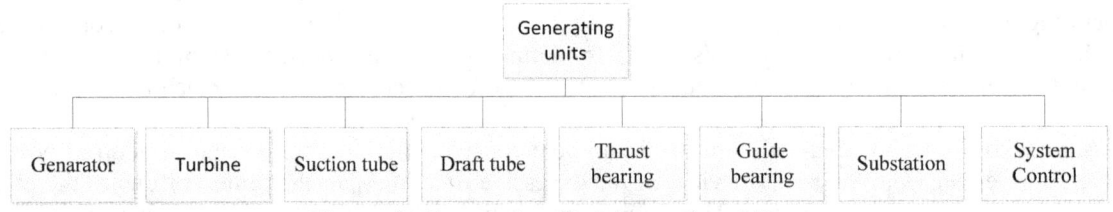

Figure 3. Generating Unit Functional Tree

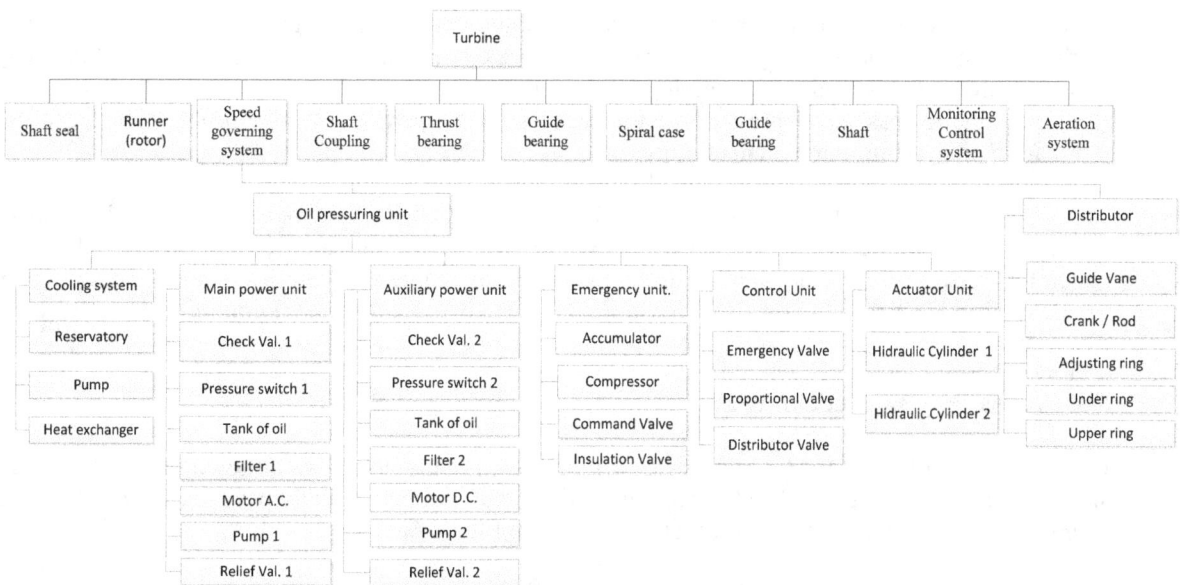

Figure 4. Hydraulic Turbine Functional Tree

The generating unit depends on some systems for monitoring and controlling its operation that includes monitoring of both active and reactive power, voltage in the generator, and temperature in some parts of the machinery, such as winding insulation and heat exchangers fluid temperature. Finally, the output of the generating unit is coupled to a substation, which connects the unit with the transmission line, having transformers that fit with the generated voltage required by the transmission line, and has a protective system which aims to protect the generating unit against faults that could occur in the transmission line.

3.2. Failure Mode and Effects Analysis (FMEA)

The analysis was performed using as initial event a failure mode of a hydro generator component. The analysis allows the study of the propagation of the failure in other subsystems or hydro generator systems, aiming at defining the loss of performance caused by the initial failure. The failure propagation effects analysis are considered in subsystems with which the component has operational relationship in accordance with the functional tree. In other words considers the propagation in the subsystems located in the higher levels of the functional tree.

For carrying on a FMEA analysis the form recommended by reference [13] is used. FMEA analysis was performed considering only the severity of the failure, defined as the loss of performance of hydro generator due to a component failure. Although the spreadsheet allows the analysis of the frequency of occurrence of the fault and even the possibility of detecting the failure in the early stages of development, depending on the application of monitoring techniques, for the present study this part of the analysis was disregarded For the present analysis the severity index is classified into three main severity levels: marginal, critical and catastrophic. Each level is split into three other sub-levels to express some variety of failure effects. A criticality scale between 1 and 9 is proposed. Values between 1 and 3 express minor effects on the turbine operation while values between 4 and 6 express significant effects on the turbine operation. Finally, failures that cause turbine unavailability or environmental degradation are classified by criticality values between 7 and 9.

FMEA analysis supports the selection of the mechanical failures that cause immediate shutdown of the hydro generator, associated with severity 9, which are: i) Failure in the turbine or generator shaft, represented by the cross section rupture due to the presence of excessive permanent deformation (due to overloads), varying its straightness, or due to crack growth associated to variable loading; ii) Failure in the turbine rotor, involving the rupture of the rotor blade due to fatigue mechanism and iii) Failure in the bearings structures, involving the rupture or permanent deformation.

Besides the catastrophic failure modes that cause equipment shutdown, the hydro generator has many auxiliary systems that are very important to support equipment performance. The FMEA analysis indicates that guides and trust bearings lubrication systems failure can cause hydro generator shutdown due to loss of oil pressure or increase of oil temperature. Those problems are caused by lubrication oil systems components failures.

The same failures for the speed governor hydraulic system will also cause the hydro generator shutdown. The failure of the refrigeration system used by the hydraulic systems and also by the generator can cause machine shutdown. Another type of failure that implies in machine shutdown is the deterioration of rotor and stator winding insulation, which operational condition must be verified over time.

3.3. Availability and reliability analysis.

For a generating unit of the hydroelectric plant, an availability analysis employing the block diagram shown in Figure 5 is carried out. From the failure data collected for two years, it was possible to characterize the reliability of the subsystems shown in Figure 5.

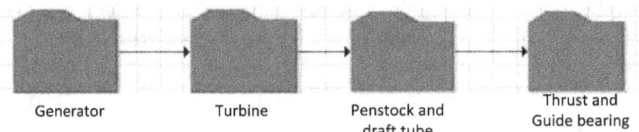

Figure 5 Simplified Block diagram of Hydroelectric

The operational data for the four generating units were registered during two consecutive years aiming at defining the failures that cause equipment unavailability (associated with corrective maintenance actions) and the expended time in programmed maintenance activities.

The operational registered data are presented in Table 1. The generating units states present in that table are defined according to IEEE 762 standard [10]. The 'in service hours' represents the number of hours that a unit is in service state. The 'reserve shutdown hours' represent the number of hours that an available unit is in reserve shutdown. The 'forced outage hours' represent the number of hours that an unit is unavailable due to the occurrence of a component failure that causes immediate removal from service. The 'basic planned outage' represents the number of hours that the unit is unavailable due to the execution of programmed maintenance activities.

Table 2 shows the number of failures associated with the basic generating unit subsystems which will be used as for reliability analysis.

From data presented in Table 2 it is interesting to observe that during two years none of the generating units presented bearing failures. It is also interesting to observe the great number of failures associated with the generator subsystem. The hydro power plant administrator only registers failures that cause interruption of power generation. According to Brazilian Electric Energy Regulatory Agency the plant administrator must report the subsystem that causes the forced outage without any root cause analysis to define the failed component.

Based on the plant operator report most of the turbine subsystem failures are associated with control guide vanes failures, including breakage of mechanical actuators, and speed governor hydraulic system failures. Regarding generator subsystem failures, most of them are caused by exciter components failures including 125 DC system failures. The penstock and draft tube subsystem failures are associated with butterfly inlet valve failures caused by hydraulic system failures.

Based on the in service hours of each generating unit and the number of failures of each subsystem the failure rate for a given subsystem is defined, as presented in Table 3. All subsystems under analysis are composed by a great number of components that must be presenting a minimum performance to guarantee the expected performance of the generating unit. As for reliability analysis each subsystem can be modeled as a series system with components presenting different reliability distributions.

Table 1. Operational data of generating units

Distribution of Hours	Generating Unit	Operational Data	
		Year 1	Year 2
In Service Hours (h)	UG1	6245.05	7699.59
	UG2	7622.36	6541.75
	UG3	5275.59	4181.21
	UG4	4442.91	3653.00
Forced Outage Hours (h)	UG1	22.55	5.41
	UG2	43.66	7.29
	UG3	0.88	10.67
	UG4	36.80	4.46
Basic Planned Outage (h)	UG1	2420.17	397.08
	UG2	1025.26	1459.02
	UG3	157.47	1383.61
	UG4	1361.70	486.81
Reserve Shutdown Hours (h)	UG1	73.23	658.82
	UG2	69.72	753.94
	UG3	3327.06	3185.51
	UG4	2919.59	4616.73
Number of Failures that cause Forced Outage	UG1	6	3
	UG2	8	3
	UG3	1	3
	UG4	7	3

Table 2. Failures distribution of generating units

Generating Unit	Subsystem	Number of failures	
		Year 1	Year 2
UG1	Generator	6	1
	Turbine	1	
	Penstock and Draft tube	0	2
	Thrust and Guide Bearings	0	0
UG2	Generator	7	1
	Turbine	1	0
	Penstock and Draft tube		2
	Thrust and Guide Bearings	0	0
UG3	Generator	1	0
	Turbine	0	1
	Penstock and Draft tube	0	2
	Thrust and Guide Bearings	0	0
UG4	Generator	3	1
	Turbine	4	0
	Penstock and Draft tube	0	2
	Thrust and Guide Bearings	0	0

Usually, for series systems with too many components, the system reliability can be modeled by an exponential distribution given that there is no frequent failure of a specific component. Once the reliability of the components are unknown and due to the small number of registered failures and diversity of failures root-causes of the subsystems it is initially recommend to adopt an exponential distribution to model their reliability distributions, [7] and [11]. The exponential distribution is defined according to the following equation:

$$R(t) = e^{-\lambda t} \qquad (3)$$

where R(t) is the subsystem reliability and λ is the subsystem failure rate.

According to Table 3 a great variability in the failure rate for the same subsystems of each generating unit is noticed. Those data indicate that although the generating units have the same design their failure rates can be influenced by possible differences in their operational conditions or even due to differences during the onsite assembly process.

The generating unit failure rate is defined as the sum of the subsystems failure rates and also presented in Table 3. Generating unit 4 presents the greater failure rate once it presented the same number of failures of units 1 and 2 with the smallest in service hours. According to the power plant generation data units 1 and 2 are preferentially used to attend the power generation demand while units 3 and 4 are used to complement generation in case of higher demand. Due to this fact generating units 1 and 2 were submitted during the two years under analysis to comprehensive scheduled maintenance (duodecennial activities) representing a mean of almost 1200 annually planned outage hours for those units. Unit 4 presented a mean of 900 annually planned outage hours. Considering that units 1 and 2 can be considered 'as good as new' after the execution of preventive maintenance tasks, their reliability were improved as shown in Table 4.

Table 3. Failures rate estimate for generating units

Generating Unit	Subsystem	Total Number of Failures	In Service Hours (h)	Subsystem Failure rate (1/h)	Generating Unit Failure rate (1/h)
UG1	Generator	7	13944.64	0.00050198	0.00071698
	Turbine	1	13944.64	0.00007171	
	Penstock and Draft tube	2	13944.64	0.00014342	
UG2	Generator	7	14163.11	0.00049424	0.00070606
	Turbine	1	14163.11	0.00007061	
	Penstock and Draft tube	2	14163.11	0.00014121	
UG3	Generator	1	9456.80	0.00010574	0.00042297
	Turbine	1	9456.80	0.00010574	
	Penstock and Draft tube	2	9456.80	0.00021149	
UG4	Generator	4	8095.91	0.00049408	0.0012352
	Turbine	4	8095.91	0.00049408	
	Penstock and Draft tube	2	8095.91	0.00024704	

For repairable systems the asymptotic availability can be used as a measure of system capacity of providing a given amount of electric power per year. The analysis is presented for two generating units, 1 and 4, representing respectively a unit recently submitted to detailed maintenance plan and a unit still submitted to triennials maintenance plan. As for reliability analysis the power generation profile shown in Figure 6 is used for units 1 and 4.

Considering the 'time to repair' database the maintainability for each of those units is calculated and based on Monte Carlo Simulation, the availability for one operational year (8760 operational hours) is calculated. The results are summarized in Table 4. It is clear that even for a greater in service hours, UG1 availability is better than UG4. This fact is associated with the execution of planned maintenance of UG1.

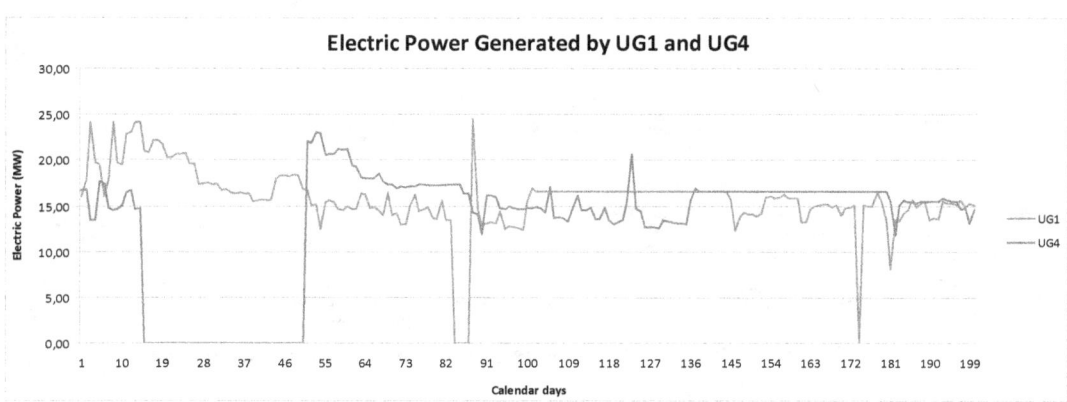

Figure 6. Power generation profile for Units 1 and 4

Table 4. Generating units availability estimate

	UG1	UG4
In service hours (h)	7450.7	4202.3
Forced outage hours	22.5	31.6
Availability	0.997	0.992

3.4. Proposals of Maintenance Policy

Taking as a basis the results of the FMEA analysis, suggestions are presented for the maintenance practices for critical equipment of the hydro power plant under study, which are defined according to the consequences of failure on the hydro generator operational condition, [12].

In accordance with RCM concepts, preventive or predictive maintenance practices must be evaluated to components whose failure modes generate consequences with severity greater than 6, in other words, those that degrade the performance or even cause shutdown of hydro generator.

Considering that most of the failures for subsystems turbine and penstock and draft tube are associated with hydraulic systems the faster the maintenance team locates a failed component the smaller will be the forced outage period. The failure diagnosis can be based on pre-defined faults that can cause interruption of hydraulic system operation. For each fault a tree must be developed to identify the possible components which failures cause the tree top event.

Those trees can be developed by engineering and maintenance crew of each power plant. In Figure 7 a tree is presented to indicate the root-cause analysis of increase in 'hydraulic oil temperature' which causes hydraulic system operational interruption. Those trees can also be used to define monitoring systems that could be able to alert about failures development in hydraulic system basic components aiming at applying predictive maintenance to improve hydro generator reliability.

This methodology can be used to develop analysis of other critical components failure modes aiming at defining monitoring systems to evaluate mechanical components degradation and to support root cause analysis as recommended by PAS 55.

The availability analysis indicates that the hydro generator maintenance practices employed by the generation company have proved effective from the point of view of maximizing the availability with reduction of corrective maintenance interventions mainly for hydro generator bearings. The use of maintenance policies selected from the application of the RCM philosophy should increase the availability of machines, minimizing the preventive and corrective interventions.

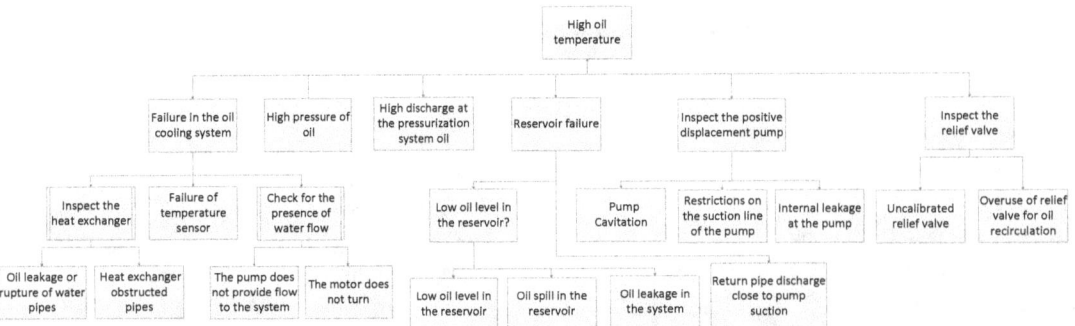

Figure 7. Example of root-cause analysis tree for failure diagnosis

The results of this analysis are:

i) It is vital the use of an on-line system for monitoring oil temperature used as a lubricating fluid in the bearings of the turbine and generator. Presently, the temperature is registered and displayed in a DCS screen but is not recorded in a database. The temperature history must be stored in a database that should be coupled to an asset management system allowing supporting bearing degradation analysis. The increase of the oil temperature may be an indicator of a failure in the cooling system or bearing failure due to oil film thickness reduction between the shaft and the pads of the bearing. Specifically in the case of monitoring system for the bearings oil temperature, if the oil temperature increases in all bearings simultaneously, there is a clear indication of oil cooling system failure. If the fault is associated with some component of the raw water feeding system an increase in temperature of the air inside the generator can also be detected, based on the indications of air temperature sensors installed near generator radiators. If the failure is associated with the oil treatment and conditioning system, it won't be observed variation of the air temperature inside the generator;

ii) The increase of oil temperature in a single bearing is an evidence of failure in cooling system (or pipes). If the failure is associated with an absence of flow of raw water, there will be an indication of water level sensor. Thus, measurement of the oil temperature is an important parameter for determining the performance of hydro-generator and may indicate the occurrence of failures in the cooling system of the hydro-generator;

iii) The temperature rise of the bearing pad indicated by the temperature sensors may be caused by an increase in oil temperature (which would also be recorded by the oil temperature sensor) or due to an improper contact between the shaft and pads, as function of the shaft vibration. The shaft vibration can be detected by proximity sensors located in the bearings;

iv) The oil analysis may assist in predicting of the occurrence of wear in the bearing components, which may be associated with a possible contact between shaft and pads, caused by the vibration of the turbine and generator shaft, or by oil contamination due to filtering system failure. A degradation analysis may be executed based on oil analysis results time history. Presently, the oil analysis is annually executed but not used for trend analysis. It is used to check the instantaneous oil condition and to program immediate corrective maintenance actions;

iv) The installation of proximity sensors in the generator to measure air gap is suggested. Such proximity sensors enable the assessment of the orbit of the generator rotor shaft, allowing the assessment of abnormalities in this orbit which characterizes the vibration of the shaft. Moreover, this sensor can monitor the evolution of a failure mode whose effect is the shaft vibration, enabling a maintenance action before a higher degradation of the operational performance of hydro generator.

It should be noted that the monitoring techniques, which support decisions associated with the implementation of predictive maintenance, are basically applied to mechanical components, for failure modes whose time dependency development is clearly defined. However, the behavior of the parameter analyzed, selected as an indicator of the development of a specific failure mode, when the machine operates in a normal condition without any performance loss, should be clearly defined, based on observations executed with a specific machine. This means that for each hydro generator should defined the "normal" behavior (associated with the optimal performance of the machine) of parameters selected for monitoring operational performance. Additionally, it should be evaluated the time evolution of these parameters considering the development of several failure modes, in order to characterize their behavior due to the occurrence of specific failure modes. Only after this survey the results of the monitoring will be used as input for decision making regarding the implementation of an intervention to execute the maintenance of a critical component of the hydro-generator. For each parameter an alarm level must defined, indicating the need to stop the machine immediately in order to avoid further damage to its components.

4. CONCLUSIONS

The methodology for implementing reliability analysis proved to be very suitable for hydro generator analysis and can support the development of asset management based on PAS 55 requirements.

The importance of carrying out a detailed functional analysis of hydro generator as an initial step in the development of reliability analysis is emphasized. This analysis should seek to define all the systems that compose a hydro-generator and the interrelationship between them, describing their functions. This analysis yields the understanding of the function of each hydro generator system and the interrelationship between components to ensure that its function is performed in accordance with a specific requirement.

Functional analysis serves as subsidy for the implementation of the Failure Modes and Effects Analysis of hydro generator, which aims to analyze the effects of the hydro generator components failures on its operational performance. The results allow the identification of critical hydro generator components, in other words those whose failure causes machine shutdown or even a severe reduction in operational performance. During the execution of this analysis it is clear that regardless of the type of hydro generator analyzed, there are some critical subsystems for ensuring the operational required performance such as the speed governing system, oil cooling system in order to control the bearing lubricating oil temperature (and hence its viscosity) and the stator winding insulation. These subsystems should be the subject to constant attention from maintenance crews which, through application of preventive or predictive tasks, should reduce the probability of unexpected failures of those components.

From the results of the FMEA analysis, the critical components or subsystems that can be submitted to predictive maintenance programs are defined, which is the main focus of the RCM philosophy. Some simple measures can be used as techniques for monitoring the operating condition of the critical components of a hydro generator, such as monitoring the oil temperature inside the bearing and in the speed governing system, monitoring the temperature of the copper conductors in the stator bars, monitoring the air temperature in the core of the generator, monitoring the temperature of bearings pads, or even checking for the presence of contaminants in the bearing lubricating oil and in the hydraulic unit of the speed governing system. The monitoring of those data already enables verification of the occurrence of anomalies in the operating condition of hydro generator, and with the aid of the FMEA tables may be used to identify possible causes of these anomalies and to predict the need for maintenance tasks before generating unit performance is reduced below a minimum value, requiring the implementation of corrective maintenance. Those variables time history can be used as part of degradation studies developed by the hydro generator operator.

To calculate the reliability and availability of the generating unit it is necessary to model the reliability and maintainability of its various components. For both analysis a database is needed where the time between failure, repair time and the causes of failure associated to each corrective intervention executed on the component are recorded in a systematic manner. Operating condition of hydro generator should be continuously recorded to correlate the occurrence of a component failure mode with its operating condition.

To demonstrate the application of the proposed methodology the reliability and availability analysis of hydro generators of a hydroelectric plant located in São Paulo, Brazil was carried out. This analysis was performed according to the steps shown in item 2 and allows the following conclusions:

i) The system for monitoring the operating temperature of some components of this hydro generator is very well designed, allowing the evaluation of abnormalities which are indicative of the development of some component failure, supporting the application of predictive maintenance techniques;
ii) The association of this temperature monitoring system database with the effects of component failures on the hydro generator operational performance, presented in the FMEA analysis, may help in the process of defining the origin of any abnormal hydro generator operational condition;
iii) The generator unit has its availability greatly affected by the number of hours employed in preventive tasks and by the required number of operational hours;
iv) The calculated availability for hydro-generators based on fault records for two years are close to the values reported by the Brazilian Association of Electric Power Generation Enterprises (ABRAGE), [1] and [2], for availability of generating units (hydro and thermal) with power range between 10MW and 30MW, which were 89.86% in 2001 and 90.95% in 2003;
v) With the increase in efficiency of maintenance tasks, focusing on predictive and preventive activities associated with items that are critical to guarantee the operating performance of generating units in accordance with the philosophy of RCM, the availability of generating units is expected to increase, affecting the operational performance of the hydroelectric plant.

Acknowledgements

The authors thanks for the financial support of Agência Nacional de Energia Elétrica (ANEEL).

References

[1] ABRAGE. *"Análise Estatística de Desempenho Unidades Geradoras Hidráulicas e Térmicas, e Equipamentos sob Responsabilidade da Geração"*, Grupo Técnico de Manutenção, Rio de Janeiro, (2001). (in portuguese)
[2] ABRAGE. *"Análise Estatística de Desempenho Unidades Geradoras Hidráulicas e Térmicas, e Equipamentos sob Responsabilidade da Geração"* Grupo Técnico de Manutenção, Rio de Janeiro, (2003). (in portuguese)
[3] Barringer, P.E. *"Reliability engineering principles"*. <http://www.barringer1.com/pdf/REP_Brochure.pdf>, (2000)
[4] Barringer, P.E and Weber, P.D. *"Life-cycle cost"*. Fifth International Conference on Process-Plant Reliability, Texas, United Stated, (1996).
[5] Blanchard, B.S., et al. *"Maintainability: a key to effective serviceability and maintenance management"*. John Wiley & sons, 1th Edition (1995).
[6] BSI- British Standards Institution. *"PAS 55: Asset Management - Part 1 & 2"*. (2008), London.
[7] Davidson J. *"The reliability of mechanical systems"*. Mechanical Engineering Publications Limited, (1998), London..
[8] DOE - Department of Energy. *"EIA-0484 International Energy Outlook 2010"*. <http://large.stanford.edu/courses/2010/ph240/riley2/docs/EIA-0484-2010.pdf>. (2010)
[9] Eti, M, et al. *"Maintenance Schemes and their Implementation for the Afam Thermal-Power Station"*. Applied Energy, Vol. 82, pp. 255-265, (2005).
[10] IEEE – The Institute of Electrical and Electronics Engineers, 1987, *"IEEE 762 Standard Definitions for Use in Reporting Electric Generating Unit Reliability, Availability, and Productivity"*. (1987), New York,
[11] Kumar D, et al. *"Reliability analysis of power-transmission cables of electric loaders using a proportional-hazard model"*. Reliability Engineering System Safety, Vol. 7, pp. 217-222, (1987).
[12] Moubray, J. *"Reliability – Centered Engineering"* John Wiley & sons, 2th Edition, (2000).
[13] US MIL-STD-1629A. *"Procedures for performing a Failure Mode, Effects and Criticality Analysis"*, (1977).

www.ingramcontent.com/pod-product-compliance
Lightning Source LLC
Chambersburg PA
CBHW081139180526
45170CB00006B/1858